Christophe B<

Fiber Bragg Gratings

Fundamentals and Applications in Telecommunications and Sensing

For a complete listing of the *Artech House Optoelectronics Library*, turn to the back of this book.

Fiber Bragg Gratings

Fundamentals and Applications
in Telecommunications and Sensing

Andreas Othonos
Kyriacos Kalli

Artech House
Boston • London

Library of Congress Cataloging-in-Publication Data
Othonos, Andreas.
   Fiber Bragg gratings : fundamentals and applications in telecommunications and
sensing / Andreas Othonos, Kyriacos Kalli.
      p. cm. — (Artech House optoelectronics library)
   Includes bibliographical references and index.
   ISBN 0-89006-344-3 (alk. paper)
    1. Optical fibers. 2. Diffraction gratings. 3. Optical detectors. I. Kalli, Kyriacos.
II. Title. III. Series.
TA1800.084 1999   99-21679
621.36'92—dc21   CIP

British Library Cataloguing in Publication Data
Othonos, Andreas
   Fiber Bragg gratings : fundamentals and applications in telecommunications and
sensing. — (Artech House optoelectronics library)
   1.Optical fiber detectors 2. Telecommunication 3. Optical fibers
   I. Title II. Kalli, Kyriacos
   621.3'692

   ISBN 0-89006-344-3

Cover design by Lynda Fishbourne

© 1999 ARTECH HOUSE, INC.
685 Canton Street
Norwood, MA 02062

International Standard Book Number: 0-89006-344-3
Cataloging-In-Publication: 99-1679

10 9 8 7 6 5 4 3

To Demetra, Anna,
&
the children

CONTENTS

Chapter 3   PROPERTIES OF FIBER BRAGG GRATINGS   95

Preface

Fiber Bragg gratings represent a key element in the established and emerging fields of optical communications and optical fiber sensing. A vast amount has been published in recent years, including a number of review papers, each with a specific emphasis on the applications arena. However, given the importance of fiber Bragg gratings to optoelectronic systems we felt that there was a need for a comprehensive book. This rapidly advancing field has made the timing of a book quite critical. This appeared viable in 1998, when we felt that the different parts to this fascinating, multi-faceted story could be presented in a complete form. Therefore, our aim was to provide a comprehensive, up-to-date overview of this subject, a foundation on which to build future work. As part of this ideal we have included more than 700 references. This work is primarily for the researcher or academic in the field of optoelectronics, however, its self-contained form is equally suitable to the engineer or graduate student, requiring only a basic knowledge of physics.

The book begins with a brief introduction to the field, followed by a detailed explanation of fiber photosensitivity. There has certainly been a need to bring together the plethora of information regarding in-fiber photosensitivity, condensing the experimental data, some of which is conflicting, into a useful format. Thus, we have dedicated Chapter 2 to this discussion. The properties and manufacturing techniques of the Bragg grating are dealt with in Chapters 3 and 4. The relatively simple strain- or temperature-induced wavelength-shift properties, described in Chapter 3, are also accompanied by a description of complex behavior, such as the fascinating aspects of ultrashort pulse propagation, and unique grating structures. Easily accessible information is also provided on the practical issues of long-term mechanical and thermal integrity. Chapter 4 informs the reader of the most important Bragg grating inscription methods and developments, whilst also providing essential details regarding the writing laser source. A condensed yet precise format for the treatment of fiber grating theory is contained within Chapter 5. All of the equations in their final form can be used to generate the examples provided in the text. Chapter 6 explores optical communications, an area where Bragg gratings have had a tremendous impact. The applications include those related to optical fiber-based components, such as fiber lasers and amplifiers, and a number of innovative filters, for example, add/drop multiplexers and comb and superstructure devices. The application to all-optical communication systems is also presented, emphasizing dense wavelength division multiplexing. An increasingly important field is optical fiber-based sensing and an expansive treatment of Bragg grating sensors is provided in Chapter 7. We investigate the multitude of passive and active wavelength-discrimination techniques, for single and multiplexed sensors. Sensors based on chirped Bragg gratings and their use as interferometric sensor elements and reflective markers are also presented. Furthermore, we present the myriad of sensing applications that Bragg gratings have impacted, from structural monitoring to medical physics. In Chapter 8 we discuss the influence of the fiber Bragg grating on the optical communications and fiber sensing markets, highlighting the areas of potential growth.

We are indebted to our families and wish to express our thanks for their support. We would also like to extend our thanks to Artech House for their support and for allowing us to format the camera-ready copies of this book. Although this greatly increased the workload of producing the book, we feel that it was ultimately worth it.

Andreas Othonos
Kyriacos Kalli

January 1999

Chapter 1

INTRODUCTION

1.1 Fiber Bragg Gratings

Following the realization of low loss optical waveguides in the 1960s, optical fibers have been developed to the point where they are now synonymous with modern telecommunication and optical sensor networks. A major drawback to the evolution of optical fiber-based networks has been the reliance on bulk optics for conditioning and controlling the guided light beam. The necessity of coupling light out of the waveguides to perform, for example, reflection, diffraction, and filtering (spatial, polarization, etc.) is an inherently lossy process. Moreover, coupling light in and out of fiber significantly increases the number of high-quality, bulk optic components, often requiring stringent tolerance on optical alignment, thus making conceptually simple systems complicated and expensive in practice. Replacing a bulk optic mirror or beam splitter with a fiber equivalent can dramatically increase system stability and portability, while reducing overall size, thus pushing laboratory-based experiments into real world environments. The most successful fiberized technology to date is the optical fiber laser and amplifier and fused tapered coupler. The intrinsic low loss nature of these components and their compatibility with integrated-optic waveguide structures have made them indispensable to the continued development of optical systems as a whole.

With the significant discovery of photosensitivity in optical fibers, a new class of in-fiber component has been developed, called the *fiber Bragg grating*. This device can perform many of the aforementioned primary functions, such as reflection and filtering, in a highly efficient, low loss manner. Fiber Bragg gratings are set to revolutionize telecommunications, and will also have a critical impact on the optical fiber sensor field. This is a comparatively simple device and in its most basic form consists of a periodic modulation of the index of refraction along the fiber core (Figure 1.1). Grating-based structures in guided wave optics have long been recognized as being very important because of their integration with fibers and the large number of device functions that they can facilitate. Surface relief grating structures, originally implemented in planar optic waveguides, have been investigated for a wide variety of filtering and coupling functions, some of which have been demonstrated in fiber and all of which have subsequently been written directly into photosensitive fiber. Ultraviolet (UV) written fiber gratings are relatively easy to fabricate; they often result in minimal perturbation of the fiber structure and are becoming increasingly inexpensive to manufacture. The advantages offered by optical fiber, such as low loss transmission, immunity to electromagnetic interference, light weight, and electrical isolation, also make the intra-core grating an ideal candidate for use in telecommunications and sensing. The versatility of the fiber Bragg grating has stimulated a number of significant innovations [1, 2].

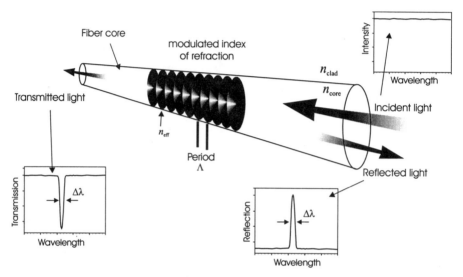

Figure 1.1   A schematic representation of an intra-core Bragg grating, with the planes of the modulated index of refraction shown along with reflected and transmitted light beams. A typical spectral response from such a Bragg grating are also shown with the peak of the spectral changes occurring at the Bragg condition ($2n_{eff}\Lambda$).

For a conventional fiber Bragg grating the periodicity of the index modulation has a physical spacing that is one half of the wavelength of light propagating in the waveguide; it is the phase matching between the grating planes and incident light that results in coherent back reflection. Reflectivities approaching 100% are possible, with the grating bandwidth ($\Delta\lambda$) tailored from typically 0.1 nm to in excess of 100 nm. These characteristics make Bragg gratings suitable for telecommunications where they are used to reflect, filter, or disperse light. Fiber lasers capable of producing light at telecommunications windows utilize Bragg gratings in forming both the high-reflectivity end mirror and output coupler to the laser cavity, realizing an efficient and inherently stable source. Moreover, the ability of gratings with nonuniform periodicity to compress or expand pulses is particularly important to high-bit-rate, long-haul communication systems. For example, grating-based dispersion compensation of 10 Gbps transmission systems over ~270 km has been demonstrated. Furthermore, the Bragg grating meets the demands of dense wavelength division multiplexing, which requires narrowband wavelength selective components, offering very high extinction between information channels. There are numerous applications that exist for low loss, fiber optic filters, including ASE noise suppression in amplified systems, pump recycling in fiber amplifiers, and soliton pulse control. Additionally, the wavelength selective properties of gratings have been used to generate true-time delays in microwave phased-array antenna systems.

The Bragg grating is also capable of coupling light from a propagating mode to another mode that has a propagation constant that matches the spatial periodicity of the grating. This may result in coupling between the forward and backward propagating core modes, or between the fundamental core mode and cladding or radiation modes. This property may be employed in fiber amplifiers to selectively outcouple unwanted wavelengths, giving uniform spectral gain.

The grating planes are subject to temperature and strain perturbations (as is the host glass material), which modify the phase matching condition and lead to wavelength dependent reflectivity. Typically, at 1.5 μm the wavelength strain responsivity is ~1 pm/nε (pico-meter per nano-strain), with a wavelength shift of 15 pm/°C for temperature excursions. Therefore, tracking the wavelength at which the Bragg reflection occurs can be related to the magnitude of an external perturbation. This functionality approaches the ideal goal of optical fiber sensors: to have an intrinsic in-line, fiber-core structure that offers an absolute readout mechanism. The reliable detection of sensor signals is critical and spectrally encoded information is potentially the simplest approach, offering simple decoding that may even be facilitated by another grating. An alternative approach is to use the grating as a reflective marker, mapping out lengths of optical fiber. Optical time domain measurements allow for accurate length or strain monitoring.

The grating may be photoimprinted into the fiber core during the fiber manufacturing process with no measurable loss to the mechanical strength of the host material. This makes it possible to place a large number of Bragg gratings at predetermined locations on the optical fiber to realize a quasi-distributed sensor network for structural monitoring with relative ease and at low cost. The basic instrumentation applicable to conventional optical fiber sensor arrays may also incorporate grating sensors, permitting the combination of both sensor types. Bragg gratings are ideal candidates for sensors that measure dynamic strain to a 1-nε resolution in aerospace applications and as temperature sensors for medical applications. They also operate well in hostile environments, such as high-pressure, borehole-drilling applications, principally as a result of the properties of the host glass material.

Finally, it is also possible to have fiber grating structures that are transmissive, coupling light between core and cladding modes. In this case the phase matching condition dictates that the grating periodicity be several hundred microns, and accordingly these components are known as long period gratings.

1.2 Historical Perspective

Fiber photosensitivity was first observed in germanium-doped silica fiber in experiments performed by Hill and coworkers [3, 4] at the Communication Research Center in Canada in 1978. During an experiment carried out to study nonlinear effects in a specially designed optical fiber, intense visible light from an argon ion laser was launched into the core of the fiber. Under prolonged exposure, an increase in the fiber attenuation was observed. It was determined that during exposure the intensity of the light back-reflected from the fiber increased significantly with time, with almost all of the incident radiation back-reflected out of the fiber. Spectral measurements confirmed that the increase in reflectivity was the result of a permanent refractive index grating being photoinduced over the 1-m fiber length–subsequently called Hill gratings. This result spawned a new interest in a previously unknown photorefractive phenomenon of optical fibers called fiber photosensitivity. In their experiment the 488 nm laser light launched into the fiber core interfered with the Fresnel reflected beam (4% reflection from the cleaved end of the fiber) and initially formed a weak standing wave intensity pattern. The high-intensity points altered the index of refraction in the photosensitive fiber core permanently. Thus, a refractive index perturbation that had the same spatial periodicity as the interference pattern was formed, with a length limited only by the coherence length of the writing radiation. This refractive index grating acted as a distributed reflector that coupled the forward to the counter-propagating light beams. The coupling of the beams provided positive feedback, which enhanced the strength of the back-reflected light and thereby increased the intensity of the interference pattern, which in turn increased the index of refraction at the high-intensity points. This process continued until the reflectivity of the grating reached a saturation level. These gratings were thus called *self-organized* or *self-induced* because they formed spontaneously within the optical fiber. The specially designed fibers were supplied by Bell Northern Research and had a small core diameter heavily doped with germanium. In these first experiments, permanent index gratings with 90% reflectivity at the argon laser writing wavelength were obtained. The change in the modulated index (Δn) was estimated to be   approximately 10^{-5} to 10^{-6}. The bandwidth of the Bragg grating, measured by stretching and temperature-tuning the fiber, was very narrow (< 200 MHz), indicating a grating length of approximately 1m. In view of the grating characteristics, their potential for applications in telecommunications was immediately recognized. The gratings only functioned, however, at the writing wavelength, in the visible part of the spectrum, which was a severe limitation for the chosen application of telecommunications.

For almost a decade after its discovery, research on fiber photosensitivity was pursued sporadically in Canada using the special Bell Northern Research fiber. During this time Lam and Garside [5] showed that the magnitude of the photoinduced refractive index change depended on the square of the writing power at the argon ion wavelength (488 nm). This suggested a two-photon process as the possible mechanism of refractive index change. The lack of international interest in fiber photosensitivity at the time was attributed to the effect being viewed as a phenomenon present only in this special fiber. Almost a decade later Stone [6] proved otherwise. Present-day researchers throughout the world are proving that photosensitivity may be a characteristic of many different types of fiber.

1.3 A New Era: Externally Inscribed Bragg Gratings

Although the discovery of photosensitivity in the form of photoinduced index changes played a key role in the advancements of optical fiber technology, devices such as self-induced gratings were not practical. This was largely because the Bragg resonance wavelength was limited to the argon ion writing wavelength (488 nm), with very small wavelength changes induced by straining the fiber. The key development that turned this phenomenon from a scientific curiosity to a mainstream tool was the side-writing technique first demonstrated at the United Technologies Research Center (this is sometimes called the transverse holographic technique). In 1989 Meltz et al. [7], following the work by Lam and Garside [5], showed that a strong index of refraction change occurred when a germanium-doped fiber was exposed to direct, single-photon, UV light close to 5 eV. This coincides with the absorption peak of a germania-related defect at a wavelength range of 240–250 nm. Irradiating the side of the optical fiber with a periodic pattern derived from the intersection of two coherent 244-nm beams in an interferometer resulted in a modulation of the core index of refraction, inducing a periodic grating. Changing the angle between the intersecting beams alters the spacing between the interference maxima; this sets the periodicity of the gratings, thus making possible reflectance at any wavelength. Even though the writing wavelength was at 244 nm, gratings could be fabricated to reflect at any wavelength thus permitting their use in modern telecommunication and sensor systems. A subtle but important point is that this method relies on an increase in refractive index that is maintained in the long wavelength region of interest (i.e., 1300–1500 nm), even though the physical phenomenon is related to the absorption of light in the ultraviolet region. This technique was steadily refined so that by 1992 index changes as large as 2×10^{-3} were reported. By 1993 it was not uncommon for publications reporting values of Δn to be commensurate with the core-cladding refractive index difference. The resulting competition between the guidance from the core-cladding refractive index difference and the diffraction/mode mixing from Δn has made possible a wide variety of linear and nonlinear optical devices.

A basic comparison of the difference in efficiency between the two-photon (associated with self-induced gratings) and the single-photon writing process may be made by comparing the fluence levels required to induce comparable index changes. For the two-photon process the photoinduced refractive index saturates after exposure to fluence levels approaching 1 GJ/cm^2; on the other hand, the single-photon process requires only 1 kJ/cm^2 for the same index change, a factor of a million times less [8]. The magnitude of the refractive index change has been shown to depend on many factors, the most important being the writing wavelength, the writing beam intensity and net dosage, and the composition of the host material and any pre-processing that the fiber may have undergone. The most commonly used light sources are KrF and ArF excimer lasers, operating at 248 and 193 nm, respectively. These lasers typically generate pulses of 1020 ns duration and at repetition rates of tens of pulses per second. A typical example indicates that exposure to laser irradiation lasting for several minutes, at intensities of 100–500 mJ/cm^2, will result in a Δn that is positive in Ge-doped, single-mode optical fiber, having a magnitude of 10^{-5} to 10^{-4}.

An issue common to internally and externally written gratings is a reflectivity dependent on the polarization of the probing light beam; that is to say, the refractive index change is birefringent. This fundamental property is relevant to understanding the photo-physics of fiber Bragg gratings and has also proven to be useful to grating applications, such as the fabrication of polarization mode converting devices or rocking filters [9].

Further developments to be discussed in detail in the chapters that follow are sensitization techniques, such as hydrogenation. A photosensitivity enhancement of an order of magnitude increase in grating reflectivity strength (Δn of 10^{-2}) has been realized for standard telecommunications fibers through hydrogenation of fibers prior to UV exposure [10]. Additionally, the use of phase masks for grating fabrication has also made a tremendous impact on the field. Reliable mass-produced gratings, the reality of commercial grating-based devices, may be realized through the use of phase masks [11]. This is a technique derived from conventional photolithography. The phase mask is a diffractive optical element that spatially modulates the UV writing beam, and it is a surface relief structure made of silica glass. Interference between the diffracted plus and minus first orders results in a periodic, near-field, high-contrast intensity pattern, having half the phase mask grating pitch. For the correct UV wavelength, the interference pattern will photoinduce a Bragg grating into the fiber core. Although the use of phase masks does not introduce any improvement in the strength of the index modulation, it does relax both the high tolerances required by grating fabrication through the aforementioned transverse holographic technique and the stringent source stability and quality conditions.

1.4 Outline of This Book

In our attempt to offer a detailed description of fiber Bragg gratings that encompasses most of their important properties and applications, we have called upon many different fields of the physical sciences. This has meant balancing a discussion of materials, science, and photophysics with the very practical issues of grating inscription and the applications-oriented aspects that are so important to telecommunications and sensing. Chapter 2 covers fiber photosensitivity and describes the multitude of fundamental processes that are responsible for this phenomenon. Chapter 3 details the basic properties and characteristics of fiber Bragg gratings. Grating inscription is presented in Chapter 4. Chapter 5 uses coupled mode theory to explain experimentally observed grating spectra and summarizes important grating design criteria. Chapters 6 and 7 offer a presentation of the most important telecommunication and sensing applications, respectively. Finally, Chapter 8 examines the place for fiber Bragg gratings in the marketplace.

References

[1]   Bennion, I., et al. "UV-written in-fibre Bragg gratings," *Optical and Quantum Electronics*, Vol. 28, 1996, pp. 93–135.

[2]   Othonos, A. "Fiber Bragg gratings," *Review of Scientific Instruments*, Vol. 68, 1997, pp. 4309-4341.

[3]   Hill, K. O., et al. "Photosensitivity in optical fiber waveguides: Application to reflection filter fabrication," *Applied Physics Letters*, Vol. 32, 1978, pp. 647–649.

[4]   Kawasaki, B. S., et al. "Narrow-band Bragg reflectors in optical fibers," *Optics Letters*, Vol. 3, 1978, pp. 66–68.

[5]   Lam, D. K. W., and B. K. Garside, "Characterization of single-mode optical fiber filters," *Applied Optics*, Vol. 20, 1981, pp. 440–445.

[6] Stone, J., "Photorefractivity in GeO$_2$-doped silica fibers," *Journal of Applied Physics*, Vol. 62, 1987, pp. 4371–4374.

[7] Meltz, G., W. W. Morey, and W. H. Glenn, "Formation of Bragg gratings in optical fibers by a transverse holographic method," *Optics Letters*, Vol. 14, 1989, pp. 823–825.

[8] Hill, K. O., et al. "Photosensitivity in optical fibers," *Annual Reviews in Material Science*, Vol. 23, 1993, pp. 12–157.

[9] Hill, K. O., et al. "Birefringent photosensitivity in monomode optical fiber: Application to the external writing of rocking filters," *Electronics Letters*, Vol. 27, 1991, pp. 1548–1550.

[10] Lemaire, P. J., et al. "High-pressure H$_2$ loading as a technique for achieving ultrahigh UV photosensitivity and thermal sensitivity in GeO$_2$ doped optical fibers," *Electronics Letters*, Vol. 29, 1993, pp. 1191–1193.

[11] Hill, K. O., et al. "Bragg gratings fabricated in monomode photosensitive optical fiber by UV exposure through a phase mask," *Applied Physics Letters*, Vol. 62, 1993, pp. 1035–1037.

Chapter 2

PHOTOSENSITIVITY
IN
OPTICAL FIBERS

2.1 Introduction to Photosensitivity

Photosensitivity in optical fiber refers to a permanent change in the index of refraction of the fiber core when exposed to light with characteristic wavelength and intensity that depends on the core material. Initially, photosensitivity was thought to be a phenomenon only associated with optical fibers having a large concentration of germanium in the core and photoexcited with 240–250-nm ultraviolet (UV) light. Following many years of research, however, photosensitivity has been observed through photoexcitation at different UV wavelengths in a wide variety of different fibers, many of which do not have germanium as the only dopant and some of which contain no germanium at all. Nevertheless, germanium-doped optical fiber remains one of the most important materials for the fabrication of devices utilizing photosensitivity. Photosensitivity in optical fibers and waveguides has significant scientific and practical importance. This phenomenon has resulted in a new class of in-fiber phase structures, of which the fiber Bragg grating is arguably the most important [1]. Fiber Bragg gratings are quietly revolutionizing modern telecommunication systems and are introducing a new optical fiber sensor that may ultimately result in its general acceptance over equivalent electrical sensor devices. Furthermore, the application of photosensitivity to standard telecommunications fibers will have an impact on the global telecommunications market, estimated to be well in excess of $100 billion. Therefore, it is important to know whether the reliability of devices fabricated using photosensitivity can place any limitations on the practical use of large scale, optical fiber networks over many decades.

We shall begin by giving a very brief historical overview of the most significant findings relating to photosensitivity in optical fibers, the details of which will be further elucidated in subsequent sections. Photosensitivity was first observed in an optical fiber that was exposed to laser light at 488 nm launched into its core [2, 3]; this phenomenon was subsequently associated with a two-photon process [4]. A transverse writing method was later used to photo-imprint Bragg gratings at a direct excitation wavelength of 240 nm [5]. The absorption band centered on this excitation (240 nm) has been related to defect centers in germanosilicate glass [6, 7]. Irradiation with a wavelength coincident with this band was shown to result in bleaching and the creation of other absorption bands, leading to a refractive index change that was described through the Kramers-Kronig relation [8]. It was then discovered that photosensitivity could be improved by up to two orders of magnitude through hydrogenation of the optical fiber core before grating inscription, and in some cases without variation of the 240-nm absorption band [9]. The latest experimental findings indicate the formation of spectral changes below 240- and 193-nm excitation of non-hydrogen-loaded, low-germanium content fiber can result in high index changes that are commensurate with the fiber core-cladding refractive index difference [10]. It has been suggested from the photoinduced index growth obtained in high- and low-germanium content fiber that photosensitivity at 193 nm obeys one-photon dynamics in high-germanium content fiber, and two-photon dynamics in low-germanium content fiber. Recently, a two-photon process has also been observed in germanosilicate glass for various UV wavelengths [11]. The current consensus explains photosensitivity as being initiated through the formation of color-centers [12] that gives way to compaction of the UV-irradiated glass [13, 14].

2.2 Photosensitivity in Silicon-Based Optical Fibers

In order to determine whether a particular dopant is necessary for a fiber to be photosensitive, many studies of the dependence on the core dopants have been carried out. The first strong indication that this was indeed the case was the observation that germanium was always present in photosensitive fibers. It was also shown that grating growth was absent in pure silica core fibers and that the OH group was not necessary for a fiber to be photosensitive [15]. These results pinpointed germanium as the essential requirement. This, however, was shown to be inaccurate, and there are numerous examples in the literature of photosensitivity in a wide range of fibers, many of which do not contain germanium as a dopant. Fibers doped with europium [16], cerium [17], and erbium:germanium [18] show varying degrees of sensitivity in a silica host optical fiber, but none are as sensitive as germania. One fiber doping that produces large index modulations (of the order of 10^{-3}) is germanium-boron co-doping [19]. Photosensitivity has also been observed in a fluorozirconate fiber [20] doped with cerium:erbium where Bragg gratings were inscribed using 246-nm radiation. From a practical point of view, the most interesting photosensitive fibers are germanium core-doped, as they are used extensively in both the telecommunications industry and optical sensor applications.

Initially, when photosensitivity was thought to occur only in germanium-doped fiber, it was believed that the germanium oxygen vacancy defects, such as a twofold coordinated neutral germanium atom (O-Ge-O or Ge_2^0 center) or the Ge-Si or Ge-Ge (the so-called *wrong bonds*), were responsible for the photoinduced index changes. However, with the demonstration of photosensitivity in most types of fiber, it is apparent that photosensitivity is a function of various mechanisms (photochemical, photomechanical, thermochemical) and the relative contribution will be fiber dependent, as well as intensity and wavelength dependent. Several models, that were proposed to describe the photoinduced refractive index changes in germanium-doped fiber share the common element of the germanium oxygen vacancy defects as precursors responsible for the photoinduced index changes. During the high-temperature gas-phase oxidation process of the modified chemical vapor deposition (MCVD) technique, GeO_2 dissociates to the GeO molecule (in other words the Ge^{2+} center) due to its higher stability at elevated temperatures. When incorporated into the glass, this molecule can manifest itself in the form of oxygen vacancy Ge-Si and Ge-Ge wrong bonds. Regardless of which particular defect causes an oxygen deficient matrix in glass, it is linked to the 240–250-nm absorption band (peaking at 242 nm) and its centers are known as germanium oxygen-deficient centers (GODCs).

Observations that give further insight to the photosensitivity of fibers are the growth dynamics of the Bragg gratings as they are exposed to UV radiation. This is a complex issue as there are many types of fibers, different UV radiation bands, and laser powers available. In spite of this, one may distinguish three distinct dynamic regimes [21]. The first regime experienced in most of the writing experiments and applicable to the internally and externally written Bragg gratings described thus far (Chapter 1) corresponds to a monotonic increase in the amplitude of the refractive index modulation. This temporal evolution of the index of refraction change is a characteristic behavior of Type I photosensitivity, and the grating formed is referred to as a Type I Bragg grating. It is commonly observed in most photosensitive fibers under either continuous wave (CW) or pulsed UV irradiation. More specifically, this classification refers to a grating produced with 100 mJ/cm^2/pulse and a cumulative fluence that is typically greater than 500 J/cm^2. This strongly depends on how photosensitive the fiber is and results in a positive refractive index change ($\Delta n > 0$). Protracted UV exposure of this grating type in some instances results in complete or partial grating erasure, followed by new spectral formation associated with a highly negative Δn (for a cumulative fluence exceeding 500 J/cm^2), and an accompanying increase in the reflecting wavelength. This form of grating is classed Type IIA. Type IIA gratings are most often observed in high GeO_2-doped fiber (> 25 mol%, high numerical aperture (NA) fibers). It is almost certain that the mechanisms responsible for Type I and Type IIA are different. Type IIA gratings have also been called Type III in the open literature. It is interesting to point out that the formation and the erasure of a Type I spectrum induces a shift in the Bragg wavelength toward the red part of the spectrum, increasing both the mean and modulated index changes. At the time of erasure, the Bragg wavelength either shifts toward the blue part of the spectrum or does not significantly shift in the course of the Type IIA grating inscription. Furthermore, when using germanosilicate fiber, a second-order diffraction spectrum appears once the first-order Type I grating has begun to saturate; the second order itself saturates at the time of the erasure of the Type I grating and fades during the growth of the Type IIA first-order grating. Phenomenological models that account for these observations usually assume that the refractive index evolution with exposure time results from two local reactions. The first of which erases and produces some defects or chemical species that lead to a positive change in the refractive index, whereas the second reaction, which is slower than the first, produces a negative change through structural reorganization.

The third dynamic regime corresponds to irradiation at energy levels greater than 1,000 mJ/cm^2, resulting in a process that allows for a single excimer light pulse to photoinduce large refractive index changes in small localized regions at the core/cladding boundary. This grating has been classed as Type II and is a result of physical damage that is limited to the fiber core, producing very large refractive index modulations estimated to be close to 10^{-2}. This allows for direct observation with a phase contrast microscope of the induced changes, which appear to constitute direct physical damage of the fiber through localized fusion. The Type II grating should not be associated with the Type IIA grating described above; there is no similarity in the underlying mechanisms responsible for these grating types. Clearly, there is some confusion with the nomenclature that arises from having many research laboratories actively pursuing different lines of research concurrently. We will attempt to maintain the classification given above, which coincides with the majority of work published in the open literature thus far. The physical properties of the above grating types may be inferred through their growth dynamics and also by measurement of thermally induced decay. The accelerated decay is different for each grating type, with Type I being the least and Type II the most stable with temperature. Type IIA falls in between. This is not surprising given that Type I has been related to local electronic defects, Type IIA to compaction, and Type II to fusion of the glass matrix.

It has been asserted that one may study the growth dynamics of UV written Bragg gratings without making a distinction between gratings fabricated using pulsed or CW lasers [22]. There is certainly evidence suggesting that the basic functional law for the growth of Type I gratings is the same under both writing conditions. It is also undoubtedly true that writing Type I gratings with pulsed laser light is more efficient than with CW radiation [23–25]. Similarly, Type IIA gratings have been written under CW and pulsed conditions [22], whereas Type II have only been fabricated on exposure to pulsed radiation. We give specific examples for each grating type, and highlight differences between them.

Patrick and Gilbert [24] have written high-reflectivity Type I Bragg gratings in Ge-doped (10 mol%) optical fiber under 244-nm CW UV light exposure with laser intensities ranging from 1.5 to 47 W/cm^2. The observed dependence of index modulation on time and intensity is in disagreement with a model for which depletion of the defect population occurs via a one-photon absorption process, which predicts similar writing efficiencies for pulsed and CW light having the same average power. Gratings having a reflectivity of 80% were obtained for a 15-minute exposure to 47 W/cm^2 of CW UV light, whereas near identical gratings, produced in the same Accutether fiber by use of a pulsed source, required only 5 W/cm^2 average intensity and a 200-second exposure time [26]. Ruling out the possibility of errors in the reported results, it is possible that this difference in writing efficiency is related to transient heating, accelerating photothermal ionization of defects, or compaction of the glass matrix as a result of the high peak intensity of the pulsed light. Figure 2.1(a) shows the peak reflectance versus time for a Type I grating exposed to an intensity of 46 W/cm^2. The prediction from a one-photon absorption process is for exponential growth, whereas that calculated from a power law of the form $\Delta n \propto t^{\alpha}$ provides the best agreement. Figure 2.1(b) shows the power law fits to three exposures at lower CW intensities. It is useful to compare this to the similar growth obtained by using a pulsed UV laser for a lower total fluence (Figure 2.2) [25]. Single pulse gratings have been reported by Askins et al. with a 2% reflectivity and modulated index amplitude of 2×10^{-5} at pulse energies of ~1 J/cm^2 [27]. A reflectivity of 65% and index modulation of 5×10^{-4} has been achieved using a boron co-doped germanosilicate fiber, exposed to a pulse energy of 25

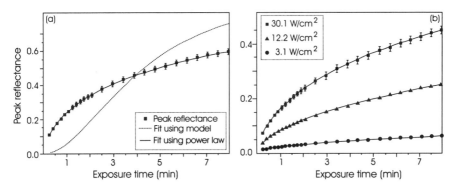

Figure 2.1   (a) Reflectance versus time for an exposure intensity of 46 W/cm² and a final index modulation of ~8x10⁻⁵. The dotted curve is a fit to R that assumes that Δn obeys the time dependence given by the model with $\Delta n_{max} = 1.71\times10^{-4}$. The solid curve is a fit that assumes that $\Delta n = Ct^b$ ($C = 4.3\times10^{-5}$, $b = 0.32$, t is in minutes).   (b) Reflectance versus time with fits assuming that $\Delta n = Ct^b$ for three gratings exposed at lower intensities (*After*: [24]).

J/cm² [28], and single pulse gratings have also been demonstrated in cerium-doped fibers [29]. The fitting parameter α increases with power density and lies between 0.25 and 0.32 for CW powers between 3 and 47 W/cm², whereas a value of 0.44 has been found for exposure to a pulsed laser [21]. Assuming a power law dependence of Δn with intensity of a form, $\Delta n \propto I^\alpha$, results in a fitting parameter α between 0.46 and 0.51 [24]. The Bragg resonant wavelength follows a similar functional form with $\alpha \sim 0.38$ [30].

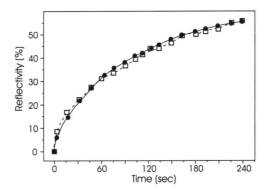

Figure 2.2   Growth rate of Bragg gratings written in AT&T Accutether fiber (10 mol% germania). The solid circles represent the growth rate of the initial Bragg grating; the squares correspond to the growth rate of the second writing after the first grating was thermally erased (*After*: [25]).

Type IIA gratings have most often been demonstrated in high Ge-content, small core fibers and have often been associated with the presence of high internal fiber stresses [31]. This is substantiated by the absence of Type IIA formation in low Ge-doped fibers at 240 nm. The lack of Type IIA behavior for gratings in hydrogen-loaded fibers implies that hydrogen treatment modifies a chemical or physical property of the fiber, which changes the conditions for the initial photosensitivity mechanisms related to the color-center and compaction models (Section 2.8). Riant and Haller have taken advantage of the photoenhanced writing efficiency at 193 nm in an attempt to fabricate Type IIA gratings in non-hydrogen-loaded, low Ge-doped, large core and hence low stress fibers, but without success [32]. A typical result for the development of Type IIA from Type I gratings is shown in Figure 2.3(a) (for an energy density lower than 330 mJ/cm^2) [33]. A pulsed laser operating at 244 nm (pulse duration 12 ns at 10 Hz) was used to record gratings in 28 mol% germanosilicate fiber. One observes a fast decrease in the transmission of the fiber, corresponding to a Bragg grating reflectivity increase of ~ 10%. The reflectivity eventually reaches 100%. Under continued irradiation the reflectivity decreases, (transmission increases) indicating the disappearance of the index modulation. Beyond this threshold another index grating appears. This complex behavior has been seen in many different fibers and it can be concluded that it does not arise from instability in the writing process. Figure 2.3(b) shows the formation of a Type IIA grating in highly Ge-doped fiber, illuminated with 110 mJ/cm^2 at 193 nm, for a repetition rate of 10 Hz [32]. This emphasizes the initial growth in the mean and modulated index with pulse number, with the mean index subsequently decreasing (or remaining constant depending on the fiber type or fluence level [33]), and the modulated index increasing. Recent experiments have shown that the formation dynamics of Type I/Type IIA grating spectra are strongly affected when the gratings are written in strained fiber [31]. Straining a fiber during grating inscription limits the Type I index modulation while accelerating the formation of the Type IIA grating [32].

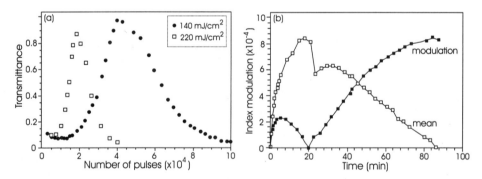

Figure 2.3   (a) Grating transmittance as a function of the number of pulses used for inscription (*After*: [33]). (b) Formation of a Type IIA grating in an unloaded, highly germanium-doped fiber (*After*: [32]).

Bragg gratings in germanosilicate fibers are found to exhibit temperature dependent decay of Δn with time after inscription. Erdogan et al. [34] reported measurements of thermally induced decay of gratings fabricated in erbium co-doped germanosilicate fiber. The decay in reflectivity is characterized by power law dependence with time, with a rapid initial decay followed by a decreasing decay rate (see Figure 2.4). Briefly, this behavior is consistent with the thermal depopulation of trapped states occupied by carriers that are photoexcited from their original band locations by UV irradiation. Thermally exciting carriers out of shallow traps causes the observed decay in the refractive index. Any residual carriers are related to the "stable" portion of the index change. A consequence of this model, which shows excellent agreement with experimental data, is that a grating may be preannealed to remove the portion of Δn that decays rapidly, leaving only the portion that has long-term stability. A noticeable difference between Type I and Type IIA gratings is their markedly different thermal behavior. Dong and Liu [35] have monitored both Type I and Type IIA temperature stability in boron co-doped germanosilicate fibers. Type I gratings were found to have reasonable short-term stability up to 300°C, whereas Type IIA gratings demonstrated excellent stability at temperatures as high as 500°C. We shall return to this aspect of grating behavior in Chapter 3 and discuss its implications to grating reliability and lifetime.

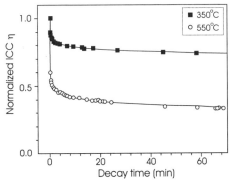

Figure 2.4   Measured integrated coupling constant normalized to starting value for two gratings heated to 350°C and 550°C as a function of decay time. The lines are fit to the data (*After*: [34]).

Figure 2.5 shows the change in the modulated index of refraction as a function of exposed energy for a germanosilicate fiber (15 mol%) to single, high-energy pulses of 248-nm UV light having duration of 20 ns [36]. Clearly there is a sharp threshold corresponding to a pulse energy of ~0.65 J/cm^2. Doubling the pulse energy from 0.45 to 0.9 J/cm^2 results in a photoinduced modulation index that is two orders of magnitude greater, whereas pulse energies to 0.56 J/cm^2 result in linear index growth. The stability of the Type II grating at elevated temperature is shown in Figure 2.6. It is stable to 800°C for periods of up to 24 hours, with no significant changes in grating reflectivity. At 900°C the grating begins to slowly decay but will remain for several hours at temperatures of 1000°C before complete erasure occurs [36]. This should be compared with a Type I grating, which is erased in seconds when exposed to temperatures of the order of 500°C. The physical damage tracks associated with the Type II grating are reminiscent to those observed during the passage of a fiber fuse, which is related to thermal breakdown in the fiber core [37]. It has been calculated that instantaneous temperature increases of several thousand degrees Celsius are responsible for Type II gratings. The resulting increase in absorption at high temperature greatly enhances the writing mechanism, leading to the sharp transition

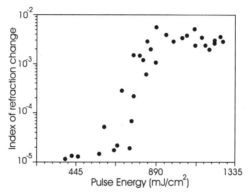

Figure 2.5   Index of refraction changes estimated for Bragg gratings inscribed in photosensitive fiber induced with a single excimer (KrF) laser pulse. The estimated refractive index change is plotted against the energy density of the laser pulse. Notice a sharp threshold at around 750 mJ/cm^2 (*After*: [36]).

observed in Figure 2.5. Gratings showing similar damage behavior have also been produced by single excimer pulse exposure through a phase mask at 193 nm [38] and 248 nm [39, 40]. The overview of results discussed above are summarized in Table 2.1, which provides typical figures for silica fiber photosensitivity for Type I, IIA, and II gratings [41]. A summary of the currently observed photosensitivity in optical fibers is shown in Figures A.1 to A.3 in the Appendix at the end of this chapter.

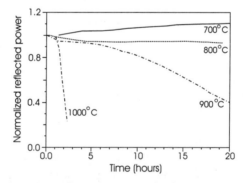

Figure 2.6   Temperature dependence of Type II gratings (*After*: [36]).

2.3 Anisotropy in the Photoinduced Index Change

Anisotropy in the photoinduced index change provides an alternative method of recovering information about the photosensitivity mechanism. That such a phenomenon exists at all immediately tells us that the defects have a preferential axis to the exciting laser beam. The argon ion and UV laser-induced refractive index change is generally anisotropic

Table 2.1   Typical Data for Germanosilicate Fiber Photosensitivity

Silicon fiber doping	Characteristics of writing sources	Fiber treatment	Change in the refractive index	Type	Erasure temperature °C	Ref.
Ge < 10% mol	240 – 262 nm Pulse or CW laser	▸ No treatment ▸ Boron doping ▸ Hydrogenation ▸ Flame brushing	$1\text{-}5\times10^{-4}$ 1×10^{-3} 3×10^{-3} 1×10^{-3}	I	100 60 23	[5, 202] [19] [139] [9]
	248 nm single pulse ~1 J/cm²		2×10^{-3}	II	800 – 1000	[36]
	193 nm 400 mJ/cm² two-photon process		10^{-3}	I	100	[10]
Ge > 15% mol	Pulse or CW under long exposure time	▸ No treatment ▸ Hydrogenation	10^{-3} at saturation of Type IIA no Type IIA	IIA	550	[33]
Ge 30% mol	334 nm CW argon ion	▸ No treatment ▸ Boron doping	0.8×10^{-4} 1×10^{-4}	I		

to a degree that is strongly dependent on the fiber type. Photoinduced birefringence in germanosilicate fibers has been studied in connection with self-organized gratings, which display a birefringence of the order of 10^{-6} [42], and externally written gratings for which the birefringence may be two orders of magnitude greater [5, 43, 44]. Similar anisotropy has been observed in the index change induced by the interference of visible light, again in germanium-doped fiber [42, 45–47]. The anisotropy of the UV induced index change has significant implications for phase gratings, offering further insight into the mechanism of photosensitivity. Exploitation of the polarization-dependent reflectivity has resulted in demonstrations of significant telecommunication components, such as the single-mode operation of an erbium-doped fiber grating laser [48], polarization mode converters and rocking filters [49], and in-fiber, in-line wave retarders [44].

It has been shown by Ouellette et al. [46] and Bardal et al. [50] that the induced birefringence is negative with respect to the polarization of the writing beam. Ouellette and co-workers [51] have extended their earlier studies of photoinduced polarization rotation in germanosilicate fiber [47] by studying the dynamic and orientational behavior of bleaching the 240-nm absorption band in fibers. In keeping with their earlier supposition, bleaching was shown to be site dependent, with the absorbing defects possessing oriented dipoles that were preferentially bleached when aligned to the laser, with the induced luminescence also being anisotropic. The role of the by-products of the bleaching, rather than the bleached defect, was emphasized as being responsible for the observed refractive index change. An important result of this work is the recognition that a site-dependent reaction rate can lead to two possible paths. A similar idea has been put forward by Hosono and co-workers [6, 7]. In the first path, the breaking of the Ge-Si wrong bond results in a GeE' center and single free electron, as supported by the evidence of Simmons et al. [52]. The second path associates the GeE' center as a reaction product between a wrong bond or GeO defect and another defect species, the reaction rate depending upon the energy transfer between the two. The latter model is favored on the basis that the defect is excited to a long-lived triplet state [6, 7], although it should be noted that both possibilities are consistent with a site-dependent reaction and may, in fact, both be active.

Erdogan and Mizrahi further characterized the anisotropic nature of the UV-induced index change in several photosensitive germanosilicate optical fibers [53]. They found that the measure of birefringence was dependent on both the fiber type and exposure conditions, growing proportionally to the total induced index change. The maximum and minimum values for the induced index change measured were 8% and 0.2%, for AT&T Accutether and hydrogenated telecommunications fiber, respectively. Erbium-doped, germanosilicate fiber suitable for use in grating-based fiber laser systems was also examined. They confirmed the findings of Poirier et al. and other workers, linking the polarization of the UV light with preferential excitation and bleaching of defects [42, 45–47, 51]. Based on this hypothesis, they were able to show that light polarized parallel to the fiber axis leads to highly isotropic gratings; however, their measurements were limited by the intrinsic birefringence of the different fibers. Figure 2.7(a) shows the measured birefringence for Accutether and erbium-doped germanosilicate fibers, respectively, under uniform UV illumination for s-polarized light ((330 mJ/cm^2)/pulse, pulse duration 15 ns at 30 Hz). The birefringence growth follows a functional dependence of the form I^α, displaying a strong dependence on the UV polarization. A reversal of birefringence is observed at early writing times [54, 55]. This has been attributed to the different effects of UV exposure on the stress-induced and geometric birefringence components that each contributes independently to the intrinsic birefringence of the fiber. Figure 2.7(b) confirms this for hydrogenated fiber, where a ratio of induced birefringence $\Delta n_s/\Delta n_p \sim 6$, even though the extinction ratio between wanted and unwanted polarizations is 10:1. Hydrogenated fiber has a very low relative birefringence of 0.2% for p-polarized UV light, leading to the fabrication of highly isotropic gratings. Reduced birefringence in hydrogenated fibers could occur if bleaching of the 240-nm band were highly polarization dependent [56].

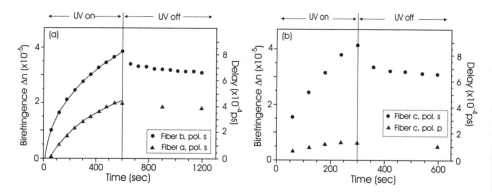

Figure 2.7   (a) Measured birefringence and estimated delay as a function of UV exposure time for Accutether [a] and [b] Er-doped fiber irradiated by a single UV beam. The solid curves are theoretical fits using a power law functional form. (b) Measured birefringence and estimated delay as a function of UV exposure time for 3 mol% hydrogenated fiber (GeO$_2$, 3 mol%) for s- and p-polarizations (*After*: [53]).

An alternative explanation for the induced birefringence has been put forward by Inniss et al. [57, 58], who used atomic force microscope (AFM) measurements to argue a case for geometric asymmetry of the side-writing process as being the major contribution, and at the same time demonstrating a new dual-exposure method of reducing birefringence. Recent measurements by Meyer and co-workers [44] on ultralow-birefringence fibers point away from the conclusions drawn in [58] in favor of preferential bleaching of orientated dipoles. By successively exposing the fiber to s- and p-polarized UV light, they achieved similar results to Erdogan and Mizrahi [53]. The ultralow-birefringence fiber exhibited very high sensitivity to s-polarized light, inducing a large birefringence component that was completely cancelled by the p-polarized light. The completely reversible nature of the mechanism contradicts [58], as does the fact that the same degree of geometric asymmetry was observed for fiber exposed to both s- (high relative birefringence) and p-polarized (low relative birefringence) light. This again points to an anisotropy produced by dipole moments with preferential bleaching of the 240-nm band. The sensitivity was found to be dependent on the total UV light dosage, with the degree of initial growth being very sensitive to the energy of the first few pulses. This fast growth process gives way to a slower process and saturation. This behavior is consistent with Poirier's [51] and Hosono's [6, 7] assertion of a dual-path reaction mechanism, with the birefringence creation and erasure through the reorientation of existing dipole moments (fast process), followed by the creation of additional dipole moments (slow process). Psaila and co-workers have investigated the effects of 193- and 240-nm radiation on photoinduced birefringence by measuring the growth of rocking filters in elliptical-core Ge-doped fiber [55]. Their results show that the photoinduced birefringence at 193 nm is larger than at 240 nm, which has implications both for the underlying physical mechanism associated with photosensitivity and for the realization of novel polarization controlling structures.

2.4 Point Defects in Silicon Glass

Defects are important to optical fibers because their absorption bands cause deleterious transmission losses; these defects are called color-centers. Ionizing radiation [59] and the fiber drawing process [60] can produce defects. Color-centers are also responsible for non-linear optical fiber transmission [61, 62], where the transmission changes in time and with light intensity and the destructive fiber fuse effect [37]. In the 1980's defects were implicated in the phenomenon of second harmonic generation [63–67], as well as the fabrication of phase gratings in optical fibers. The one-photon process that triggers the photoinduced change at 240 nm is well below the band gap at 146 nm, thereby implying that point defects in the ideal glass tetrahedral network are responsible for the observation of photosensitivity. In 1956 Weeks [68] reported a narrow resonance in the electron spin resonance (ESR) spectra of neutron irradiated crystalline quartz and silica from a species termed the E' center [69]. This was the first study of point defects in amorphous silica, and many other defects in silica and germanosilicate glass have since been characterized. Point defects in optical fibers are important to various phenomena, yet their origin, chemical structure, and role are uncertain. To optimize or control these defects it is essential that their nature be understood.

Using ESR, three intrinsic defects in silica (Figure 2.8) have been identified: the E' center, the nonbonding oxygen hole center (NBOHC, $\equiv$ O.), and the peroxy (or superoxide) radical ($\equiv$ O-O.) [70]; all have been correlated with absorption bands. A fourth significant diamagnetic defect exists that was originally identified by Kaiser [60] in silica optical fibers by a characteristic absorption band at 630 nm. It is known as the drawing induced defect (DID). Kaiser noted a fiber loss of 500 dB/km at this wavelength. Several variants of the generic E' center have been identified in the ESR hyperfine structure of quartz [69, 71] and amorphous silica [70, 72], which have a common structure of the form Si. An unpaired electron (which renders the center ESR active) occupies a tetrahedral (sp³) orbital of an under-coordinated silicon atom (bonded to only three oxygen atoms). The most common and thermally stable variant, the SiE' center, is shown in Figure 2.8. In oxygen-deficient silica, the Si-Si wrong bond is the dominant precursor defect, so SiE' defects are common. While the basic Si structure was postulated as early as 1960 [69, 71], the asymmetric nature of the SiE' was not realized until 1975 [73].

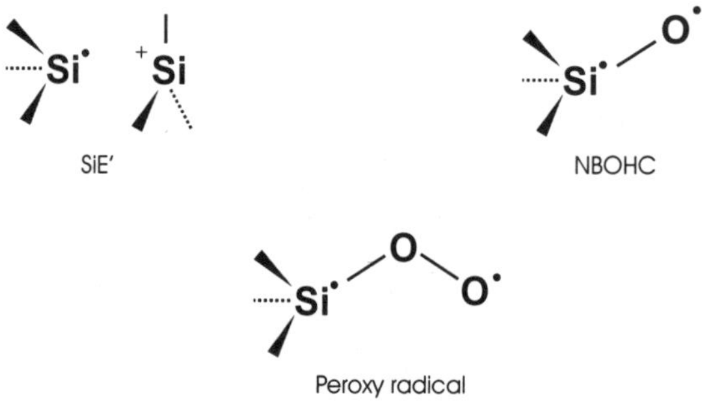

Figure 2.8   The structure of point defects in silica.

The E' center is considered to be the fundamental defect center in SiO_2, and it is associated with aging effects and radiation degradation. Yip et al. [73] characterized the defect electron, through ESR data, as a nonbinding tetrahedral hybrid orbital centered on silicon and pointing in a direction normally associated with a Si-O bond. A relaxed oxygen vacancy model was proposed for the E' center by Fowler and co-workers [74, 75].

Several absorption bands arise from intrinsic defects. In silica 160, 173, 215, 245, 260, and 630 nm [70] are most prominent. Not all bands have been definitively assigned; there is consensus over the SiE' band at 215 nm, the preoxy radical at 160 nm, the NBOHC at 260 nm, and the Si-Si wrong bond at 245 nm, but not for the DID band at 630 nm [60].

2.5 Point Defects in Germanium-Doped Silica (Germanosilicate) Glass

Since germanosilicate fiber cores are the most important photosensitive fibers, it is useful to look at the point defects in germanosilicate glass. Ge, unlike Si, has two moderately stable oxidation states, +2 and +4; thus, germania can be expected as both GeO_2 and GeO in glass. From thermodynamic considerations one expects that concentrations of GeO_2 will be proportional to the GODC concentration in the glass. It is well known that the suboxide GeO becomes more stable than GeO_2 at high temperature [76, 77]. This is important in fiber preform fabrication since the incorporation of GeO produced during the high-temperature, gas phase oxidation process of MCVD results in an oxygen-deficient matrix. Although GeO is sometimes considered to exist as discrete molecules in the germanosilicate matrix [78], it will likely manifest itself in the form of Ge-Si wrong bonds, which are suspected defect precursors. Indeed, it has been found by Jackson et al. [77] that maintaining glass at an elevated temperature ($\sim 1600°C$) followed by rapid cooling to room temperature results in the formation of a large number of defects. This parallels the drawing of optical fiber from a glass preform; thus, one may anticipate that the fiber will possess a defect distribution related to the drawing process, in addition to any variance in fiber or preform stress distributions.

Early work on photoinduced paramagnetic defect centers of germanosilicate fibers was performed by Friebele et al. [79] in 1974 on low loss (10 dB/km) Corning Glass Work's optical fiber. ESR spectra performed on γ-irradiated germanosilicate fibers with a Ge core concentration of 10 mol% identified four overlapping spectra for defects, denoted Ge(0), Ge(1), Ge(2), and Ge(3), in addition to silicon E' centers. The damage centers were directly related to the Ge content in the core. The notation Ge(n) centers is used for diamagnetic Ge sites that can trap an electron to become Ge(n)⁻ centers. The Ge(0) and Ge(3) centers have since been identified with GeE', the center having the deepest electron trap depth [80, 81]. Further studies of radiation effects in Ge-doped silica identified the paramagnetic GeE' defect center as a by-product of the GODC bleaching, induced by exposure to visible light and radiation [82]. This behavior was accounted for by considering the GODC to be a Ge atom with a next-nearest-neighbor Si atom. The breakage of the bond between this pair formed the GeE' center and released an electron that facilitated the formation of new defect centers, such as the paramagnetic Ge(1)⁻ and Ge(2)⁻, through charge retrapping elsewhere in the network [80].

While the Ge(n)⁻ centers were initially thought to be variants of the GeE' center, further ESR studies [83] suggested they were composed of an electron trapped at a normal four coordinated Ge atom, with distortion of the tetrahedral structure. Tsai et al. [80, 84] proposed such structures for the Ge(1) and Ge(2) centers. The Ge(1) center is an electron trapped at a Ge atom coordinated to four O-Si next-nearest-neighbor atoms, while the Ge(2) center is an electron trapped at a Ge atom coordinated to one O-Ge (≡ Ge-O-Ge ≡) and three O-Si next-nearest-neighbor atoms (Figure 2.9). Ge(n)⁻ centers are likely to be sites of the tetrahedral network characterized by large distortions of the O-Ge-O angles from 109.4 degrees, which mix in the Ge d-orbitals for the efficient capture of electrons into deep traps. Other ESR investigations indicate that the GeE' center is axially symmetric. The structures of the Ge-Si and GeE' centers are also shown in Figure 2.9. Of significance is the large bond length increase between the Ge and Si atoms following ionization of the Ge-Si wrong bond. This is driven by the change of hybridization of the Si atom from sp^3 to sp^2, which draws the Si atom toward the plane of the three oxygen atoms to which it is bonded. There are implications to such large structural changes, both for stress relief and for compaction models. The ESR studies are complemented by absorption and luminescence spectroscopy measurements. The wide variety of defects in germanosilicate glass often results in the transfer of energy between them, which can be observed using spectroscopy. However, there is some disagreement over specific band assignments.

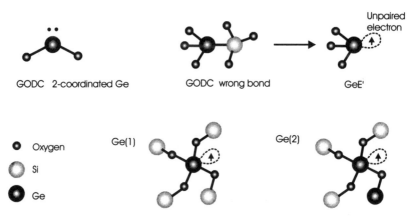

Figure 2.9   Possible GODC candidates. The GeE' center and the Ge(1) and Ge(2) electron trap centers.

A number of Ge-doped silica absorption bands have been reported [78, 82] principally at 180, 213, 240, 281, 325, and 517 nm (which obscure some of the silica bands), and the DIDs at 630 nm. The 281-nm (4.4 eV) and 213-nm (5.8 eV) bands have been assigned to the Ge(1) and Ge(2) centers [82]. The strong 240-nm band and the weak 325-nm band (which scales 1000 times smaller than the 240-nm band) have been assigned to the wrong bond defect [61, 82], although the original assignment was to discrete GeO molecules [78]. (The GeO is a reactive species and is more likely to be incorporated as an oxygen vacancy in the germania silica network, producing the Ge-Si wrong bond.) The 240-nm band is attributed to a singlet-to-singlet transition (S_0 to S_1), and the 325-nm band to the partially forbidden singlet-to-triplet transition (S_0 to T_1). The absorption spectrum in the UV range is shown in Figure 2.10(a). In the course of UV irradiation, the absorption spectrum changes as shown in Figure 2.10(b) [25], with the 240-nm band decreasing together with the fluorescence intensity. At the same time, the absorption increases both for longer and shorter wavelengths. The band at 517 nm has been tentatively assigned to an O hole center [82]. It is interesting that the omnipresent GeE' defect has only recently been reliably correlated to any absorption band, namely the 195-nm band, by Hosono et al. [85], through isochronal annealing experiments (Figure 2.11(a)). Prior to this, the only evidence that assigned a band to this defect was a weak feature at 340 nm associated with an emission band at 400 nm in an isolated report by Kashiwazaki et al. [86]. Tsai and Griscom [84] have predicted a 4.4-eV absorption band for the GeE' center in Ge-doped silica, as observed in GeO_2 glass [87], but as yet this remains experimentally unidentified. The results of Atkins et al. [88], who measured the UV induced absorption change down to 165 nm, indicated that the main absorption changes in germanium-doped silica are due to bands at 195 nm (6.35 eV) and 242 nm (5.1 eV). They identified additional weak bands at 256 nm (4.85 eV),

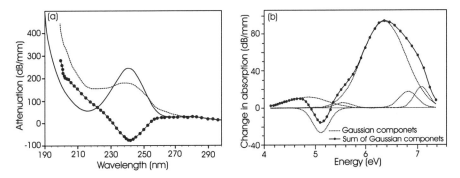

Figure 2.10 (a) UV absorption spectra before (solid line) and after (dashed line) writing an 81% peak reflectivity grating in AT&T Accutether single-mode fiber. The change in attenuation (solid circles) is also shown (*After*: [88]). (b) Changes in the UV absorption spectra for 3 mol% germania MCVD fiber preform core: data points (solid circles), six Gaussian components of fit (dashed line), and sum of Gaussian components (solid line) (*After*: [25]).

224 nm (5.64 eV), 183 nm (6.82 eV), and 175 nm (7.1 eV). These weak bands are typically an order of magnitude weaker than the main band at 195 nm, and they contribute to only a small fraction of the total index change.

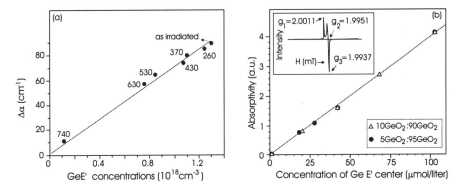

Figure 2.11 (a) Correlation between concentrations of laser-induced GeE' centers and absorption coefficients of the 6.4-eV band. The data were taken from isochronal annealing experiments. Numbers in figure denote annealing temperature and solid line a least square fit to the data (*After*: [85]). (b) Correlation between concentrations UV-induced GeE' centers and the absorptivity of UV-bleached 5.06-eV band. The inset is the ESR signal of the UV-induced GeE' center (*After*: [6]).

Luminescence spectroscopy has not been used as extensively as ESR and absorption measurements for investigating defects, although it has received increased attention recently. Most germanosilicate luminescence bands occur at similar wavelengths to those of silica [89], and there is a suspicion that some silica bands may result from Ge impurities [90]. Such luminescence bands have been reported at 275, 295, 400, 413, 450, and 650 nm

[91]. For example, absorption at 260 nm and luminescence at 300 and 413 nm have been correlated with a germanium oxygen vacancy, whereas absorption at 245 nm and luminescence at 295 and 400 nm with a silicon oxygen vacancy [89, 92]. Williams et al. made UV loss measurements on germanosilicate fiber samples and associated the fluorescence at 413 nm with the 242-nm absorption band [93]. Poirier et al. have linked the blue luminescence at 400 nm to site-dependent bleaching of the 240-nm band [51]. Furthermore, Malo et al. [94] have shown that a photoexcited fiber exhibits a thermoluminescence signal when heated. Additional bands from germanosilicate glass have been reported at 560, 620, and 680 nm [89]. As with absorption studies, considerable controversy exists over the band assignments. Furthermore, luminescence spectra can, in analogy with their absorption counterparts, be described in terms of direct and indirect transitions. The 290- and 400-nm bands are considered to be S_1 to S_0 and T_1 to S_0 transitions, respectively. Therefore, these bands and their associated absorption result from the same GODC defect.

Mizrahi and Atkins questioned the validity of correlating the 242-nm absorption band in germanosilicate fibers to the 400-nm luminescence band. They observed that the partial bleaching of the 242-nm band by light at the same wavelength was not accompanied by any changes in luminescence on an equivalent time scale. This significant result indicates that at least two different defects are absorbing light at 5.1 eV [26]. Hosono and co-workers reported comparable observations and arrived at a similar conclusion [6]. The time-dependent emission spectrum from the optical fiber core shows that UV-induced changes are fast and large at the beginning, followed by a slower and weaker response [51]; therefore, one must consider the role of structural defects [6]. Duval et al. [95] have shown that Δn is proportional to the fluorescence change that is associated with the 240-nm absorption at the beginning, but then the photoluminescence levels off, whereas Δn continues to increase. Finally, it has been confirmed that the bleaching rate of the 240-nm absorption band is dependent on both the UV excitation intensity and the Ge concentration. Therefore, decreasing the UV power results in some of the bleached defects returning to their initial state [96].

Cathodoluminescence (CL) is an alternative technique used to study the spatial distribution of defects in optical fiber preforms [97]. Sharing similarities with luminescence, it relies on an electron beam rather than photons for sample excitation. CL and luminescence bands are ordinarily identical, but whereas luminescence bands can be excited individually by selecting a discrete excitation wavelength, all CL bands are excited together. Atkins et al. [97] have applied CL to MCVD- and vapor phase axial deposition (VAD-) manufactured preforms with an electron beam of energy 20 keV. CL has also been applied to optical fibers [98, 99] and Atkins and co-workers assigned the GeE' and Ge DID centers to CL bands at 425 nm and 650 nm, respectively. It should be noted that the 650-nm band has been independently related to DIDs by Dianov et al. through laser-induced luminescence [100]. The CL from silica glass degrades upon exposure to the electron beam [101], but it has been shown that if the beam current is small, the degradation is negligible during the time required to acquire spectra and intensity images [102]. One must consider the possibility that CL does more than passively probe the sample, given the comparatively high-energy electron beam.

2.5.1 The Defect Precursors

The complex electron transfer phenomena triggered by irradiation with 240-nm UV light remains obscure. Two possible candidates, the Ge-Ge or Si-Ge wrong bonds (neutral oxygen monovacancy) or the Ge^{2+} ions coordinated by two oxygen atoms and having two lone pair electrons (neutral oxygen divacancy) [6], are believed to induce the 5-eV band in germanosilicate glass. The latter candidate has alternatively been proposed to be a divacancy-type oxygen-deficiency center (ODC) [103]. These issues are discussed below, which is analogous to the GODC, for germanosilicate glass preforms and optical fibers.

2.5.1.1 GeO_2: SiO_2 Preforms

The 5-eV optical absorption band in SiO_2 (the B_2 band) was first proposed to be an oxygen-deficient center by Arnold [104]. Tohmon et al. later identified two types of B_2 bands, $B_{2\alpha}$ and $B_{2\beta}$ [105]. The former band has an optical absorption center at 5 eV that is always accompanied by a band at 7.6 eV, strong luminescence at 4.3 eV, and a weak band at 2.7 eV. The $B_{2\beta}$ band has its absorption centered at 5.16 eV, with weak and strong luminescence bands at 4.3 and 3.1 eV, respectively. The 4.3-eV luminescence excited by the 7.6-eV band decays in a time scale shorter than 10 ns and implies a direct singlet-to-singlet transition (a relaxed S_0 to S_1 ODC), whereas the 4.3-eV luminescence excited by the $B_{2\alpha}$ band (5 eV) has a decay time of 100 μs. This is indicative of a ground singlet-to-triplet transition (an unrelaxed S_0 to T_1 ODC). Nevertheless, these two distinct bands are caused by the same oxygen vacancy. Mizrahi and Atkins found evidence that the 5-eV absorption band is correlated to more than one defect [26].

There is very strong evidence that the photoinduced refractive index changes in germanosilicate glass, through CW Ar^+ laser (at 488 nm) or filtered Xe/Hg lamp irradiation, originate from the photochemical conversion of GODCs (5-eV absorption band) into GeE' centers, which are linked to an intense absorption band peaking at 6.4 eV [85]. The conversion of the wrong bond defect (a possible GODC) to GeE', GeO_3^+ (or SiO_3^+) and the release of an electron has been confirmed by Hosono et al. [6]. Kashyap et al. have given further credence to free electron movement by measuring carrier induced changes in the photocurrent of Ge- and P-doped silica waveguides under UV irradiation [106]. Optical and ESR measurements on chemical vapor deposition (CVD) prepared germanosilicate glass preforms [6] have resulted in the decomposition of the 5-eV absorption band into two closely spaced components. The formation of the GeE' center increases in parallel with the bleaching of the band at 5.06 eV. This is shown in Figure 2.11(b), where the oscillator strength is estimated to be ~0.4. The linear relationship in Figure 2.11(b) provides a basis for attributing the 5.06-eV band to the precursor site of the GeE' center. The precursor is a neutral oxygen vacancy (NOV) coordinated with two Ge ions (a wrong bond). The second component at 5.16 eV remains unbleached when excited with low light levels at 248 nm, but displays intense luminescence at ~400 nm, along with a weak signal at 290 nm. This second component has been assigned to a Ge^{2+} coordinated to two oxygen atoms [6]. This suggests that alternative photochemical reactions may be present in germanosilicate glasses that depend on the UV power density.

We shall report in some detail the work relating to the possibility of two different reaction pathways, thus pointing out that the validity of relating the 5.16-eV absorption band to a two-coordinated Ge (Ge^{2+}) is uncertain. Tsai and co-workers [103] have presented a strong argument in favor of an ODC defect that is consistent with Tohmon [105]. A reaction path proposed between two-coordinated defects and atomic hydrogen results in paramagnetic centers of the form, X.-H, where X may be either Si or Ge and . denotes a nonbonding electron. When X = Ge, a 11.9-mT ESR doublet is proposed to exist, making it possible to directly correlate between temperature induced changes in the 5.16-eV band and the corresponding changes in the UV induced, hydrogen-associated paramagnetic defects [84, 107]. To test this hypothesis, the optical absorption coefficients of the 5.16-eV absorption band of silica samples containing various concentrations of Ge (up to 180 ppm) were measured along with ESR spectra [103]. The 5.16-eV band was related to the presence of Ge, increasing as the square root of the Ge concentration, rather than the anticipated unitary dependence, as would be expected for a two-coordinated Ge defect. The square root dependency is indicative of a divacancy-type ODC. Furthermore, an association with a two-coordinated Ge defect would be apparent in ESR data, with the concentration of the 11.9-mT doublets being proportional to the product of the 5.16-eV absorption coefficient and OH concentration. No such linearity is observed, suggesting an alternative defect to the two-coordinated Ge. At the high temperatures required for glass manufacture, two GODC divacancies are in thermal equilibrium with a GeO_2 in silica (Figure 2.12). Therefore, the thermochemical equilibrium constant relates the square of the ODC concentration with linear proportionality to GeO_2, as observed. This is a very strong argument in favor of a divacancy-type ODC [6]. Poirier et al. have reported photobleaching of the luminescence associated with this band, which is consistent with this conclusion [51].

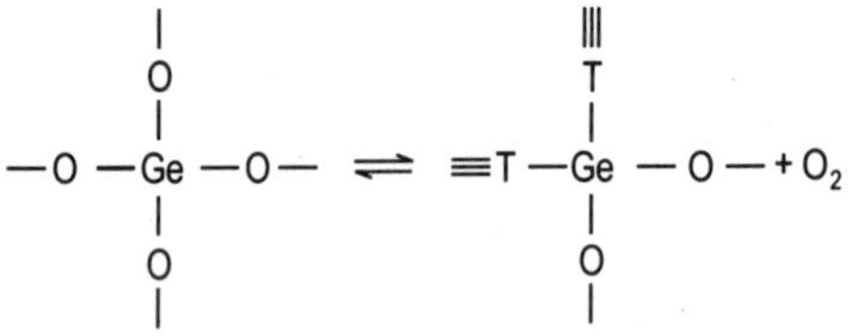

Figure 2.12   Two oxygen-deficient centers of a Ge divacancy in thermochemical equilibrium with a GeO_2 in silica, where T is Ge or Si.

To further study the photochemical reactions in GeO_2: SiO_2 glass, Nishii et al. [7] exposed VAD-prepared germanosilicate glass to UV radiation from a Hg discharge lamp (4.9 eV) and a KrF excimer laser (5 eV), confirming the existence of the two photochemical reaction paths [7] (discussed above). Exposure to Hg lamp radiation (16 mW/cm^2) induced GeE' centers and bleaching of the absorption band due to oxygen deficient defects near 5 eV and the emergence of an intense band near 6.4 eV; this is a one-photon process. However, irradiation with a KrF laser (10 mJ/cm^2/pulse, 20 ns) generated two types of paramagnetic defects: electron trapped centers associated with fourfold coordinated Ge ions, so-called Ge-electron centers (GEC = Ge(1)$^-$ and Ge(2)$^-$), and a self-trapped hole center (STH: a bridging oxygen trapping a hole), which is a two-photon process. The two possible reactions proceed independently and depend on the power densities of UV photons. The formation of GECs is saturated by irradiation with KrF laser pulses, finally being converted to GeE' centers under prolonged irradiation. Figure 2.13 shows ESR measurements of 10GeO$_2$: 90SiO$_2$ glass irradiated with the Hg lamp and KrF laser. The formation of GeE' centers was confirmed after exposure to the Hg lamp radiation, as shown in Figure 2.13(a). The estimated concentration of GeE' centers was approximately 4×10^{15} spins/cm^3, apparently increasing linearly with decreasing absorptivity of the 5-eV band. An intense ESR signal was observed near the resonance signal of the GeE' center after irradiation with 100 KrF laser shots. The signals in the region of g < 2 can be assigned to electron trapped centers associated with Ge(1) and Ge(2) [80, 82, 83]. The resonance signal at g = 2.01 is attributed to the STH [108]. Irradiation with KrF laser pulses exceeding 10^3 shots (Figure 2.13(c)), leaves the total spin concentration unchanged, whereas the shape of the resonance signal in the region of g > 2 differs completely from curve b. This points to a structural change of the GECs under continuous irradiation. The spectrum of signal c minus b (Figure 2.13(d)), gives the typical line shape of the GeE' center. The g values are identical to the GeE' center induced with the Hg lamp [6]; therefore, there is evidence that continuous laser irradiation partially converts the total number of GECs to a GeE' center. The majority of paramagnetic centers are formed quickly at the initial stage of irradiation ($<10^3$ shots), saturating at 2.6×10^{17} spins/cm^3. The photochemical conversion of GECs to GeE' is also observed in absorption spectra.

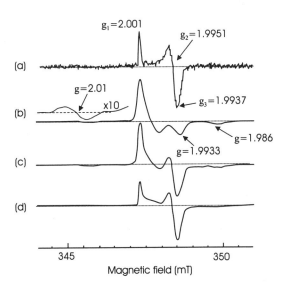

Figure 2.13   ESR spectra of 10GeO$_2$: 90SiO$_2$ glass irradiated with a Hg lamp and a KrF laser: (a) illuminated with Hg lamp (230 h), (b) irradiated with 10^2 shots KrF laser, (c) irradiated with 3×10^4 KrF laser shots, and (d) difference spectrum between (b) and (c) (*After:* [7]).

Figure 2.14 shows the difference spectra before and after irradiation with the Hg lamp and KrF laser. Irradiation with the Hg lamp (Figure 2.14(a)) leads to bleached and induced bands centered at 5.1 and 6.4 eV, respectively [85]. Independent formation of color-centers occurs in the glass matrix depending on the power densities of the UV photons, as is indicated by the reproduction of curve c (the addition of curves a and b) and shown in curve d. Friebele and Griscom [82] report that the absorption bands of Ge(1) and Ge(2) induced in a germanosilicate optical fiber by irradiation with γ-rays are located at 4.4 and 5.8 eV, respectively. As shown in the inset of Figure 2.14, the absorption bands due to these GECs can be recognized in the initial stage of irradiation with KrF laser pulses ($<10^2$ shots), with the measured spectrum separated into two main Gaussian bands centered at 4.4 and 5.8 eV, strongly suggesting the simultaneous formation of Ge(1) and Ge(2) with a KrF laser. A third, unidentified peak, at ~4.8 eV, is often found to improve the fitting [25, 82, 85]. A difference absorption band has recently been identified and measured at 4.92 eV by Takahashi et al. [109]: it is related to the 5.06-eV band, but its physical significance remains obscure. A two-photon process explains the formation of GECs and STHs, considering that the laser intensities are seven orders of magnitude greater than the Hg lamp and that the optical band gap energy of the glass used at ~7.1 eV exceeds the laser photon energy.

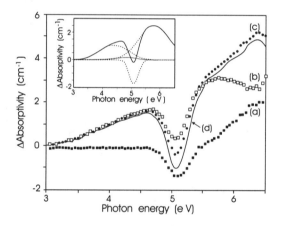

Figure 2.14   Difference absorption spectra of $10GeO_2$: $90SiO_2$ glass before and after irradiation with (a) Hg lamp (230 h), (b) KrF laser (10^2 shots), (c) Hg lamp (230 h) + KrF laser (10^2 shots), and (d) the sum of (a) + (b). Inset: difference absorption spectra of $10GeO_2$:$90SiO_2$ glass before and after irradiation with 60-KrF laser shots. Dashed lines are individual Gaussian components of solid curve (*After*: [7]).

The difference absorption spectra (background subtracted) of 10 mol% germanosilicate glass for KrF laser irradiation shows near-saturated spectral changes below 5 eV (10^3 shots); the absorptivity around 6.4 eV gradually increases with pulse number (3×10^4) (Figure 2.15). These spectral changes are explained by the conversion of GECs to GeE' centers by continuous irradiation with the KrF laser. The color-center giving an intense absorption near 6.4 eV is closely related to the GeE' centers [85]. Figure 2.16

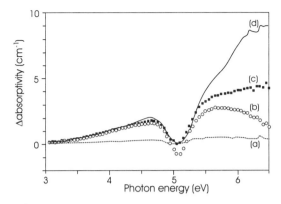

Figure 2.15 Difference absorption spectra of $10GeO_2$: $90SiO_2$ glass before and after irradiation with KrF laser: (a) 10 shots, (b) 60 shots, (c) 10^3 shots, and (d) 3×10^4 shots (*After:* [7]).

shows a proposed energy level diagram that follows from the UV-induced, photochemical reactions in GeO_2: SiO_2 glass. Hg lamp irradiation proceeds through a one-photon absorption process that is rapidly saturated because of the low concentration of neutral oxygen monovacancies ($\sim 10^{16} cm^{-3}$) [6]. This leads to (2.1)

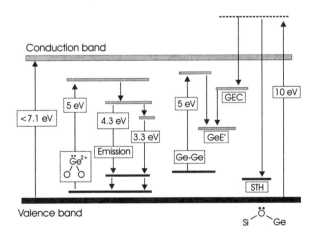

Figure 2.16 Schematic energy diagram showing relevant defect levels and photochemical reactions in $10GeO_2$: $90SiO_2$ glass caused by irradiation with UV light (*After:* [7]).

$$-Ge - Ge- \ (\text{or Ge-Si}) \xrightarrow{h\nu} GeE' + GeO_3^+ (\text{or } SiO_3^+) + e^- \qquad (2.1)$$

The simultaneous formation of GECs and STHs leads to

$$-\overset{|}{\underset{|}{\text{Ge}}}- \;+\; \overset{\cdot\cdot}{\underset{\diagdown}{\text{O}}} \; \xrightarrow{\;2h\nu\;}\; \text{GEC} + \text{STH} \qquad (2.2)$$

The lone pair electrons on bridging oxygens are excited to the conduction band via two-photon absorption processes. The conversion of GECs to GeE' centers can be described by the following reaction:

$$\text{GEC} \longrightarrow \text{GeE}' + \text{O (nonbridging oxygen)} \qquad (2.3)$$

This conversion of a GEC with a negative charge into a GeE' with no nominal charge may be regarded as structural relaxation. The driving force is electrostatic repulsion between a trapped electron and the neighboring oxygen atoms; the structural change restores local electro-neutrality around the Ge ions by detaching a nonbridging oxygen (with a negative charge) from its coordination sphere. Therefore, the concentration of GeE' centers in glass, for long UV exposure times, exceeds pre-existing Ge-related ODC concentrations because GECs, which are the precursors of the GeE' centers in this case, are created from the intrinsic structure. Hosono et al. [11] give a quantitative relation between GEC formation in germanosilicate glass and photon densities of various excimer lasers. Excimer laser radiation at ArF (6.4 eV), KrF (5 eV), and XeCl (4 eV) suggests that the formation of Ge electron centers, in 10-mol% VAD-prepared germanosilicate glass occurs via two-photon absorption processes [11]. Although the wavelength of the KrF laser corresponds to the absorption band of existing GODCs, no significant difference in the formation efficiency is seen between ArF and KrF laser light (Figure 2.17). This implies that the GECs are not derived from the precursor defects in glass but from intrinsic GeO_4 members. The formation of the GeE' center from the ODCs coordinated by two Ge ions occurs by trapping a positive hole following a structural relaxation of the pyramidal GeO_3^+ into a planar form (Figure 2.18), where h^+ denotes a positive hole. On the other hand, such a

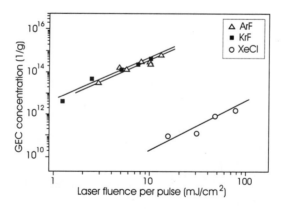

Figure 2.17 Concentrations of GEC created per pulse, of various excimer lasers as a function of laser fluence per pulse (*After* [11]).

distinct displacement of relevant atoms is not involved in the formation process of the GECs from GeO$_4$ units. It is therefore considered that the formation of GECs is energetically favorable over GeE' formation from the ODCs (Figure 2.18). The efficiency for XeCl laser light is smaller by four orders of magnitude compared with ArF or KrF light. The defect formation efficiencies for the case where the two-photon energy is close to the optical band gap, k(308), are much lower than for that when the two-photon energy is enough to exceed the band gap, k(248) and k(193).

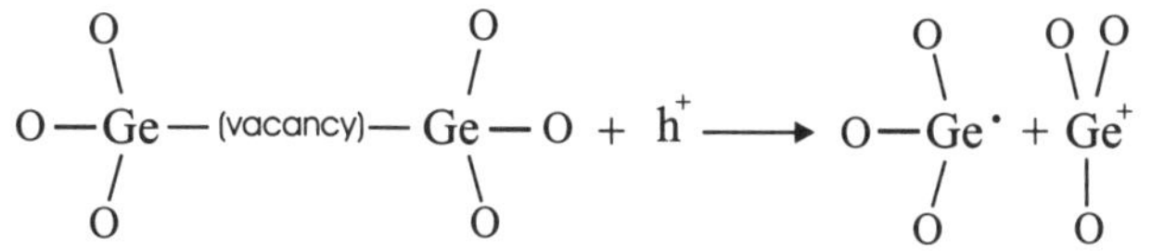

Figure 2.18   The formation of the GeE' center from the ODCs coordinated by two Ge ions occurs by trapping a positive hole following a structural relaxation of the pyramidal GeO$_3^+$ into a planar form (*After*: [11]).

The evidence presented thus far gives an indication of how UV irradiation produces different defect precursors. It is also known that physical compaction of the glass results in similar behavior to UV irradiation [110]. St. J. Russell et al. [8] proposed that compaction triggered by bond breakage could result in the observed cumulative refractive index changes. Compaction has received much attention recently as an alternative explanation to the color-center model (Section 2.8). Dianov and co-workers [111] have looked for evidence that point defects may be the precursors to the UV-induced densification, even though they cannot be considered as the direct cause. Densification is expected to have an impact on any residual defects, changing their spectroscopic properties as a result of the network rearrangement. To differentiate between the effects of UV irradiation and glass densification on defect transformations, the influence of mechanical pressure on the spectroscopic properties of GODCs of germanosilicate glass has been examined. Preform samples prepared with MCVD and VAD methods (10 and 7 mol% GeO$_2$, absorption coefficient at 5 eV: 180 and 30 dB/mm, respectively) and hydrostatically pressurized at 9 Gpa and 300°C results in significant densification without crystallization. UV absorption and luminescence spectra were measured under excitation by N$_2$ and KrF laser radiation (3.68 and 5 eV, respectively) along with ESR spectra of the initial and densified samples. Hydrostatic pressure treatment results in a densification of $16 \pm 2\%$ (MCVD) and $19 \pm 2\%$ (VAD), corresponding to the same level of densification measured for hydrostatically pressurized silica glass [110]. Absorption spectra of initial and densified VAD-prepared glass, with accompanying Gaussian decompositions identifying three different GODCs, is shown in Figure 2.19. Considerable modification in the spectra are observed and reflect changes in the defect concentration (areas), structure (band displacement), and the glass network (broadening). A significant growth of a metastable absorption band at 7.3 eV is observed, but only its tail is resolvable; therefore, a large error may be associated with the position of this band. It grows, however, by a factor of 15. There is qualitative agreement

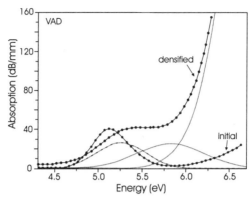

Figure 2.19   Absorption spectra of initial and densified germanosilicate glass and their decomposition into three Gaussian components (*After*: [111]).

with UV-induced changes in germanosilicate glass, with large absorption induced under densification in the range 5.5–6.5 eV (paramagnetic Ge centers Ge(2) and GeE' have UV irradiated absorption bands at 5.8 and 6.4 eV, respectively). However, ESR measurements indicate that no Ge paramagnetic defects are created during densification of the glass, with only GeE' centers observed in both initial and densified samples with an unchanged concentration of $5 \times 10^{15} cm^{-3}$. Therefore, in contrast to UV irradiation, pressure treatment does not induce free charge carriers and radicals with unpaired spins, indicating a possible link with diamagnetic centers in the glass.

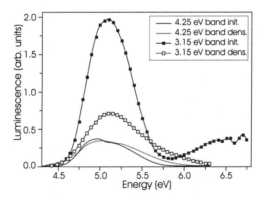

Figure 2.20   Luminescence excitation spectra in initial and densified germanosilicate glass, with excitation at 5.1 eV (*After*: [111]).

The luminescence spectra of initial and densified glass feature 4.25-eV (UV) and 3.15-eV (blue) bands excited by 5-eV photons; only the blue luminescence is observed under 3.68-eV irradiation. The blue luminescence band in densified glass decreases fourfold at both excitation wavelengths, whereas the UV band decreases by a factor of 1.5. Figure 2.20 confirms the difference of the influence of densification on glass in the UV and blue luminescence bands, with the former changing marginally, whereas the efficiency of excitation in the band near 5.1 eV decreases threefold, implying that these bands have different defect assignments. The band beyond 6.5 eV also disappears in the densified sample. The luminescence kinetics also varies with glass densification [112]. Figure 2.21 gives such an example for MCVD glass, the behavior of VAD prepared glass being the same. Under excitation with 3.68 and 5 eV, the luminescence reaches a maximum between 2 and 8 μs before decaying. Following Gallagher and Osterberg [113], two exponential functions may be used to fit the luminescence kinetics: $I_{lum} = A_1\exp(-t/\tau_1) - A_2\exp(-t/\tau_2)$. The first term represents conventional decay of a T_1 state with intensity A_1 at $t = 0$. The second term is a delayed excitation of the T_1 state and results from energy transfer with another defect. The difference $(A_1 - A_2)$ is the initial population of T_1, and the ratio $(A_1 - A_2)/A_1$ is a measure of the fractional GODC luminescence. As the 3.68-eV excitation can also give rise to 3.15-eV luminescence, the energy transfer must occur between the same states of two defects. For nondensified preforms, typically 65% of the maximum luminescence arises from the initial intensity, with another ~35% transferred from another defect. Densification reduces this portion of the energy transfer to 30%, and, as we have seen, the overall luminescence decreases by a factor of four. We have also seen that the absorption decrease at 5 eV cannot account for this. Given the measurement accuracy one cannot draw definitive conclusions, but this may imply that the band at 330 nm (3.76 eV) is not restricted to an S_0 to T_1 GODC transition, which only accounts for 25% of the 3.15-eV luminescence intensity.

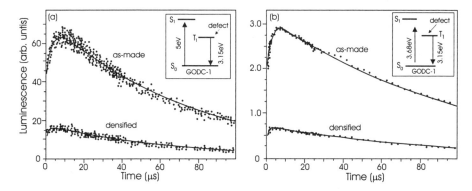

Figure 2.21 Kinetics of 3.15-eV luminescence excited by 5.0 (a) and 3.68 eV (b) photons in as-made and densified MCVD germanosilicate glass. Insets: diagrams of GODCs T_1 state excitation including energy transfer from another defect (*After*: [112]).

It is generally accepted that the UV luminescence band results from a singlet-to-singlet (S_1 to S_0) transition, and the blue band from a triplet-to-singlet (T_1 to S_0) transition inside the same defect. However, the difference in behavior of these bands under glass densification is too large to conform to this interpretation. Given that the blue luminescence lifetime changes only slightly, no significant nonradiative relaxation for the T_1 state arises in densified glass. Furthermore, given that the decrease in blue luminescence is not accompanied by an increase in the UV luminescence, the rate of the S_1 to T_1 conversion is not greatly affected by the glass densification, which further implies that the blue and UV luminescence bands belong to different defects. The decrease in the blue luminescence is usually observed in the writing of Bragg gratings and is associated with the bleaching of GODC. Clearly, densification can decrease this further. Therefore, the combined action of GODC bleaching and glass densification results in blue luminescence decay.

2.5.1.2 Germanosilicate Optical Fibers

The evidence presented thus far can only be considered unequivocal for bulk glass. Although there is ample confirmation that fibers and preforms share essentially the same absorption spectra, they possess different stress and defect distributions. In preforms these arise from thermal expansion inhomogeneities, whereas in optical fibers one must also consider the role of drawing induced defects. That the Ge-Si wrong bonds are responsible for the photosensitivity of MCVD germanosilicate fibers is widely accepted, it must also be kept in mind that Ge-Si bonds cannot be the only triggers. The wrong bonds can be ruptured as the glass cools on drawing because of the different thermal expansion coefficients for doped and undoped glass. The subsequent freeing of an electron results in the E' center production, relaxing the lattice around the defect as the silicon atom retracts into a trigonal planar configuration with an accompanying atomic displacement [73]. For small displacements the recombination of the E' center with an electron forms the wrong bond; large displacements result in a DID center [91]. These defects can also be annealed out at higher temperatures reverting back to wrong bonds. Griscom [70] and Robertson [114] have reported wrong bond ruptures. It has been observed by Hanafusa et al. [115] that the concentration of SiE' and GeE' centers produced by fiber drawing increases with the drawing speed (which increases tension and drawing temperature through thermoelastic

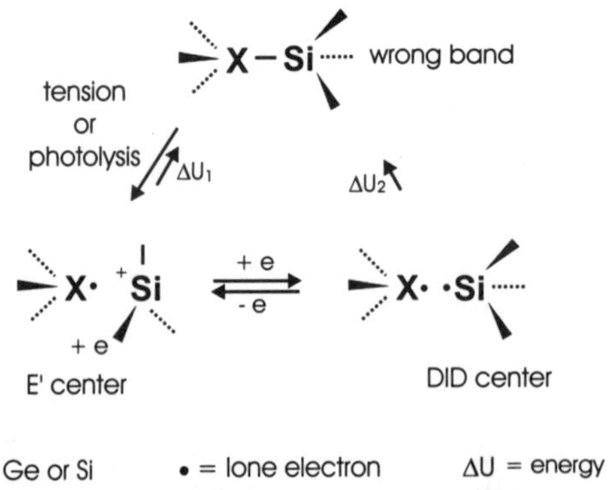

Figure 2.22   The transformations between point defects (wrong bonds, E' centers, and drawing induced defects) in germanosilicate fibers associated with photolytic processing.

stress) [116]. Atkins et al. [97] have viewed the equilibrium between the E' and DID centers as an oxidation reduction reaction, suggesting that this balance is altered by changing the electron properties of co-dopants and oxygen content in the glass. A reaction scheme (Figure 2.22) proposed by Atkins et al. [97] shows that the precursors of the DIDs are the Ge-Si and Si-Si wrong bonds, and not the Ge-O-Si and Si-O-Si bonds, which would yield NBOHC defects [60, 117, 118]. The formation of NBOHCs has been contradicted by Friebele et al. [119] who found that for drawn, pure silica fibers an additional diamagnetic defect was responsible for the 630-nm band. Annealing experiments led to the conclusion that the 630-nm absorption resulted from a diamagnetic electron trap [120]. These experiments reveal a distinction between the GeE' center and the DID center in germanosilicate fibers. Profiling of the GeE' center across a germanosilicate core preform shows an almost identical profile to the DID center, with maximum concentrations at the core center and core/cladding interface being very similar, indicating that the centers are closely related but not identical.

There have been a number of recent studies investigating the relationship between the ubiquitous GeE' center and the photoinduced refractive index changes. By comparing optical and ESR data, Simmons et al. [52] have found a similar wavelength dependence for the GeE' defect center concentration and the formation of self-organized gratings, under 488-nm excitation, in accord with the UV absorption characteristics of oxygen-deficient germania. Therefore, grating formation results from oxygen vacancies located at substitutional Ge sites that, on exposure to UV light, result in ionized defect band bleaching, liberating an electron and creating a GeE' hole trap. This is supported by the temperature dependence of GeE' centers (GeE'+GeE'$_{d1}$+GeE'$_{d2}$) and the reflectivity of a typical Bragg grating formed with pulsed UV light in a Ge-doped silica fiber (Figure 2.23) [121]. Up to 650°C the grating reflectivity is well correlated to the total concentration of GeE' centers. This indicates that GeE' defects cannot be charge trapping sites for internal-electric-field Bragg gratings. For this to be true, the individual modulation should be proportional to the square of the concentration of GeE' centers (due to the dc Kerr effect), with the reflectivity of the grating changing with the fourth power of the GeE'

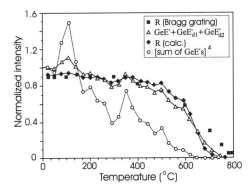

Figure 2.23   Normalized concentrations of the sum of GeE', GeE'$_{d1}$, GeE'$_{d2}$, the reflectivity R from a typical grating, and the calculated reflectivity as a function of annealing temperature in Ge-doped silica (*After*: [121]).

concentration. Such a relationship does not exist. Rather, one observes a unitary correlation between GeE' centers and reflectivity, which implies that GeE' centers are directly responsible for both Bragg and self-organized gratings, at least for the Ge- and Ge/P-doped fibers investigated.

Tsai et al. [107] have made a direct behavioral study of photoinduced color-centers and their correlation with the index modulation associated with the Type I Bragg grating. The fluence dependence of the photoinduced GeE' center, its thermal annealing behavior, and its reaction with hydrogen are similar to that of the index modulation generated in both hydrogen-loaded and unloaded germanosilicate fibers. There was also strong evidence of a diamagnetic structure, possibly densification contributing to the index modulation. One may hypothesize that if GeE' centers are associated with Bragg gratings in germanosilicate fibers, then their growth rate should be much higher in hydrogen-loaded fibers and should be similar to the growth rate of gratings written in the same fiber [9, 122–124]. A saturated exponential growth of GeE' is anticipated, as reported for Δn in hydrogenated fibers [125]. Furthermore, the reaction of GeE' centers with hydrogen above room temperature implies that the reflectivity of gratings in germanosilicate fibers can be modified by heat treatment in the presence of hydrogen, as has been reported [126]. ESR spectra of hydrogenated and standard fibers irradiated with a KrF laser, with 30 mJ/pulse for 270 shots, show that the GeE' generated in the former is greater than that in the latter, in accordance with the much higher photosensitivity observed for hydrogen-loaded fibers. Figure 2.24 shows that the fluence dependence of the Ge(1) is similar for both loaded and unloaded fibers, despite that the photosensitivity of the former is much larger.

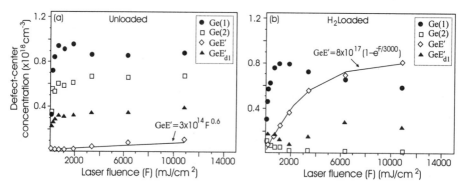

Figure 2.24 Paramagnetic defect-center concentration versus UV photon fluence (0-11J/cm^2) in (a) unloaded and (b) H$_2$-loaded Ge-SiO$_2$ fibers (*After*: [107]).

The fluence dependence of Ge(2), although similar to Ge(1) in unloaded fibers, is more complex in hydrogen-loaded fiber. Growth at low fluences is followed by saturation and is completely uncorrelated with the growth of Δn. This growth dynamic can be understood by the reaction of Ge(2) centers with hydrogen, forming the hydrogen-associated Ge(1) center, as characterized by a 11.9-mT doublet [84]. It is important to note that no correlation has been observed between the index modulation and Ge(1) and Ge(2) centers. The electron diffusion length under UV excitation is >2 μm, and Ge(1)/Ge(2) centers will likely contribute to the mean index component through their optical absorption through the Kramers-Kronig relation [82]. The power-law growth of GeE' in unloaded fibers and the saturating exponential growth in hydrogen-loaded fibers for an accumulated fluence of 12 J/cm^2 [127] is in agreement with the grating growth behavior in hydrogenated and standard fibers [24, 30, 125]. This result, and the similar photosensitivity and thermal stability [121] of GeE' centers and grating reflectivity, provides evidence that GeE' centers are associated with the modulated component of the grating at these fluence levels. Raman spectroscopy [128] suggests that Δn can be expressed as the sum of index changes per unit concentration of GeE' and GeH for gratings written in loaded fibers, in addition to any photoinduced structure. Thermal annealing in the presence of hydrogen will result in a GeE' center being replaced with GeH, which leads to [107]

$$\Delta n = K_{GeH}([GeE']_0 + [GeH]_0) + K_X[X] + (K_{GeE'} - K_{GeH})[GeE'] \qquad (2.4)$$

where X is any photoinduced structure (other than GeE' and GeH), associated with the grating. K_I are the index changes per unit concentration of I, and $[I]_0$ is the initial concentration of I, where I = GeE', GeH, X. Measuring Δn versus GeE' for hydrogen-loaded and unloaded fibers results in identical gradients $K_{GeE'} - K_{GeH}$ 1.5x10^{-23} cm^3, which is derived from (2.4) [107]. It is found that GeE' is approximately the same for both fibers because of the different exposure times for hydrogen-loaded (30 seconds) and unloaded (5 minutes) fibers. The value of Δn for gratings fabricated in hydrogen-loaded fiber is double that measured in unloaded fibers; there is an additional contribution of GeH to Δn. Optical absorption associated with GeH has been reported by Krol et al. [129]. An independent estimation has been made for $K_{GeE'}$ and K_{GeH} by thermally annealing gratings fabricated in unloaded fiber without postexposure to hydrogen impregnation. Without hydrogen, GeH is much smaller than GeE' and contributes negligibly to Δn. In this case Δn is linearly related to GeE' with a proportionality constant being the index change per unit concentration of GeE'. Isochronal annealing of Δn and GeE' in fibers without postexposure

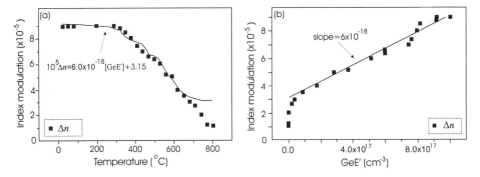

Figure 2.25   Isochronal thermal annealing of Δn and GeE' center concentration in unloaded fibers without postexposure H$_2$ loading: (a) Δn versus temperature and fit to linear relationship between Δn and [GeE']; (b) Δn versus [GeE']. Laser fluence is 300 J/cm^2 (*After*: [107]).

loading demonstrates that Δn follows GeE' up to ~650°C (Figure 2.25(a)). From Figure 2.25(b), the index change per unit concentration of GeE' centers is found to have a value of 6×10^{-23} cm³; thus, for a [GeE'] of 10^{17} cm⁻³, the index modulation arising from GeE' centers is ~6×10^{-6}. The index change per unit concentration for GeH is 4.5×10^{-23} cm³; therefore, for [GeH] of 10^{17} cm⁻³, Δn owing to GeH is 4.5×10^{-6}. Figure 2.25 shows that Δn continues to decrease above 650°C, whereas GeE' and other paramagnetic centers are no longer observed, suggesting a diamagnetic contribution to the photoinduced structure of the grating, such as densification, as observed by AFM [23, 130]. The 30% contribution to Δn by this structure, estimated from the data of Figure 2.25, is comparable to that reported in [23] (refer to Section 2.8.7).

This approach has recently been extended and the correlation between the thermal stability of GeE' centers and Type IIA gratings has been made [131]. A comparison was made of the photoinduced defect center concentrations in high NA, heavily doped, germanosilicate fibers with the complex dynamics of the refractive index changes associated with gratings written in the same fiber. It was found that stretching the fiber leads to significant changes in the accumulated fluence dependence of the photoinduced defect center concentration. Taunay et al. [131] exposed 28 mol% GeO₂-doped fiber (1.8-µm core diameter and cut-off wavelength at 0.8 µm) to a KrF laser through a phase mask. The grating reflectivity and Bragg wavelength were monitored in the course of grating writing, as varying degrees of strain were applied to the fiber. The formation of paramagnetic defect centers was monitored by ESR measurements for fibers exposed to the same fluence levels as used to write gratings. Figure 2.26 shows the growth of the refractive index with accumulated fluence for a mean fluence per pulse of 375 mJ/cm² and a 20-Hz repetition rate for the formation of Type IIA gratings. It is apparent that increasing the strain from ~0.4×10^{-3} to 9.9×10^{-3} results in the negation of the Type I grating, whereas the efficiency of the Type IIA grating increases. These features are similar to the results reported in [31]. Figure 2.27 shows the concentration of paramagnetic defect centers [GeE' and GEC (Ge(1)+Ge(2))] as a function of accumulated fluence. The concentration of the UV-induced GeE' centers increases with fluence, while the GEC centers quickly saturate

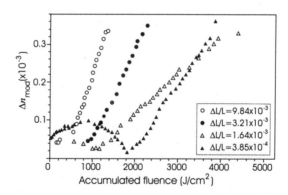

Figure 2.26   Growth of the UV-induced refractive index modulation as a function of the accumulated fluence for gratings written in fibers exposed to various strain levels (*After*: [131]).

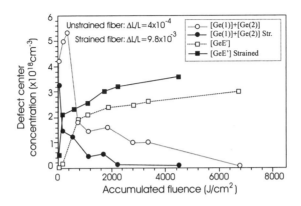

Figure 2.27 Evolutions of the paramagnetic centers, GeE', Ge(1)+Ge(2), in the course of a uniform UV exposure, before and after applied strain (*After*: [131]).

(<400 J/cm^2). This is the case for an accumulated fluence an order of magnitude less than that reported in [127]. The ESR spectra further indicate that the fluence dependence and the total concentration of the UV-induced GeE' centers are actually greater in the strained fiber. The concentration of GeE' centers in the unstrained fiber is noticeable at an accumulated fluence level for which the Type I modulated refractive index saturates and decreases at 800 J/cm^2. Straining the fiber causes bleaching of the GEC centers; their concentration saturates at 3.3×10^{18}cm^{-3} and disappears at 2 kJ/cm^2, compared with 5.3×10^{18}cm^{-3} for the unstrained fiber. The disappearance of the Type I grating in strained fibers has been related to the increased rate of photobleaching of the GEC centers and the decrease in the rate of densification. This result clearly differs from the correlation of GeE' centers with Type I gratings reported earlier, where fluences were <12 J/cm^2. Although the GeE' defect is predominant in the strained fiber (along with an increased Type IIA photosensitivity), there is no direct correlation evident in the growth dynamics. GeE' centers display saturating exponential growth in both strained and unstrained fibers.

Type I and Type IIA Bragg gratings have also been fabricated in Ge-free, N$_2$-doped silica-core fibers, with a change in refractive index $\sim10^{-3}$ under 193-nm illumination without hydrogen loading (Figure 2.28) [132]. The growth dynamics are similar to their Ge-doped counterparts for both hydrogen-loaded and unloaded cases. Whereas Type IIA behavior can be induced in unloaded fibers with fluence levels of several kJ/cm^2, hydrogen-loaded examples display a monotonic increase in the index modulation in complete agreement with hydrogen-loaded, high Ge content fibers [32]. Temperature-induced changes in reflectivity also parallel the previous findings for Type I/Type IIA gratings, with the latter being more stable. Interestingly, an increase in reflectivity between 350°C and 600°C is observed for the Type IIA gratings and is consistent with Dong and Liu [35]. The mechanism for this fiber's photosensitivity may be related to the two-photon absorption mechanism proposed by Albert et al. for lightly Ge-doped fiber exposed to 193-nm irradiation [133]. As yet, this remains unresolved.

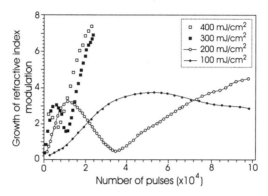

Figure 2.28 Growth of refractive index modulation versus pulse number for Bragg gratings written in N_2-doped silica fiber exposed to 193-nm laser light (*After*: [132]).

The GODC singlet-singlet (S_0-S_1) transition band at 242 nm also has a weak absorption band with a maximum at 330 nm associated with a singlet-triplet (S_0-T_1) transition [76]. Dianov and co-workers have proposed that direct triplet-state excitation, by near-UV 330-nm light, should result in the same changes as observed at 240 nm, if it is involved in the photosensitivity of germanosilicate glass. The strong photobleaching of the GODC triplet-singlet luminescence that occurs, accompanied by luminescence at 650 nm associated with DIDs, supports this hypothesis [100]. Recently, Starodubov et al. [134] have found that photoexcitation at 334 nm has induced index changes in germanosilicate glass. However, the absorption of the S_0-T_1 band is three orders of magnitude less than the S_0-S_1 transition, and writing intensities of several kW/cm^2 are typically required to give an index change of 10^{-4}. Under 334-nm irradiation photoionization of defects is not required; rather, the excitation of a triplet-state of the GODC defect activates structural transformations in the glass matrix. To evaluate the role of Ge, Bragg gratings were written to saturation with near-UV light in germanosilicate fiber with different Ge-core concentrations. The change in refractive index varies linearly with the Ge concentration (Figure 2.29), indicating that the defect responsible for the near-UV photosensitivity contains a single Ge atom (a Ge-Si wrong bond, a twofold coordinated Ge, or an associated GODC). A schematic for the different possible pathways for near- and mid-UV light initiating a refractive index change in Ge-doped fiber is shown in Figure 2.30. Light at 240 nm excites a GODC from its ground singlet state S_0 to its excited state S_1 from which it can ionize spontaneously or by absorbing another 240-nm photon–such ionization is thought to be necessary for an index change [135]. However, a GODC excited to S_1 can also relax to the long-lived triplet-state T_1. From there the defect undergoes a metamorphosis and changes its structure to a DID. Light at 330 nm can excite the triplet-state directly from the ground state. It is proposed that structural rearrangement of the GODC into the DID is the principal cause of the light-induced refractive index change. Ce co-doping acts as an efficient electron donor and should enhance the index formation, if photoionization plays a role in near-UV photosensitivity. Direct excitation of the Ce^{3+} absorption band centered at 320 nm will be many orders of magnitude more efficient at ionizing Ce^{3+} than a GODC [76]. However, simply releasing

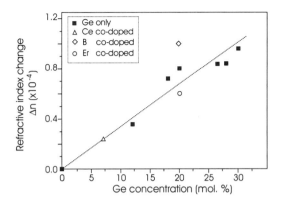

Figure 2.29 Dependence of light-induced index change on fiber's Ge concentration under near-UV laser illumination. Gratings fabricated with 334-nm light for a ~250 kJ/cm^2 exposure. Boron co-doping enhances the grating strength but Ce co-doping does not (*After*: [134]).

more electrons by photoionization does not necessarily create a larger refractive index change (Figure 2.29). Therefore, since the E' center concentration, which is correlated to the index change, is not modified by the presence of Ce^{3+}, it appears that photoionization has no role to play in the index change. On the other hand, co-doping with B does increase the photosensitivity, which supports the idea that a structural transformation lies behind the enhancement mechanism for this approach [19]. B facilitates a structural modification by softening the glass, making it easier for the embedded GODC to undergo the transformation from its triplet state. Evidence of this is found in annealing experiments, which confirm that fibers containing B lose their gratings at lower temperatures than fibers without. Finally, the same argument suggests that gratings can be efficiently fabricated in germanosilicate fibers co-doped with Er^{3+} under near-UV light illumination, irrespective of the Er^{3+} concentration.

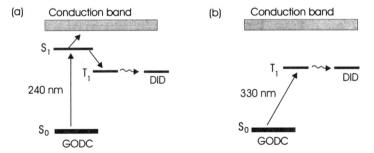

Figure 2.30 Energy-level diagram of a germanium oxygen-deficient defect, showing proposed pathways excited with (a) pulsed mid-UV light and (b) CW near-UV light. The excited triple state relaxes into a new defect state, a drawing-induced defect. This has been proposed to cause the index change that makes the Bragg gratings (*After*: [134]).

Permanent changes in the refractive index have also been induced in pure GeO_2 bulk glass by exposure to the third harmonic (3.51 eV) of a Q-switched Nd:YAG laser [136]. The spatial profile of Δn varies with the third power of the irradiance. The formation of the GeE' centers (6.4 eV) by below band gap irradiation assumes that the GeE' center is generated by two reaction processes: First, the formation of a metastable defect species associated with Ge via two-photon absorption; and second, the structural relaxation of the metastable defect into a GeE' center through a one-photon absorption process. It was proposed that the metastable defect is a GEC, given that it is known to relax to GeE' through a thermal stabilization process [137] and the associated Ge(1) can absorb the UV irradiation. The GEC is the dominant species in germanosilicate glass irradiated at 193 nm [11], which corresponds to photon energy of 6.5 eV. This is comparable to the two-photon energy used in this study, 2x3.51 eV = 7.02 eV. Taken as a whole, these results are consistent with the proposed two-photochemical-reaction model for the formation of the GeE' center, which would lead to the measured positive refractive index change [7].

Finally, Albert et al. [10] have compared the growth rates of fiber Bragg gratings written in two standard optical fibers with different Ge doping as a function of the 193-nm ArF laser pulse energy density. The grating growth rate was found to be linearly proportional to the laser pulse energy density for fibers with high Ge doping (a one-photon process (OP)), but proportional to the square of the pulse energy density for standard telecommunications fibers with low Ge concentration (a two-photon process (TP)). An ArF laser, operated at 50 pulses/s with a fluence per pulse of 60 mJ/cm^2, inscribed Bragg gratings using a phase mask in highly doped fiber (high Ge-doped fiber: Alcatel's bend insensitive fiber, 8-mol% Ge-doped) and in lightly doped fiber (Corning SMF-28, Ge 3 mol%). The growth rate for the high Ge-doped fiber (Figure 2.31(a)), is very fast, with the index modulation saturating at a value of 3.6×10^{-4}, which agrees with earlier published results [133]. Much slower growth rates are observed for the low Ge-doped fiber (Figure 2.31(b)), with the index modulation saturating at an unexpectedly high value of 0.91×10^{-3} for high fluences (almost three times higher). There was no evidence of the Type II

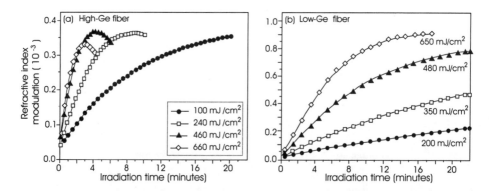

Figure 2.31   (a) Growth of refractive index modulation amplitude in high Ge-doped fiber (Alcatel bend-insensitive) resulting from irradiation through a phase mask with ArF laser for pulses of different energy density. (b) Same as (a) in low Ge-doped fiber (Corning SMF-28) (*After*: [10]).

damage mechanism reported by Dyer et al. [38]. The difference may be explained in two ways. The two processes create either similar numbers of color-centers with different polarizability, leading to a higher index change in the TP case, or the same kinds of color-centers, with the TP effect creating more of them. To test the hypotheses, a comparative isochronal annealing experiment was carried out on gratings fabricated using OP and TP processes at 193 nm, as well as at 248 nm in SMF-28 fiber. Approximately 50% of the initial index modulation is preserved after 30 minutes at 600°C; therefore, it appears that similar color-centers are formed in the three cases. Differences in the maximum index change, or in the index growth rate, result from the different efficiency of 5-, 6.4-, and (2x6.4) 12.8-eV photons at driving the photochemical reactions in which the color-centers are produced.

2.6 Enhanced Photosensitivity in Silica Optical Fibers

Since the discovery of photosensitivity and the first demonstration of grating formation in germanosilicate fibers, there has been considerable effort in understanding and increasing the photosensitivity in optical fibers. Standard single-mode telecommunication fibers, doped with 3% germania, typically display index changes of $\sim 3 \times 10^{-5}$. Generally increasing the dopant level or subjecting the fiber or preform to reducing conditions at high temperatures can result in large index changes of $\sim 5 \times 10^{-4}$ (there is one atypical report as high as 1.2×10^{-3} [138]). It is more often desirable, however, to fabricate photoinduced devices in standard optical fibers for compatibility with existing systems. Sensitization techniques have been developed for writing high reflectivity gratings in these fibers. An increase the photoinduced index modulation to values of the order of 10^{-3} and higher have been realized via low temperature hydrogen loading (hydrogenation) [9] or flame brushing [139].

2.6.1 Hydrogen Loading (Hydrogenation)

Lemaire and co-workers [9] were first to report a simple, alternative, but highly effective approach for achieving very high UV photosensitivity in optical fibers using low temperature hydrogen treatment prior to the UV exposure. Fibers are soaked in hydrogen gas at temperatures ranging from 20–75°C and pressures from ~20 atm to more than 750 atm (typically 150 atm), which results in diffusion of hydrogen molecules into the fiber core. In excess of 95% equilibrium solubility at the fiber core can be achieved with room temperature treatment. Permanent changes in the fiber core refractive index of 0.01 are possible. One advantage of hydrogen loading is the fabrication of Bragg gratings in any germanosilicate and germanium-free fibers. Additionally, in unexposed fiber sections the hydrogen diffuses out, leaving negligible absorption losses at the important optical communication windows.

In experiments [9] using standard AT&T single-mode optical fiber (3-mol% germania loaded with 3.3% hydrogen), gratings having bandwidths of 4 nm and peak changes

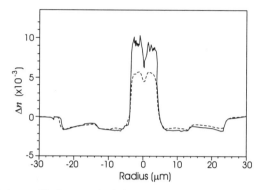

Figure 2.32 Refractive index profile for a standard single-mode fiber with 3% GeO$_2$, and for a grating that was UV written in the same fiber after loading with 3.3% hydrogen (solid curve) (*After*: [9]).

in the modulated index of ~6x10^{-3} were written using the transverse holographic technique by exposure to pulsed irradiation at 241 nm with a fluence of ~300 mJ/cm^2 (30 Hz) for 10 minutes. A comparison of the refractive index profile at the midpoint of a grating with an untreated fiber is shown in Figure 2.32. The average core index has increased by ~3.4x10^{-3}. Similar results were obtained for both MCVD- and VAD-drawn fibers containing ~3% germania. The mechanism is therefore not dependent on fiber or preform processing, but rather on the interactions between germania and hydrogen molecules, coupled with the UV exposure conditions. Figure 2.33 shows the absorption spectrum changes in the infrared (IR) for a germanosilicate fiber exposed to 1 atm pressure of hydrogen gas at 100°C. The sharp absorption peak at 1.24 μm, which is due to molecular hydrogen, is saturated after 10 hours. The absorption band due to OH formation is comprised of two closely spaced

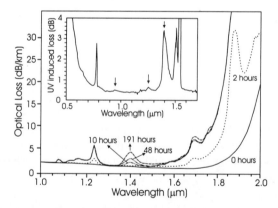

Figure 2.33 Absorption spectrum changes in the IR for a germanosilicate fiber exposed to 1 atmospheric pressure of hydrogen gas at 100°C (*After*: [140]). Inset: UV-induced losses in ~5-mm long grating in fiber with 9% GeO$_2$. Features at 770 and 1500 nm are due to the gratings: marked peaks at 0.95, 1.24, and 1.39 μm are due to OH (*After*: [9]).

peaks at 1.39 μm (Si-OH) and 1.41 μm (Ge-OH) [140]. This indicates that hydrogen molecules react with germanosilicate glass and form OH absorbing species. On the other hand, UV irradiation of untreated samples shows no OH formation [9, 122]. Figure 2.33 (inset) shows the UV-induced loss changes that occur in response to writing a strong grating in a 9% germania fiber loaded with 4.1% hydrogen. The OH ion concentration is consistently close to the germania content of the fiber. Optical absorption spectra of a germanosilicate preform rod heated in a hydrogen atmosphere at 500°C for different times are shown in Figure 2.34(a) [141]. The growth of the broad 240-nm absorption band is clearly seen in this figure. This indicates that the reaction of hydrogen molecules at Ge sites produces GODCs assigned to the broad 240-nm absorption band. Figures 2.33 and 2.34(a) suggest that the GODCs and OH species are formed from thermally driven reactions between hydrogen and Si-O-Ge glass sites. The inscription of Bragg gratings in hydrogen-loaded fiber undoubtedly involves both thermal and photolytic mechanisms. Atkins and co-workers [122] investigated thermal effects by exposing germanosilicate glass (loaded 1 mol % hydrogen) to a CO_2 laser (CW mode) for 10 seconds, which resulted in a glass temperature of 600°C. The UV absorption spectrum, shown in Figure 2.34(b), clearly shows growth of the GODC band near 240 nm, from 20 dB/mm before heating to 380 dB/mm after heating. IR spectra of the sample indicate the formation of ~980 ppm OH. The tail of the OH broadband absorption peak at 1.39 and 1.41 μm introduces losses that are often unacceptable to designers of telecommunications network systems. However, by loading fiber with deuterium instead of hydrogen, the UV-induced absorption peak is shifted to longer wavelengths, out of the erbium amplifier band of 1.55 μm [124].

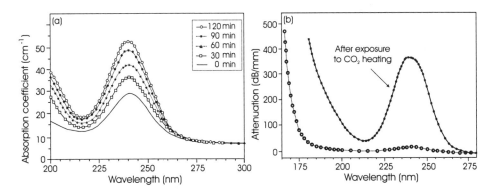

Figure 2.34   (a) Optical absorption spectra of a germanosilicate preform rod heated in a H_2 atmosphere at 500°C for different times (*After*: [141]). (b) UV spectra of 1 mol% H_2-loaded germanosilicate glass before (dashed) and after (solid) 10-second exposure to a CO_2 laser beam (*After*: [122]).

Atkins et al. have also measured short wavelength absorption spectra between 165 and 300 nm for hydrogen-loaded and unloaded 3 mol% GeO_2-doped preform samples [122]. Hydrogenation and UV exposure introduces extremely large short wavelength increases in absorption. A notable difference in the spectra of UV-exposed hydrogenated samples is

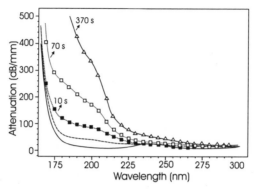

Figure 2.35 UV spectrum of 2.4 mol% H$_2$-loaded germanosilicate glass exposed to a KrF laser operating at 248 nm, 120 mJ/cm^2, and 30 Hz. Spectrum before exposure (solid line), untreated sample after 360 seconds (dashed line), spectrum after 370 seconds (triangle), 70 seconds (square) and 10 seconds (solid square) (*After*: [122]).

the absence of the 240-nm band, and any blue fluorescence disappears in a matter of seconds. In untreated samples both phenomena are clearly evident (Figure 2.35). Heating a hydrogenated sample to 800°C (following 370 seconds of UV exposure) removes almost all of the UV absorption below 220 nm, whereas a band at 240 nm develops with an attenuation (120 dB/mm) six times that of the band prior to UV exposure.

Substantial index changes have also been achieved by rapidly heating hydrogen-loaded fibers [9, 122, 126]. Dramatic increases have been measured in the growth rate of the Si-OH overtone peak at 1.39 μm with a combination of UV exposure and heating in hydrogenated germanosilicate fibers (3.5 mol% GeO$_2$). This shows that reactions can occur at every tetrahedrally coordinated Ge in the glass, with the OH level approaching the germania content of the fiber, as with UV exposure [126]. Loading P$_2$O$_5$-doped fiber with D$_2$ and applying heat treatment has resulted in the first reported evidence for 248-nm photosensitivity for this fiber [126]. Atkins et al. [122] have confirmed that a combination of thermal and photolytic sample excitation increased the OH levels at 190 ppm OH (by weight) compared to 60 ppm OH for UV laser excitation alone, with no OH formation for non-hydrogen-loaded samples. The thermal treatment initiates oxygen-deficient defect formation, while the UV irradiation bleaches the defects leading to the observed index changes. However, there is ample data to indicate that this is oversimplified; a more complex scenario is likely given by the following observations. First, the rapid bleaching of the 240-nm absorption band suggests a direct reaction of hydrogen with the defect sites. Second, the 240-nm band is stronger after thermally annealing a UV-irradiated hydrogenated sample compared with non-hydrogen-loaded samples. Evidence that links the UV-induced spectral changes in hydrogen-treated samples to the laser repetition rate indicates a temperature dependence. What is clear is that the strong UV light dissociates the hydrogen molecules in the glass, leading to the formation of Si-OH and Ge-OH, in addition to the formation of oxygen-deficient centers. A photochemical model in which UV photons disrupt a GeO bond at a tetrahedrally coordinated Ge, followed by an irreversible reaction of the excited bond with a H_2 molecule, appears to explain the photosensitivity observed in both Ge- and P_2O_5-doped cores. UV losses rise with increased H_2 reaction and temperature [9, 122], which offers an explanation for the enhanced reaction rates in heated fibers. Given that hydrogen-initiated reactions continue even after the 240-nm band has disappeared, it should be possible to initiate hydrogen reactions with sufficiently energetic photons regardless of whether the UV wavelength overlaps the 240-nm band or not [122, 126]. Albert and co-workers [133] have confirmed this using 193-nm writing wavelength in germanosilicate glass (refer to Section 2.7.1).

Malo et al. [142] have studied the effects of hydrogenation on the guiding properties of silica-based optical waveguides and fibers by monitoring the reflectivity spectrum of gratings over several days. Two gratings were inscribed in weakly photosensitive Corning SMF-28 fiber prior to and after hydrogen treatment referred to as grating 1 and 2, respectively. Grating 1 undergoes a maximum refractive index increase of 0.05% before returning to its original value, a result that is undoubtedly related to hydrogen absorption and effusion. The behavior of grating 2 is more complex and is principally governed by the reaction of UV-induced hydrogen dissociation forming Ge-OH bonds in the core region. The grating center wavelength follows changes in the hydrogen concentration that result from diffusion and reactions with defects in the glass, finally returning to a Bragg wavelength close to its original value. Using standard Fickian diffusion equations for a cylindrical geometry explains the data (Figure 2.36). The normalized concentration of molecular hydrogen in the core was well correlated to the shifts in the Bragg wavelength. Farries et al. [143] also measured very similar behavior for hydrogen-treated fiber gratings. They measured a mean wavelength shift of 1 nm over a 30-day period, which correlated to the time the hydrogen was allowed to diffuse from the fiber prior to UV exposure. Outgassing prior to UV exposure substantially increased grating stability. However, it is not possible to reliably predict the magnitude of any wavelength drift, as it certainly relates to the residual concentration of hydrogen at the time of UV exposure and the degree of depletion within the core. It has been shown that postexposure annealing can stabilize gratings against long-term wavelength variations [144].

Attempts have also been made to associate hydrogenation with the paramagnetic

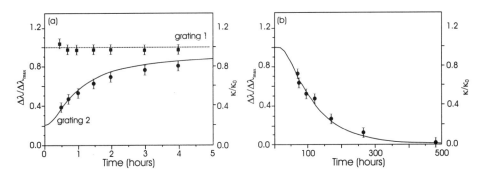

Figure 2.36 Short-term (a) and long-term (b) evolution of grating resonance wavelength λ_B and a normalized molecular hydrogen concentration (κ/κ_0) (*After:* [142]).

defects believed to be responsible for the index change [107]. It was shown that gratings written in hydrogenated fibers may have Δn expressed as a sum of contributions from GeE' and GeH centers:

$$\Delta n = K_{GeE'}[GeE'] + K_{GeH}[GeH] + K_X[X] \tag{2.5}$$

(for definition of symbols see Section 2.5.1.2). It is found that the index change per unit concentration of GeH contributed to an index change of 4.5×10^{-6}. This should be compared with an index contribution from the GeE' centers of 6×10^{-6}. Therefore, a significant contribution to Δn arises from GeH centers. It has been shown that similar concentrations of GeE' centers are induced in photosensitive and SMF-28 hydrogenated fibers, even though the 240-nm band in SMF-28 is an order of magnitude less than in the photosensitive fiber [145]. This implies that the GeE' center formation in hydrogen-loaded fiber does not necessarily result from the breakage of wrong bonds. That photogeneration of GeE' centers occurs at low fluence levels, as opposed to photobleaching in unloaded fibers, suggests that GeE' centers in the hydrogenated fibers are photoinduced by hydrogen reacting with Ge-SiO_2 to form GeH. The growth of the index modulation of gratings in hydrogenated fiber is correlated with that for GeE'. GECs are found to have similar saturation concentrations in hydrogenated and unloaded fibers, this suggests that GECs are not relevant to the photoinduced reaction scheme and result from electron trapping [80].

Recent measurements using a CW Ar^+ laser have been made on hydrogenated fibers excited at 334 nm . Exposure of a standard fiber to near-UV light results in the excitation spectrum of a GODC triplet state (T_1), the transition occurring directly or through an intermediate GODC singlet state, with a characteristic shoulder at 3.7 eV (335 nm). However, very little sensitivity is observed at 351 nm, corresponding to photon energy of 3.5 eV, as reported in [146]. The spectrum associated with hydrogenated fiber shows a shift to 4.5 eV, implying that the excitation of a GODC triplet state is no longer the key to the index changing mechanism (Figure 2.37). The maximum photosensitivity is now coincident with the dissociation energy of hydrogen molecules. It has been proposed that because the initial and final states share the same symmetry, 4.5 eV may not itself be enough energy to break the H-H bond [147]. The presence of a suitable GODC or Ge(1) center can transfer energy to the hydrogen molecule to form GeH or OH. This is indicated by luminescence studies, where there is a decrease in the lifetime of an excited triplet state of a GODC in hydrogen-loaded fibers compared with unloaded fiber [147]. The efficient formation of mainly GeOH bands has been detected for 334-nm irradiation, with the saturated concentration of OH bonds being twice the concentration of molecular hydrogen, as is observed for 240-nm excitation. This implies that mainly OH bonds are formed. The actual index growth rate depends linearly on the Ge concentration and its saturated level on hydrogen concentration. The OH growth shows a quadratic dependence and this implies that there is another contributing mechanism to the refractive index change.

Finally, the growth dynamics for the manufacture of rocking filters in both hydrogen-loaded and unloaded fibers at 193 nm have been examined by Psaila et al. [55]. Their results indicate that hydrogenation leads to large rocking angles (up to 8 degrees have been measured), corresponding to a photoinduced birefringence of 2.4×10^{-5} and allowing for 95% coupling between the polarization states.

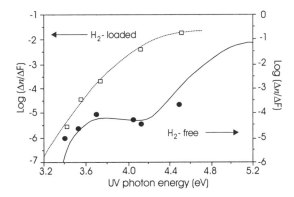

Figure 2.37 H$_2$-loaded fiber's photosensitivity versus UV writing energy (squares) approximated by a Gaussian fit (solid line). H$_2$-free fiber index change (circles) normalized by fluence with known GODC absorption spectrum (solid line) (*After*: [147]).

2.6.2 Flame Brushing

Flame brushing is a simple and effective technique for enhancing the photosensitivity in germanosilicate fiber [139]. The photosensitization introduced through flame brushing is achieved with a negligible loss at the important high-transmission communications windows. The region of the optical waveguide to be photosensitized is brushed repeatedly by a flame fueled with hydrogen and a small amount of oxygen, reaching a temperature of ~1700°C. The photosensitization process takes approximately 20 minutes. At these temperatures, the hydrogen diffuses into the fiber core very quickly and reacts with the germanosilicate glass to produce GODCs, creating a strong absorption band at 240 nm and rendering the core highly photosensitive. The flame-brush technique has been used to increase the photosensitivity of standard telecommunications fiber by a factor greater than 10, achieving changes in the index of refraction >10^{-3} [139]. A comparison may be made between standard and flame-brushed fiber, for which index changes of 1.6x10^{-4} and 1.75x10^{-3} have been realized, respectively, under similar laser writing conditions. Increased levels of photosensitivity are also exhibited in planar waveguides. The index step of a Ge-doped layer on silica was measured to increase by 15% after 10 minutes of flame-brush treatment, with an accompanying increase in the absorption coefficient of the waveguide reaching 700 dB/mm at 240 nm. This change is best described by the data of Figure 2.38. A comparison of curves (a) and (b) clearly shows the large changes that flame brushing can induce. A measure of the photoinduced change in the absorption coefficient is shown in curve (c) for the flame-brushed waveguide layer on exposure to UV light. Strong bleaching of the absorption band at 240 nm was achieved by irradiation at 248 nm for 40 minutes with a fluence of 112 mJ/cm^2 per pulse at 50-Hz repetition rate. The bleaching was accompanied by an increase in the bands at 213 and 281 nm, which are normally assigned to the Ge(2) and Ge(1) centers, respectively. With higher fluences of 400 mJ/cm^2 per pulse, gratings of 90% reflectivity have been written in Ge:silica on silicon waveguide, corresponding to a core index modulation of 9.5×10^{-4}. Temperature stability measurements on a photoinduced Bragg grating show that the refractive index modulation maintains 40% of its original value after the sample was held at 500°C for 17 hours.

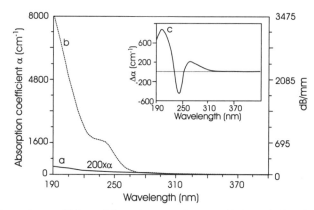

Figure 2.38   Absorption coefficient of the substrate and waveguide layer before flame-brush treatment multiplied by 200 (curve a) and of the waveguide layer after flame-brush treatment (curve b). Curve c is the photoinduced change in the absorption coefficient of the flame-photo-sensitized waveguide layer after exposure to UV light (*After:* [139]).

The enhanced photosensitivity techniques of flame brushing and hydrogen loading follow the same concept. In both cases, hydrogen is used in a chemical reaction with germanosilicate glass to form GODCs that are responsible for the photosensitivity. The formation of Bragg gratings in flame-brushed germanosilicate fibers undoubtedly involves both thermal and photolytic mechanisms, except that in this case the thermally driven chemical reactions occur simultaneously as the hydrogen diffuses into the core at elevated temperatures. Subsequent UV irradiation bleaches the GODC band giving rise to index changes. For fluence levels exceeding 250 mJ/cm^2 per pulse, a surface relief structure is also written at the silicon/silica boundary of a planar waveguide structure. It is believed that these surface relief gratings result from light-induced melting of the silicon combined with stress relaxation at the interface. There are several advantages in enhancing fiber photosensitivity by flame brushing. The increased photosensitivity in the fiber is permanent, as opposed to hydrogen loading where the fiber loses its photosensitivity as the hydrogen diffuses out of the fiber. It allows strong Bragg gratings to be fabricated in standard telecommunications fibers that typically exhibit no intrinsic photosensitivity. Localization of photosensitivity due to the relatively small flame can be used to brush the fiber. However, one major drawback of this technique is that the high temperature flame weakens the fiber, which has serious implications for the long-term stability of any device fabricated using this approach.

2.6.3 Co-doping

The addition of various co-dopants in germanosilicate fiber has also resulted in photosensitivity enhancement. In particular, B co-doping can lead to a saturated index change ~4 times larger than that obtained in pure germanosilicate fibers [19]. These fibers were fabricated using the MCVD technique. A comparison of the relative photosensitivity of four different types of fibers including boron co-doping is given in Table 2.2.

Table 2.2 Relative Photosensitivity for Four Different Fibers

Fiber type	Fiber Δn	Saturated index modulation	Maximum reflectivity for 2-mm gratings	Time for reflectivity to saturate
Standard fiber ~ 4 mol% Ge	0.005	3.4×10^{-5}	1.2%	2 hours
High index fiber ~20 mol% Ge	0.03	2.5×10^{-4}	45%	~ 2 hours
Reduced fiber ~ 10 mol% Ge	0.01	5×10^{-4}	78%	~ 1 hour
Boron co-doped fiber ~ 15 mol% Ge	0.003	7×10^{-4}	95%	~ 10 min

After: [19].

The fibers were irradiated with a modest power intensity of 1 W/cm^2 from a frequency-doubled CW argon ion laser until the grating reflectivity saturated. The results showed that the fiber containing boron had an enhanced photosensitivity. This fiber was much more photosensitive than the fiber with higher germanium concentration and without boron co-doping. In addition, saturated index changes were higher and were achieved faster than for any of the other fibers. This suggests that there is an additional mechanism operating in the boron co-doped fiber that enhances the photoinduced refractive index changes. The germanium-boron co-doped fiber was fabricated with a germanium composition of 15 mol%. In the absence of boron this fiber would have a refractive index difference of 0.025 between the core and cladding. However, when the preform was drawn into fiber, the measure value for Δn dropped to 0.003. It appears that the addition of boron reduces the core index of refraction. This result is not surprising, as it is known that the addition of boron oxide to silica can result in a compound glass that has a lower index of refraction than that of silica [148]. Studies have shown that boron-doped silica glass system results in lower refractive index values when the glass is quenched, while subsequent thermal annealing causes the refractive index to increase. This is consistent with the fact that Δn dropped from 0.025 to 0.003 when the preform was drawn into fiber, since fibers are naturally quenched during the drawing process. This effect is assumed to be due to a build-up in thermo-elastic stresses in the fiber core resulting from the large difference in thermo-mechanical properties between the boron-containing core and the silica cladding. It is well known that tension reduces the refractive index through the stress-optic effect. Ultraviolet absorption measurements of the fiber between 200 and 300 nm showed only the characteristic GODC peak at 240 nm. The boron co-doping did not affect the peak absorption at 240 nm nor the shape of the 240-nm peak, and no other absorption peaks were observed in this wavelength range [19]. The absorption measurements suggest that boron co-doping does not enhance the fiber photosensitivity through production of GODCs as in the case of hydrogen-loading and flame-brushing techniques. Instead, it is believed that boron co-doping increases the photosensitivity of the fiber by allowing photoinduced stress relaxation to occur. In view of the stress-induced refractive index changes known to occur in boron-doped silica fibers, it seems likely that the refractive index increases through photoinduced stress relaxation initiated by the breaking of the wrong bonds by UV light.

2.7 Photosensitivity at Other Writing Wavelengths

The preceding sections have shown that some form of externally induced enhancement can reverse the lack of natural photosensitivity in standard telecommunications fiber. It has recently been demonstrated that Bragg grating devices can also be inscribed in telecommunications fibers using the different laser wavelength of the ArF excimer vacuum UV (VUV) radiation at 193 nm [38, 133, 149] and with far greater writing efficiency. An immediate advantage of using 193 nm is a reduction of laser-induced damage when using a phase mask and the higher spatial resolution in diffraction-limited applications, such as point-by-point writing. More recently, photosensitivity has been demonstrated in fibers exposed to high-energy 351-, 334-, and 157-nm laser wavelengths [134, 136, 146, 150].

2.7.1 ArF Excimer Vacuum UV Radiation at 193 nm

We have concentrated thus far on showing how photosensitivity, and hence the inscription of Bragg gratings in optical fibers, can to some degree be associated with the bleaching of an absorption band located near 5.0 eV (245-nm band) [88]. We now report in some detail the salient points regarding photo-excitation at 193 nm, for which the previous statement no longer holds. Albert and co-workers have fabricated fiber Bragg gratings using KrF (248 nm) and ArF (193 nm) excimer laser pulses (via the phase mask technique) for comparative purposes. Bragg gratings fabricated at 193-nm irradiation appear to develop much stronger reflectivity than at 248 nm under similar excitation conditions. The decision to investigate photosensitivity at shorter wavelengths was provoked by an observation that there was no 5-eV absorption in some hydrogenated germanosilicate glass, despite the greatly increased photosensitivity as a result of hydrogenation [122]. Furthermore, there was mounting evidence that the dominant spectroscopic absorption changes associated with photosensitivity were occurring at wavelengths shorter than 200 nm [88, 122].

Absorption measurements from 190–400 nm indicate that bleaching is induced with

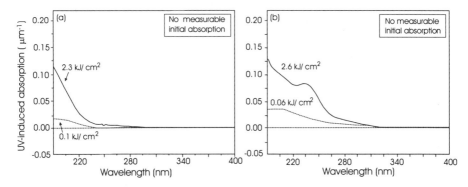

Figure 2.39 Spectra of absorption changes induced by UV light in initially transparent H$_2$-loaded Ge-doped silica: (a) at 248 nm, (b) at 193 nm. The two curves in each graph correspond to different cumulative UV doses (*After*: [133]).

an excimer laser operating at 50 pulses/s with pulse energy densities of 120 mJ/cm^2 at 248 nm and 40 mJ/cm^2 at 193 nm, respectively. Bragg grating fabrication was performed in a non-hydrogen-loaded, germanosilicate Alcatel fiber (GeO$_2$ 7 mol%), similar to that used in [88]. Measurements were also made on planar optical waveguides, consisting of Ge-doped silica on synthetic silica substrates, photolytically sensitized via hydrogenation [9]. Figure 2.39 shows the spectra for the hydrogenated (<7 days) Ge-doped silica waveguides. There is no detectable absorption from the dissolved molecular hydrogen, and any UV absorption predominantly results from the tail of the silica UV band gap. Nevertheless, strong absorption changes occur in the doped layer on exposure to UV light. At 248 nm the absorption increase occurs below 190 nm, which lacks an absorption band near 242 nm (Figure 2.39(a)). Conversely, 193-nm excitation results in a distinct absorption band near 242 nm, with an accompanying but weaker feature near 210 nm, in addition to the tail

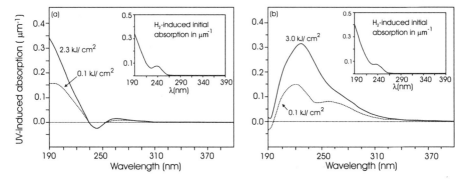

Figure 2.40 Spectra of absorption changes induced by UV light in H$_2$-loaded Ge-doped silica with initial absorption bands (shown in the insets): (a) at 248 nm, (b) at 193 nm. The two curves in each graph correspond to different cumulative UV doses (*After*: [133]).

of a strong absorption below 190 nm (Figure 2.39(b)). The formation of the 242-nm band occurs only in cases of high dosage exposure. Samples hydrogenated for longer time periods give markedly different results (Figure 2.40). Absorption bands near 242 nm and at wavelengths shorter than 190 nm are clearly evident (see insets). The initial absorption spectra in these cases are similar to those observed in standard optical fiber [88]. Exposure at 248 nm (Figure 2.40(a)), also follows the trend observed in optical fibers: the 242-nm band is quickly and completely bleached after a relatively small UV dose, but the absorption continues to increase significantly at shorter wavelengths with further exposure to the bleaching radiation. Exposure to 193-nm light (Figure 2.40(b)), results in different absorption changes with two bands appearing early in the exposure, centered near 220 and 260 nm, and merging into a strong 225-nm peak at a higher UV dose. Little bleaching of the high initial absorption at 193 nm is detected. The strongest absorption increases are achieved in the case shown in Figure 2.40(a) (i.e., bleaching at 248 nm when there is an initial strong 242-nm band). Doses greater than 40 kJ/cm^2 proved necessary to saturate the UV absorption increase in the case of Figure 2.39(a). Writing efficiencies for in-fiber Bragg gratings using 193-nm and 248-nm light were also compared, with the gratings exposed to doses of 300 mJ/cm^2 per pulse at 248 nm and 120 mJ/cm^2 per pulse at 193 nm. For total UV doses of 1.4 kJ/cm^2 (4-minute exposure), the reflectivity of a grating exposed at 193 nm reached 80%, whereas a grating of comparable length exposed at 248 nm reached only 20% after a total dose of 5.4 kJ/cm^2 (6 minute exposure). In contrast to 193 nm, the reflectivity of the gratings written at 248 nm continued to increase slowly for at least 60 minutes of irradiation. The photoinduced absorption changes observed under 193-nm illumination show some similarity to those in pure silica, where 220- and 260-nm absorption bands are induced under similar conditions [138, 151]. In the case of silica, two-photon absorption or the excitation of a band between 7 and 8 eV [151] is believed to be responsible for the rearrangement of defects. Isochronal annealing experiments revealed no difference in the thermal stability of gratings fabricated at the two wavelengths–in both cases, the refractive index modulation dropped by 10% of its initial value at 300°C and by

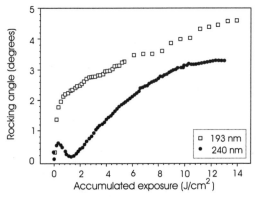

Figure 2.41   Comparison of rocking filter growth dynamics fabricated in a hydrogenated fiber at 193 and 240 nm. At 1.3 J/cm^2 the rocking filter fabricated at 240 nm passes though a minimum, attributed to a change in sign in the photoinduced birefringence (*After*: [55]).

50% near 600°C after 30 minutes. The experimental findings of Psaila et al. [55] point to the possible existence of two different mechanisms for the creation of photoinduced birefringence at these wavelengths. The writing efficiency at 193 nm is significantly larger than at 240 nm and produces rocking angles that are up to three times greater. Additionally, the growth dynamics at these wavelengths are dissimilar (Figure 2.41), suggesting that the different wavelengths result in photosensitivity via different mechanisms.

Following the work of Albert, Dyer et al. [38] likewise reported the formation of high reflectivity gratings in optical fiber using a 193-nm ArF laser irradiated phase mask. The authors, however, were clearly working in a different energy density regime, linking the observed changes to a damage mechanism (Type II) at the core-cladding interface under single-pulse, high-energy laser excitation [36]. This mechanism was enhanced by the larger absorption in the doped fiber core at this wavelength. The greater absorption was coupled with an incubation effect in which the damage threshold was reduced while the absorption grew with increasing number of pulses, which permitted the rapid formation (~10 pulses) of damage gratings at modest fluences (~400 mJ/cm²). The fluence distribution from the phase mask had dominant fluence peaks spaced at a distance d (i.e., the phase mask periodicity), and not $d/2$, which in a threshold process leads to the mask imprinting d-spaced features [152]. Malo et al. [39] reported a similar result for damage fiber gratings. Whereas similar experiments using a 248-nm KrF laser showed only a narrow range of fluence between successful damage grating production and catastrophic failure of the fiber, this was not the case at 193 nm. The larger UV absorption at the core-cladding interface and the strong incubation effect permitted the gratings to be written at a significantly lower fluence, without fiber failure.

There is a plethora of new data regarding the 193-nm inscription mechanism, with Type I and Type II gratings being formed. Dong et al. [149] discovered that the fast formation (~1500 pulses at ~1 (J/cm²)/pulse) of Type IIA fiber gratings at 193 nm, which produces highly negative index modulations of $\sim -3 \times 10^{-4}$, is also possible. It was found that a grating

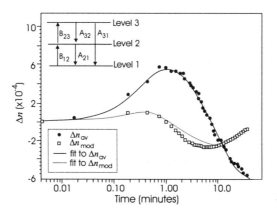

Figure 2.42 Growth of index modulation and average index when writing with a KrF excimer laser at 0.31 J/cm²/pulse, together with the fits from the proposed three-level energy model. The inset shows the energy diagram of the proposed model (*After:* [149]).

with positive index modulation was first formed, followed by a strong negative grating. A three energy level phenomenological system was used to model the photosensitivity, with all the model parameters determined from a single growth measurement of the average index change. This approach produces an excellent match between predicted fits and measured index modulation growth. Negative index gratings were written in boron-doped germanosilicate fiber and proved to be more stable than the positive index gratings formed at the early stage of grating growth [22]. In terms of the model, this is attributed to a higher decay energy barrier for the level associated with the negative index change. A similar type of grating growth and stability was observed when writing at 248 nm by Niay et al. [22]. Dong et al. [149] also observed that grating growth at 193 nm was much faster than at 248 nm, which agrees with all the data presented thus far for both Type I and Type II gratings. Figure 2.42 shows the growth of the index modulation and the average index change (0.31J/cm², 20 Hz) and covers all the growth features. The index modulation reaches a positive maximum after ~0.5 minutes with a strength of ~1x10⁻⁴, followed by a negative maximum at ~7 minutes with a strength of ~ −2.5x10⁻⁴. The index modulation decreases slowly toward zero after the negative maximum. The average index change reaches its positive maximum after ~1 minute and saturation was reached after ~40 minutes. The curves in Figure 2.42 are supported by theoretical fits obtained from the model. It is found that Type II behavior can be excluded, given that the maximum negative index modulations are independent of pulse intensity. The inverse of the time required to reach the negative index modulation maximum varies linearly with the pulse intensities, depending only on the total fluence of the exposure. The three-level energy system model explains a positive index change by the depletion of a ground level (level 1) populating a lower level (level 2). A negative index change results from a higher level (level 3) populated by depletion of level 2. The model predicts the linear dependence of growth rate on intensity and the constant saturated negative index changes at the different pulse intensities. The maximum negative index modulation is predicted to increase linearly with the writing fringe contrast

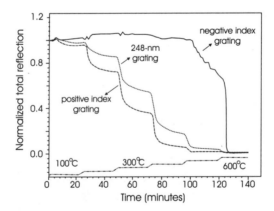

Figure 2.43   The stability of three gratings. The positive index grating was written until the positive index-modulation maximum. The negative index grating was written until the negative index-modulation maximum (*After:* [149]).

up to 95%, giving a negative index modulation of $\sim -5.2 \times 10^{-4}$. The same three-level model is equally applicable to 248 nm, as growth at this wavelength is similar but slower (relevant growth rate parameters would need adjusting). Phenomenological models can certainly predict aspects of grating growth but cannot address the fundamental mechanisms that result in the growth. The results of thermal stability tests conducted on a positive and negative index grating are shown in Figure 2.43. A third positive index grating written with a 248-nm excimer laser ($\sim$40 minutes at 20 Hz and $\sim$0.3 J/cm^2/pulse) is also shown. The grating reflectivities were 77%, 98%, and 60%, respectively. Clearly, the negative index grating is far more stable than the other two grating types.

2.7.2 Near-UV Irradiation at 334 and 351 nm

The first indications of a near-UV interaction with glass were observed by Niguchi et al. in 1986 [153]. A UV-induced absorption band at 630 nm and a luminescence band at 650 nm arose when a multimode germanosilicate fiber was exposed to an intensity of 10^3 W/cm^2 of 351-nm light. Investigations into the bleaching dynamics of 400-nm luminescence under near-UV excitation (266 nm) by Poirier et al. [51] brought to light that a triplet-state photochemical reaction could be responsible for the observed exponential luminescence decay. Dianov and co-workers have since proposed that if a triplet-state excitation is involved in photosensitivity of germanosilicate glass, the direct excitation of this state by near-UV 330-nm light should result in the same changes as observed at 240 nm. However, 330 nm cannot directly ionize the defect (334 nm is three times as efficient for writing Bragg gratings as 351 nm). The observation of 650-nm luminescence and bleaching of the GODC blue luminescence by CW 351-nm radiation confirmed this [100]. Atkins et al. have assigned the luminescence at 650 nm to DIDs [97]. Indeed, the increase in red luminescence has demonstrated a better correlation to the index change than the decrease in blue luminescence, while the relative correlation between the two bands is maintained over a range of excitation intensities, in contrast to 5-eV excitation [154]. The existence of two mechanisms for GODC photodestruction explains the discrepancy between the blue luminescence and refractive index behavior. Direct evidence of a near-UV index change in glass was reported in 1996 with the fabrication of a long period grating in Ge-doped fiber [155]. Starodubov and co-workers have since shown that efficient Bragg grating fabrication at 1550 nm in germanosilicate fibers is possible by use of near-UV light [134]. Using 334-nm light, the side writing of Bragg gratings with an index change of $\sim 10^{-4}$ in Ge-doped fiber was demonstrated. No hydrogenation of the fibers was required. These gratings were shown to have the same temperature stability as gratings fabricated with 240-nm light. An enhancement of the 334-nm photosensitivity in boron co-doped fibers is observed, suggesting that B facilitates a structural transformation of the glass.

A refractive index grating has been side-written in the core of a hydrogenated germanosilicate optical fiber, following a brief exposure to a CO$_2$ laser, using a phase mask and the 351-nm wavelength from an Ar ion laser [146]. The index change was 2×10^{-4}. Permanent changes in the refractive index have also been induced in pure GeO$_2$ bulk glass by exposure to the third harmonic (3.51 eV) of a Q-switched Nd:YAG laser [136]. The poor writing efficiency at this wavelength meant that the laser light had to be focused to a tight 28-μm beam diameter. A pulse beam energy of 7 μJ (4-ns pulse) at a repetition frequency of 10 Hz resulted in a permanent positive refractive index change after a prolonged exposure of several minutes. Using the z-scan technique, Δn was found to be proportional to the irradiation time during the first 10 minutes of exposure, saturating at a value of $\sim 10^{-4}$ at 30 minutes. The formation of the paramagnetic GeE' defect center, with a maximum spin density of 3.1×10^{18} spins/cm^3, was also confirmed. Taking account of the defect density and absorption coefficient, the refractive index change through the Kramers-Kronig transformation was $\sim 10^{-4}$, providing semi-quantitative agreement with the measured result.

2.7.3 Photosensitivity at 157 nm in Germanosilicate Fiber

Recently, Herman et al. [150] have demonstrated a new photosensitivity response of optical fibers and slab waveguides to light at 157 nm, from a F_2 excimer laser. Strong photosensitivity responses were anticipated because of the close proximity of 7.9-eV photons to the band gap of the germanosilicate glass at 7.1 eV. The photosensitivity response of two single-mode fiber types was examined: first, a high Ge-doped fiber (8%-GeO$_2$), and second, standard telecommunication fiber (3%-GeO$_2$ Corning SMF-28). Both fiber types were also soaked in 3-atm hydrogen in excess of 2 weeks to provide a total of four samples. The fibers were exposed to F_2 irradiation in ambient argon to provide VUV transparency, thereby permitting real-time characterization of the laser-induced index changes at 1.55 μm. Up to 9000 exposures were made for single-pulse fluence values of 25, 100, and 450 mJ/cm^2 (Figure 2.44). It was found that rates of index change are several times larger at fluences of 100–450 J/cm^2 and orders of magnitude faster at lower fluence, in comparison with the results of Albert et al. for 193-nm irradiation [10]. The 157-nm fluence dependence suggests a single-photon dependence of index change in departure with the two-photon response noted at 193 nm in Albert's work for the same fiber type. The 157-nm induced index changes were ~10-fold faster for hydrogen-loaded fiber, while higher

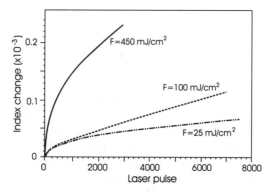

Figure 2.44 Index of refraction changes observed in Corning SMF-28 fiber exposed to 157-nm radiation. Values were inferred using a Michelson interferometer configuration (*After*: [150]).

germanium content had little effect on the rates. All fiber types yielded to a nonlinear damage mechanism at higher fluence, especially for the highly doped or hydrogen-loaded fibers, which is a result of the strong material absorption occurring under exposure to VUV wavelengths.

2.8 Photosensitivity Mechanisms

The precise origins of photosensitivity and the accompanying refractive index change have yet to be fully understood. It is clear that no single model can explain all the experimental results, as there are several microscopic mechanisms at work. There is substantial experimental evidence supporting the mechanism put forward by Hand and Russell [12]. The resultant color-centers are responsible for changes in the UV absorption spectrum of the glass, and the refractive index change follows the Kramers-Kronig relationship [12]. Many experiments [25, 52, 88, 93, 121] support the GeE' center model for photosensitivity, which is certainly the mechanism responsible for the original self-organized gratings [2]. The color-center model, however, does not completely explain all the experimental observations [22, 33], and an alternative model based on glass densification induced by photoionization of the Ge defects [156] also has experimental support [157]. We have seen earlier that the influence of the laser-writing wavelength, as well as power, fiber types, and processing, leads to many possible reaction pathways.

Measurements of the spectral changes accompanying UV irradiation and grating inscription have shown bleaching of the 240-nm band and the growth of absorption features at shorter wavelengths [25, 85, 88, 93], in particular at 195 nm. Kramers-Kronig analysis of these data yields values for the refractive index change in close agreement with those inferred from measurements on photoinduced gratings, which provides support for the color-center model. It is also consistent with the same model that the bleaching of the 240-nm band can be reversed subsequently by heating to 900°C [25]; a grating written, thermally erased, and rewritten in the same section of fiber exhibited essentially the same properties each time. Malo et al. have shown [158], however, that annealing standard germanosilicate telecommunications fiber in air at 1200°C can remove its photosensitivity irrecoverably. Conversely, Cordier et al. have presented the results of a tunneling electron microscope (TEM) investigation of gratings UV-written in a fiber preform. These show microstructural changes aligned with the grating fringes, which are interpreted as densification resulting from strain relaxation induced by the creation of the GeE' centers [159]. These authors argue that a greater spectral absorption range than the ~165- to 300-nm range considered by Atkins et al. [25] must be included in Kramers-Kronig analysis if the photoinduced refractive index is to be accurately determined. In the sections that follow, we will describe the various mechanisms associated with the changes in the index of refraction. After accumulating all the experimental findings, it becomes obvious that there are two main mechanisms that are involved in photosensitivity (at least for the most common germanosilicate fiber). These are described by the color-center model and the compaction model.

2.8.1 Color-Center Model

We have highlighted how a change in the index of refraction in a germanosilicate fiber triggered by a single, sub-band gap photon implies that point defects in the ideal glass tetrahedral network can be responsible for this process. Ironically, a great deal of research has been directed toward minimizing the formation of these color-center defects. With their connection to fiber Bragg gratings, however, the role of the defects in optical fibers has changed dramatically. The color-center model has received a great deal of attention, and while its contribution to explaining photosensitivity in germanosilicate fibers is considered complementary to other phenomena, there is increasing evidence that it is most applicable to hydrogenated germanosilicate fibers where the formation of microscopic defects occurs. Any change in the refractive index (i.e., through the formation of a grating), is associated with the photoinduced change in absorption through the Kramers-Kronig relation, expressed as [8]

$$\Delta n_{\text{eff}}(\lambda) = \frac{1}{2\pi^2} P \int_0^\infty \frac{\Delta\alpha_{\text{eff}}(\lambda)}{1-(\lambda/\lambda')^2} d\lambda \qquad (2.6)$$

where P is the principle part of the integral, λ is the wavelength, and $\Delta\alpha_{\text{eff}}(\lambda)$ is the effective change in the absorption coefficient of the defect, given by

$$\Delta\alpha_{\text{eff}}(\lambda) = (1/L) \int_0^L \Delta\alpha(\lambda, z) dz \qquad (2.7)$$

where L is the sample thickness. This takes into account the fact that the bleaching beam is strongly attenuated as it passes through the sample, and thus bleaching does not occur uniformly with increasing depth. $\Delta\alpha_{\text{eff}}(\lambda)$ may be modeled as a Gaussian distribution. Equation 2.6 may be used to calculate the index change that is induced by bleaching of the absorption bands. The boundaries are set to λ_1 and λ_2, the limits of the spectral range within which absorption changes take place and λ' is the wavelength for which the refractive index is calculated. The validity of (2.6) requires that λ' is much greater than the upper and lower bound limits. This relationship arises from the causality condition for the dielectric response–that there can be no index change before the application of a perturbation–and demonstrates that the index change produced in the infrared/visible region of the spectrum by the photoinduced processing results from a change in the absorption spectrum of the glass in the UV/far-UV spectral region. Measuring the Bragg grating reflectivity enables one to evaluate the effective index change, which may then be compared to the value calculated from (2.6).

In this model, first proposed by Hand and Russell [12], photoinduced changes in the material properties of the glass introduce new localized electronic excitations and transitions of defects. It is precisely these color-center defects, because of their strong optical absorption, that are proposed to give rise to the change in the refractive index associated with photosensitivity. The bleachable wrong bond defects, which initially absorb the light, are transformed into defects that are more polarizable by virtue of the fact that their electronic transitions take place at longer wavelengths (e.g., Ge(1) centers), or that they have stronger transitions (e.g., Ge(2), GeE') [91]. The observation of weak birefringence induced in low birefringence fibers, by two-photon absorption, indicates that oriented defects are produced [42, 50] in accord with this model. Further to this, the color-center model presumes that the refractive index at a particular point is related only to the number density and orientation of defects in that region, determined purely by their electronic absorption spectra. Any nuclear displacement arising from the photoinduced process is limited to a few atoms and is only weakly coupled to long-range displacements of the atoms in the network. Thus, only the electronic properties of the defects produced are important. The permanence and thermal annealing properties of the photoinduced index change are attributed to the slow kinetics of the reverse process.

Poyntz-Wright et al. [8, 61, 62] developed a quantitative color-center model primarily from their studies of photoinduced absorption at 488 nm. They proposed the mechanism whereby Ge-Si wrong bonds are transformed into GeE' centers by two-photon absorption and trapped electrons in Ge(n) centers, which give the absorption tails in the visible region. Using the parameters of the Hand and Russell model for the electronic transitions of the defect centers, this change is three orders of magnitude too small to explain the observed value of 2×10^{-4} under the same experimental conditions. If one first considers the change in absorption in the range 200–400 nm on a bulk sample before and after irradiation, one finds that a value of 1000 cm^{-1} is needed to agree with observed changes. This implies that accounting for large absorption arising in the deep UV may lead to a reasonable agreement [8]. It is noted that the model did not include a contribution from the GeE' center, which at that time had not been correlated to an absorption band, despite it being the dominant ESR active defect produced.

Matters were further complicated by the results of Williams et al. [160] who measured large differences between the UV absorption spectra of fibers and preforms. These results were later refuted by the same authors who used an improved technique, and by Gallagher et al. [161], and independently by Atkins [162] who found optical fiber and preform absorption spectra to be in general agreement. Atkins and Mizrahi [88] reported on the absorption spectrum between 200 and 300 nm of a fiber core before and after inscribing a Bragg grating with 81% reflectivity (see Figure 2.10(a)). The band at 241 nm was partially bleached and new absorption bands arose. Using the Kramers-Kronig relation, the observed changes in the absorption spectrum indicated that only 16% of the index change (inferred from the reflectivity of the grating), could be accounted for. A strong increase in the absorption edge in the region between 190 and 200 nm was considered responsible for a substantial part of the index change. The origin of this absorption was not known but was tentatively associated with the GeE' center. Atkins et al. [25] extended this work and reported a study of absorption changes in optical fiber preform cores. They measured the absorption changes between 165 and 300 nm, for a 3-mol% germania MCVD preform before and after UV exposure. The absorption changes observed in the region of 200–300 nm were consistent with that previously reported [88] for phase gratings written in optical fibers. The 241-nm band was bleached and a strong and broad absorption band centered at 195 nm appeared. Again this absorption band was linked to the ubiquitous GeE' center. Kramers-Kronig calculations of the absorption changes gave an index change in good agreement with that estimated from a Bragg grating written in fibers of similar composition and under the same UV exposure. The induced absorption changes could be completely reversed by heating the fiber at $900^{\circ}C$ for 60 minutes and duplicated by re-exposure at the same intensity and duration. This thermal reversibility of grating inscription supports the validity of the color-center model and cannot be explained through the stress-relief model (Section 2.8.6). Figure 2.2 shows the growth of a Bragg grating written in AT&T Accutether fiber with UV exposure and upon thermal erasure by heating from a CO_2 laser. The bleaching of the absorption band and the subsequent creation of new absorption bands agrees with the redistribution of defects first suggested by Hand and Russell [12]. The fact that the absorption changes reverse as a grating is heated is consistent with mechanisms of grating formation in which the absorption changes play a major role. It is interesting to point out that these results are in conflict with other reported data [163] in which each cycle of writing and erasing a grating was found to reduce the fiber's photosensitivity. Nevertheless, there is strong evidence linking the mechanism of refractive index changes, at least in part, to the color-center model.

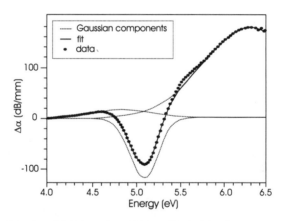

Figure 2.45   Measured $\Delta\alpha$ in a 0.2-NA preform (10.2 mol% GeO_2), after an exposure of 20 minutes at 50 mJ/cm²/pulse, fitted with three Gaussian functions (*After:* [135]).

Recently, Hosono et al. [85] have found a strong correlation between the optical absorption band peaking at 6.4 eV and the GeE' centers, with an oscillator strength of 0.5 (Figure 2.11(a)). These results point to compatibility with the Kramers-Kronig mechanism. This particular defect has been strongly linked with photosensitivity in hydrogenated fibers [107]. Tsai et al. [121] reported similar thermal stability between photoinduced GeE' centers and Δn, suggesting an association through the Kramers-Kronig relation and the color-center model (Figure 2.23) [25, 135]. This is supported by the power-law growth of both Δn [24,30] and the concentration of GeE' centers [127]. By taking into account the contributions from deep UV absorption bands, there is strong evidence [25, 135] that this model can explain a large part of the measured magnitude of the index change for Type I gratings. Dong and co-workers [135] have presented a rigorous evaluation of the photoinduced absorption change in MCVD-prepared germanosilicate preforms. Their results show that large photoinduced absorption changes between 165 and 300 nm (~700 decibels per millimeter (dB/mm) at 195 nm) account for ~3×10^{-4} of the index change expected at 1.5 μm for Type I gratings, through the Kramers-Kronig relation [25, 93]. It was found that the change in the strength of the 242-nm band peak is typically half that of the 195-nm band peak (6.35 eV), contributing a negative index change that is ~24% of the positive index change caused by the 195-nm band. The measured absorption changes used for the refractive index calculations, are shown in Figure 2.45 along with Gaussian fits centered at 242, 195, and 256 nm, showing good agreement. A pulsed KrF excimer laser at 248.5 nm and 20 Hz was used for all the UV exposures. The refractive index change at 1.5 μm may be expressed in terms of the fitted absorption curves as

$$\Delta n = (2.34\Delta\alpha_{242} + 4.96\Delta\alpha_{195} + 5.62\Delta\alpha_{256}) \times 10^{-7} \qquad (2.8)$$

where $\Delta\alpha$ terms are in dB/mm.

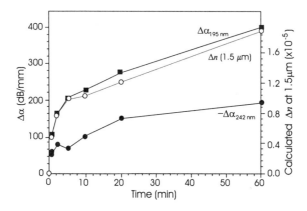

Figure 2.46 Growth of $\Delta\alpha_{242}$ and $\Delta\alpha_{195}$ in a 0.18-NA preform (8.3 mol% GeO_2) at 90 mJ/cm²/pulse and calculated Δn at 1.5 μm (*After*: [135]).

Figure 2.46 shows the evolution of the two main absorption bands in seven samples containing 8.3 mol% germania (0.18 NA) as functions of exposure times at 90 mJ/cm²/pulse. The calculated index change at 1.5 μm was ~2×10^{-4} for the 60-minute exposure. The 195-nm band increased by a very significant 0.4 dB/mm. It should also be noted that the ratio $\Delta\alpha_{195}/\Delta\alpha_{242} \approx -2$ is maintained for all the different exposure times, indicating a close relationship between the bands that is independent of the pulse energy densities and different germania concentrations. The thermal annealing dynamics of the photoinduced absorption changes resemble those of fiber gratings, decaying at temperatures as low as 200°C and totally disappearing at 800°C, giving further support to the color-center model. Additionally, for the same exposure (20 minutes at 60 mJ/cm²/pulse) the two main bands increase approximately linearly with the germania

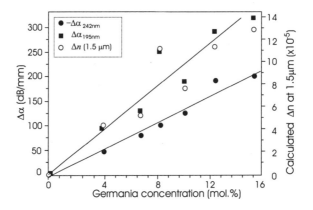

Figure 2.47   Dependence of the UV-induced changes on germania concentrations after exposure to 60 mJ/cm²/pulse for 20 minutes (*After*: [135]).

concentration (Figure 2.47)–a preform containing 15.5 mol% germania (0.25 NA), having attained a calculated index change of 1.4×10^{-4} (unsaturated) at 1.5 μm. It is interesting to note that $\Delta\alpha_{195}/\Delta\alpha_{242} \approx -2$ is approximately maintained. The pulse energy dependence of the absorption change for samples with 8.3 mol% germania further confirms maintenance of the relation between the bands (Figure 2.48). Above 50 mJ/cm²/pulse, the absorption change shows signs of saturation with the pulse energy density for an exposure time of 20 minutes. To simulate actual grating writing conditions, a focused excimer beam at 248.5 nm was used. Figure 2.49 shows the absorption changes obtained in time for a sample with 12.5 mol% germania (0.22 NA), for 250 mJ/cm²/pulse. The absorption change at 248.5 nm is measured by monitoring the transmission of the focused beam through the sample. The

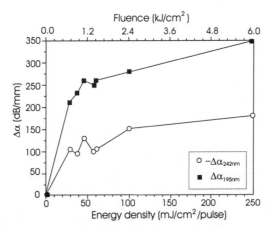

Figure 2.48   Pulse energy density dependence of absorption change in a 0.18-NA preform (8.3 mol% GeO₂) after 20 minutes (*After*: [135]).

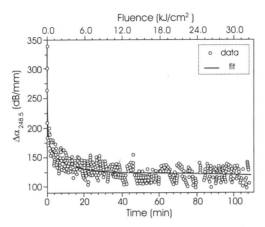

Figure 2.49 Dynamics of the absorption change at 248.5 nm at 250 mJ/cm²/pulse in a 0.22-NA preform (12.5 mol% GeO₂) (*After:* [135]).

total fluence (250 mJ/cm²/pulse for 2 hours) would have resulted in Type IIA gratings if a fiber were used in place of the sample. A stretched exponential function proves a good fit to the data

$$\alpha_{248.5}(t) = \alpha_{248.5}(\infty) + [\alpha_{248.5}(0) - \alpha_{248.5}(\infty)]\exp[-(t/t_0)^\beta] \qquad (2.9)$$

where $\alpha_{248.5}(0) = 340$ dB/mm, $\alpha_{248.5}(\infty) = 120$ dB/mm, $t_0 = 0.55$ minutes, and $\beta = 0.315$. The Type IIA grating should only occur after ~1 hour of exposure with a well-saturated absorption change. For the measured ratios $\Delta\alpha_{195}/\Delta\alpha_{242} \approx -2$ and $\Delta\alpha_{256}/\Delta\alpha_{242} \approx -0.15$, the estimated $\Delta\alpha_{242}$, $\Delta\alpha_{195}$, and Δn (1.5 μm) are ~360 dB/mm, 720 dB/mm and 3×10^{-4}, respectively, at their saturation level. For the original sample, $\alpha_{242} = 455$ dB/mm, therefore, ~80% of the original 242-nm band is bleached after the exposure. These results show that an appreciable component of the refractive index change in Type I gratings can be explained by the Kramers-Kronig relation. However, it is difficult to explain the ~10^{-3} index change obtained by Limberger [138] using this model.

Comparing Figures 2.46 and 2.48, one can see that for the same fluence but for different pulse energy densities, almost the same absorption changes are obtained. This indicates that the UV-induced absorption change is a linear mechanism driven by single-photon absorption. That the relationship $\Delta\alpha_{195}/\Delta\alpha_{242} \approx -2$ is maintained, regardless of fluence level and germania content, indicates a close relationship between these bands, as do the results described in [6, 85]. The dynamics of the absorption at 248.5 nm is very well fitted with a stretched exponential function. In two recent studies of UV-induced blue fluorescence in germanosilicate fibers [51, 164], the bleaching of the fluorescence with UV exposure was found to have a similar dependence with values of $\beta = 0.3$. A stretched exponential time dependence of energy or electronic transfer was also obtained when donor-acceptor pairs were simulated with a random distributed separation [165]. This is clear evidence of a direct link between the bleaching of the 242-nm band, the decay of the blue fluorescence, and the index change. Given that Type IIA gratings are anticipated in a region in which the absorption change at 248.5 nm is well saturated, different mechanisms need to be considered to explain the change in the refractive index, as we have highlighted earlier. Leconte et al. [166] have presented a modified Kramers-Kronig calculation with UV-induced absorption changes taken as functions of the total pulse fluence (exposure interval and pulse energy density) and wavelength for hydrogenated and unloaded fibers [166]. Finally, Digonnet has presented a mathematically rigorous method to compute the UV-induced refractive index change [167].

The color-center model cannot satisfactorily explain the behavior of all fiber types and their dopants. Silica fibers doped with P, Sn, or Ta display evidence that the corresponding changes in the UV absorption spectra affect grating formation. This contrasts with observations made on hydrogen-loaded, rare-earth-doped, aluminosilicate fibers, where the Kramers-Kronig analysis performed over the wavelength range 190–800 nm fails to account for the refractive index changes. It is useful to explain where discrepancies arise when computing the index change. One requires the knowledge, in principle, of the entire spectrum from zero to infinite wavelength. This information is not, of course, available and theoretical extrapolation has to be used. Given that the color-center model is strictly local, its assessment through the Kramers-Kronig relation requires that during grating fabrication one measures the UV-induced loss spectrum for each exposure interval and at each place along the grating for the spectral range covering the color-centers' absorption. Assuming that these measurements can be realized with a spatial resolution high enough to match the grating pitch, the Kramers-Kronig analysis would then give, for each exposure, the true form of the periodic refractive index change along the fiber axis. Subsequent description of the change in refractive index via a Fourier series would give both the mean and modulated index changes corresponding to the grating. The calculation is performed on a macroscopic scale; however, the following considerations should be noted. The total index change is the sum of contributions from a mean (Δn_{mean}, dc) and modulated (Δn_{mod}, ac) component. The absorption bands generated from these two components are used in the Kramers-Kronig calculation. Clearly, a calculation using the contribution from both absorption bands will result in a total refractive index change comprised of the sum of the two associated indices. A precise calculation of Δn_{mod} when $\Delta n_{mean} \neq 0$, is not simple to perform unless one knows exactly the contribution from the dc absorption-related bands.

2.8.2 The Electron Charge Migration Model

The electron charge migration model is derived from the color-center model and follows a classical approach that has been applied to explain photorefractivity in $Bi_{12}SiO_{20}$ or $Bi_{12}GeO_{20}$ materials [168]. An electric field created by charge migration acts on the index either by the Pockel's effect, where the index is proportional to the electric field, or by the Kerr effect, where the index is proportional to the square of the electric field. Payne proposed that charges can be extracted from defects by the action of light in an irradiated zone and trapped in deep levels in the dark zone [169]. This photo-excitation of defects leads to strong periodic, longitudinal electric fields within the fiber, which through the

Pockel's effect creates a second-order nonlinear $\chi^{(2)}$. Fermann has pointed to the formation of Ge(1) and Ge(2) defects in GeO$_2$-doped fiber when exposed to light at 488 nm [170]. It was shown that the Ge(2) defect can be photo-excited, releasing an electron. The spatial nonuniformity of the optical field in the optical fiber causes the subsequent diffusion of the photo-excited electrons from points of high to low optical intensity. This invokes photoconductivity that has actually been observed [171]. Equilibrium is reached when the electric field created by the charge transfer compensates the photoinduced concentration gradient. This electric field reaches about 10^7V/m and yields through the Pockel's effect an index grating amplitude of 7×10^{-7} taking $\chi^{(2)} = 7\times10^{-14}$ V/m (the second-order nonlinear coefficient deduced from second harmonic generation (SHG) experiments). This is clearly a very weak effect. For self-organized gratings, however, the reflection coefficient, calculated over a half-meter length of optical fiber, is 55%, which is in close agreement with the results observed experimentally by Hill and co-workers [2]. Nevertheless, this theory cannot explain results obtained through grating fabrication via the transverse holographic technique and the use of pulsed UV laser sources.

2.8.3 Permanent Electric Dipole Model

The dipole model is based on the formation of built-in periodic space-charge electric fields by the photo-excitation of defects. Photo-ionization of the GODCs, Ge-Si, or Ge-Ge, creates positively charged GeE' hole-centers and free electrons. The defect is fixed to the matrix whereas the electron has enough energy to escape, diffusing away from and getting trapped at neighboring Ge(1) and Ge(2) sites to form negatively charged Ge(1)$^-$ and Ge(2)$^-$ electron traps, respectively [172]. The GeE' hole traps and Ge(1)$^-$ and Ge(2)$^-$ electron traps result in electric dipoles with spacing of the order of several angstroms. Each resulting dipole will produce a static dc polarization field that extends many molecular lengths. For example, at a distance of 1 nm away from the dipole center the field will be approximately $E = 145$ V/mm, for a dipole spacing of 0.3 nm. These static electric fields induce local index of refraction changes proportional to E^2 through the dc Kerr effect. During the writing process of a Bragg grating, when the fiber is exposed to a UV interference pattern, the free electrons in the high intensity regions will diffuse until they are trapped by defects in the low intensity regions. This redistribution of charges within the fiber will create periodic space-charge electric fields. The periodic refractive index change is proportional to $\chi^{(3)}E^2$, where $\chi^{(3)}$ is the third-order nonlinear coefficient, and E is the electric field of the dipole source. The change in refractive index detected by a guided mode is then the refractive index change averaged over the defect volume. The dipole model was partly inspired by the photorefractivity models in crystals, where there is a $\pi/2$ phase shift of the index change relative to the interference pattern of the UV light, as shown in Figure 2.50. Although this mechanism works very well for photorefractive crystals, it might be difficult to justify in the case of photosensitive fibers due to the large number density of dipoles required. Hand and Russell [12] estimate that 10^{26}m^{-3} dipoles are needed for a Δn of 2×10^{-4}. Given that the molecular density is 2.2×10^{28}m^{-3}, the number of defects must be several orders of magnitude higher than estimated from experiments.

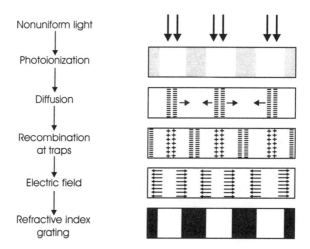

Figure 2.50   Response of a photorefractive material to a sinusoidal spatial light pattern.

2.8.4 Ionic Migration Model

It is also possible that ionic migration may be induced following diffusion, leading to a redistribution of the dopant responsible for the index change. There are two possible paths depending on whether the dopant profile is necessary in the transport process. Bonds may be excited in response to photon absorption, freeing a radical or an ion. The force acting on the species is proportional to the gradient of chemical potential, that is, the concentration gradient (Fick's law). The other possibility is that the beam acts directly on the material, initiating a microscopic mechanism of motion [173]. Although this model provides an alternative explanation for the index change, this effect has yet to be observed experimentally.

2.8.5 Soret Effect

Another form of ionic migration is the movement induced by the application of a laser-induced temperature gradient, called the Soret effect [174]. This has received detailed treatment by Miotello and Kelly [174] for silicon-implanted silica. It should be noted that this phenomenon could, in principle, be induced by any excitation laser wavelength, given there is a strong absorption of the photon energy into heat. A plethora of experimental evidence indicates otherwise. Furthermore, the fact that silica fiber, which is strongly absorbing in the mid-UV, does not exhibit photosensitivity provides strong evidence against this model.

2.8.6 Stress-Relief Model

The stress-relief model [91, 175] is based on the hypothesis that the refractive index change arises from the alleviation of built-in thermo-elastic stresses in the fiber core. The fiber optic core in a germanosilicate fiber is under tension because of the difference in the thermal expansion of the core and the cladding as the glass is cooled below the fictive temperature (glass transition temperature) during fiber drawing. This means that because the temperature of the fiber decreases rapidly there is a point where the structure (including defects) is frozen in. Through the stress-optic effect, it is known that tension reduces the refractive index, and it is therefore expected that stress-relief will increase the refractive index. It is proposed that during UV irradiation the wrong bonds break and promote relaxation in the tensioned glass hence reducing frozen-in thermal stresses in the core [91]. Although there is an abundance of breakable wrong bonds in germanosilicate core fibers, this is not the case for pure silica core fibers, which are not photosensitive in the UV (in agreement with the model).

The index change produced by the relaxation of the above stress can be calculated. It is useful to consider this stress-optic effect in terms of the Kramers-Kronig relation, which attributes it to the shift of the electronic band gap with stress. The glass can be modeled as a continuous, random, tetrahedral network of covalent bonds. Tension stretches the bonds, decreases the wavelength of the gap, and this leads to a decrease in the refractive index as described by the Kramers-Kronig relationship. The index change with stress for x-polarized light is characterized by the stress-optic coefficients c_1 and c_2:

$$n_x = n_0 - c_1 \sigma_x - c_2 (\sigma_y + \sigma_z) \tag{2.10}$$

where n_0 is the index of the unstressed material and σ_x, σ_y, and σ_z are the stress coefficients along the respective axes. The dispersion of c_1 has been measured to be quite weak because it arises from a small change over that of pure silica [91], as expected from the small shift of the band gap. Substitution of the stresses into the above equation gives the magnitude of the index change due to the thermo-elastic stress relaxation in optical fibers. It is demonstrated that they can be as large as 10^{-3} for highly stressed fibers (they are of the same magnitude as the typical photoinduced index change). Recently, Fonjallaz et al. [176] have reported the measurement of axial stress modifications in fiber Bragg gratings and have shown that tension in the core of single-mode germanosilicate fibers is greatly increased during Bragg grating formation. A strong increase of the tension has also been observed in [13], which contradicts the stress-relief model [91]. Finally, the thermal reversibility of grating inscription cannot be explained by this model [88].

2.8.7 Compaction/Densification Model

The compaction model is based on laser irradiation-induced density changes that result in refractive index changes. Irradiation by laser light at 248 nm at intensities well below the breakdown threshold has been shown to induce thermally reversible, linear compaction in amorphous silica, leading to refractive index changes. Fiori and Devine [110, 177] used a

KrF excimer laser to irradiate thin-film amorphous silica samples grown on Si wafers. Figure 2.51 shows the variation of this oxide thin-film thickness as a function of accumulated UV dose for a nominally 100-nm oxide sample. At an accumulated dose of 2 kJ/cm², there is an obvious reduction in the film thickness (approximately 16%) and a corresponding evolution of refractive index during laser irradiation. After annealing for 1 hour in a vacuum of 10^{-6} torr at 950°C, the compaction disappears and the original thickness and pre-irradiated refractive index value is recovered. Continued accumulation of UV irradiation beyond this reversible compaction regime leads to irreversible compaction, until the film is entirely etched after a total accumulated dose of 17 kJ/cm². An approximately linear relationship has been found between the index of refraction and the density change. The linear compaction, $\Delta t/t$, is translated to a volume change by transforming through, $\Delta V/V = 3(\Delta t/t)/(1+2\sigma)$ where σ is the Poisson's ratio. The reversibility in compaction, coincident with the creation of defects, conforms to results taken from implanted fused silica. Figure 2.51 shows the evolution of the refractive index: a rapid increase approaching 20% of the equilibrium value. The Lorentz-Lorenz relation for the refractivity (derived from the Clausius Mausotti relation) is

$$R = \frac{(n^2-1)}{\rho(n^2+2)} \tag{2.11}$$

with ρ being the specific gravity and n the refractive index. One obtains by differentiation

$$\Delta n = -\frac{(n^2+2)(n^2-1)}{6n}\left[1 - \frac{\Delta R}{(R\,\Delta V/V)}\right]\frac{\Delta V}{V} \tag{2.12}$$

The Lorentz-Lorenz relation can also be expressed in terms of density, and one obtains, with $\rho = 2.2$ (cgs units) for silica, $\Delta n = 0.267\Delta\rho$. Similar empirical laws for SiO₂

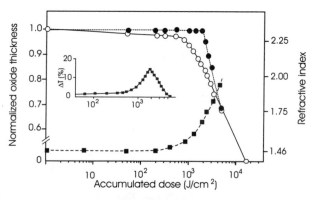

Figure 2.51   Observed compaction of a 100-nm thick oxide as a function of accumulated UV dose (open circle); the same sample after annealing for 1hour at 950°C in vacuum (solid circle). Inset: difference between these two results. Evolution of the refractive index during irradiation (solid square) (*After:* [110]).

($\Delta n = 0.25\Delta\rho$) have been found [177]. If $(V/R)\Delta R/\Delta V < 1$, then compaction results in the observed increase in refractive index. The first term in parentheses relates to structural volume changes; the second term, at least as far as SiO_2 is concerned, relates to the volume of the oxygen ion. The material density can also be described in terms of the number of polarizable oscillators per unit volume, whereas changes in the glass molar refractivity can be expressed through the material's macroscopic polarizability (i.e., the sum of the polarizabilities of these oscillators). Therefore, (2.12) may be expressed as $\Delta n \propto (\Delta V/V - \Delta\alpha/\alpha)$ [178]. Clearly there is competition between these two terms, which affects the refractive index. For example, the refractive index increase caused by a volume expansion (density decrease) would be larger in the absence of a simultaneous increase in the material polarizability. The measured and predicted values of the refractive index change are found to agree to within 10%, linking index variations to laser-induced structural variations. Fiori and Devine hinted at the possibility of a breakage of high- to low-order membered ring structures in the glass, and hence volume, with a reduction in the mean ring sizes to a limit of two-membered rings. This hypothesis has been supported by Raman spectra of compressed amorphous silica (Section 2.8.8) [179]. Beyond the two-member limit, reversible compaction gives way to irreversible compaction and etching, which results from sub- and direct-band gap network defects [180, 181]. Fiori and Devine [177] also measured the refractive index variation in hydrostatically compressed silica, producing results in very good agreement with laser-compacted, amorphous silica. This confirmed their hypothesis that laser and hydrostatically induced compaction arise though similar physical mechanisms, with compaction of amorphous silica proceeding through internal structural rearrangements and not primarily through a process of defect creation.

There is an important connection to photosensitivity given that compaction of silica at a free surface occurs upon laser irradiation. The effects of pulsed excimer laser irradiation at 193 nm examined by Rothschild et al. [182] identified both color-center formation and stress-induced birefringence, which are related to laser-induced microscopic volume changes of the solid. The absorption at 215 nm (correlated to the E' center) and birefringence effects were found to be strongly correlated, both increasing quadratically with increasing fluence (indicative of a two-photon absorption across the band gap at 146 nm) and displaying similar low temperature behavior. Figure 2.52 shows that the birefringence and color-center absorption initially increase linearly with the number of pulses (50 mJ/cm^2/pulse) to 10^6 pulses. Above this regime divergence occurs until 1.5×10^6 pulses, thereafter reducing the fluence to 12 mJ/cm^2/pulse results in a partial bleaching of the absorption band, while the birefringence remains unchanged. Application of an external, radial compressive stress completely eliminates the radiation-induced birefringence, implying that the unexposed areas are under tensile stress caused by compaction of the exposed volume. The density increase is accompanied by a refractive index change ($\sim 5 \times 10^{-5}$) unrelated to the color-centers, which remains unchanged even after 40% bleaching of the color-centers. One may therefore conclude that two separate effects are evident: the first resides as E' center absorbers and the second as a matrix change leading to compaction and the refractive index change. Even though the band gap of amorphous silica is 9 eV, there is ample evidence that sub-band [183] radiation can couple energy into fused silica, causing point defect formation. Excimer laser-induced

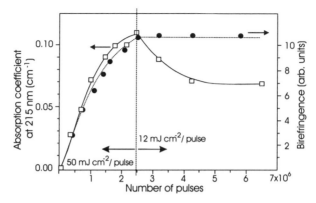

Figure 2.52   Effect of irradiation history on the 215-nm absorption coefficient and stress birefringence (at 633 nm). Three regimes are noted: (a) a linear increase at high (50 mJ/cm²/pulse) fluence, up to ~10⁶ pulses; (b) a sublinear increase at the high fluence for the next ~1.5x10⁶ pulses; and (c) partial bleaching but no change in birefringence upon reducing the fluence to 12 mJ/cm²/pulse (*After*: [182]).

densification of fused silica at 193 nm [184] is well known, and the use of excimer lasers for optical lithography puts great demands on optical materials to resist radiation damage. The densification of fused silica follows a power law in dose. More importantly, densification obeys reciprocity (i.e., it depends on NI^2 but not separately on pulse number N and fluence per pulse I).

Several microscopic mechanisms can lead to a volume change: a phase transformation (a change of the long-range order), a change in the polymerization of the glass (a change in the medium-range order), or a change of coordination (a change in the short-range order). X-ray absorption spectroscopy by means of synchrotron radiation, specifically shows that in Ge-containing glass exhibiting photosensitivity, Ge has a fourfold and sixfold structure. A change of the short-range order environment will induce a change in the long-range order, but the reverse is not true. There is evidence that under UV excitation, bond breaking leads to local structural changes that change the coordination number. The disordered nature of glass gives a medium-range order that results from intermolecular bonding. Structural voids allow molecular elements to rotate, leading to a variation in the degree of polymerization. For example, in SiO_2-P_2O_5, the phosphorus coordinance remains constant [185], but one oxygen can be doubly instead of simply bonded. Therefore, although the short-range order is the same, the medium range is different. Thus, the hardness decreases with P, as is observed. Photoinduced densification utilizes the change in coordination structure around a dopant, which traps an electron or hole when UV irradiated. A drastic coordination change can be expected from a tetrahedrally coordinated to a three-coordinated state [186]. The relaxation of a GEC to a GeE' center under prolonged KrF laser irradiation corresponds to this type of coordination change [85].

Raman measurements in Ge doped glass suggest that the irradiation and hydrostatic pressure can change the rate of fourfold, fivefold, and sixfold rings present in the silica network [187] (discussed further in Section 2.8.8). Long-range order is based on inducing crystallization or the redistribution of aggregates or precipitates. High pulse densities that result in material fusion are changes of this type [40]. In this case, as the photon energy density is so high, multiple step absorption can arise. Absorption initially causes photoconductivity [171] and additionally causes broadband absorption. Then, a second photon is absorbed in the conduction band and all its energy is released under heat because of the high density of electron states available in the conduction band. This leads to fusion.

Thus far, we have illustrated the densification process in bulk Si samples. However, this may be extended to describe the UV-induced densification processes in germanosilicate optical fibers, which is critical to understanding the formation dynamics of fiber Bragg gratings. Poumellec [130] investigated Bragg grating formation in germanosilicate glass using an optical microscope via interferometric microscopy, and concluded that densification occurs and accounts for 7% of the UV-induced refractive index increase. The MCVD-manufactured preform was composed of a B_2O_3- (4 mol%) and P_2O_5- (0.3 mol%) doped silica-clad and germanium-doped core (11 mol%). Gratings were written at 243 nm for a pulse energy density of 100 mJ/cm^2 and peak power density 4 MW/cm^2 (17 ns at 40 Hz). The evolution of the refractive index for both the fiber and preform is shown in Figure 2.53, with the relative contributions from the mean and modulated components shown. Comparison of the modulated index for the fiber and the preform shows similar results, supporting the hypothesis that preform behavior under UV irradiation can be strongly correlated to the behavior of optical fibers. Above 60 kJ/cm^2 the index change reduces tending towards Type IIA grating formation. For the microscopy measurements the interference pattern was set to 8.3 μm and the UV dosage limited to the regime relevant to Type I gratings (10 kJ/cm^2). One should be aware that this is fiber dependent and there are conflicting reports on the exact value of this limit [131]. Figure 2.54 shows the grating profile, with the grating preferentially formed to one side of the core. The profiles for a Type I grating (10,000 pulses) show the longitudinal (a) and transverse (b) perspectives. If compaction contributes to 7% to the refractive index

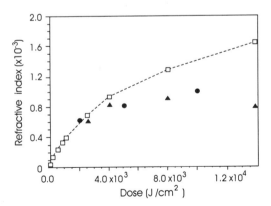

Figure 2.53 Refractive index variations for fiber and preform as a function of dose of UV-pulsed irradiation. Preform (circle): refractive index modulation. Fiber (triangle): refractive index modulation. Fiber (square): mean refractive index (*After*: [130]).

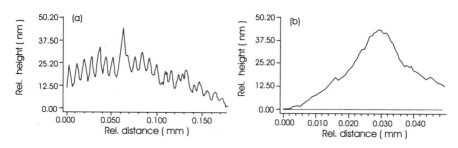

Figure 2.54 Grating profiles. (a) Longitudinal: 12.5-nm maximum modulation and 23-nm mean value at maximum. (b) Transverse profile passing between grating fringes (*After:* [130]).

change, there should be greater temperature stability for this component compared with contributions that arise from the color-center model. Limberger et al. [13] studied compaction and photoelastic-induced index changes in fiber Bragg gratings. They found that the tension on the core of single-mode fibers is strongly increased by the formation of a Bragg grating, in contradiction with the stress-relief model [91]. This tension increase lowers the refractive index because of the photoelastic effect. On the other hand, the compaction of the core network results in an increased refractive index. The two contributions were evaluated from axial stress measurements, from the determined index modulation amplitude, and from the mean index change of the Bragg gratings. The total Bragg grating index modulation had a positive mean value, explained by a structural modification of the germanosilicate core network into a more compact configuration. The mean index change was observed to be at least 20% smaller than the index modulation amplitude. The total Bragg grating index modulation was found to be smaller than the compaction-induced index modulation by 30%–35% because of the photoelastic effect. It was argued that given that the color-center model cannot account for index changes larger than $\sim 4 \times 10^{-4}$ [135], a structural modification leading to compaction of the glass matrix must be the main contribution (inelastic) to the observed index change. Figure 2.55 shows as an example the Bragg grating index modulation amplitude, Δn_{mod}, and the mean axial stress increase at the core center, $\Delta \sigma_{z,0,mean}$. The similarity between these two curves shows that there is a linear relation between Δn_{mod} and $\Delta \sigma_{z,0,mean}$ where the proportionality factor is $(0.8 \pm 0.2) \times 10^{-4} \text{mm}^2/\text{kg}$ [176]. Figure 2.55 also depicts the evolution of Δn_{mean} obtained from the Bragg wavelength shift. Δn_{mean} is positive since the Bragg resonance shifts to longer wavelengths during the irradiation and corresponds to 80% of Δn_{mod} at the end of the irradiation. Mean values smaller than the index amplitudes were observed in contrast to results earlier published [43, 130]. There have been a number of recent works dealing with Bragg grating investigation via optical microscopy and AFM [23, 57]. Cordier and co-workers [188] performed the first TEM measurements regarding densification on optical fibers. Type I gratings were examined, in both hydrogen-treated and untreated fibers, and it was concluded that fringes associated with density changes were only observed in untreated fibers, which points to an explanation through other mechanisms, such as the defect mechanisms in the color-center model. Where density changes were observed, it

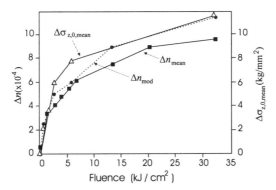

Figure 2.55 Index modulation amplitude, Δn_{mod}, and axial stress modification at the core center, $\Delta \sigma_{z,0,mean}$, for fiber gratings fabricated with different fluences. Δn_{mean} is the evolution of the mean index change for the grating written with the largest dose (*After*: [13]).

was found that the fringe periodicity mimicked the UV irradiance profile, implying that glass compaction was induced at the high points of the UV irradiation pattern.

Douay et al. have produced a comprehensive study into densification and its involvement in photosensitivity in silica-based optical fibers and glass [41]. A comparison of densification in hydrogen-loaded and non-hydrogen-loaded preform slices was carried out to determine whether hydrogen contributes to the densification process. Through this study several conclusions were reached. It was found that the modulated refractive index was thermally reversible (unlike the mean core index), in most non-hydrogen-loaded germanosilicate and aluminosilicate fibers, except for highly doped germanosilicate fibers. Photoelastic densification could account for 40% of the photosensitivity of non-hydrogen-loaded germanosilicate or aluminosilicate plates. Unlike the enhancement of UV photosensitivity via hydrogen loading, no increase in densification following hydrogenation was observed in germanosilicate plates. Given that hydrogenation considerably increases the UV-induced excess loss below 220 nm without strong saturation effects, it was concluded that the color-center model accounts for a large part of the photosensitivity in the hydrogenated germanosilicate plates. The refractive index modulation in non-hydrogen loaded germanosilicate was thermally reversible, whereas the mean index was not. If one assumes that the heating-induced increase in the mean refractive index arises from thermal compaction of the germanosilicate fiber core and that densification accounts for a non-negligible part of the UV-induced refractive index modulation, then thermal reversibility of the modified change implies that the heating-induced compaction of the core does not prevent further UV-induced densification. Poumellec [21] has shown that by starting with a densification annealing model one can obtain similar behavior to that given by Ergodan et al. [34], who based their calculations on a charge transfer model for the decay mechanism, whereby carriers excited during writing are trapped in a broad distribution of trap states and the rate of thermal depopulation is an activated function of the trap depth. Poumellec's modified model is applicable to photoinduced structural changes.

It is not quite clear when and for what parameters compaction is important. Under certain experimental conditions compaction plays a major role in the UV-induced index of refraction, a role that was previously assumed to be played by the defect in the color-center model. It is the belief of the authors that compaction is certainly one of the major mechanisms in explaining UV photosensitivity. However, its exact contribution under various experimental conditions has to be investigated further.

2.8.8 Raman Spectroscopy of Germanosilicate Fibers

Raman spectroscopy can be used to probe macroscopic structures having large range order, such as optical fibers, while offering sufficient resolution to identify changes associated with UV-induced compaction [187]. The silica fiber Raman spectrum is generally characterized by having a main peak at 450 cm^{-1} identified with the symmetric stretching motion of the bridging oxygen atoms. Other peaks, at ~490 and ~600 cm^{-1}, have controversial assignments, labeled as defect lines D1 and D2, respectively. They are related to structural defects in the vitreous state and are also attributed to fourfold and threefold rings involving the Si-O base unit. They provide information regarding the thermal and mechanical history of the glass. Given this, it is highly likely that the fiber-drawing parameters (tensile stress and furnace temperatures) play a key role in influencing the number, relative orientations, and shapes of these structures. Certainly, Si-O coordination changes have been documented under similar demanding conditions [185]. UV-irradiation-induced structural transformation of germanosilicate glass fibers measured using Raman spectroscopy has been investigated by Dianov and co-workers [189]. Significant changes were observed in the Raman spectra after UV irradiation, indicating long-range transformation of the glass structure. Single-mode, germanosilicate fibers (core composition 20–25 mol% GeO$_2$ and 2 mol% P$_2$O$_5$) were exposed to 248-nm KrF excimer laser pulses, with a fluence per pulse of 484 mJ/cm^2, for a total fluence of ~41 kJ/cm^2. Bragg gratings were fabricated in the same fiber with a total fluence equal to that of the homogenous irradiation. Raman fiber spectra were recorded with a spectral resolution of better than 1 cm^{-1} during excitation with an Ar$^+$ laser operating at 514.5 nm.

The reflectivity and Bragg wavelength as a function of the total laser fluence are shown in Figure 2.56. The fluence dependence of the reflectivity and the Bragg wavelength are typical for Type I and Type IIA Bragg gratings. Figure 2.57 shows the Raman (background subtracted) spectra for samples, before and after being UV irradiated at fluence levels corresponding to Type IIA gratings. Before irradiation, a peak with the maximum at 435 cm^{-1} dominates the Raman spectrum, after irradiation this shifts to higher frequencies. The peak shift is accompanied by an intensity increase of the red luminescence band, possibly explained by the existence of different forms of nonbridging oxygen-hole centers in the glass [190], or the creation of DID centers [100]. A general deformation of this band occurs, decreasing in intensity of its low frequency side while increasing on its high frequency side. The peak is associated with the symmetric stretching modes of bridging oxygen v$_s$ (Si-O-Si), v$_s$ (Si-O-Ge), and v$_s$ (Ge-O-Ge) of sixfold rings of SiO$_4$ and GeO$_4$ tetrahedra in glass. The changes in the Raman band near 435 cm^{-1} are associated with the decrease

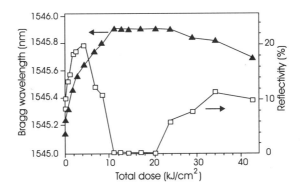

Figure 2.56   Reflectivity and Bragg wavelength of a grating written in a fiber with 25 mol% GeO₂ in the core as a function of the total irradiation fluence (*After*: [189]).

of order of the tetrahedron rings in the glass. The weak shoulder of the Raman spectrum near 480 cm⁻¹ is associated with the same modes of fourfold rings, whereas any asymmetry in the main peak results from the presence of fivefold and sevenfold rings. The shoulder near 580 cm⁻¹ (threefold rings), showing significant growth, along with the peak near 670 cm⁻¹, is assigned to other vibrational modes of glass, v_δ (Ge-O-Ge) and v_δ (Ge-O-Si). The low frequency Boson peak at 50 cm⁻¹, due to the quasi-localized vibration modes associated with the middle order of the glass, shifts by 10 cm⁻¹ into the high frequency range. The significant changes in the fiber Raman spectra, after UV excitation, indicate changes in the density of the oscillation vibrational states of germanosilicate glass. UV irradiation may induce breaking of large (sixfold or more) rings (350–470 cm⁻¹ frequency) and the formation of fourfold, threefold, or twofold (greater than 480 cm⁻¹ frequency) rings. The increase in the number of lowfold rings in germanosilicate glass after UV irradiation has been interpreted by the destruction of the GeO₂ deficiency centers and the conversion of the broken bonds formed during UV irradiation into O atoms with the formation of Ge-O-Ge and possibly Si-O-Ge(Si) bonds [189].

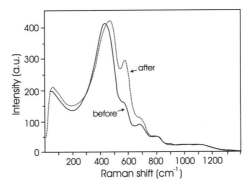

Figure 2.57   Raman spectra (background subtracted) of 75-mol% SiO₂: 25-mol% GeO₂ optical fiber core, before and after UV irradiation (*After*: [189]).

A possible scheme for such a glass structure transformation in which three coordinated O atoms and lowfold rings are formed has been calculated [189]. An increase in the concentration of lowfold rings in the glass structure should result in an increase in its density and refractive index, as observed in [110]. A correlation between the refractive index change and stress change in the fiber core during Bragg grating formation by UV irradiation was reported in [13]. The changes in the Raman spectra cannot be realized by preform fabrication or during glass drawing of germanosilicate core fibers and certainly result from exposure to UV light.

2.9 Photosensitivity: Co-dopants and Other Fiber Types

The photosensitivity of standard germanosilicate fibers with moderate concentrations of germanium in the core (5–7 mol% GeO_2) is not large and does not allow for the efficient writing of refractive index gratings. To increase the writing efficiency one may increase the germanium concentration and/or load the fiber with hydrogen. The former results in a reduction of the modal field diameter and additional losses when the fiber is spliced with a standard telecommunication fiber; whereas the latter raises losses in the IR region owing to the OH group absorption, unless D_2 is used. Therefore, optimization of the grating writing technology and the development of novel types of photosensitive fiber are now a priority. In fact, photosensitivity has been observed in co-doped germanosilicate fibers and germanium-free fibers. We briefly describe some of the more important results in the sections that follow and highlight these results in Table 2.3. It is also noteworthy that the important Er^{3+}-doped fiber, possessing a relatively high germanium content also exhibits significant UV photosensitivity [18].

2.9.1 Nitrogen Co-doping

Dianov et al. [191] have fabricated a germanosilicate fiber (7 mol% GeO_2) co-doped with nitrogen by the surface-plasma chemical-vapor deposition (SPCVD) process. The fiber has been found to be far more photosensitive than similar N_2-free fibers. They have demonstrated a photoinduced refractive index change of 2×10^{-3} at an irradiation dose of 75 kJ/cm^2 and a wavelength of 244 nm without hydrogen loading, and an index change of 1×10^{-2} after hydrogen loading. The increase in photosensitivity is attributed to a greater concentration of GODCs in this glass. Hydrogen loading magnifies the photosensitivity of N_2-doped germanosilicate fibers. The waveguide parameters of such fibers can be made to closely match standard telecommunication fibers; therefore, gratings written in N_2-doped, germanosilicate fibers may find many applications in fiber optic communications.

Table 2.3 The Influence of Co-dopants on Fiber Photosensitivity

Nature of co-doping	Characteristics of writing sources	Fiber treatment	Change in the refractive index	Type	Erasure temperature °C	Ref.
P	248 nm pulsed	‣ Hydrogenation followed by heating up to 400°C	7×10^{-4}			[126]
P	193 nm pulsed	‣ Hydrogenation	2×10^{-4}	I	100	[192]
P+Yb^{3+}, Er^{3+}	193 nm pulsed	‣ Hydrogenation	10^{-3}			[203]
P, Al, Yb^{3+}, Er^{3+}	248 nm pulsed	‣ Hydrogenation	2.5×10^{-5}	I		[204]
P, Sn	248 nm pulsed	‣ No treatment	5×10^{-4}	I	100	[196]
Ge Sn	248 nm pulsed ~400 mJ/cm²	‣ No treatment	2×10^{-3}	I		[205]
	244 nm CW 30mW/cm² or pulsed 200 mJ/cm²		10^{-3}	I IIA	100 500	[206]
P, Ge	193 nm pulsed ~200 mJ/cm²		2.6×10^{-4}	IIA		[207]
N	193 nm pulsed	‣ Hydrogenation does not improve the fibers photosensitivity	8.5×10^{-4}	IIA	700	[132]

2.9.2 Germanium-Free Silicon Oxynitride Fiber

Dianov et al. [132] have demonstrated photosensitivity in germanium-free, nitrogen-doped silica fiber. Such fibers are known to be more resistant to γ-radiation than normal germanosilicate fibers and have radiation resistance comparable to results obtained for pure silica core fibers. As a result, inscription of Bragg gratings in such fibers opens up unique possibilities for the development of various sensors intended for use in hazardous radiation environments. Their fiber was fabricated by hydrogen-free, reduced-pressure SPCVD. This fiber was also hydrogen loaded at 100 atm for 20 days at room temperature. In their experiments Bragg gratings were inscribed using the phase-mask technique with 193-nm UV light. Typical time evolutions of the refractive index modulation calculated from the grating reflectivity are shown in Figure 2.28 for four writing fluences. Further details are given in Section 2.5.1.2. The dynamics of growth of the Bragg gratings in the non-hydrogen-loaded fiber closely resembles Type IIA gratings in silicate fibers doped with a high concentration of germanium. The temperature stability of the gratings is very promising: heating the fiber to 1200°C for 30 minutes decreased to refractive index modulation by only a factor of two. Together with the good radiation resistance of N$_2$-doped silica fibers, this allows for unique possibilities of many optoelectronic systems intended for use in a hostile environment.

2.9.3 Phosphorus-doped Fibers

Although the majority of research on the photosensitivity of glass has been directed towards forms of germanium doping, many applications demand co-doping with rare-earth ions. Germanosilicate glass is not an ideal host, the solubility of rare-earth ions is low. A better host is achieved by doping the silica with phosphorus and small amounts of aluminum, rare-earth ions, and/or with tin, boron, or germanium. It is well known, however, that the presence of phosphorus bleaches the absorption band centered on 240 nm and thus reduces the photoinduced index change. Photolytic grating writing in phosphorus-doped waveguides has recently been possible only with substantial hydrogen loading [192]. Recently, Canning et al. [193] have demonstrated photosensitivity at 193 nm in phosphosilicate fibers fabricated by the flash condensation technique [194]. Strasser et al. [195] showed for the first time that strong (>3 nm spectral width) UV-induced Bragg gratings can be written in P-doped silica material. This grating was produced by 10-minute exposure at 30 Hz and 90 mJ/cm^{2}pulse. The three types of dynamic behavior have been observed for phosphosilicate fibers. Type II gratings have been induced through the exposure of a tin-doped fiber to light at 248 nm at a fluence per pulse higher than 400 mJ/cm^2 [196]. Lowering the fluence to 250 mJ/cm^2 resulted in the formation of Type I gratings. There is little information regarding the thermal stability of gratings written in phosphosilicate fibers; however, thermal erasure of a grating written in a phosphosilicate channel waveguide at 193 nm has been demonstrated [192].

2.9.4 Rare-Earth-doped Fibers

In view of the important applications of rare-earth-doped fibers in fiber laser technology, the direct fabrication of Bragg gratings in such fiber was an extremely important development. Fiber gratings have been written into rare-earth-doped fibers [43, 197] following hydrogen loading of the fiber to achieve photosensitivity. Permanent refractive index gratings have been observed in several Eu^{3+}-, Pr^{3+}-, and Er^{3+}-doped oxide glasses [198]. Dong et al. [29] have conducted a comparison study between a Ge-doped optical fiber and Ce^{3+}-doped aluminosilicate optical fiber and concluded that Ce^{3+}-doped fiber gratings have performance similar to those germanosilicate fiber gratings in terms of both grating strength and thermal stability. The important applications relating to rare-earth-doped optical fibers will be discussed in later chapters.

2.9.5 ZBLAN Fluoride Glass Optical Fibers

Fluoride-based glass optical fibers were initially not thought to be photosensitive. However, some recent experiments have proven otherwise. Poignant et al. [199] and Taunay et al. [200] recently wrote permanent gratings in both bulk samples and optical fibers of Ce-doped fluorozirconate glass. Photosensitivity in fluorozirconate glass fiber is not very well understood, and far less is known than the photosensitivity encountered in germanosilicate fiber. Recently, however, Williams et al. [201] have performed several experiments to elucidate the mechanism of photosensitivity in rare-earth-doped fluorozirconate glass. Glass doped with Ce, Tb, Tm, and Pr was studied. Permanent holographic gratings were written in bulk samples using 248-nm light, with the strongest gratings observed in Ce:ZBLAN. UV-induced changes in both absorption and ESR spectra were observed. In the Ce-doped glass the grating formation dynamics were recorded as a function of writing beam intensity and Ce concentration. The results indicated that the mechanism of photosensitivity involves color-center creation through a stepwise two-photon excitation of a Ce ion. The color-centers can subsequently be bleached by one-photon at 248 nm.

2.9.6 Aluminosilicate Optical Fibers

Thus far we have concentrated on photosensitivity germanosilicate glass; however, aluminosilicate glass has also exhibited photosensitivity in recent experiments. Table 2.4 shows that photosensitivity in aluminosilicate fibers has only been demonstrated under exposure to pulsed light, with the highest refractive index modulations obtained through hydrogen loading and co-doping the fiber with rare-earth ions. To date Ce dopants have proven to be most efficient. The grating dynamics are classed as Type I. Figure 2.58 shows the growth rate of gratings written in non-hydrogen-loaded Ce^{3+}-doped aluminosilicate fiber. Contrary to observations in germanosilicate fibers, the rate of grating growth in strained aluminosilicate fiber is not reduced by the application of strain. In contrast, gratings written in either hydrogen- or non-hydrogen-loaded aluminosilicate fibers, show similar thermal stability to gratings written in non-hydrogen-loaded germanosilicate

Table 2.4 Photosensitivity for Doped Aluminosilicate Fibers

Nature of co-doping	Characteristics of writing sources	Fiber treatment	Change in the refractive index	Type	Erasure temperature °C	Ref.
$Al + Eu^{2+}$	248 nm pulsed	▸ No treatment	2.5×10^{-5}	I	300	[16]
$Al + Ce^{3+}$	292 nm pulsed 265 nm pulsed	▸ No treatment	2.5×10^{-5} 3.7×10^{-4}	I	23–150	[17, 41]
$Al, P, + Ce^{3+}$	248 nm single pulse 266 nm pulsed	▸ No treatment	5×10^{-5} 1.4×10^{-4}	I	150	[29, 208]
$Al + Ce^{3+}$ $Al + Tb^{3+}$ $Al + Er^{3+}$ $Al + Tm^{3+}$ $Al + Yb^{3+}; Er^{3+}$	240 nm pulsed 240 nm pulsed 235 nm pulsed 235 nm pulsed 193 nm pulsed	▸ Hydrogenation	1.5×10^{-5} 6×10^{-4} 5×10^{-5} 8×10^{-5} 5×10^{-4}	I	100	[41]

fibers. However, gratings written by exposure to a small-cumulated fluence (less than 1 kJ/cm^2) counter this observation since they spontaneously bleach (partially) at room temperature [41].

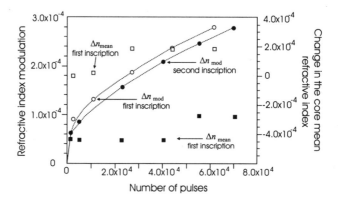

Figure 2.58   Growth rate of gratings written in non-hydrogen-loaded Ce^{3+}-doped aluminosilicate fiber. Initial writing (open symbols) and second writing (black symbols) after the first grating was thermally erased at 900°C for 15 minutes. Δn_{mean} (squares) and Δn_{mod} (circles) (*After*: [41]).

2.10   Maintaining the Index Change

We close this chapter by briefly discussing some of the issues related to stabilizing the UV-induced refractive index changes. Once the index change associated with photosensitivity has occurred, the major contributing factor to erasure of this change (other than energy coinciding with the UV-writing band) is temperature, strain, and external sources of radiation. The factor most often encountered is temperature-related erasure. For example, the energy traps associated with defect centers can be thermally excited, leading to partial or total removal of the index change. The thermal energy can also result in a structural change of the medium- and long-range order of the glass. However, the strain needed to erase the index change in an optical fiber far exceeds the breaking strain of the host material, and therefore, cannot significantly alter the permanence of the UV-induced index change. In the case of external radiation, one must be extremely careful to avoid exposure to high radiation doses that can lead to changes in all the material properties (defect distribution and bond rupture). Maintaining photoinduced index changes will essentially depend on the following three factors: the initial writing wavelength, the presence of co-dopants, and the strain distribution in the fiber. The relative contributions determine the underlying mechanism that dictates the physical properties that govern the in-fiber photosensitivity and will, of course, determine what grating type is written.

References

[1]   Bennion, I., et al. "UV-written in-fibre Bragg gratings," *Optical and Quantum Electronics*, Vol. 28, 1996, pp. 93–135.

[2]   Hill, K. O., et al. "Photosensitivity in optical fiber waveguides: Application to reflection filter fabrication," *Applied Physics Letters*, Vol. 32, 1978, pp. 647–649.

[3]   Kawasaki, B. S., et al. "Narrow-band Bragg reflectors in optical fibers," *Optics Letters*, Vol. 3, 1978, pp. 66–68.

[4]   Lam, D. K. W., and B. K. Garside, "Characterization of single-mode optical fiber filters," *Applied Optics*, Vol. 20, 1981, pp. 440–445.

[5]   Meltz, G., W. W. Morey, and W. H. Glenn, "Formation of Bragg gratings in optical fibers by a transverse holographic method," *Optics Letters*, Vol. 14, 1989, pp. 823–825.

[6]   Hosono, H., et al. "Nature and origin of the 5-eV band in SiO_2: GeO_2 glasses," *Physical Review B*, Vol. 46, 1992, pp. 11445–11451.

[7]   Nishii, J., et al. "Photochemical reactions in GeO_2-SiO_2 glasses induced by ultraviolet irradiation: Comparison between Hg lamp and excimer laser," *Physical Review B*, Vol. 52, 1995, pp. 1661–1665.

[8]   St. J. Russell, P., et al. "Optically-induced creation, transformation and organisation of defects and colour-centres in optical fibres," International Workshop on Photoinduced Self-Organization Effects in Optical Fiber, Quebec City, Quebec, May 10–11, *Proceedings SPIE*, Vol. 1516, 1991, pp. 47–54.

[9]   Lemaire, P. J., et al. "High-pressure H_2 loading as a technique for achieving ultrahigh UV photosensitivity and thermal sensitivity in GeO_2 doped optical fibres," *Electronics Letters*, Vol. 29, 1993, pp. 1191–1193.

[10]   Albert, J., et al. "Comparison of one-photon and two-photon effects in the photosensitivity of germanium-doped silica optical fibers exposed to intense ArF excimer laser pulses," *Applied Physics Letters*, Vol. 67, 1995, pp. 3529–3531.

[11]   Hosono, H., H. Kawazeo, and J. Nishii, "Defect formation in SiO_2: GeO_2 glasses studied by irradiation with excimer laser light," *Physical Review B*, Vol. 52, 1996, pp. R11921–R11923.

[12]   Hand, D. P., and P. St. J. Russell, "Photoinduced refractive-index changes in germanosilicate fibers," *Optics Letters*, Vol. 15, 1990, pp. 102–104.

[13]   Limberger, H. G., et al. "Compaction- and photoelastic-induced index changes in fiber Bragg gratings," *Applied Physics Letters*, Vol. 68, 1996, pp. 3069–3071.

[14]   Poumellec, B., et al. "The UV-induced refractive index grating in Ge: SiO_2 preforms: Additional CW experiments and the microscopic origin of the change in index," *Journal of Physics D: Applied Physics*, Vol. 29, 1996, pp. 1842–1856.

[15]   Stone, J., "Photorefractivity in GeO_2-doped silica fibers," *Journal of Applied Physics*, Vol. 62, 1987, pp. 4371–4374.

[16]   Hill, K. O., et al. "Photosensitivity in Eu^{2+}: Al_2O_3 doped core fiber: preliminary results and applications to mode converters," Conference on Optical Fiber Communication, Technical Digest Series (Optical Society of America, Washington, DC), Vol. 14, 1991, pp. 14–17.

[17]   Broer, M. M., R. L. Cone, and J. R. Simpson, "Ultraviolet-induced distributed-feeback gratings in Ce^{3+}-doped silica optical fibers," *Optics Letters*, Vol. 16, 1991, pp. 1391–1393.

[18]   Bilodeau, F., et al. "Ultraviolet-light photosensitivity in Er^{3+}-Ge-doped optical fiber," *Optics Letters*, Vol. 15, 1990, pp. 1138–1140.

[19]   Williams, D. L., et al. "Enhanced UV photosensitivity in boron codoped germanosilicate fibres," *Electronics Letters*, Vol. 29, 1993, pp. 45–47.

[20]   Niay, P., et al. "Fabrication of Bragg gratings in fluorozirconate fibers and application to fiber lasers," Conference on Laser and Electro-Optics, Technical Digest Series (Optical Society of America, Washington, DC), Vol. 8, Paper CPD 91/21, 1994.

[21]   Poumellec, B., and F. Kherbouche, "The photorefractive Bragg gratings in the fibers for telecommunications," *Journal of Physics III France*, Vol. 6, 1996, pp. 1595–1624.

[22]   Niay, P., et al. "Behavior of spectral transmissions of Bragg gratings written in germania-doped fibers: Writing and erasing experiments using pulsed or CW UV exposure," *Optics Communications*,

Vol. 113, 1994, pp. 176–192.

[23]   Poumellec. B., et al. "UV induced densification during Bragg grating inscription in Ge: SiO$_2$ preforms," *Optical Materials*, Vol. 4, 1995, pp. 441–449.

[24]   Patrick, H., and S. L. Gilbert, "Growth of Bragg gratings produced by continuous-wave ultraviolet light in optical fiber," *Optics Letters*, Vol. 18, 1993, pp. 1484–1486.

[25]   Atkins, R. M., V. Mizrahi, and T. Erdogan, "248-nm induced vacuum UV spectral changes in optical fibre preform cores: Support for a colour centre model of photosensitivity," *Electronics Letters*, Vol. 29, 1993, pp. 385–387.

[26]   Mizrahi, V., and R. M. Atkins, "Constant fluorescence during phase grating formation and defect band bleaching in optical fibres under 5.1-eV laser exposure," *Electronics Letters*, Vol. 28, 1992, pp. 2210–2211.

[27]   Askins, C. G., et al. "Fiber Bragg reflectors prepared by a single excimer pulse," *Optics Letters*, Vol. 17, 1992, pp. 833–835.

[28]   Archambault, J. L., L. Reekie, and P. St. J. Russell, "High reflectivity and narrow bandwidth fibre gratings written by single excimer pulse," *Electronics Letters*, Vol. 29, 1993, pp. 28–29.

[29]   Dong, L., et al. "Bragg gratings in Ce^{3+}-doped fibers written by a single excimer pulse," *Optics Letters*, Vol. 18, 1993, pp. 861–863.

[30]   Anderson, D. Z., et al. "Production of in-fibre gratings using a diffractive optical element," *Electronics Letters*, Vol. 29, 1993, pp. 566–568.

[31]   Niay, P., et al. "Bragg grating photoinscription within various types of fibers and glasses," *Proceedings of Topical Meeting on Photosensitivity and Quadratic Nonlinearity in Glass Waveguides: Fundamentals and Applications*, Portland, OR, Sept. 9–11, Technical Digest Series (Optical Society of America, Washington, DC), Vol. 22, , Paper SuA1, 1995, pp. 66–69.

[32]   Riant, I., and F. Haller, "Study of the photosensitivity at 193 nm and comparison with photosensitivity at 240 nm influence of fiber tension: Type IIA aging," *IEEE Journal of Lightwave Technology*, Vol. 15, 1997, pp. 1464–1469.

[33]   Xie, W. X., et al. "Experimental evidence of two types of photorefractive effects occurring during photoinscriptions of Bragg gratings written within germanosilicate fibres," *Optics Communications*, Vol. 104, 1993, pp. 185–195.

[34]   Erdogan, T., et al. "Decay of ultraviolet-induced fiber Bragg gratings," *Journal of Applied Physics*, Vol. 76, 1994, pp. 73–80.

[35]   Dong, L., and W. F. Liu, "Thermal decay of fiber Bragg gratings of positive and negative index changes formed at 193 nm in a boron-codoped germanosilicate fiber," *Applied Optics*, Vol. 36, 1997, pp. 8222–8226.

[36]   Archambault, J. L., L. Reekie, and P. St. J. Russell, "100% reflectivity Bragg reflectors produced in optical fibres by single excimer laser pulses," *Electronics Letters*, Vol. 29, 1993, pp. 453–455.

[37]   Hand, D. P., and P. St. J. Russell, "Solitary thermal shock waves and optical damage in optical fibers: The fiber fuse," *Optics Letters*, Vol. 13, 1988, pp. 767–769.

[38]   Dyer, P. E., et al. "High reflectivity fibre gratings produced by incubated damage using a 193 nm ArF laser," *Electronics Letters*, Vol. 30, 1994, pp. 860–862.

[39]   Malo, B., et al. "Single excimer-pulse writing of fiber gratings by use of a zero order nulled phase mask: Grating spectral response and visualization of index perturbations," *Optics Letters*, Vol. 18, 1993, pp. 1277–1279.

[40]   Mihailov, S. J., and M. C. Gower, "Periodic cladding surface structures induced when recording fiber Bragg reflectors with a single pulse from a KrF excimer laser," *Applied Physics Letters*, Vol. 65, 1994, pp. 2639–2641.

[41]   Douay, M., et al. "Densification involved in the UV-based photosensitivity of silica glasses and optical fibers," *IEEE Journal of Lightwave Technology*, Vol. 15, 1997, pp. 1329–1342.

[42]   Parent, M., et al. "Proprietes de polarisation des reflecteurs de Bragg induits par photosensibilite dans les fibres optiques monomodes," *Applied Optics*, Vol. 24, 1985, pp. 354–357.

[43]   Meltz, G., and W. W. Morey, "Bragg grating formation and germanosilicate fiber photosensitivity," International Workshop on Photoinduced Self-Organization Effects in Optical Fiber, Quebec City,

Quebec, May 10–11, *SPIE Proceedings*, Vol. 1516, 1991, pp. 185–199.

[44] Meyer, T., et al. "Reversibility of photoinduced birefringence in ultralow-birefringence fibers," *Optics Letters*, Vol. 21, 1996, pp. 1661–1663.

[45] St. J. Russell, P., and D. P. Hand, "Rocking filter formation in photosensitive high birefringence optical fibres," *Electronics Letters*, Vol. 52, 1990, pp. 1846–1848.

[46] Ouellette, F., D. Gagnon, and M. Poirier, "Permanent photoinduced birefringence in a Ge-doped fiber," *Applied Physics Letters*, Vol. 58, 1991, pp. 1813–1815.

[47] Lauzon, J., et al. "Dynamic polarization coupling in elliptical-core photosensitive optical fiber," *Optics Letters*, Vol. 17, 1992, pp. 1664–1666.

[48] Mizrahi, V., et al. "Stable single-mode erbium fiber-grating laser for digital communications," *IEEE Journal of Lightwave Technology*, Vol. 11, 1993, pp. 2021–2025.

[49] Hill, K. O., et al. "Birefringent photosensitivity in monomode optical fiber: Application to the external writing of rocking filters," *Electronics Letters*, Vol. 27, 1991, pp. 1548–1550.

[50] Bardal, S., A. Kamal, and P. St. J. Russell, "Photoinduced birefringece in optical fibers: A comparative study of low-birefringence and high-birefringence fibers," *Optics Letters*, Vol. 17, 1992, pp. 411–413.

[51] Poirier, M., et al. "Dynamic and orientational behavior of UV-induced luminescence bleaching in Ge-doped silica optical fiber," *Optics Letters*, Vol. 18, 1993, pp. 870–872.

[52] Simmons, K. D., et al. "Correlation of defect centers with wavelength-dependent photosensitive response in germania-doped silica optical fibers," *Optics Letters*, Vol. 16, 1991, pp. 141–143.

[53] Erdogan, T., and V. Mizrahi, "Characterization of UV-induced birefringence in photosensitive Ge-doped silica optical fibers," *Journal of the Optical Society of America B*, Vol. 11, 1994, pp. 2100–2105.

[54] Psaila, D. C., F. Ouellette, and C. Martijn de Sterke, "Characterization of photoinduced birefringece change in optical fiber rocking filters," *Applied Physics Letters*, Vol. 68, 1996, pp. 900–902.

[55] Psaila, D. C., C. Martijn de Sterke, and F. Ouellette, "Fabrication of rocking filters at 193 nm," *Optics Letters*, Vol. 21, 1996, pp. 1550–1552.

[56] Albert, J., et al. "Dichroism in the absorption spectrum of photobleached ion-implanted silica," *Optics Letters*, Vol. 18, 1993, pp. 1126–1128.

[57] Inniss, D., et al. "Atomic force microscopy study of UV-induced anisotropy in hydrogen-loaded germaosilicate fibers," *Applied Physics Letters*, Vol. 65, 1994, pp. 1528–1530.

[58] Vengsarkar, A. M., et al. "Birefringence reduction in side-written photoinduced fiber devices by a dual-exposure method," *Optics Letters*, Vol. 19, 1994, pp. 1260–1262.

[59] Friebele, E. J., et al. "Optical fiber waveguides in radiation environments II," *Nuclear Instruments and Methods in Physics Research B*, Vol. 1, 1984, pp. 355–369.

[60] Kaiser, P., "Drawing-induced coloration in vitreous silica fibers," *Journal of the Optical Society of America*, Vol. 64, 1974, pp. 475–481.

[61] St. J. Russell, P., D. P. Hand, and L. J. Poyntz-Wright, "Frequency doubling, absorption and grating formation in glass fibres: Effective defects or defective effects?" Fiber Laser Sources and Amplifiers II, *Proceedings SPIE*, Vol. 1373, 1990, pp. 126–139.

[62] Poyntz-Wright, L. J., M. E. Fermann, and P. St. J. Russell, "Nonlinear transmission and color-center dynamics in germanosilicate fibers at 420–540nm," *Optics Letters*, Vol. 13, 1988, pp. 1023–1025.

[63] Osterberg, U., and W. Margulis, "Dye laser pumped by Nd:YAG laser pulses frequency doubled in a glass optical fiber," *Optics Letters*, Vol. 11, 1986, pp. 516–518.

[64] Stolen, R. H., and H. W. K. Tom, "Self-organized phase-matched harmonic generation in optical fibers," *Optics Letters*, Vol. 12, 1987, pp. 585–587.

[65] Anderson, D. Z., "Efficient second-harmonic generation in glass fibers: The possible role of photo-induced charge redistribution," Nonlinear Properties of Materials, Proceedings SPIE, Vol. 1148, 1989, pp.186–196.

[66] Osterberg, U., and W. Margulis, "Experimental studies on efficient frequency doubling in glass optical fibers," *Optics Letters*, Vol. 12, 1987, pp. 57–59.

[67] Anoikin, E. V., et al. "Photoinduced second-harmonic generation in gamma-ray-irradiated optical fibers," *Optics Letters*, Vol. 15, 1990, pp. 834–835.

[68]   Weeks, R. A., "Paramagnetic resonance of lattice defects in irradiated quartz," *Journal of Applied Physics*, Vol. 27, 1956, pp. 1376–1381.

[69]   Nelson, C. M., and R. A. Weeks, *Journal of the American Ceramic Society*, Vol. 43, 1960, pp. 396–399.

[70]   Griscom, D. L., "Defect structure of glasses: Some outstanding questions in regard to vitreous silica," *Journal of Non-Crystalline Solids*, Vol. 73, 1985, pp. 51–77.

[71]   Weeks, R. A., and C. M. Nelson, *Journal of the American Ceramic Society*, Vol. 43, 1960, pp. 399–404.

[72]   Griscom, D. L., "Characterization of three E'-center variants in X- and γ-irradiated high purity a-SiO_2," *Nuclear Instruments and Methods in Physics Research B*, Vol. 1, 1984, pp. 481–488.

[73]   Yip, K. L., and W. B. Fowler, "Electronic structure of E'$_1$ centers in SiO_2," *Physical Review B*, Vol. 11, 1975, pp. 2327–2338.

[74]   Yip, K. L., and W. B. Fowler, "Electronic structure of SiO_2. I. Theory and sample calculations," *Physical Review B*, Vol. 10, 1974, pp. 1391–1399.

[75]   Yip, K. L., and W. B. Fowler, "Electronic structure of SiO_2. II. Calculations and results," *Physical Review B*, Vol. 10, 1974, pp. 1400–1408.

[76]   Neustruev, V. B., "Colour centres in germanosilicate glass and optical fibres," *Journal of Physics, Condensed Matter*, Vol. 6, 1994, pp. 6901–6936.

[77]   Jackson, J. M., et al. "Preparation effects on the UV optical properties of GeO_2 glasses," *Journal of Applied Physics*, Vol. 58, 1985, pp. 2308–2311.

[78]   Yuen, M. J., "Ultraviolet absorption studies of germanium silicate glasses," *Applied Optics*, Vol. 21, 1982, pp. 136–140.

[79]   Friebele, E. J., D. L. Griscom, and G. H. Sigel, Jr, "Defect centers in a germanium-doped silica-core optical fiber," *Journal of Applied Physics*, Vol. 45, 1974, pp. 3424–3428.

[80]   Tsai, T. E., D. L. Griscom, and E. J. Friebele, "On the structure of Ge-associated defect centers in irradiated high purity GeO_2 and Ge-doped SiO_2 glasses," *Diffusion and Defect Data*, Vol. 53–54, 1987, pp. 469–476.

[81]   Tsai, T. E., et al. "Radiation induced defect centers in high-purity GeO_2 glasses," *Journal of Applied Physics*, Vol. 62, 1987, pp. 2264–2268.

[82]   Friebele, E. J., and D. L. Griscom, "Color centers in glass optical fiber waveguides," *Materials Research Society Symposium Proceedings*, Vol. 61, 1986, pp. 319–331.

[83]   Kawazoe, H., "Effects of modes of glass-formation on structure of intrinsic or photon induced defects centered on III, IV or V cations in oxide glasses," *Journal of Non-Crystalline Solids*, Vol. 71, 1985, pp. 231–243.

[84]   Tsai, T. E., and D. L. Griscom, "Defect centers and photoinduced self-organization in Ge-doped silica core fiber," International Workshop on Photoinduced Self-Organization Effects in Optical Fiber, Quebec City, Quebec, May 10–11, *Proceedings SPIE*, Vol. 1516, 1991, pp. 14–28.

[85]   Hosono, H., et al. "Correlation between GeE' centers and optical absorption bands in SiO_2:GeO_2 glasses," *Japanese Journal of Applied Physics*, Vol. 35, 1996, pp. L234–L236.

[86]   Kashiwazaki, A., et al. "Effects of sintering atmosphere on defects in SiO_2: GeO_2 VAD fiber," *Materials Research Society Symposium Proceedings*, Vol. 88, 1987, pp. 217–223.

[87]   Purcell, T. A., and R. A. Weeks, "Electron spin resonance and optical absorption in GeO_2," *Journal of Chemical Physics*, Vol. 43, 1965, pp. 483–491.

[88]   Atkins, R. M., and V. Mizrahi, "Observations of changes in UV absorption bands of single mode germanosilicate core optical fibres on writing and thermally erasing refractive index gratings," *Electronics Letters*, Vol. 28, 1992, pp. 1743–1744.

[89]   Kohketsu, M., et al. "Photoluminescence in VAD SiO_2: GeO_2 glasses sintered under reducing or oxidizing conditions," *Japanese Journal of Applied Physics*, Vol. 28, 1989, pp. 622–631.

[90]   Levy, P. W., "Reactor and gamma-ray induced coloring in crystalline quartz and Corning fused silica," *Journal of Chemical Physics*, Vol. 23, 1955, pp. 764–765.

[91]   Sceats, M. G., G. R. Atkins, and S. B. Poole, "Photo-induced index changes in optical fibers," *Annual Reviews in Material Science*, Vol. 23, 1993, pp. 381–410.

[92] Gallagher, M., and U. Osterberg, "Spectroscopy of defects in germanium-doped silica glass," *Journal of Applied Physics*, Vol. 74, 1993, pp. 2771–2778.

[93] Williams, D. L., et al. "Direct observation of UV induced bleaching of 240nm absorption band in photosensitive germanosilicate glass fibres," *Electronics Letters*, Vol. 28, 1992, pp. 369–371.

[94] Malo, B., et al. "Fiber mode converters: Point-by-point fabrication of index gratings, visualization using thermoluminescence, and applications," Optical Fiber Communications Conference, San Diego, CA, Paper WL3, 1991.

[95] Duval, Y., et al. "Correlation between ultraviolet-induced refractive index change and photoluminescence in Ge-doped fiber," *Applied Physics Letters*, Vol. 61, 1992, pp. 2955–2957.

[96] Stepanov, D. Y., F. Ouellette, and G. R. Atkins, "Changes in spatial distribution of UV-excited luminescence in Ge-doped fibre preforms during UV exposure," *Electronics Letters*, Vol. 29, 1993, pp. 1975–1977.

[97] Atkins, G. R., et al. "Control of defects in optical fibers: A study using cathodoluminescence spectroscopy," *IEEE Journal of Lightwave Technology*, Vol. 11, 1993, pp. 1793–1801.

[98] Atkins, G. R., et al. "Defects in optical fibres in regions of high stress gradients," *Electronics Letters*, Vol. 27, 1991, pp. 1432–1433.

[99] Williams, D. L., M. J. Wilson, and B. J. Ainslie, "Spectral and spatial study of photosensitive optical fibre preforms by cathodoluminescence," *Electronics Letters*, Vol. 28, 1992, pp. 1744–1746.

[100] Dianov, E. M., D. S. Starodubov, and A. A. Frolov, "UV argon laser induced luminescence changes in germanosilicate fibre preforms," *Electronics Letters*, Vol. 32, 1996, pp. 246–247.

[101] Skuja, L., and W. Entzian, "Cathodoluminescence of intrinsic defects in glassy silica, thermal silica films and a-quartz," *Physica Status Solidi A*, Vol. 96, 1986, pp. 191–198.

[102] Atkins, G. R., M. G. Sceats, and S. B. Poole, "Beam current effects in cathodoluminescence spectroscopy," Symposium on Optical Fiber Measurements, Boulder, CO, 1992.

[103] Tsai, T. E., et al. "Structural origin of the 5.16eV optical absorption bands in silica and Ge-doped silica," *Applied Physics Letters*, Vol. 64, 1994, pp. 1481–1483.

[104] Arnold, G. W., "Ion-implantation effects in noncrystalline SiO_2," *IEEE Transactions in Nuclear Science*, Vol. NS-20, 1973, pp. 220–223.

[105] Tohmon, R., et al. "Correlation of the 5.0- and 7.6-eV absorption bands in SiO_2 with oxygen vacancy," *Physical Review B*, Vol. 39, 1989, pp. 1337–1345.

[106] Kashyap, R., G. D. Maxwell, and D. L. Williams, "Photoconduction in germanium and phosphorus doped silica waveguides," *Applied Physics Letters*, Vol. 62, 1993, pp. 214–216.

[107] Tsai, T. E., G. M. Williams, and E. J. Friebele, "Index structure of fiber Bragg gratings in Ge-SiO₂ fibers," *Optics Letters*, Vol. 22, 1997, pp. 224–226.

[108] Griscom, D. L., "Self-trapped holes in amorphous silicon dioxide," *Physical Review B*, Vol. 40, 1989, pp. 4224–4226.

[109] Takahashi, M., et al. "Thermal equilibrium of Ge-related defects in a GeO_2-SiO_2 glass," *Applied Physics Letters*, Vol. 72, 1998, pp. 1287–1289.

[110] Fiori, C., and R. A. B. Devine, "Evidence for a wide continuum of polymorphs in a-SiO_2," *Physical Review B*, Vol. 33, 1986, pp. 2972–2974.

[111] Dianov, E. M., et al. "Optical absorption and luminescence of germanium oxygen-deficient centers in densified germanosilicate glass," *Optics Letters*, Vol. 22, 1997, pp. 1089–1091.

[112] Dianov, E. M., et al. "Rise and decay of 3.15-eV luminescence in germanosilicate glass: Influence of glass densification," Conference on Bragg Gratings, Photosensitivity, and Poling in Glass Fibers and Waveguides: Applications and Fundamentals, Williamsburg, VA, Oct. 26–28, Technical Digest Series (Optical Society of America, Washington, DC), Vol. 17, 1997, pp. 172–174.

[113] Gallagher, M., and U. Osterberg, "Time resolved 3.10-eV luminescence in germanium-doped silica glass," *Applied Physics Letters*, Vol. 63, 1993, pp. 2987–2989.

[114] Robertson, J., "Defect mechanisms in a-SiO_2," *Philosophy Magazine B*, Vol. 52, 1985, pp. 371–377.

[115] Hanafusa, H., Y. Hibino, and F. Yamamoto, "Formation mechanism of drawing-induced E' centers in silica optical fibers," *Journal of Applied Physics*, Vol. 58, 1985, pp. 1356–1361.

[116] Hibino, Y., and H. Hanafusa, "ESR study on E'-centers induced by optical fiber drawing process,"

Japanese Journal of Applied Physics, Vol. 22, 1983, pp. L766–L768.

[117]   Lemaire, P. J., and M. D. de Coteau, "Optical spectra of silica core optical fibers exposed to hydrogen," *Materials Research Society Symposium Proceedings*, Vol. 88, 1987, pp. 225–232.

[118]   Kawazoe, H., "Drawing or photon induced defects in silica based waveguides," *Materials Research Society Symposium Proceedings*, Vol. 88, 1987, pp. 193–199.

[119]   Friebele, E. J., G. H. Sigel, Jr., and D. L. Griscom, "Drawing-induced defect centers in a fused silica core fiber," *Applied Physics Letters*, Vol. 28, 1976, pp. 516–518.

[120]   Friebele, E. J., D. L. Griscom, and M. J. Marrone, "The optical absorption and luminescence bands near 2 eV in irradiated and drawn synthetic silica," *Journal of Non-Crystalline Solids*, Vol. 71, 1985, pp. 133–144.

[121]   Tsai, T. E., E. J. Friebele, and D. L. Griscom, "Thermal stability of photoinduced gratings and paramagnetic centers in Ge- and Ge/P-doped silica optical fibers," *Optics Letters*, Vol. 18, 1993, pp. 935–937.

[122]   Atkins, R. M., et al. "Mechanisms of enhanced UV photosensitivity via hydrogen loading in germanosilicate glasses," *Electronics Letters*, Vol. 29, 1993, pp. 1234–1235.

[123]   Maxwell, G. D., et al. "Hydrogenated low loss planar silica waveguides," *Electronics Letters*, Vol. 29, 1993, pp. 425–426.

[124]   Mizrahi, V., et al. "Ultraviolet laser fabrication of ultrastrong optical fiber gratings and of germania-doped channel waveguides," *Applied Physics Letters*, Vol. 63, 1993, pp. 1727–1729.

[125]   Strasser, T. A., et al. "Ultraviolet laser fabrication of strong, nearly polarization-independent Bragg reflectors in germanium-doped silica waveguides on silica substrates," *Applied Physics Letters*, Vol. 65, 1994, pp. 3308–3310.

[126]   Lemaire, P. J., et al. "Thermally enhanced ultraviolet photosensitivity in GeO_2 and P_2O_5 doped optical fibers," *Applied Physics Letters*, Vol. 66, 1995, pp. 2034–2036.

[127]   Tsai, T. E., et al. "Defect centers induced by harmonic wavelength of 1.06 μm light in Ge/P-doped fibers," *Electro-Optics and Non-Linear Optics in Ceramic Transactions*, Vol. 14, 1990, pp. 127–136.

[128]   Greene, B. I., et al. *Journal of Non-Crystalline Solids*, Vol. 168, 1994, pp. 195.

[129]   Krol, D. M., R. M. Atkins, and P. J. Lemaire, "Photoinduced second-harmonic generation and luminescence of defects in Ge-doped silica fibers," International Workshop on Photoinduced Self-Organization Effects in Optical Fiber, Quebec City, Quebec, May 10–11, *Proceedings SPIE*, Vol. 1516, 1991, pp. 38–46.

[130]   Poumellec. B., et al. "UV induced densification during Bragg grating inscription in Ge:SiO₂ preforms: Interferometric microscopy investigations," *Optical Materials*, Vol. 4, 1995, pp. 404–409.

[131]   Taunay, T., et al. "Growth kinetics of photoinduced gratings and paramagnetic centers in high NA, heavily Ge-doped silica optical fibers," Conference on Bragg Gratings, Photosensitivity, and Poling in Glass Fibers and Waveguides: Applications and Fundamentals, Williamsburg, VA, Oct. 26–28, Technical Digest Series (Optical Society of America, Washington, DC), Vol. 17, 1997, pp. 181–183.

[132]   Dianov, E. M., et al. "Grating formation in a germanium free silicon oxynitride fibre," *Electronics Letters*, Vol. 33, 1997, pp. 236–238.

[133]   Albert, J., et al. "Photosensitivity in Ge-doped silica optical waveguides and fibers with 193-nm light from an ArF excimer laser," *Optics Letters*, Vol. 19, 1994, pp. 387–389.

[134]   Starodubov, D. S., et al. "Bragg grating fabrication in germanosilicate fibers by use of near-UV light: a new pathway for refractive-index changes," *Optics Letters*, Vol. 22, 1997, pp. 1086–1088.

[135]   Dong, L., et al. "Photoinduced absorption changes in germanosilicate preforms: Evidence for the color-center model of photosensitivity," *Applied Optics*, Vol. 34, 1995, pp. 3436–3440.

[136]   Watanabe, Y., et al. "Permanent refractive-index changes in pure GeO_2 glass slabs induced by irradiation with below-gap light," Conference on Bragg Gratings, Photosensitivity, and Poling in Glass Fibers and Waveguides: Applications and Fundamentals, Williamsburg, VA, Oct. 26–28, Technical Digest Series (Optical Society of America, Washington, DC), Vol. 17, 1997, pp. 77–79.

[137]   Watanabe, Y., et al. "Structure and mechanism of formation of drawing- or radiation-induced defects in SiO_2: GeO_2 optical fiber," *Japanese Journal of Applied Physics*, Vol. 25, 1986, pp. 425–431.

[138]   Limberger, H. G., P. Y. Fonjallaz, and R. P. Salathe, "Spectral characterisation of photoinduced highly

efficient Bragg gratings in standard telecommunications fibres," *Electronics Letters*, Vol. 29, 1993, pp. 47–49.

[139] Bilodeau, F., et al. "Photosensitization of optical fiber and silica-on-silicon/silica waveguides," *Optics Letters*, Vol. 18, 1993, pp. 953–955.

[140] Noguchi, K., et al. "Loss increase for optical fibers exposed to hydrogen atmosphere," *IEEE Journal of Lightwave Technology*, Vol. LT-3, 1985, pp. 236–243.

[141] Awazu, K., H. Kawazoe, and M. Yamane, "Simultaneous generation of optical absorption bands at 5.14 and 0.452 eV in 9SiO$_2$: GeO$_2$ glasses heated under an H$_2$ atmosphere," *Journal of Applied Physics*, Vol. 68, 1990, pp. 2713–2718.

[142] Malo, B., et al. "Effective index drift from molecular hydrogen diffusion in hydrogen-loaded optical fibres and its effect on Bragg grating fabrication," *Electronics Letters*, Vol. 30, 1994, pp. 442–444.

[143] Farries, M. C., et al. "Fabrication and performance of packaged fibre gratings for telecommunications," *IEEE Colloquium Optical Fibre Gratings*, London, England, Jan. 1995.

[144] Lemaire, P. J., "Enhanced UV photosensitivity in fibers and waveguides by high pressure hydrogen loading," Conference on Optical Fiber Communication (OFC'95), San Diego, CA, Feb/March, (Optical Society of America, Washington, DC), Paper WN5, 1995.

[145] Tsai, T. E., and E. J. Friebele, "Kinetics of defect centers formation and photosensitivity in Ge-SiO$_2$ fibers of various compositions," Conference on Bragg Gratings, Photosensitivity, and Poling in Glass Fibers and Waveguides: Applications and Fundamentals, Williamsburg, VA, Oct. 26–28, Technical Digest Series (Optical Society of America, Washington, DC), Vol. 17, 1997, pp. 101–103.

[146] Atkins, R. M., and R. P. Espindola, "Photosensitivity and grating writing in hydrogen loaded germanosilicate core optical fibers at 325 and 351 nm," *Applied Physics Letters*, Vol. 70, 1997, pp. 1068–1069.

[147] Grubsky, V., D. S. Starodubov, and J. Feonberg, "Mechanisms of index change induced by near-UV light in hydrogen-loaded fibers," Conference on Bragg Gratings, Photosensitivity, and Poling in Glass Fibers and Waveguides: Applications and Fundamentals, Williamsburg, VA, Oct. 26–28, Technical Digest Series (Optical Society of America, Washington, DC), Vol. 17, 1997, pp. 98–100.

[148] Camlibel, I., D. A. Pinnow, and F. W. Dabby, "Optical aging characteristics of borosilicate clad fused silica core fiber optical waveguides," *Applied Physics Letters*, Vol. 26, 1975, pp. 185-187.

[149] Dong, L., W. F. Liu, and L. Reekie, "Negative-index gratings formed by a 193-nm excimer laser," *Optics Letters*, Vol. 21, 1996, pp. 2032–2034.

[150] Herman, P. R., K. Beckley, and S. Ness, "157-nm photosensitivity in germanosilicate waveguides," Conference on Bragg Gratings, Photosensitivity, and Poling in Glass Fibers and Waveguides: Applications and Fundamentals, Williamsburg, VA, Oct. 26–28, Technical Digest Series (Optical Society of America, Washington, DC), Vol. 17, 1997, pp. 159–161.

[151] Awazu, K., and H. Kawazoe, "O$_2$ molecules dissolved in synthetic silica glasses and their photochemical reactions induced by ArF excimer laser radiation," *Journal of Applied Physics*, Vol. 68, 1990, pp. 3584–3591.

[152] Dyer, P. E., et al. "Study and analysis of submicron-period grating formation on polymers ablated using a KrF laser irradiated phase mask," *Applied Physics Letters*, Vol. 64, 1994, pp. 3389–3391.

[153] Noguchi, K., N. Uesugi, and K. Suzuki, "Visible-region optical absorption property for germanium-doped silica optical fibres induced by UV light propagation," *Electronics Letters*, Vol. 22, 1986, pp. 519–520.

[154] Svalgaard, M., "Dynamics of ultraviolet induced luminescene and fiber Bragg grating formation in high fluence regime," *Proceedings of Topical Meeting on Photosensitivity and Quadratic Nonlinearity in Glass Waveguides: Fundamentals and Applications*, Portland, OR, Sept. 9–11, Technical Digest Series (Optical Society of America, Washington, DC), Vol. 22, 1995, pp. 160–164.

[155] Dianov, E. M., et al. "Refractive-index gratings written by near-ultraviolet radiation," *Optics Letters*, Vol. 22, 1997, pp. 221–223.

[156] Bernadin, J. P., and N. M. Lawandy, "Dynamics of the formation of Bragg gratings in germanosilicate optical fibers," *Optics Communications*, Vol. 79, 1990, pp. 194–199.

[157] Chiang, K. S., M. G. Sceats, and D. Wong, "Ultraviolet photolytic-induced changes in optical fibers:

The thermal expansion coefficient," *Optics Letters*, Vol. 18, 1993, pp. 965–967.

[158]   Malo, B., et al. "Elimination of photoinduced absorption in Ge-doped silica fibres by annealing of ultraviolet colour centres," *Electronics Letters*, Vol. 28, 1992, pp. 1598–1599.

[159]   Cordier, P., et al. "TEM characterization of structural changes in glass associated to Bragg grating inscription in a germanosilicate optical fiber preform," *Optics Communications*, Vol. 111, 1994, pp. 269–275.

[160]   Williams, D. L., et al. "Ultraviolet studies on photosensitive germanosilicate preforms and fibers," *Applied Physics Letters*, Vol. 59, 1991, pp. 762–764.

[161]   Gallagher, M. D., and U. L. Osterberg, "Ultraviolet absorption measurements in single-mode optical glass fibers," *Applied Physics Letters*, Vol. 60, 1992, pp. 1791–1793.

[162]   Atkins, R. M., "Measurement of the ultraviolet absorption spectrum of optical fibers," *Optics Letters*, Vol. 17, 1992, pp. 469–471.

[163]   Fertein, E., et al. "Shifts in resonance wavelengths of Bragg gratings during writing or bleaching experiments by UV illumination within germanosilicate optical fibre," *Electronics Letters*, Vol. 27, 1991, pp. 1838–1839.

[164]   Gilbert, S. L., and H. Patrick, "Comparison of UV-induced fluorescence and Bragg grating growth in optical fiber," Conference on Laser and Electro-Optics, Technical Digest Series (Optical Society of America, Washington, DC), Vol. 8, 1994, p. 244.

[165]   Byers, J. D., et al. "Efficient numerical simulation of the time-dependence of electronic-energy transfer in polymers: Short-range transfer and foster trapping," *Macromolecules*, Vol. 23, 1990, pp. 4835–4844.

[166]   Leconte, B., et al. "Analysis of color-center-related contribution to Bragg grating formation in Ge: SiO_2 fiber based on a local Kramers-Kronig transformation of excess loss spectra," *Applied Optics*, Vol. 36, 1997, pp. 5923–5930.

[167]   Digonnet, M. J., "A Kramers-Kronig analysis of the absorption change in fiber gratings," *Proceedings SPIE*, Vol. 2841, 1996, pp. 109–120.

[168]   Attard, A. E., "Fermi level shift in $Bi_{12}SiO_{20}$ vis photon-induced trap level occupation," *Journal of Applied Physics*, Vol. 71, 1992, pp. 933–937.

[169]   Payne, F. P., "Photorefractive gratings in single-mode optical fibres," *Electronics Letters*, Vol. 25, 1989, pp. 498–499.

[170]   Fermann, M. E., "Characterisation techniques for special optical fibres," Ph.D. Dissertation, University of Southampton, United Kingdom, 1988.

[171]   Bagratashvili, V. N., et al. "Direct observation of ultraviolet laser induced photocurrent in oxygen deficient silica and germanosilicate glasses," *Applied Physics Letters*, Vol. 68, 1996, pp. 1616–1618.

[172]   Williams, D. L., et al. "Photosensitive index changes in germania doped silica fibres and waveguides," Photosensitivity and Self-Organization in Optical Fibers and Waveguides, *Proceedings SPIE*, Vol. 2044, 1993, pp. 55–68.

[173]   Lawandy, N. M., "Light induced transport and delocalization in transparent amorphous systems," *Optics Communications*, Vol. 74, 1989, pp. 180–184.

[174]   Miotello, A., and R. Kelly, "Laser irradiation effects in Si^+-implanted SiO_2," *Nuclear Instruments and Methods in Physics Research B*, Vol. 65, 1992, pp. 217–222.

[175]   Wong, D., S. B. Poole, and M. G. Sceats, "Stress-birefringence reduction in elliptical-core fibers under ultraviolet irradiation," *Optics Letters*, Vol. 17, 1992, pp. 1773–1775.

[176]   Fonjallaz, P. Y., et. al. "Tension increase correlated to refractive-index change in fibers containing UV-written Bragg gratings," *Optics Letters*, Vol. 20, 1995, pp. 1346–1348.

[177]   Fiori, C., and R. A. B. Devine, "Ultraviolet irradiation induced compaction and photoetching in amorphous thermal SiO_2," *Materials Research Society Symposium Proceedings*, Vol. 61, 1986, pp. 187–195.

[178]   Bazylenko, M. V., D. Moss, and J. Canning, "Complex photosensitivity observed in germanosilica planar waveguides," *Optics Letters*, Vol. 23, 1998, pp. 697–699.

[179]   Grimsditch, M., "Polymorphism in amorphous SiO_2," *Physical Review Letters*, Vol. 52, 1984, pp. 2379–2381.

[180] O'Reilly, E. P., and J. Robertson, "Theory of defects in vitreous silicon dioxide," *Physical Review B*, Vol.27, 1983, pp. 3780–3795.

[181] Fiori, C., and R. A. B. Devine, "High resolution ultraviolet photoablation of SiO_x films," *Applied Physics Letters*, Vol. 47, 1985, pp. 361–362.

[182] Rothschild, M., D. J. Ehrlich, and D. C. Shaver, "Effects of excimer laser irradiation on the transmission, index of refraction, and density of ultraviolet grade fused silica," *Applied Physics Letters*, Vol. 55, 1989, pp. 1276–1278.

[183] Arai, K., et al. "Two-photon processes in defect formation by excimer lasers in synthetic silica glass," *Applied Physics Letters*, Vol. 53, 1988, pp. 1891–1893.

[184] Allan, D. C., et al. "193-nm excimer-laser-induced densification of fused silica," *Optics Letters*, Vol. 21, 1996, pp. 1960–1962.

[185] Weeding, T. L., et al. "Silicon coordination changes from 4-fold to 6-fold on devitrification of silicon phosphate glass," *Nature*, Vol. 318, 1985, pp. 352–353.

[186] Hosono, H., "Defects associated with photosensitivity in GeO_2-SiO_2 glasses," Conference on Bragg Gratings, Photosensitivity, and Poling in Glass Fibers and Waveguides: Applications and Fundamentals, Williamsburg, VA, Oct. 26–28, Technical Digest Series (Optical Society of America, Washington, DC), Vol. 17, 1997, pp. 166–168.

[187] Gabriagues, J. M., and H. Fevrier, "Analysis of frequency-doubling processes in optical fibers using Raman spectroscopy," *Optics Letters*, Vol. 12, 1987, pp. 720–722.

[188] Cordier, P., et al. "Evidence by transmission electron microscopy of densification associated to Bragg grating photoimprinting in germanosilicate optical fibers," *Applied Physics Letters*, Vol. 70, 1997, pp. 1204–1206.

[189] Dianov, E. M., et al. "UV-irradiation-induced structural transformation of germanosilicate glass fiber," *Optics Letters*, Vol. 22, 1997, pp. 1754–1756.

[190] Munekuni, S., et al. "Various types of nonbridging oxygen hole center in high-purity silica glass," *Journal of Applied Physics*, Vol. 68, 1990, pp. 1212–1217.

[191] Dianov, E. M., et al. "Highly photosensitive nitrogen-doped germanosilicate fibre for index grating writing", *Electronics Letters*, Vol. 33, 1997, pp. 1334–1336.

[192] Malo, B., et al. "Photosensitivity in phosphorus-doped silica glass and optical waveguides" *Applied Physics Letters*, Vol. 65, 1994, pp. 394–396.

[193] Canning, J., and Sceats, M. G., "Transient gratings in rare-earth doped phosphosilicate optical fibres through periodic population inversion," *Electronics Letters*, Vol. 31, 1995, pp. 576–578.

[194] Cater, A. L. G., M. G. Sceat, and S. B. Poole, "Flash condensation technique for the fabrication of high phosphorous-content rare earth doped fibers" *Electronics Letters*, Vol. 28, 1992, pp. 2009–2010.

[195] Strasser, T. A., A. E. White, and M. F. Yan, Optical Fiber Conference, San Jose, CA, p. 159, 1995.

[196] Dong, L., et al. "Strong photosensitive gratings in tin-doped phosphosilicate optical fibers," *Optics Letters*, Vol. 20, 1995, pp. 1982–1984.

[197] Ball, G. A., W. W. Morey, and J. P. Waters, "Nd^{3+} fiber laser utilizing intra-core Bragg reflectors" *Electronics Letters*, Vol. 26, 1990, pp. 1829–1830.

[198] Broer, M. M., A. J. Bruce, and W. H. Grodkiewicz, "Photoinduced refractive-index changes in several Eu^{3+}-, Pr^{3+}-, and Er^{3+}-doped oxide glasses," *Physical Review B*, Vol. 45, 1992, pp. 7077–7083.

[199] Poignant, H., et al. "Efficiency and thermal behavior of cerium doped fluorozirconate glass fiber gratings," *Electronics Letters*, Vol. 40, 1994, pp. 1339–1341.

[200] Taunay, T., et al. "Ultraviolet-induced permanent Bragg gratings in cerium-doped ZBLAN glass or optical fibers," *Optics Letters*, Vol. 19, 1994, pp. 1269–1271.

[201] Williams, G. M., et al. "Photosensitivity of Rare Earth doped ZBLAN Fluoride Glasses", *IEEE Journal of Lightwave technology*, Vol. 15, 1997, pp. 1357–1362.

Selected Bibliography

[202]   Patrick, H., et al. "Annealing of Bragg gratings in hydrogen-loaded optical fiber," *Journal of Applied Physics*, Vol. 78, 1995, pp. 2940–2945.

[203]   Canning, J., et al. "Transient and permanent gratings in phoshosilicate optical fibers produced by the flash condensation technique," *Optics Letters*, Vol. 20, 1995, pp. 2189–2191.

[204]   Archambault, J. L., et al. "High reflectivity photorefractive Bragg gratings in germania-free optical fibers," Conference on Lasers and Electro-Optics, Technical Digest Series (Optical Society of America, Washington, DC), Vol. 8, 1994, pp. 242–243.

[205]   Dong, L., et al. "Enhanced photosensitivity in tin-codoped germanosilicate optical fibers," *IEEE Photonics Technology Letters*, Vol. 7, 1995, pp. 1048–1050.

[206]   Douay, M., et al. "Annealing of gratings photowritten in tin-codoped germanosilicate preform plates and fibers," Optoelectronics '97:Integrated Devices Applications on Photosensitive Optical Materials and Devices, *Proceedings SPIE*, San Jose, CA, Feb. 8–14, 1997.

[207]   Dong, L., et al. "Grating formation in a phosphorus-doped germanosilicate fiber," Conference on Optical Fiber Communication (OFC 1996), Paper TuO2, 1996, pp. 82–83.

[208]   Dong, L., et al. "Photosensitivity in Ce^{3+}-doped optical fibers," *Journal of the Optical Society of America B*, Vol. 10, 1993, pp. 89–93.

Appendix: A Summary of Photosensitivity in Optical Fibers

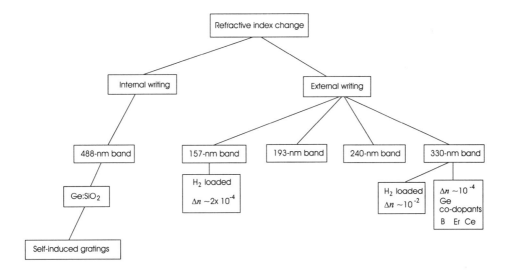

Figure A.1   A summary of the current knowledge regarding photosensitivity in optical fibers: overview.

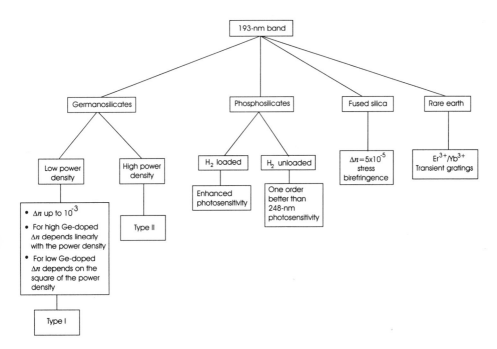

Figure A.2   A summary of the current knowledge regarding photosensitivity in optical fibers: 193-nm band excitation.

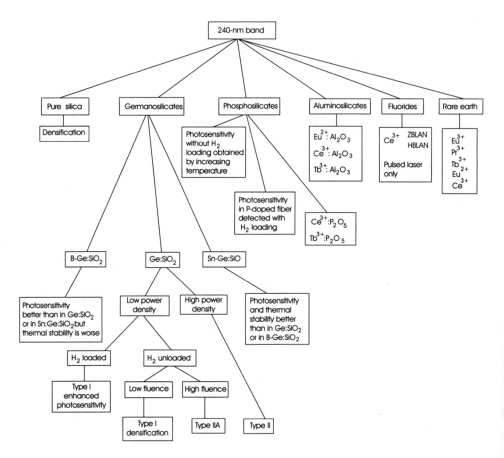

Figure A.3   A summary of the current knowledge regarding photosensitivity in optical fibers: 240-nm band excitation.

Chapter 3

PROPERTIES
OF
FIBER BRAGG GRATINGS

In this chapter we will describe in detail the various properties that are characteristic of fiber Bragg gratings. We will begin by examining the measurable wavelength-dependent properties, such as the reflection and transmission spectral profiles, for a number of simple and complex grating structures. The dependence of the grating wavelength response to externally applied perturbations, such as temperature and strain, is also investigated. Pulse propagation through the grating structure is used to characterize its dynamic response and associative nonlinear optical phenomena. Practical issues of longevity and reliability are examined through thermally initiated grating decay and mechanical strength degradation resulting from optical fiber exposure to intense UV irradiation. Finally, we look at the impact of loss effects, such as short wavelength, radiation mode coupling and incoherent scattering losses.

3.1 Simple Bragg Gratings

In its simplest form a fiber Bragg grating consists of a periodic modulation of the refractive index in the core of a single-mode optical fiber. These types of uniform fiber gratings, where the phase fronts are perpendicular to the fiber's longitudinal axis and with grating planes having constant period (Figure 3.1), are considered the fundamental building blocks for most Bragg grating structures. Light, guided along the core of an optical fiber, will be scattered by each grating plane. If the Bragg condition is not satisfied, the reflected light from each of the subsequent planes becomes progressively out of phase and will eventually cancel out. Additionally, light that is not coincident with the Bragg wavelength resonance will experience very weak reflection at each of the grating planes because of the index mismatch–this reflection accumulates over the length of the grating. As an example, a 1-mm grating at 1.5 μm with a strong Δn of 10^{-3} will reflect ~ 0.05% of the off-resonance incident light. Where the Bragg condition is satisfied, the contributions of reflected light from each grating plane add constructively in the backward direction to form a back-reflected peak with a center wavelength defined by the grating parameters. This is

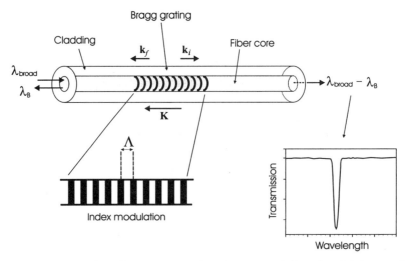

Figure 3.1   Illustration of a uniform Bragg grating with constant index of modulation amplitude and period. Also shown are the incident, diffracted, and grating wave vectors that have to be matched for momentum to be conserved.

analogous to a volume hologram or a crystal lattice diffracting X-rays.

The Bragg grating condition is simply the requirement that satisfies both energy and momentum conservation. Energy conservation ($\hbar\omega_f = \hbar\omega_i$) requires that the frequency of the incident radiation and the reflected radiation is the same. Momentum conservation requires that the incident wavevector, k_i, plus the grating wavevector, K, equal the wavevector of the scattered radiation k_f. This is simply stated as

$$k_i + K = k_f \tag{3.1}$$

where the grating wavevector, K, has a direction normal to the grating planes with a magnitude $2\pi/\Lambda$ (Λ is the grating spacing shown in Figure 3.1). The diffracted wavevector is equal in magnitude, but opposite in direction, to the incident wavevector. Hence, the momentum conservation condition becomes

$$2\left(\frac{2\pi n_{\text{eff}}}{\lambda_B}\right) = \frac{2\pi}{\Lambda} \tag{3.2}$$

which simplifies to the first-order Bragg condition

$$\lambda_B = 2n_{\text{eff}}\Lambda \tag{3.3}$$

where the Bragg grating wavelength, λ_B, is the free space center wavelength of the input light that will be back-reflected from the Bragg grating, and n_{eff} is the effective refractive index of the fiber core at the free space center wavelength.

3.2 Uniform Bragg Grating Reflectivity

Consider a uniform Bragg grating formed within the core of an optical fiber with an average refractive index n_0. The index of refractive profile can be expressed as

$$n(z) = n_0 + \Delta n \cos\left(\frac{2\pi z}{\Lambda}\right) \tag{3.4}$$

where Δn is the amplitude of the induced refractive index perturbation (typical values 10^{-5} to 10^{-3}), and z is the distance along the fiber longitudinal axis. Using the coupled mode theory of Lam and Garside [1] that described the reflection properties of a Bragg grating, the reflectivity of a grating with constant modulation amplitude and period is given by the following expression:

$$R(l,\lambda) = \frac{\Omega^2 \sinh^2(sl)}{\Delta k^2 \sinh^2(sl) + s^2 \cosh^2(sl)} \tag{3.5}$$

where $R(l,\lambda)$ is the reflectivity that is a function of the grating length l and wavelength λ. Ω is the coupling coefficient, $\Delta k = k - \pi/\lambda$ is the detuning wavevector, $k = 2\pi n_0/\lambda$ is the propagation constant, and $s^2 = \Omega^2 - \Delta k^2$. The coupling coefficient, Ω, for sinusoidal variation of index perturbation along the fiber axis is given by

$$\Omega = \frac{\pi \Delta n}{\lambda} M_p \tag{3.6}$$

where M_p is the fraction of the fiber mode power contained by the fiber core. On the basis that the grating is uniformly written through the core, M_p can be approximated by $1 - V^{-2}$, where V is the normalized frequency of the fiber. $V = (2\pi/\lambda) a (n_{co}^2 - n_{cl}^2)^{1/2}$ where a is the core radius, n_{co} and n_{cl} the core and cladding indices, respectively. At the Bragg grating center wavelength there is no wavevector detuning and $\Delta k = 0$; therefore, the expression for the reflectivity becomes

$$R(l,\lambda) = \tanh^2(\Omega l) \tag{3.7}$$

The reflectivity increases as the induced index of refraction change increases. Similarly, as the length of the grating increases so does the resultant reflectivity. A calculated reflection spectrum as a function of the wavelength is shown in Figure 3.2. The side lobes of the resonance are due to multiple reflections to and from opposite ends of the grating region. The sine spectrum arises mathematically through the Fourier transform of a harmonic signal having finite extent; an infinitely long grating would transform to an ideal delta function response in the wavelength domain. A general expression for the approximate full-width-half maximum bandwidth of a grating is given by [2]

$$\Delta\lambda = \lambda_B s \sqrt{\left(\frac{\Delta n}{2n_0}\right)^2 + \left(\frac{1}{N}\right)^2} \tag{3.8}$$

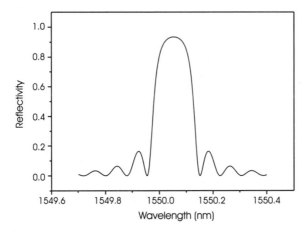

Figure 3.2   Reflection spectrum of a Bragg grating as a function of wavelength.

where N is the number of the grating planes. The parameter $s \sim 1$ for strong gratings (with near 100% reflection) whereas $s \sim 0.5$ for weak gratings.

3.3 Strain and Temperature Sensitivity of Bragg Gratings

The Bragg grating resonance, which is the center wavelength of back-reflected light from a Bragg grating, depends on the effective index of refraction of the core and the periodicity of the grating. The effective index of refraction, as well as the periodic spacing between the grating planes, will be affected by changes in strain and temperature. Using (3.3) the shift in the Bragg grating center wavelength due to strain and temperature changes is given by

$$\Delta\lambda_{\mathrm{B}} = 2\left(\Lambda \frac{\partial n_{\mathrm{eff}}}{\partial l} + n_{\mathrm{eff}} \frac{\partial \Lambda}{\partial l} \right)\Delta l + 2\left(\Lambda \frac{\partial n_{\mathrm{eff}}}{\partial T} + n_{\mathrm{eff}} \frac{\partial \Lambda}{\partial T} \right)\Delta T \qquad (3.9)$$

The first term in (3.9) represents the strain effect on an optical fiber. This corresponds to a change in the grating spacing and the strain-optic induced change in the refractive index. The above strain effect term may be expressed as [3]

$$\Delta\lambda_{\mathrm{B}} = \lambda_{\mathrm{B}}(1 - p_{\mathrm{e}})\varepsilon_{z} \qquad (3.10)$$

where p_{e} is an effective strain-optic constant defined as

$$p_{\mathrm{e}} = \frac{n_{\mathrm{eff}}^{2}}{2}[p_{12} - \nu(p_{11} + p_{12})] \qquad (3.11)$$

p_{11} and p_{12} are components of the strain-optic tensor, and ν is the Poisson's ratio. For a typical germanosilicate optical fiber $p_{11} = 0.113$, $p_{12} = 0.252$, $\nu = 0.16$, and $n_{eff} = 1.482$. Using these parameters and the above equations, the anticipated strain sensitivity at ~1550 nm is a 1.2-pm change as a result of applying 1 $\mu\varepsilon$ to the Bragg grating. Experimental results of a Bragg center wavelength shift with applied stress on a 1548.2-nm grating are shown in Figure 3.3(a).

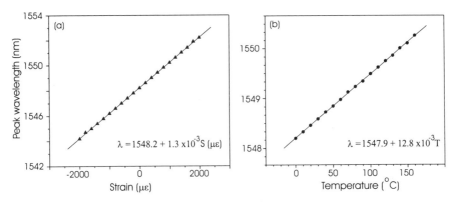

Figure 3.3 Peak reflection from the Bragg grating (a) under applied stress, and (b) at different temperatures. The Bragg grating formed the output coupler of an erbium-doped fiber laser. The grating peak reflectivity was centered at 1548.2 nm under zero stress at room temperature.

The second term in (3.9) represents the effect of temperature on an optical fiber. A shift in the Bragg wavelength due to thermal expansion changes the grating spacing and the index of refraction. This fractional wavelength shift for a temperature change ΔT may be written as [3]

$$\Delta\lambda_B = \lambda_B(\alpha_\Lambda + \alpha_n)\Delta T \tag{3.12}$$

where $\alpha_\Lambda = (1/\Lambda)(\partial\Lambda/\partial T)$ is the thermal expansion coefficient for the fiber (approximately 0.55×10^{-6} for silica). The quantity $\alpha_n = (1/n_{eff})(\partial n_{eff}/\partial T)$ represents the thermo-optic coefficient, which is approximately equal to 8.6×10^{-6} for the germania-doped, silica-core fiber. Clearly the index change is by far the dominant effect. From (3.12) the expected sensitivity for a 1550-nm Bragg grating is approximately 13.7 pm/°C. Figure 3.3(b) shows experimental results of a Bragg grating center wavelength shift as a function of temperature. It now becomes apparent that any change in wavelength, associated with the action of an external perturbation to the grating, is the sum of strain and temperature terms. Therefore, in sensing applications where only one perturbation is of interest, the deconvolution of temperature and strain becomes necessary (this is discussed further in Chapter 7).

3.4   Other Properties of Fiber Gratings

When a grating is formed under conditions for which the modulated index change is saturated under UV exposure, then the effective length will be reduced as the transmitted signal is depleted by reflection. As a result, the spectrum will broaden appreciably and depart from a symmetric sinc or Gaussian shape spectrum, whose width is inversely proportional to the grating length. This is illustrated in Figure 3.4 (a) and (b). In addition, the cosinusoidal shape of the grating will distort into a waveform with steeper sides. A second-order Bragg line (Figure 3.4(c)) will appear from the new harmonics in the Fourier spatial spectrum of the grating [4].

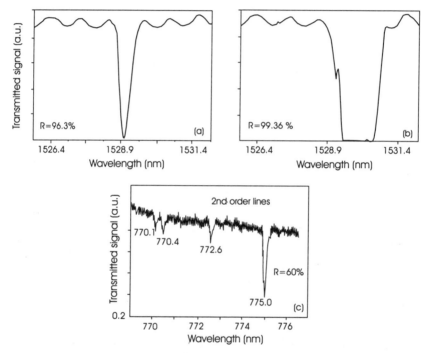

Figure 3.4   A strongly reflecting grating with a large index change (a) becomes saturated and (b) the spectrum broadens under continuous exposure because the incident wave is completely reflected before reaching the end of the grating. The strongly saturated grating is no longer sinusoidal, and the peak index regions are flattened, whereas the valleys in the perturbation index distribution are sharpened. As a result, second-order Bragg reflection lines (c) are observed at about one-half the fundamental Bragg wavelength and at other shorter wavelengths for higher order modes (*After:* [4]).

Another interesting feature, which is observed in strongly reflecting gratings with large index perturbations, is the small-shape spectral resonance on the short wavelength side of the grating centerline. This is due to self-chirping from $\Delta n_{eff}(z)$. Such features do not occur if the average index change is held constant or adjusted to be constant by a second exposure of the grating. A Bragg grating will also couple dissimilar modes in reflection and transmission, provided the following two conditions are satisfied: phase matching and sufficient mode overlap in the region of the fiber that contains the grating. The phase matching condition, which ensures a coherent exchange of energy between the modes, is given by [4]

$$n_{\text{eff}} - \frac{\lambda}{\Lambda_z} = m_{\text{e}} \qquad (3.13)$$

where n_{eff} is the modal index of the incident wave and m_{e} is the modal index of the grating-coupled reflected or transmitted wave. It should be pointed out that the above equation allows for a tilted or blazed grating by adjusting the grating pitch along the fiber axis Λ_z. Figure 3.5 shows the mode-coupling, phase-matching requirement graphically. This illustration indicates that five different types of interaction can occur, depending on the ratio of the wavelength to the period of the grating. Normal bound-mode propagation occurs when the effective index of the wave lies between the cladding and the core values. A grating that reflects a like mode couples waves between the upper, forward-propagating branch of the dispersion relation to its lower negative-going, mirror image. This situation occurs when the grating has a period sufficiently fine that the Bragg condition is obeyed. However, this same grating will also couple to other modes at shorter wavelengths; some will be reflected and/or absorbed, whereas others will be radiated away from the fiber (this is discussed in the next section). These interactions are observed as a series of transmission dips in the spectrum, at wavelengths that are less than the Bragg wavelength (Figure 3.6(a)). In the case where the grating is tilted or radially nonuniform, the interactions will take place between symmetric and asymmetric modes (Figure 3.6(b)).

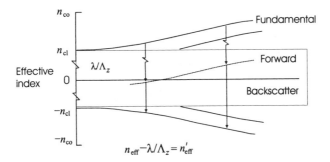

Figure 3.5   A schematic of the phase-matching conditions to achieve synchronous mode coupling with a fiber Bragg grating. The ratio of the wavelength λ and the grating pitch Λ_z along the fiber axis determines which type of mode (cladding or bound, backward or forward propagating) is excited by the incident, forward-propagating LP_{01} fundamental mode.

It should also be mentioned that a useful situation arises if the grating period becomes much coarser. In this case the fundamental mode exchanges energy in a resonant fashion with a forward-going cladding mode. The effect is similar to mode coupling in a two-core fiber or between modes in a multimode waveguide. These gratings can be made easily with a simple transmission mask because the required pitch is a few hundred microns, in contrast with the fine submicron pitch that is required to reflect a bound mode. Gratings made with a periodicity that coincides with the phase-matching resonance condition between the forward-propagating guided mode and cladding are known as long period gratings (Section 3.12).

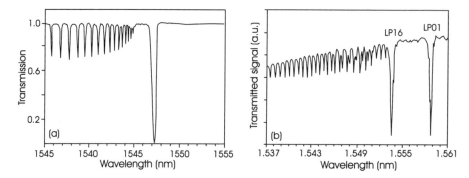

Figure 3.6   (a) Schematic transmission profile for a strong fiber Bragg grating, showing loss to radiation modes on the short wavelength side, sharply modified by the cladding mode structure. (b) Transmission spectrum of blazed grating (4.36 degree effective tilt angle within the core) in depressed-cladding fiber, showing pronounced coupling into the asymmetric LP_{16} backward-propagating cladding mode (*After:* [4]).

3.4.1 Cladding and Radiation Mode Coupling

Bragg gratings written in a highly photosensitive fiber, or in fiber that has been hydrogenated, have a very pronounced transmission structure on the short wavelength side of the Bragg peak (see Figure 3.6(a)). This feature is only observable in the transmission spectrum (the main peak appears only when viewed in reflection); therefore, this structure must result from light leaving the side of the fiber–to analyze it one must take into account radiation mode coupling. Usually radiation mode coupling, which is routinely observed from surface relief gratings made by physically etching the core of a polished optical fiber, is a smooth function of wavelength. However, the transmission spectrum of the Bragg grating (as seen in Figure 3.6(b)) consists of multiple sharp peaks that modulate this coupling, which is a direct consequence of the cylindrical cladding-air interface. Dipping the cladding into glycerin, which in effect eliminates the cladding-air interface, may eliminate this effect. Nevertheless, the cladding mode radiation-related problems become very serious with large excess losses at wavelengths shorter than the peak reflection wavelength. As a result, highly reflective chirped gratings have lower reflectivity at shorter wavelengths when the signal is coupled from the longer wavelength side of the fiber grating. There are several approaches to avoiding the radiation mode effect. One proposed method to counter this problem is the suppression of the normalized refractive index modulation, associated with this coupling by having a uniform photosensitive region across the cross-section plane of the optical fiber [5]. From the orthogonality principle of the modes, the overlap of the modal fields and the grating index modulation would be zero in this case. The LP_{01} (fundamental) mode will therefore not couple into any of the cladding modes. Since the LP_{01} mode only has significant field distribution over the core and the part of the cladding immediately next to the core, it is usually sufficient to have only this part of the optical fiber photosensitive. Although it is possible to introduce a photosensitive cladding around a photosensitive core, it is, however, very difficult to obtain the same photosensitivity over both cladding and core.

The second proposed method is to use a high-NA fiber [6]. This increases the gap between the main grating band and the next cladding mode coupling band to approximately 7 nm for high-NA fiber, a value that is far less than the requirements of most applications. Recently Dong and co-workers [7] proposed and demonstrated a new method for suppressing the coupling from guided optical modes into cladding modes. A depressed cladding, with appropriate index and thickness, is added between the photosensitive core and the normal cladding. This has proven to be very effective in reducing the cladding mode field strength over the core region of the optical fiber, minimizing this form of intermodal coupling. This method can also be combined with the photosensitive cladding method to achieve a further suppression of the coupling. Haggans et al. [8] compared the cladding and radiation mode loss characteristics of gratings written in fibers having a narrow-depressed cladding with the properties of gratings written in matched clad, photosensitive clad, and wide-depressed clad fibers under varying grating azimuthal asymmetry conditions. Fibers with narrow-depressed claddings were found to have lower cladding mode losses for small degrees of asymmetry than for fiber designs previously proposed for cladding-mode rejection. Depending of the degree of azimuthal asymmetry present in a given grating-writing process it was found that for large effective grating tilts ($\theta > \sim 1.3$ degrees) a matched clad design was superior. For negligible tilt or the equivalent asymmetry ($\theta < \sim 0.25$ degrees), the wide-depressed clad of Dong et al. [7] and the photosensitive clad of Delevaque et al. [5] gave superior suppression of cladding modes. For intermediate asymmetries ($\theta \sim 0.75$ degrees), a value that may occur for strong gratings, a narrow-depressed inner cladding fiber was shown to have lower cladding mode losses than the aforementioned fiber types.

3.4.2 Apodization of Fiber Gratings

The main peak in the reflection spectrum of a finite length Bragg grating with uniform modulation of the index of refraction is accompanied by a series of sidelobes at adjacent wavelengths. It is important in some applications to lower and, if possible, eliminate the reflectivity of these sidelobes, or to apodize the reflection spectrum of the grating. For example, in dense wavelength division multiplexing (DWDM) it is important to have very high rejection of the nonresonant light in order to eliminate cross talk between information channels, and therefore, apodization becomes absolutely necessary. Apodized fiber gratings can have very sharp spectral responses, with channel spacings down to 100 GHz. For applications such as add-drop filters, or in demultiplexers, the grating response to less than −30 dB from the maximum reflection is important. Another benefit of apodization is the improvement of the dispersion compensation characteristic of chirped Bragg grating, for which the group delay becomes linearized and the modulation associated with the presence of sidelobes is eliminated. For a 10-Gbps signal the level of the modulation should be less than 100 ps, ±10 ps is considered an acceptable value. In practice, apodization is accomplished by varying the amplitude of the coupling coefficient along the length of the grating. The apodization of fiber Bragg gratings using a phase mask with variable diffraction efficiency has been reported by Albert et al. [9]. Bragg gratings, with sidelobe levels 26 dB lower than the peak reflectivity, were fabricated in standard tele-communication fibers. This represents a reduction of 14 dB in the sidelobe levels compared to uniform gratings with the same bandwidth and reflectivity. Figure 3.7 shows the spectral reflection response of an apodized and an unapodized fiber Bragg grating reflector reported by the same group [10], where a 20 dB reduction in the sidelobe levels was achieved.

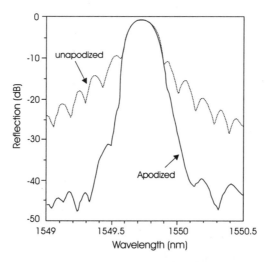

Figure 3.7   Reflection spectrum of fiber Bragg gratings photo-imprinted with a uniform diffrating phase mask and with a phase mask having a Gaussian profile of diffraction efficiency (*After:* [10]).

A cosine apodization technique obtained by repetitive, symmetric longitudinal stretching of the fiber around the center of the grating while the grating was written has been recently reported by Kashyap et al. [11]. This apodization scheme is applicable to all types of fiber gratings, written by direct replication by a scanning or a static beam, or by use of any other interferometer and is independent of length. The authors estimate that the bandwidth of the apodized gratings (0.4 nm) is sufficient to compensate for the dispersion of a 10-Gbps transmission system at a dispersion of 17 ps/nm/km over a distance of ~85 km. The simplicity of this technique allows the rapid production of fiber gratings required for wavelength division multiplexed (WDM) systems and dispersion compensation. Albert et al. [12] have demonstrated an efficient apodization technique, whereby a Moiré technique is used in the fabrication of a diffractive phase mask by electron beam lithography. The phase mask has a variable diffraction efficiency that results in apodization when illuminated with a uniform UV beam . An apodized grating with a channel width of 0.5 nm (transmission bandwidth < −30 dB) and a minimum channel spacing of 1.8 nm (bandwidth needed for the reflection to drop below −20 dB) has been demonstrated resulting in a device with a 28% bandwidth utilization factor.

3.5 Types of Fiber Bragg Gratings

There are a several distinct types of fiber Bragg grating structures: the common Bragg reflector, the blazed Bragg grating, and the chirped Bragg grating. These fiber Bragg gratings are distinguished either by their grating pitch (spacing between grating planes) or tilt. The most common fiber Bragg grating is the Bragg reflector, which has a constant pitch. The blazed grating has phase fronts tilted with respect to the fiber axis; that is, the angle between the grating planes and the fiber axis is less than 90 degrees. The chirped grating has an aperiodic pitch, displaying a monotonic increase in the spacing between grating planes. A brief overview of these Bragg gratings along with some of their applications is presented solely for the purpose of establishing the gratings' properties. Chapters 6 and 7 will deal with applications in greater detail.

3.5.1 Common Bragg Reflector

The common Bragg reflector, the simplest and most used fiber Bragg grating, is illustrated in Figure 3.1. Depending on parameters such as grating length and magnitude of induced index change, the Bragg reflector can function as a narrowband transmission or reflection filter or a broadband mirror. In combination with other Bragg reflectors, these devices can be arranged to function as band-pass filters. Two such configurations are shown in Figure 3.8. The first configuration modifies a broadband spectrum, utilizing the Bragg gratings to remove discrete wavelength components, whereas the second arrangement incorporates the Bragg gratings as highly reflecting mirrors to construct a fiber Fabry-Perot cavity.

Bragg reflectors are considered to be excellent strain and temperature sensing devices because the measurements are wavelength encoded. This eliminates the problems of amplitude or intensity fluctuations that exist in many other types of fiber-based sensor systems. Each Bragg reflector can be designated its own wavelength-encoded signature; therefore, a series of gratings can be written in the same fiber, each having a distinct Bragg resonance signal. This configuration can be used for wavelength division multiplexing or quasi-distributed sensing [13]. Gratings have also proven to be very useful components in tunable fiber or semiconductor lasers [14, 15], serving as one or both ends of the laser cavity (depending on the laser configuration). Varying the Bragg resonance feedback signal to the grating tunes the laser wavelength. Using this approach Ball and Morey [16] demonstrated a continuously tunable, single-mode, erbium fiber laser with two Bragg reflectors configured in a Fabry-Perot cavity. Continuous tunability without mode hopping was achieved when both the gratings and enclosed fiber were stretched uniformly. Bragg grating fiber lasers can also be used as sensors where the Bragg reflector serves the dual purpose of tuning element and sensor [17]. A series of Bragg reflectors having distinct wavelength-encoded signatures can be multiplexed in a fiber laser sensor configuration for multipoint sensing [18, 19].

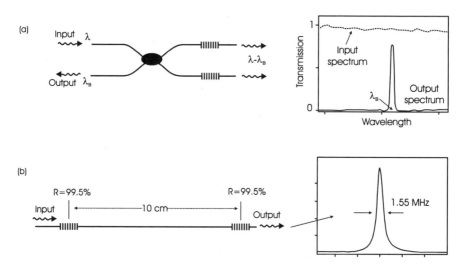

Figure 3.8   Fiber optic band-pass filters using Bragg reflectors with resonant wavelength at λ_B: (a) filter arranged in a Michelson type configuration and (b) filter arranged in a Fabry-Perot type configuration.

3.5.2 Blazed Bragg Grating

Tilting (or blazing) the Bragg grating planes at angles to the fiber axis (Figure 3.9(a)), will result in light that is otherwise guided in the fiber core being coupled into loosely bound, guided-cladding or radiation modes. The tilt of the grating planes and strength of the index modulation determines the coupling efficiency and bandwidth of the light that is tapped out. The criterion to satisfy the Bragg condition of a blazed grating is similar to that of the Bragg reflector that was analyzed earlier. Figure 3.9(b) illustrates the vector diagram of the Bragg condition (energy and momentum conservation) for the blazed grating. Here the wavevector of the grating $\mathbf{K}$ is incident at an angle θ_b with respect to the fiber axis. The magnitudes of the incident, $\mathbf{v}_i$, and the scattered, $\mathbf{v}_s$, wavevectors must be equal ($v = |\mathbf{v}_i| = |\mathbf{v}_s|$). Simple trigonometry shows that the scattered wavevector must be at an angle $2\theta_b$ with respect to the fiber axis. Applying the law of cosines to the momentum diagram gives

$$|\mathbf{v}_i|^2 + |\mathbf{v}_s|^2 - 2|\mathbf{v}_i||\mathbf{v}_s|\cos(\pi - 2\theta_b) = |\mathbf{K}|^2 \qquad (3.14)$$

which reduces to $\cos(\theta_b) = |\mathbf{K}| / 2v$ and shows that the scattering angle is restricted by the Bragg wavelength and the effective refractive index. It is clear from (3.14) that for blazed gratings not only different wavelengths emerge at different angles, but different modes of the same wavelength also emerge at slightly different angles due to their different propagation constants.

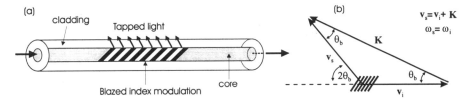

Figure 3.9   (a) A schematic diagram of a blazed grating. (b) A vector diagram for the Bragg condition of a blazed grating.

Figure 3.10 shows the output coupling of 488- and 514.5-nm light from an argon ion laser. The green argon ion wavelength has two modes and the blue wavelength has three modes that propagate in the fiber. These wavelengths and their modes are well separated and resolvable; thus, the grating tap acts as a spectrometer and mode discriminator. Meltz and Morey [20] have achieved out-coupling efficiencies as high as 21% at 488 and 514.5 nm.

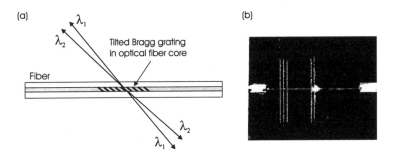

Figure 3.10   (a) Illustration of separated wavelength tapped out at different angles. (b) Image of the radiation out-coupled at 488 and 514.5 nm from a fiber Bragg grating tap (*After:* [20]).

Erbium-doped fiber amplifiers are now an integral part of long-haul, high-bit-rate communication systems and are finding applications in areas of wide bandwidth amplification. Kashyap et al. [21] demonstrated the use of multiple blazed gratings to flatten the gain spectrum of erbium-doped fiber amplifiers. A gain variation of ±1.6 dB over a bandwidth of 33 nm in a saturated erbium-doped fiber amplifier was reduced to ±0.3 dB. This is important in fiber communications that use several signals at different wavelengths and gives a uniform signal-to-noise ratio at the receiver output. Another interesting application of blazed gratings is in mode conversion. Mode converters are fabricated by inducing a periodic refractive index perturbation along the fiber length with a periodicity that bridges the momentum mismatch between the modes to allow phase-matched coupling between the selected modes. Different grating periods are used for mode conversion at different wavelengths. Hill et al. [22] demonstrated efficient mode conversion between forward propagation LP_{01} and LP_{11} modes.

3.5.3 Chirped Bragg Grating

One of the most interesting Bragg grating structures with immediate applications in telecommunications is the chirped Bragg grating. This grating has a monotonically varying period, as illustrated schematically in Figure 3.11.

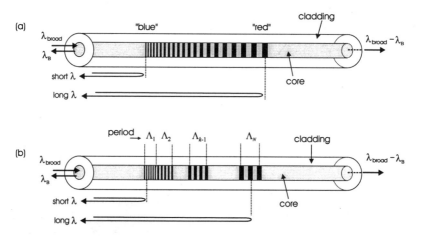

Figure 3.11   (a) A schematic diagram of a chirped grating with an aperiodic pitch. For forward-propagating light as shown, long wavelengths travel further into the grating before being reflected. (b) A schematic diagram of a cascade of several gratings with increasing period that are used to simulate long, chirped gratings.

There are certain characteristic properties offered by monotonically varying the period of gratings that are considered advantages for specific applications in telecommunications and sensor technology, such as dispersion compensation and the stable synthesis of multiple wavelength sources [23, 24]. These types of gratings can be realized by axially varying either the period of the grating Λ or the index of refraction of the core or both. From (3.3) we have

$$\lambda_B(z) = 2n_{\text{eff}}(z)\Lambda(z) \tag{3.15}$$

The simplest type of chirped grating structure is one where the variation in the grating period is linear:

$$\Lambda(z) = \Lambda_0 + \Lambda_1 z \tag{3.16}$$

where Λ_0 is the starting period and Λ_1 is the linear change (slope) along the length of the grating. Thus, one may consider such a grating structure made up of a series of smaller length uniform Bragg gratings increasing in period. If such a structure is designed properly, one may realize a broadband reflector. Typically, the linear chirped grating has associated with it a chirped value/unit length (Λ_1) and the starting period. For example, a chirped grating 2 cm in length may have a starting wavelength at 1550 nm and a chirped value of 1 nm/cm. This implies that the end of the chirped grating will have a wavelength period corresponding to 1552 nm. Chirped gratings have been written in optical fibers using various methods [25–27]. These fabrication techniques will be discussed in the next chapter.

In optically amplified, long-haul, high-bit-rate communication systems the main limitation to data transmission is pulse broadening caused by chromatic dispersion. The pulse broadening can be eliminated by incorporating an element having a dispersion of opposite sign and equal magnitude to that of the optical fiber link. Traditionally, optical fibers displaying the correct negative dispersion characteristics have been incorporated into telecommunication lines. However, the modal field diameter of the compensating fiber rarely matches that of the standard guide; therefore, splicing between different fiber sections requires pre-fusion preparation (heating of the fiber to produce diffusion of the dopants in the guiding core until the modal field overlap is optimized). This can be achieved with a single fusion-splicing device; however, it is advantageous if this can be avoided, particularly if a suitable in-line, in-fiber component is available. In a chirped grating the resonant frequency is a linear function of the axial position along the grating, so that different frequencies present in the pulse are reflected at different points and thus acquire different delay times (Figure 3.11). It is now possible to compress temporally broadened pulses. Figure 3.12(a) gives examples of the chirped grating transmission and reflection profiles [28]; an example of the group delay as a function of wavelength for a chirped grating is shown in Figure 3.12(b) [11].

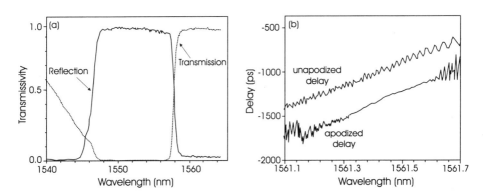

Figure 3.12 (a) Typical transmission and reflection profiles of a chirped Bragg grating (*After:* [28]); (b) Group delay for unapodized and apodized chirped grating (*After:* [11]).

Kashyap and co-workers have demonstrated concatenated, apodized gratings called "super-step" gratings. The gratings having arbitrary length were constructed for a predetermined wavelength span. A 13.5-ns delay for a 1.3-m super-stepped, chirped grating over 10 nm was achieved [29, 30]. Furthermore, in telecommunication systems residual pump light emitted from an optical fiber amplifier can limit receiver performance through an increase in excess noise and receiver saturation. The fiber amplifier performance may be improved by reflecting back the unabsorbed pump light at the amplifier output. Farries et al. [31] demonstrated the use of broadband chirped fiber Bragg grating for pump rejection and recycling of unabsorbed pump light from an erbium-doped fiber amplifier. In that work an amplifier was pumped with a 980-nm diode laser, and a broadband chirped fiber Bragg filter centered at 980 nm was used to reject and recycle the unabsorbed pump light.

3.6 Photosensitivity Types of Fiber Bragg Gratings

In Chapter 2 we classified the different types of Bragg gratings into three distinct categories of photosensitivity types, Type I, Type IIA, and Type II. The final grating type depended upon the initial writing conditions (laser power and wavelength, CW or pulsed energy delivery) and fiber properties. We elaborate on the earlier information, but what follows throughout this chapter should be taken as being complementary to the data discussed in Chapter 2.

3.6.1 Type I Bragg Gratings

Type I Bragg gratings refer to gratings that are formed in normal photosensitive fibers under moderate intensities. The growth dynamics of the Type I grating is shown in Figure 2.1 and is characterized by a power law with time of the form $\Delta n \propto t^{\alpha}$ [32]. A typical spectral response of a uniform period, Type I Bragg grating is shown in Figure 3.13.

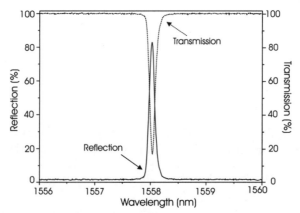

Figure 3.13   A typical spectral response of a uniform period, Type I Bragg grating. Transmission and reflection spectral of a broadband light source in the region of where the Bragg condition is satisfied.

It is interesting to note that the reflection spectra of the guiding mode is complementary to the transmission signal, implying that there is negligible loss due to absorption or reflection into the cladding. This is a fundamental characteristic of a Type I Bragg grating. Furthermore, due to the photosensitivity type of the Bragg grating, the grating itself has a characteristic behavior with respect to temperature erasure. Type I gratings can be erased at relatively low temperatures (approximately 200°C). Nevertheless, Type I gratings are the most utilized Bragg gratings and operate effectively from −40° to +80°C, a temperature range that satisfactorily covers most telecommunications and sensor applications.

3.6.2 Type IIA Bragg Gratings

Type IIA fiber Bragg gratings appear to have the same spectral characteristics as Type I gratings. The transmission and reflection spectra are complimentary, rendering this type of grating indistinguishable from Type I in a static situation. However, because of the different mechanism involved in fabricating these gratings there are some distinguishable features noticeable under dynamic conditions observed either in the initial fabrication or the temperature erasure of the gratings. Type IIA gratings are inscribed through a long process, following Type I grating inscription, as Figure 2.3 shows [33, 34]. At approximately 30 minutes of exposure (depending of the fiber type and exposure fluence) the Type IIA grating is fully developed. Clearly, Type IIA gratings are not very practical to fabricate. Although the mechanism of the index change is different from Type I, occurring through compaction of the glass matrix, the behavior subject to external perturbations is the same for both grating types. Irrespective of the subtleties of the index change on a microscopic scale, the perturbations act macroscopically and, therefore, the wavelength response remains the same. However, when the grating is exposed to high ambient temperature, a noticeable erasure is observed only at temperatures as high as 500°C (Section 3.9 and figures therein). A clear advantage of the Type IIA over the Type I gratings is the dramatically improved temperature stability of the grating, which proves very useful if the system is exposed in high ambient temperatures (as may be the case for sensor applications).

3.6.3 Type II Bragg Gratings

Type II Bragg gratings are the most distinct of all grating types and are formed under very high, single-pulse fluence (>0.5 J/cm^2)[35]. A typical transmission and reflection spectrum of a Type II grating is shown in Figure 3.14. The reflection appears to be broad, and several features over the entire spectral profile are believed to be due to nonuniformities in the excimer beam profile that are strongly magnified by the highly nonlinear response mechanism of the glass core. Type II gratings pass wavelengths longer than the Bragg wavelength, whereas shorter wavelengths are strongly coupled into the cladding, as is observed for etched or relief fiber gratings [36], permitting their use as effective wavelength selective taps.

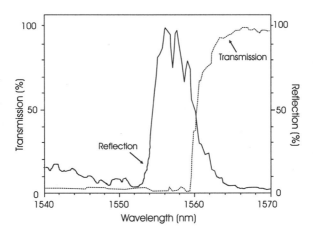

Figure 3.14   Reflection and transmission spectra of a Type II grating. At wavelengths below the Bragg wavelength of 1556 nm the light is strongly coupled into the cladding (*After:* [35]).

Type II gratings were first demonstrated in an experiment carried out to study the relationship between pulse energy and grating strength, where a series of single-pulse gratings were produced with a UV excimer laser beam [35]. The index of refraction change was estimated from the reflection spectrum using coupled mode theory (the result is summarized in Figure 2.5). It is apparent that there is threshold at pulse energy of 650 mJ/cm^2, above which the induced index modulation increases dramatically. Below the threshold point, the index modulation grows linearly with energy density, whereas above this point the index modulation appears to saturate. The gratings formed with low index of refraction modulation were Type I gratings. The behavior observed in Figure 2.5 suggests that there is a critical level of absorbed energy, which triggers off a highly nonlinear mechanism, initiating dramatic changes in the optical fiber. Examination of a Type II grating with an optical microscope revealed a damaged track at the core-cladding interface that is unique to this grating type, strongly suggesting that it may be responsible for the large index change. The fact that this damage is localized on one side of the core suggests that most of the UV light has been strongly absorbed locally.

Results of stability tests have shown Type II gratings to be extremely stable at elevated temperatures (see Figure 2.6) [35]. At 800°C over a period of 24 hours, no degradation in grating reflectivity was evident. At 1000°C most of the grating disappears after 4 hours, implying that the localized fusion has been thermally "washed out." This superior temperature stability can be utilized for sensing applications in hostile environments. One of the most attractive features of Type II gratings is that (as with Type I gratings) highly reflective gratings can be formed in just a few nanoseconds, the duration of a single excimer pulse. This is of great practical importance for large-scale mass production of strong gratings during the fiber drawing process before application of the protective polymer coating. Although the concept of fabrication of single-pulse Type I and Type II Bragg gratings during the fiber drawing process has been successfully demonstrated

[37, 38], the quality of in-line gratings must be improved. One distinct advantage of producing fiber Bragg gratings during the drawing process is that in-line fabrication avoids potential contact with the pristine outer surface of the glass. Off-line fabrication requires a section of the fiber to be stripped off its UV absorbing polymer coating in order for the grating to be exposed. This drastically weakens the fiber at the site of the grating due to surface contamination, even if the fiber is subsequently recoated.

3.7 Novel Bragg Grating Structures

In this section we will include several atypical grating structures that have immediate applications to telecommunications and sensing devices. These novel Bragg grating structures include superimposed multiple Bragg gratings, superstructure Bragg gratings, and phase-shifted Bragg gratings.

3.7.1 Superimposed Multiple Bragg Gratings

Othonos and co-workers [39] demonstrated the inscription of several Bragg gratings at the same location on an optical fiber. This is of interest as a device in fiber communications, lasers, and sensor systems, because multiple Bragg gratings at the same location basically perform a comb function that is ideally suited for multiplexing and demultiplexing signals. All the gratings are written at the same location of the fiber, which makes this approach well suited to optical integrated technology, where the issue of size is always a concern. This can also be used for material detection where the multiple Bragg lines can be designed to match the signature frequencies of a given material. A narrow linewidth KrF excimer laser was used in an interferometric setup to inscribe the different Bragg gratings on the same fiber location (Figure 3.15). The fiber (AT&T Accutether) was first hydrogen loaded to enhance its photosensitivity. Figure 3.15 shows the reflectivity for seven superimposed

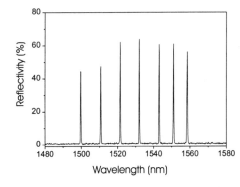
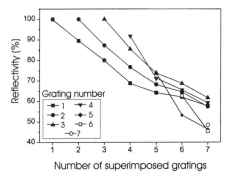

Figure 3.15   Superimposed multiple Bragg gratings at the same location on photosensitive optical fiber. The plot on the right shows reflectivities for each of the Bragg gratings, as a function of the number of gratings, superimposed on the same location (*After:* [39]).

Bragg gratings. The first grating was written at 1550.05 nm and reached a reflectivity of ~100% within 15 seconds (at 30 Hz) of UV exposure and had a linewidth of 0.25 nm. After adjusting the interferometer to write at a different Bragg wavelength, the second grating was written at 1542.6 nm with approximately the same characteristics. Each time a new grating was inscribed, the reflectivity of the existing gratings was reduced. Nevertheless, even after superimposing five gratings, the individual grating reflectivities were higher than 60%. Additionally, the center wavelength of the existing Bragg gratings shifted to longer wavelengths each time a new grating was inscribed because of the change of the effective index of refraction (e.g. the first grating shifted to 1550.975 nm by the time the last grating was inscribed). The shift in wavelength of the first grating after writing all seven gratings corresponds to an effective index of refraction increase of 0.86×10^{-3}.

3.7.2 Superstructure Bragg Gratings

The superstructure Bragg grating refers to a grating fiber structure fabricated with a modulated exposure over the length of the grating [40]. One such approach used by Eggleton et al. [40] was to translate the UV writing beam along a fiber and phase mask assembly while the intensity of the beam was modulated. An excimer-pumped dye laser with a frequency doubler was used to produce 2.0 mJ at 240 nm. Hydrogenated, single-mode, boron-codoped fiber was placed in near contact with a phase mask, and the ultraviolet light was focused through the phase mask into the fiber core by a cylindrical lens, exposing a length of approximately 1 mm. To fabricate a 40-mm long superstructure, the excimer laser was periodically triggered at intervals of 15 seconds to produce bursts of 150 shots at a repetition rate of 10 Hz, while the ultraviolet beam was translated at a constant velocity of 0.19 mm/s along the mask. The resulting period of the grating envelope was approximately 5.65 mm, forming seven periods of the superstructure. The reflection spectrum of this grating structure is shown in Figure 3.16. There is strong reflection at five discrete wavelengths corresponding to the spatial frequencies of the grating, with reflectance varying from 30% to 95%. These superstructure gratings can be used as comb filters for signal processing and for increasing the tunability of the fiber laser-grating reflector.

3.7.3 Phase-Shifted Bragg Gratings

Bragg gratings generally act as narrowband reflection filters centered at the Bragg wavelength because of the stop band associated with a one-dimensional periodic medium. Many applications, such as channel selection in a multichannel communication system, would benefit if the fiber grating could be designed as a narrowband transmission filter. Although techniques based on Michelson and Fabry-Perot interferometers have been developed for this purpose [41], their use requires multiple gratings and may introduce additional losses. A technique commonly used in distributed feedback (DFB) semiconductor lasers [42] can be used to tailor the transmission spectrum to suit specific

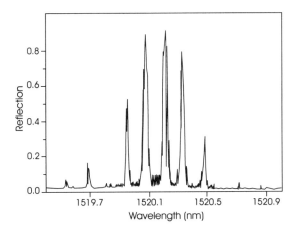

Figure 3.16 Reflection spectrum from a grating superstructure fabricated by translating the UV writing beam along a fiber and phase mask assembly while the intensity of the beam was modulated (*After:* [40]).

requirements. This approach relies on introducing a phase shift across the fiber grating, whose location and magnitude can be adjusted to design a specific transmission spectrum. This is a generalization of an idea first proposed by Haus and Shank [43] in 1976. The principle of the phase shift was demonstrated by Alferness et al. [44] in periodic structures made from semiconductor materials where a phase shift was introduced by etching a larger spacing at the center of the device. This forms the basis of single-mode, phase-shifted semiconductor DFB lasers. A similar device may be constructed in optical fibers using various techniques:

1. Phase masks, in which phase shift regions have been incorporated into the mask design [45];
2. Post-processing of a grating by exposure of the grating region to pulses of UV laser radiation (Figure 3.17) [46];
3. Post-fabrication processing using localized heat treatment [47].

Such processing produces two gratings out of phase with each other, which act as a wavelength selective Fabry-Perot resonator, allowing light at the resonance to penetrate the stop band of the original grating. The resonance wavelength depends on the size of the phase change. One of the most obvious applications includes production of very narrow-band transmission and reflection filters. Moreover, multiple phase shifts can be introduced to produce other devices, such as comb filters, or to obtain single-mode operation of DFB fiber lasers.

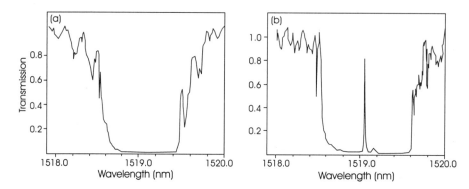

Figure 3.17   Normalized transmission spectra (a) before and (b) after a phase shift is introduced to a Bragg grating structure, using UV post-processing (*After:* [46]).

3.8   Pulse Propagation in Gratings

In view of the importance of ultrashort pulse applications in telecommunications and pulse propagation in optical fiber, we now discuss the effects of pulse propagation through Bragg gratings and review some of the most recent work. We consider pulse propagation where the spectral bandwidth of the pulse is narrower and larger than the grating response bandwidth. High-intensity pulse propagation through uniform gratings is also reviewed.

3.8.1   Ultrashort Pulse Propagation Through Fiber Gratings

In high-bit-rate, telecommunications applications ultrashort pulse propagation through fiber Bragg gratings is of primary importance, where the grating can act as a narrowband reflector or pulse compressor. It is therefore of interest to obtain a detailed understanding of the exact behavior of ultrashort pulse propagation through fiber Bragg gratings. Initially, we consider the case where the spectral bandwidth of the pulse is narrower than the grating response bandwidth. Taverner et al. [48] used ps (10^{-12}s) pulses from a figure-eight fiber soliton laser to study the effects of fiber gratings on short pulses. The grating was mounted in a jig that permitted strain tuning of the peak reflection wavelength. The authors investigated the pulse response of the grating in the temporal and spectral domains by measuring the autocorrelation function and the optical spectrum of the reflected and transmitted pulses. The source permitted tuning of the pulse duration (and associated optical bandwidth) over the range of 2.3–6 ps. Two fiber Bragg gratings (Type I) were examined. Grating A had a peak reflectivity R = 0.86 at 1532 nm, with a spectral half-width $\Delta\lambda_B$ = 0.78 nm; for grating B, R = 0.47 at 1531 nm, and $\Delta\lambda_B$ = 1.46 nm. Initially, the reflection and transmission of the gratings were investigated as a function of wavelength separation $\Delta\lambda_{offset}$, between the peak of the grating response and the peak of the pulse spectrum. Figure 3.18 shows plots of the observed pulse half-width broadening factors versus $\Delta\lambda_{offset}$ for the reflection of 5.6-ps hyperbolic-secant pulse from grating A ($\Delta\lambda_{pulse}$ / $\Delta\lambda_B = 0.56$, where $\Delta\lambda_{pulse}$ corresponds to the pulse spectral half-width). Additionally, 2.9-ps hyperbolic-secant pulses from grating B ($\Delta\lambda_{pulse}$ / $\Delta\lambda_B = 0.59$) were observed in both the temporal and spectral domains . Minimum temporal broadening factors of 35% for grating B and 50% for grating A, at wavelength offset of $\Delta\lambda_{offset} = 0$, were observed. Increased broadening and pulse deformation were observed as the reflection peak moved from the pulse center until $\Delta\lambda_{offset}$ / $\Delta\lambda_B \approx 1$, where the pulse was observed to become multipeaked. Superimposed upon the plots in Figure 3.18 are the theoretical pulse-broadening factors obtained by fitting the grating response function to the solutions of coupled mode theory for a uniform sinusoidal grating (Chapter 5). The theory was found to be in good agreement with the experimentally observed results. Furthermore, Taverner et al. [48] obtained a plot of broadening factor as a function $\Delta\lambda_{pulse}$ / $\Delta\lambda_B$ by varying the duration of the pulses incident on the gratings. They were able to vary $0.2 < \Delta\lambda_{pulse}$ / $\Delta\lambda_B < 1.5$ with the soliton spectrum centered on the grating in all instances, demonstrating significant dispersive and pulse-shaping effects for $\Delta\lambda_{pulse}$ / $\Delta\lambda_B > 0.2$.

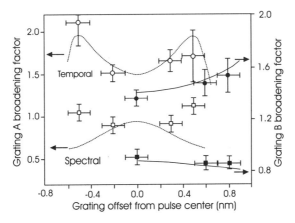

Figure 3.18   Experimental (points) and theoretical (curves) autocorrelation half-width and spectral half-width broadening factors for 5.6-ps hyperbolic-secant pulse reflection from grating A and 2.9-ps hyperbolic-secant pulse reflection from grating B (*After:* [48]).

We now consider the case where the spectral bandwidth of the incident pulse is larger than that of the grating response. Although there is relatively little work in this area, it is of interest from both a fundamental and applications viewpoint. Pulse propagation though fiber gratings can be modeled using the coupled mode theory, even though these equations (Chapter 5) are derived in the context of a monochromatic frequency source (CW signal) for linear propagation. The response to an input pulse can be obtained by considering each spectral component separately and integrating over the entire spectrum of the incident pulse. The reflected pulse field amplitude can be obtained by multiplying the appropriate grating frequency response with the input pulse spectrum, and the corresponding pulse behavior is recovered by taking the inverse Fourier transform. The reflection, transform-limited, picosecond Gaussian pulses from several different uniform gratings, varying in strength, are shown in Figure 3.19 [49].

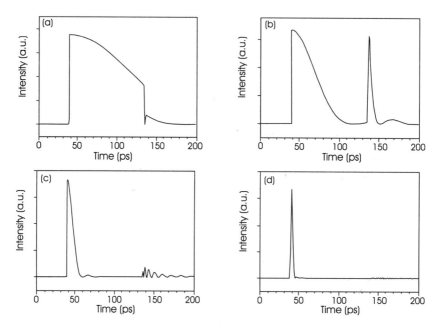

Figure 3.19   Reflected pulses from uniform gratings of length $L = 1.0$ cm and grating strengths (a) 3×10^{-5}, (b) 8×10^{-5}, (c) 3×10^{-4}, and (d) 1×10^{-3}, which are referred to in the text as weak, medium, strong, and very strong, respectively. The input is a transform-limited 1-ps Gaussian pulse (*After:* [49]).

Clearly, the transform-limited Gaussian pulse has undergone a considerable shape modification. For a weak index modulation grating (3×10^{-5}), the reflected pulse has a square-like shape with a gradual fall-off in intensity. As the grating strength increases, there is a separation of the reflected pulse into two distinct components; a main reflection peak and transient subpulses. The main reflection peak and the start of the transient sub-pulses are separated in time by the round-trip propagation time through the grating. Furthermore, the duration and the number of oscillations in the transient subpulses both increase with increasing grating strength, as seen in Figure 3.19(a). A single picosecond (ps) pulse travels approximately 0.2 mm in the fiber over the period equal to its full width half maximum (FWHM), which is considerably shorter than the length of the Bragg grating. Clearly, at any given time the pulse interacts only with a fraction of the total length of the grating. Furthermore, as a result of the weak UV-induced index of refraction modulation, the pulse will propagate well into the grating, continuously generating a reflected signal. When the input pulse finally reaches the end of the grating, the last reflection must travel back through the grating. Therefore, the total duration of the reflected pulse is precisely the round-trip propagation time through the grating. The overall reflected pulse is a coherent sum of the reflected components generated as the input pulse propagates through the grating. The observed long duration of the overall reflected pulse is consistent with its narrow spectrum. The decrease in the intensity of the reflected pulse with time is mainly due to the intensity loss of the incoming pulse as it propagates through the grating (Figure 3.19(a)). Furthermore, the sharp rise and fall time of the reflected pulse is also consistent with the broadband reflection that occurs when the input pulse enters or exits the grating. For stronger uniform gratings (Figure 3.19) there is a separation of the reflected pulse into two distinct components: a main reflected peak, which is primarily due to frequencies in the stop band of the CW grating response, and the transient subpulses arising from the sidelobe frequencies. This pulse-grating behavior can be explained as follows. Due to the strong index modulation, and thus the strong reflection, the frequencies in the stop band are primarily reflected by a short segment at the beginning of the grating. This interaction results in the short duration of the main reflection peak. Frequencies that lie in the sidelobes of the CW grating response, however, will propagate to the end of the grating and thus contribute to the transient pulses. Looking at the reflected delay time response of the grating it is clear that the frequencies of the stop band are reflected before those in the sidelobes. The difference in the reflected time delay between frequencies in the stop band and those in the sidelobes corresponds to the propagation time through the grating, resulting in the separation between the main reflection peak and transient subpulses. The oscillatory nature of the transient subpulses arises from the beating of the sidelobes frequencies, which are symmetrically spaced about the Bragg resonance wavelength. For this particular example in Figure 3.19(c), the period of oscillations in the transient subpulses for the strong grating is approximately 5 ps, which corresponds to a frequency spacing of approximately 1.6 nm.

A rather surprising result is that following the main reflection peak, there are no significant reflections from the grating until the transient subpulses appear after a time equal to the round-trip propagation time through the grating. This phenomenon can be explained using the effective medium description [50] where the grating is replaced by an effective medium with no grating, but rather with a frequency-dependent refractive index. In this model the sidelobe frequencies that propagate in the grating do not reflect until they reach the discontinuity at the end of the grating between the effective medium and the unmodified fiber core. Once reflected, they will only appear at the front of the grating after having traveled back through the medium. The minimum total distance traveled is twice the grating length, corresponding to the round-trip propagation time through the grating. In addition, since the refractive index of the effective medium is frequency dependent (Figure 3.20), different wavelengths will propagate with different speeds within the medium. The temporal spreading in the transient subpulses is primarily due to this effect.

Let us next consider the case of linearly chirped gratings where the amplitude of the modulated index is assumed constant. For a weak index of refraction variation ($\sim 3 \times 10^{-5}$) the reflected pulse has a square shape; as the refractive index variation becomes stronger, the transients are more pronounced (Figure 3.21). There was also no difference in the reflected pulse shape from a positively or negatively chirped grating of the same magnitude. The shaping in the reflected pulses from chirped grating is attributed to the position dependence of the resonance wavelength. For a linear chirped grating, the interaction between the pulse and the grating is similar to the unchirped case, the only

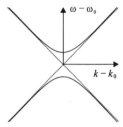

Figure 3.20   Dispersion relation for a linear periodic structure solved from the coupled mode equations, where n_0 is the refractive index of the unmodified fiber core and k is the wave number.

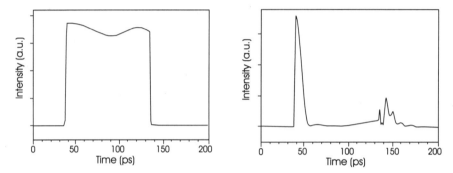

Figure 3.21   Reflected pulses from linearly chirped gratings of length $L = 1.0$ cm (a) 3×10^{-5} and (b) 3×10^{-4}. The input pulse is a 1-ps transform-limited Gaussian pulse (*After:* [49]).

difference being the effect of the shift in the resonance wavelength. As the input pulse propagates through the structure, each subgrating segment (due to its different resonance frequency) interacts differently with the pulse. Therefore, the reflected frequencies vary, and, as the pulse spectrum is approximately constant over this range of frequencies, the reflected pulse components will experience a smaller decrease in intensity compared with the unchirped case; the overall reflected pulse will have a square-like shape. Following this interpretation, it is clear that the reflected pulse is independent of the sign of the chirp.

Recently, experimental confirmation of the theoretical simulations have been provided by measuring the temporal response of transform-limited picosecond Gaussian pulses reflected and transmitted from a fiber grating [51]. Figure 3.22 shows the experimentally measured reflected and transmitted pulses.

3.8.2   High-Intensity Pulse Propagation in Bragg Gratings

Nonlinear optics is a rich area of intense theoretical and experimental study [52, 53], yielding a wealth of physical phenomena. Examples of nonlinear effects that have been studied and exploited include those involving the second- and third-order nonlinear

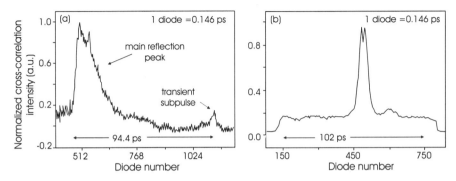

Figure 3.22 The experimentally measured cross-correlation of the (a) reflected and (b) transmitted pulses from a fiber grating, with the input pulse tuned to the Bragg wavelength of the grating (*After:* [51]).

susceptibilities and the nonlinear inelastic scattering phenomena. Because the generation of a nonlinear optical response requires a strong interaction of the electromagnetic field with the medium, the observation of nonlinear effects in bulk media requires a combination of high intensity and long interaction length. In long distance communications, a third-order nonlinear effect of great significance is self-phase modulation [54], which together with anomalous dispersion can result in the formation of bright temporal optical solitons [55, 56]. The small group velocity dispersion of optical fibers is such that the distance over which temporal solitons form is typically on the order of kilometers. To obtain soliton-like effects over short interaction lengths, researchers have studied spatial solitons, where the optical nonlinearity (Kerr effect) is balanced by diffraction, rather than by dispersion, and which have been observed in samples only 5 mm in length [57]. Nonlinear properties of periodic media were first investigated by Winful et al. [58], who showed theoretically that such media exhibit bistability, multistability, and switching behavior. This prediction has been confirmed experimentally in a variety of geometries, such as in semiconductors. Clearly, similar nonlinear effects may also be observed in Bragg gratings fabricated in optical fibers. In Bragg grating solitons the grating dispersion is balanced by the nonlinear Kerr effect. The dispersion of a chirped Bragg grating can be up to five orders of magnitude larger than that of an equivalent length of UV-unprocessed fiber [59, 60]. Grating solitons may be observed in gratings of only a few centimeters in length [61], thus rivaling the compactness of spatial soliton geometries. This implies that the required optical intensities to generate Bragg grating solitons are correspondingly higher than in uniform media. At the Bragg wavelength λ_B, and for wavelengths close to it, gratings reflect strongly. In fact, for such a grating the reflectivity is high over the range of wavelengths $\Delta\lambda$ given by

$$\frac{\Delta\lambda}{\lambda_B} = \frac{\Delta n}{n_{\text{eff}}} \tag{3.17}$$

where the wavelength band $\Delta\lambda$ is referred to as the photonic band gap [62]. For example, a grating with $\Delta n \sim 3 \times 10^{-4}$ and $\lambda_B = 1550$ nm results in a photonic band gap width of $\Delta\lambda = 0.3$ nm (Figure 3.23(a)). Let us consider the dispersive properties of Bragg gratings for wavelengths close to the Bragg wavelength λ_B, but not so close so as to be within the photonic band gap. In this case the group velocity v_g of the light is strongly affected by the grating because of multiple Fresnel reflections off the periodic structure. The group velocity varies between 0 and c/n_{eff} over a range of wavelengths of order $\Delta\lambda$. Close to the Bragg resonance, gratings exhibit strong quadratic dispersion that can be many orders of magnitude larger than for UV-unprocessed fiber [59, 60]. For $\lambda > \lambda_B$ red wavelengths travel faster than blue; therefore, we have normal dispersion. In contrast, for $\lambda < \lambda_B$ the dispersion is anomalous. It is also interesting to point out that dispersion is only larger over a wavelength range that is of the order of $\Delta\lambda$. Moreover, since the quadratic dispersion itself is a strong function of wavelength close to the Bragg resonance, gratings also exhibit comparatively strong higher order dispersion. Therefore, at wavelengths close to λ_B but outside the photonic band gap, low-intensity light pulses propagate through a grating far more slowly than through UV-unprocessed fiber; however, the strong dispersion rapidly broadens the pulses.

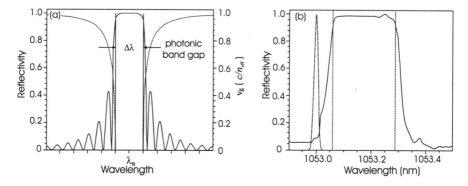

Figure 3.23   (a) Reflectivity (solid line) of a uniform Bragg grating as a function of wavelength; the Bragg wavelength λ_B, and the width of the photonic band gap $\Delta\lambda$, the edges of which are indicated by the vertical lines. Typically, $\Delta\lambda \approx 0.2$ nm. The dashed line indicates the group velocity in units of the speed of light in the unprocessed fiber. (b) Measured reflection spectrum of a uniform grating. Vertical dashed lines: approximate extent of the photonic band gap. Curved dashed line: idealized emission spectrum of the laser (*After*: [61]).

To discuss the properties of the gratings under high intensities one has to consider the optical nonlinearity of the glass. In UV-unprocessed fiber, the nonlinearity, in combination with anomalous dispersion, can give rise to solitons [54]. We have also discussed how fiber gratings support grating solitons. Since the dispersion must be anomalous in a uniform grating, solitons only occur for wavelengths below λ_B. Several key differences between grating solitons and conventional solitons can immediately be recognized. First, the larger grating dispersion means that in grating solitons the effect of the nonlinearity is also greater than in conventional solitons. That the strength of the nonlinearity is not affected by the grating implies that grating solitons require higher intensities than conventional solitons.

Second, because the soliton period (the propagation length over which solitons effects become apparent) is inversely proportional to the dispersion, the relevant length scales for grating solitons are much smaller than for conventional solitons. Third, higher order dispersion is more important in gratings than in UV-unprocessed fiber; therefore, it can generally not be treated perturbatively– an approximation that is usually allowed for conventional solitons.

It was recently demonstrated that approximately 15 GW/cm^2 are required to form grating solitons for a grating with length of 35 mm. At low intensities one expects to observe linear optical effects. Martijn de Sterke et al. [61] launched 70-ps pulses with an intensity of 1 MW/cm^2 into gratings and measured the transmitted pulse shapes for (a) the input tuned far from the band gap and (b) the input launched only about 30 pm outside the band gap edge. There were three important differences observed between these two cases. The key point is that in case (a), the detuning of light pulse from the Bragg resonance is sufficiently large so that the grating can be ignored; this is not true in case (b).

1. The total transmitted energy in case (b) is smaller than in (a) because of the higher grating reflectivity close to the Bragg wavelength (Figure 3.23(b)).
2. In case (b) the pulse was about 67 ps delayed with respect to that in (a). This was due to the smaller group velocity close to the Bragg wavelength.
3. In case (b) the transmitted pulse is roughly 20 ps wider than in (a). This is due to the strong dispersion close to the Bragg wavelength.

3.8.3 Weak Nonlinear Regime, Intermediate Intensity

When 80-ps pulses with peak intensity of 20 GW/cm^2 were launched into the grating, some interesting results were noted. The measured pulse shapes are shown in Figure 3.24(a). The solid line corresponds to the transmitted pulse when the input pulse was tuned far from the Bragg resonance. This clearly represents a case where the pulse is unaffected by the grating. When the input wavelength was tuned just below the Bragg resonance, where the dispersion was anomalous, the transmitted pulse is compressed to approximately 25 ps, and the peak intensity is enhanced. The 45-ps delay of the compressed pulse, relative to the transmitted pulse far from resonance, indicates that the pulse is propagating at an average group velocity of $\sim 0.74 c/n_{eff}$. Winful first predicted the compression of high-intensity pulses in periodic structures theoretically in 1985 [63]. Eggleton et al. [64] have shown that the compressed pulse is associated with the formation of a Bragg grating soliton.

3.8.4 Strong Nonlinear Regime, High Intensity

Further increasing the input pulse intensity leads to unexpected results. Rather than observing higher peak intensity for the transmission pulse, the pulse appears to split inside the grating. This is illustrated in Figure 3.24(b) for which the incoming pulse has peak intensity of 65 GW/cm^2, a width of 80 ps, and a detuning below λ_B of 0.15 nm. Similar behavior is observed in numerical calculations in which the central frequency of the incoming pulse was inside the photonic band gap [65–67]. However, because of the high reflectivities of the grating at such frequencies, the threshold for the instability was found to be high. In contrast, when the input pulse has frequency outside the photonic band gap, the threshold is much lower. The intensity behavior can be understood in terms of modulation instability inside the grating. This usually applies to fields with constant amplitude in a nonlinear medium with anomalous dispersion, in which this instability leads to the formation of a train of pulses. However, fields with a sufficiently slow-varying envelope should exhibit similar behavior. The effect described here may form the basis for an all-optical pulse generator based on a fiber grating [64]. By optimizing the grating and pulse parameters, it may be possible to reduce the instability threshold considerably.

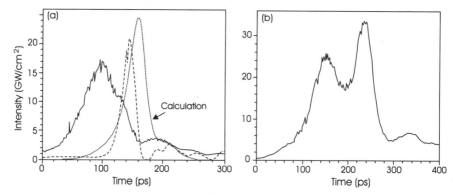

Figure 3.24 (a) Solid line corresponds to the measured transmitted intensity versus time when the incoming pulse is detuned far from the Bragg resonance. The measured pulse width with the incoming pulse frequency close to the Bragg resonance is shown by the dash line. (b) Measured transmitted intensity versus time for pulses with a peak input intensity of 65 GW/cm², a width of 80 ps, and a detuning below λ_B of 0.15 nm (*After:* [61]).

3.9 Lifetime and Reliability of Fiber Bragg Gratings

It is envisaged that optical fiber gratings will play a key role in the next generation of optical fiber communication systems, sensor networks, laser sources, and switching devices. Regardless of whether the optical fiber has undergone any form of pretreatment prior to UV excitation, some thermal decay of the grating occurs over time, even at room temperature. We touched upon this in Chapter 2 when the grating properties were briefly discussed. Long-term grating stability has serious implications given the large number of potential applications that are device-critical to telecommunications and optical fiber sensors. Whether the intended use is in DWDM systems, resonant fiber structures, or sensors located on an aircraft or around a nuclear power plant, the stability of a device that can provide "absolute" wavelength-encoded information is critical. Additionally, the stability of the host material must also be accounted for. If it is found that the inscription process can lead to long-term instability, the impact could be significant–the last 20 years has seen an escalation in the number of optical fiber cables that have been installed worldwide for telecommunications purposes. This trend is set to continue. There has also been much recent interest in optical fiber sensors for measuring strain, temperature, and pressure, with the primary use directed towards civil engineering structures. Therefore, the effect of mechanical degradation in optical fibers resulting from exposure to UV laser radiation has important consequences for the reliability of devices incorporating fiber Bragg gratings. Although the losses of individual gratings may be considered small, the cumulative loss over a large network can place stringent requirements on losses. Similarly, in resonant grating structures (e.g., Fabry-Perot filters, fiber lasers, and switches) grating losses can be detrimental to the cavity Q. In strong gratings (>50%) coupling to radiation modes on the short wavelength side of the band gap is another unwanted loss mechanism, as is incoherent scattering losses.

3.9.1 Thermal Decay of Fiber Bragg Gratings

In Chapter 2 we discussed how grating formation revolves around laser-induced excitation of glass into a metastable state; as a result, the grating will decay over time at elevated temperatures. The extent to which this occurs depends on the fiber and grating type, whereas all grating types written in nonhydrogenated fiber are stable at room temperature over many years. One approach to stabilize the grating is to preanneal at a temperature that exceeds the anticipated serviceable temperature of the grating-based device–called accelerated aging. There is, however, no single, universal stabilization anneal for all grating types since the refractive index stability depends on the fiber properties and the laser writing conditions. The thermally induced decay under accelerated test temperatures implies that the UV-induced defects are not thermodynamically stable, having sites that are reversible in nature followed by decay of the associated refractive index. Each individual site has an associated activation energy barrier E_{a1} that must be overcome in order to cause the index decay (Figure 3.25). The amorphous nature of glass results in a distribution of activation energies, the components of which can decay with discrete time constants. Low activation energy sites such as E_{a1} decay faster than high activation energy sites E_{a3}. A net

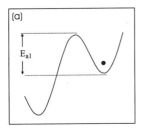
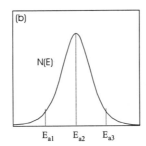
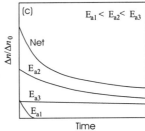

Figure 3.25   (a) The refractive index modulation in optical fibers results from UV-induced defect sites, (b) the amorphous nature of glass leads to a distribution of possible activation energies for these sites, and (c) the net decay of the grating is a composite of individual components with various time constants.

grating decay that is a summation of the individual components generally follows thermal excitation. One observes an initial rapid decay followed by a slowing but non-zero rate. Accelerated aging experiments [68, 69] are required to accurately simulate long-term grating stability, regardless of the particular grating type classification. Thus far two approaches have emerged for analyzing and predicting grating decay using accelerated aging data, the so-called "aging curve" and "power law" approaches. Both are based on similar physical principles, although it has recently been shown [69] that the former approach is generally more applicable.

3.9.1.1 Power Law Description of Bragg Grating Decay

In 1994 Erdogan et al. [68] provided the first detailed study of grating stability and proposed a model to explain the thermal degradation characteristics of fiber Bragg gratings written in germanium and erbium-germanium-co-doped silica fibers. The model showed that the decay of the UV-induced index change could be described by a power law function of time with a small exponent ($\ll 1$). This has a similar form to that reported for grating growth [32] and the exponential decay of UV-induced luminescence bleaching in Ge-doped fiber [70]. The authors phenomenologically described the observed slowing in the decay following the rapid initial decay through a decay mechanism in which carriers initially excited during writing are trapped in a broad energy distribution of trap states (this is consistent with the amorphous nature of glass). The rate of thermal depopulation was an activated function of the trap depth. The current consensus among researchers is that the power law model adequately explains the observed behavior with different fitting coefficients for the different fibers used. Indeed, this model also confirms the greater thermal stability of strong gratings written at higher UV intensities with correspondingly higher core temperatures that results from the partial thermal anneal during the UV writing process. However, Patrick et al. [71], and more recently Baker et al. [72], have evidence that the gratings written in hydrogen-loaded fibers are less stable than their standard fiber counterparts and cannot be accurately described by the power law, whereas Egan et al. [73] have reported otherwise.

The study of Erdogan et al. [68] was particularly important as Er^{3+}-doped germanosilicate fiber, used for making short, single-frequency fiber lasers, was investigated. The fiber core was doped with 14 mol% Ge and 600 ppm Er^{3+} co-doped with aluminum to minimize concentration-quenching effects. Gratings were written with UV light at 244 nm for a peak index change of $\sim 2 \times 10^{-4}$. The gratings were heated to elevated temperatures and their transmission spectra were monitored in real time. Figure 2.4 shows examples of decay data at $350°$ and $550°C$, indicative of power law dependence with time, for the integrated coupling constant (ICC). This quantity is proportional to the UV-induced refractive index change even for nonuniform gratings [68], and it is defined by

$$ICC = \tanh^{-1}\left(\sqrt{1 - T_{\min}}\right) \qquad (3.18)$$

This quantity does not account for any contribution from the mean index change. Furthermore, quantity η is defined as the ICC normalized to its value at $t = 0$ and given the proportionality between the ICC and the UV-induced index change, η is essentially a normalized index change, $\Delta n(t) / \Delta n_0$, with the form

$$\eta = \frac{1}{1 + A(t/t_1)^\alpha} \qquad (3.19)$$

where t is time, and t_1 is a unit factor (set to 1 minute) that renders the term dimensionless. This form for grating decay results for specific initial distributions of activation energies, for which the relationship resembles an exponential form for the rising part of the distribution curve. A and α are dimensionless, temperature-dependent terms.

Figure 3.26(a) shows the dependence of the power law factor A on temperature, obtained from curve fits to (3.19), in which both A and α are allowed to vary freely. The circles denote a presumed linear dependence of α with time, $\alpha = T/T_0$ with $T_0 = 5250$K, for which there is only reasonable agreement above 500K. This indicates that A can be written in the form $A = A_0 \exp(\alpha T)$, where T is in Kelvin and A_0 and α in K^{-1}. By fitting the accelerated aging data for any grating at a given temperature to (3.19), the terms of A and α may be obtained for that temperature. Repeating this procedure as a function of temperature leads to the relation between A, α, and T, from which the grating stability can be predicted using (3.19) for a given temperature and time. Figure 3.26(b) shows that the dependence of the exponent α on temperature displays a linear relation only above 600K, with a large statistical variation below this point. The large errors primarily depend on the large uncertainty in the ICC for small measurement errors of T_{min}, along with the substantially smaller grating decay at low temperatures, on which α is very dependent.

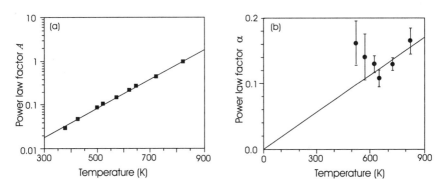

Figure 3.26 (a) The power law factor A obtained from the curve fits to (3.19) with α varied freely (error bars) and a fixed to $\alpha = T/5250$ K (open circles), and (b) α versus temperature for the data obtained in fitting to (3.19) (*After:* [68]).

A model that is similar to that used to describe dispersive carrier transport in amorphous semiconductors may be modified to predict the measured behavior [68]. Carriers are excited by the UV irradiation from the single homogeneously broadened absorption spectrum by GODCs at 5 eV. The carriers are then assumed trapped in a continuous distribution of energy states, rather than at a single trap level. If we consider that it is the breakage of the wrong bonds that creates defects and frees electrons (Chapter 2), then it is plausible to assume that thermal excitation of these electrons from traps can lead to the reformation of the wrong bond, simultaneously removing the UV-induced index change. The dissociation energy E_a of the trap follows a simple Arrhenius law of the form

$$\frac{\Delta n(t)}{\Delta n_0} = \exp(-\nu t) \qquad (3.20)$$

where ν is a frequency factor related to E_a, temperature T, and Boltzmann's constant K_B via $\nu(T) = \nu_0 \exp(-E_a / K_B T)$. Clearly, this is not the case for standard or, as we will show, hydrogenated fibers, as the Arrhenius law predicts that at a constant temperature the trap depopulation, as well as the index modulation, should decay exponentially to zero, rather than reaching a plateau [68]. Therefore, the correct description is one in which almost all electrons up to a given trap depth, rather than a fraction of electrons at a single trap, are wiped out by the decay process; hence, the decay history is important in determining the subsequent behavior. Figure 3.27 shows a simplified diagram for this mechanism. $E = 0$ is the point where electrons are free (the conduction band minimum). The thermal releasing of electrons results in the reoccupation of the original deep level occupied prior to UV excitation. Evidence of this has been reported in [74]. The essence of the theoretical picture is the separation of the distribution of trapped carriers into energies above and below a demarcation energy E_d that depends on time and temperature. It is therefore convenient to interpret the experimental data in terms of E_d, where $E_d = K_B T \ln(\nu t)$, where ν is obtained by acquiring multiple data sets as a function of temperature and fitting them together.

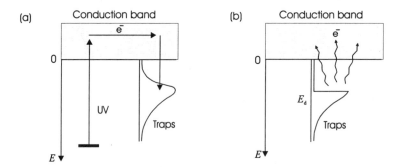

Figure 3.27 Diagram of the physical model for which (a) electrons excited by UV irradiation are trapped in a continuous distribution of traps, and (b) thermal depopulation of the traps at a given time and temperature approximately corresponds to shallower traps ($E < E_d$) being emptied and deeper traps ($E > E_d$) remaining full.

In Figure 3.28 the experimentally determined values for η (assumed proportional

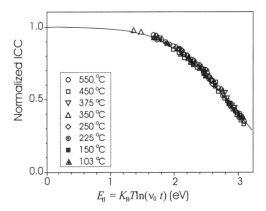

Figure 3.28 Plot of the normalized integrated coupling constant as a function of the demarcation energy E_d (aging curve). The solid line is the fit using (3.21) (*After:* [68]).

to N, the total number of trapped electrons remaining at a given time) are plotted versus the demarcation energy E_d. The different symbols correspond to the different decay experiments performed at various temperatures. Plotting the change in refractive index modulation as a function of the demarcation energy, which is equivalent to the aging parameter, gives the aging curve. The parameter η may also be written as

$$\eta = \frac{1}{1 + \exp\left[\left(E_d - \Delta E\right)\Big/ K_B T_0\right]} \tag{3.21}$$

This presents a simple analytical form for the dependence of the normalized UV-induced index change on the demarcation energy E_d, and, hence, the aging of the grating may be established. From the experimental data of Figure 3.28 and the corresponding fit, the energy distribution of the traps (ΔE) is ~2.8 eV deep, with a width of 1.6 eV. Figure 3.28 immediately suggests which combinations of time and temperature are capable of erasing a given fraction $1 - \eta$ of the initial UV-induced index change. Each value of η has a unique demarcation energy associated with it, and it is this that validates its use for determining accelerated aging. To show the importance of the decay history, a grating was first heated to 350°C, cooled and then reheated to 550°C (Figure 3.29). The grating decayed to ~33% of its initial value, regardless of whether the grating started this cycle with its initial strength (Figure 2.4), or with 75% of its initial strength (Figure 3.29), with the first 25% erased by the 350°C cycle. A single-activation-energy process would result in 33% of 75%, or 25%, of the initial grating strength to be left after the 550°C cycle in Figure 3.29. This suggests that given an arbitrary temporal segment of a grating's decay history, it is possible to ascertain how much decay has occurred prior to that segment. This also verifies the possibility of accelerated aging of the UV-induced refractive index change pre-annealing a grating to keep only the stable component of the index. Finally, we give a comparison of a power law prediction for a grating operated at a temperature of 80°C for 25 years, written in boron-germanium–co-doped or Ge-doped fiber. The former will decay to 88% of its original UV-induced modulated index value compared with 92% for the latter. Thus, the predicted result is consistent with the experimental observation [75, 76] that B co-doping reduces the thermal stability of the fiber grating, which further validates this model.

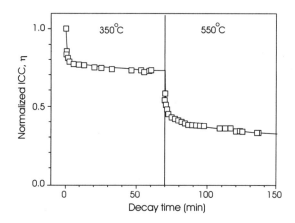

Figure 3.29   Decay of a single grating heated to 350°C and then to 550°C. The data are consistent with the treatment of the 350°C segment as decelerated aging of the 550°C decay, as shown by the curve prediction (*After:* [68]).

3.9.1.2 Stability of Type I Bragg Gratings in Hydrogenated and Unloaded Fiber

Annealing studies made on Type I Bragg gratings written in hydrogen-loaded and unloaded fiber (9 mol% Ge-doped) using either a CW or pulsed laser source, have indicated that gratings in unloaded fibers are more robust at elevated temperatures (Figure 3.30(a)) [71]. For example, 110°C (176°C) for 10 hours produced no measurable change (5%) in the index modulation for the grating in unloaded fiber, whereas hydrogenated fiber decreased by more than 20% (40%) under the same conditions. For the unloaded fiber grating a temperature change of 400°C would be required to produce a 40% increase. Figure 3.30(b) compares the normalized index modulation after 10 hours of annealing as a function of annealing temperature for gratings written using a pulsed or CW source. It is interesting that the thermal stability of the index modulation is essentially the same under both writing conditions. This indicates that the stability of the index modulation for Type I gratings is independent of the peak intensity of the writing source and confirms that essentially the same grating type has been written with both sources. For annealing temperatures between 100° and 350°C the hydrogenated fiber shows larger decreases in the index modulation, although above this temperature gratings in both fiber types tend to the same value of $\Delta n / \Delta n_0$. That the index change in hydrogenated fiber first decreases rapidly, before reaching a residual level comparable to that observed in unloaded fibers, implies that the species responsible for the index change in hydrogen-loaded fibers is less stable than in unloaded fiber. It can be associated with a broader trap distribution. Annealing at low temperatures removes that part associated with hydrogenation, leaving a more stable component behind. This is consistent with the notion that hydrogenation produces defect centers that contribute to already existing defect centers intrinsic to the fiber [77].

The long-term room temperature stability of gratings written in hydrogenated fiber is inferior to unloaded fiber, often showing a 15% reflectivity decrease over 6 months compared to no change for unloaded fiber gratings. This indicates that the thermal history of the grating is far less important to its erasure than the maximum operating temperature experienced, which has implications if the gratings are written in hydrogenated fiber for high temperature applications. Additionally, strong gratings in hydrogenated fiber (Δn_0 of 10^{-3} or larger) are no more stable at elevated temperatures than moderately reflecting gratings (Δn_0: 5×10^{-5} to 1×10^{-4}), unlike gratings in unloaded fiber that undergo a partial anneal with the high power UV writing conditions.

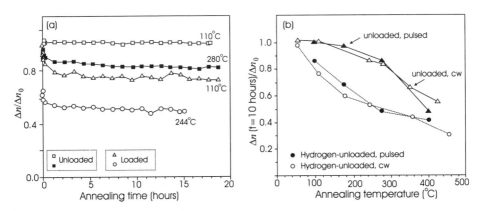

Figure 3.30   Normalized index modulation $\Delta n/\Delta n_0$ versus time for gratings written in (a) hydrogenated and unloaded fiber, using a CW UV source, (b) after 10 hours of annealing as a function of annealing temperature. The circles represent gratings written using the pulsed (open) and CW (closed) source in the unloaded fiber. The squares are data from gratings written in the hydrogenated fiber (*After:*[71]).

Finally, annealing of the UV-induced SiOH/GeOH absorption bands in hydrogenated fiber displays no correlation with grating decay; the OH absorption band is stable at 180°C, unlike the associated grating, and is erased only when exposed to temperatures as high as 630°C for several hours.

3.9.1.3   Stability of Type IIA Bragg Gratings

Dong and Liu [78] have examined the thermally induced decay of positive (Type I) and negative (Type IIA) index gratings, written at 193 nm in B–co-doped germanosilicate fiber, characterizing the energy levels of the system and predicting grating lifetimes. It should be noted that Type IIA gratings have not been observed in hydrogenated fibers. Figure 3.31 shows that the decay of gratings with varying levels of positive and negative index changes represented by the reflection growth profile in the inset. The Type I grating, A, decays rapidly under heating. The Type IIA grating, F, is stable to 300°C. The curves C, D, and E confirm that heating can cause grating growth since the balance between the positive and the negative index contributions changes at 300°C.

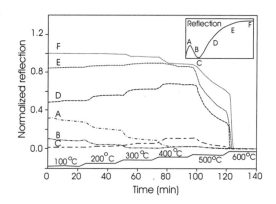

Figure 3.31   Decay of gratings with varying levels of positive and negative index change (*After:* [78]).

The decay of the positive index gratings with temperature is shown in Figure 3.32(a). Figure 3.32(b) shows the decay of gratings associated with a negative index change, showing excellent temperature stability to 300°C.

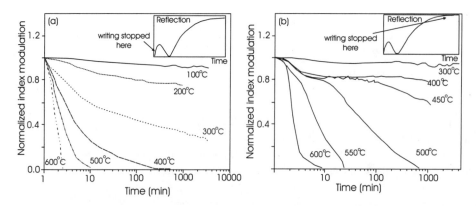

Figure 3.32   (a) Decay of gratings formed mainly by positive index changes at different temperatures and power law fittings to the data. (b) Decay of gratings formed mainly by negative index changes at different temperatures and power law fittings to the decay (*After:* [78]).

3.9.1.4 "Log time" Model for Gratings Written in Hydrogenated Fiber

Baker et al. [72] have studied gratings written in hydrogen-treated and untreated, B co-doped germanosilicate fibers using a CW source at 244 nm, over several thousand hours. A "log time" model best describes gratings written in hydrogenated fibers. Pre-annealing the gratings at 85°C for 48 hours prior to measurements produced a 20% reduction in grating strength, which is consistent with [71], whereas the measured decay was found not to produce a significant decrease in reflectivity but rather a reduction in the grating bandwidth. When η is plotted against log time for the hydrogen-loaded fiber, the decay slope is constant for all temperatures, with a variable time delay before degradation begins. Therefore, for hydrogenated fiber η can be modified as

$$\begin{aligned} \eta &= 1 - k\log(t/\tau) &&\text{for } t > \tau \\ \eta &= 1 &&\text{for } t \leq \tau \end{aligned} \tag{3.22}$$

with k being constant and t being temperature dependent. This can be fitted to data for hydrogenated fiber with similar mean errors and variance as the power law fit to unloaded fibers. The temperature dependence of τ is approximately linear with $(1/T)$, which allows for lifetime predictions to be made, while following an Arrhenius relationship similar to that for the aging of glass [79].

The log time model also supports the proposal that hydrogen-loaded fiber gratings are associated with a broadening trap population. The rate of change of η with log time is linear after time τ and the gradient is equal for all temperatures, implying that the population of traps (activation energies) is constant over the range of energies explored. This is a special case of E_d where the population of traps approximates to a top hat functional shape. Considering a worst case operating temperature of 80°C for 25 years, the power law model predicts that B-Ge–co-doped fiber will decay to 44% of the dehydrogenated value (46% using the log time variation model), compared with 88% for nonhydrogenated fiber. Hydrogenation has reduced the thermal stability and thermal stabilization is required. A typical requirement for a grating device is an operating lifetime of over 25 years for a temperature range of −40° to +80°C. Using the decay characteristics, a temperature/time combination can be determined to accelerate the aging of the fiber grating. If the grating is held at an accelerated temperature T_2 for a time t_2, this will give the equivalent accelerated decay to that obtained at the generating temperature T_1 for a time t_1. For the power law model,

$$t_2 = \exp\left\{ aT_0\left[\left(\frac{T_1}{T_2}\right) - 1\right] \right\} t_1^{(T_1/T_2)} \tag{3.23}$$

for B-Ge–co-doped fiber, $a = 13.1 \times 10^{-3} \text{K}^{-1}$ and $T_0 = 2941$ K; and for the log time model,

$$t_2 = t_1\left(\frac{\tau_2}{\tau_1}\right) \tag{3.24}$$

The application of the appropriate annealing process can completely negate grating degradation. For example, aging a grating to an equivalent time of 80°C of 2000 years will result in the following degradation of the modulated index change over 25 years:

1. Germanium-doped silica fiber < 0.01%;
2. Boron and germanium co-doped silica fiber < 0.05%;
3. Hydrogenated boron and germanium co-doped silica fiber < 0.3%.

Clearly, accelerating the aging must be carefully applied and tailored to suit the device operating conditions while accounting for what pretreatment the fiber underwent to increase its photosensitivity.

3.9.1.5 Aging Curve Approach

Kannan et al. [69] have also examined the validity of using the power law description for gratings written in hydrogenated fiber via the aging curve approach. A comparison between the two techniques showed that the aging curve offers advantages over the power law technique. In the aging curve approach data obtained at different temperatures are combined to give a composite data curve from which reliability predictions can be made. The extent of accelerated aging on a grating at temperature T and time t may be described by an aging parameter, or equivalently by the demarcation energy E_d. The Erdogan et al. [68] power law analysis of accelerated aging data relies on satisfying the conditions relating A and α to temperature, which is often not the case. Figure 3.33 compares experimentally observed decay to that predicted by the two techniques, showing that the aging curve approach is validated against the power law technique. Any deviation between the power law predictions and measured results, as indicated in Figure 3.33, results from

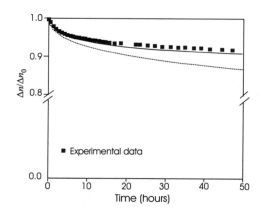

Figure 3.33 Comparison of the aging curve approach with the power law technique. The continuous line represents the aging curve prediction, whereas the broken line corresponds to the power law prediction for a grating at 200°C. The symbols represent the experimentally measured data (*After:* [69]).

assumptions made regarding the initial defect distribution, irrespective of how this may be modified under a stabilizing anneal. Any modification in distribution from a stabilizing anneal would   result in a truncated distribution at which lower energies are no longer assumed by the power law, as illustrated in Figure 3.34. If curve (a) is the energy distribution of an as-written grating, then upon stabilization, the distribution is modified to (d). The power law applied to such a truncated distribution results in an overestimate in the predicted decay, as was found in Figure 3.33. Likewise, underestimates in the predicted decay can arise if the energy distribution of sites in destabilized gratings follows (b) and (c), with the low energy end of the distribution producing more sites than assumed by the power law. The aging curve approach proves to be valid for all the aforementioned cases, regardless of the actual distribution. In [68] a good linear fit of α to T was obtained at >600 K, in contrast to the poor fit at <600K. However, this approach of using high temperatures to obtain linearity can result in the introduction of decay mechanisms that will not be observed in practice. Therefore, cautious interpretation of data that rely on extrapolating the linear α versus T relationship obtained at high temperature to the far lower operating temperature is required. It is best to confirm that α versus T is indeed linear in the realm of the proposed operating temperature before relying on this approach.

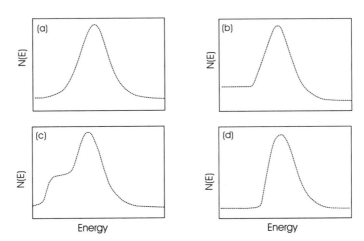

Figure 3.34 Different energy distributions of sites and applicability of the power law to them: (a) the distribution assumed by Erdogan et al. for which the power law is valid; (b) and (c) situations for which the power law could underestimate decay; and (d) the case of truncated distribution, for which the power law overestimates the amount of decay.

3.9.2 Mechanical Strength of Fiber Bragg Gratings

The mechanical resistance of optical fibers has been well characterized with regard to humidity, chemical agents, stripping methods, and fiber splicing. It has proved difficult, however, to arrive at conclusive data for measuring the mechanical degradation of UV-exposed optical fibers. The results in the literature that point to a UV-induced weakening of the glass are contradicted by evidence for gratings written in fiber during the glass drawing stage, just prior to jacketing with a plastic buffer. Any flaws in the optical fiber are the result of external (surface) or internal defects (on a molecular bond scale) that can act as sources of stress fracture, breaking the material bonds and causing catastrophic failure of the device. From Weibull's empirical law, the flaw distribution is given by the following equation [80]:

$$\log\left(\ln\frac{1}{1-F(\sigma)}\right) = m\log\left(\frac{\sigma}{\sigma_0}\right) \tag{3.25}$$

where σ is the stress applied to the fiber during the fatigue test and σ_0 and m are constants characteristic of the material. $F(\sigma)$ is the cumulative failure probability and is defined as the probability of breakage below stress level σ; therefore, $1 - F(\sigma)$ is the probability of survival. For a group of N samples, the cumulative probability of failure is calculated by, $F = (i - 0.5)/N$, with $i = 1, 2, 3,...,N$. Equation 3.25 gives the coordinates of the axes of the Weibull curve. The slope of the curve is called the Weibull shape parameter m, which represents the sharpness of the distribution and is used in lifetime models of optical fibers [81]. An equivalent m-value for pristine fibers is in the range 60–160. An equivalent Weibull plot shows the breaking stress σ against the cumulative failure probability F. The advantage of the former is that it allows the calculation of the m-value, while the latter gives absolute values of the breaking stresses.

Feced et al. [82] measured the mechanical (tensile) strength degradation of UV-exposed optical fibers on exposure to pulsed UV radiation at 193 and 248 nm. Their data led them to conclude that significant statistical differences exist between the mean breaking stress of fibers exposed to 193 nm compared 248 nm, the latter wavelength causing greater damage. Varelas et al. [83] extended this study by comparing the mechanical strength degradation of 22 mol% Ge-doped silica fibers resulting from exposure to pulsed (248 nm) and CW (240 nm) laser sources. In both cases results were obtained for total irradiation doses of 0.5 and 1 kJ/cm². Figure 3.35 shows the Weibull plot for CW and pulsed irradiated fibers, with the Weibull distribution of pristine fiber given as

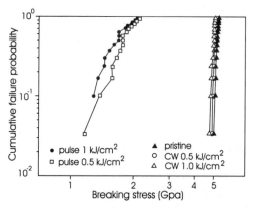

Figure 3.35 Weibull plot comparing CW and pulsed irradiation influence on fiber mechanical resistance (*After:* [83]).

Table 3.1 Comparative Table of Mechanical Performance of Different Irradiation Parameters

	F_{pulse} mJ/cm^2	F_{total} kJ/cm^2	Breaking strength max GPa	median GPa	min GPa	m-value
No irradiation	0.0	0.0	5.18	5.13	4.98	112
CW		0.5	5.15	5.03	4.88	84.6
		1	5.08	4.96	4.79	72.8
Pulsed	150	0.5	2.09	1.76	1.16	7.9
		1	2.09	1.68	1.16	7.1

After: [83].

a reference. The median breaking strengths and m-values are summarized in Table 3.1. The CW case shows a small increase in median breaking stress compared with pristine fiber. In the pulsed case the total dose influence was five times more pronounced for the median breaking strength compared with the CW case. The total dose dependence is also less pronounced for the CW irradiation. The same authors have also examined the median breaking stress of irradiated fibers against total irradiation dose and fluence/pulse (Figure 3.36) [84]. The data are qualitatively the same against total dose regardless of the pulse energy densities and also show a reduction in median breaking stress from 4.7 GPa to < 2.5 Gpa at 1 kJ/cm^2. Above 1.5 kJ/cm^2 the median breaking stress reaches a higher value of 2.7 Gpa that remains almost constant. On increasing the energy density/pulse, the median breaking stress is lowered (error bars equivalent to ± 0.63 GPa). The behavior was explained as follows: the first stage corresponds to an increase in flaws from UV-induced stresses on the fiber surface that increase with the fluence/pulse and total UV dose. To

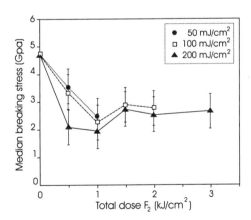

Figure 3.36 Median breaking stress of irradiated fibers against total irradiated dose and fluence per pulse (*After:* [84]).

counter this, the high repetition rate of the irradiation increases the surface temperature, thermally annealing some of the flaws induced in the first stage, and this increases the median breaking stress of the fiber. Equilibrium between the two processes can explain the apparently constant median breaking stress above a dose of 1.5 kJ/cm².

In contrast to [82], Askins et al. [85] have reported, that fiber Bragg gratings written using a KrF excimer laser on the draw tower suffered no degradation in strength as a result of high-intensity laser exposure and exhibited tensile strength comparable to theoretical estimates for silica fibers. Figure 3.37 shows the Weibull plots of the fiber breaking strengths for fiber drawn directly from the tower, compared with the stripped SMF-28 fiber, as reported by Feced et al. [82]. The stripped fiber samples were exposed for 24 hours to 50% relative humidity at 24°C prior to laser exposure or proof testing. The jacketed fiber in Figure 3.37 displayed no indication that exposure to pulsed 248-nm irradiation of 1 J/cm² decreased the tensile fiber strength. The data for UV radiation exposed fiber compare favorably to the unirradiated reference fiber. A comparison of the results for the stripped (solid circles) and jacketed (solid line) demonstrates that the stripped fiber is weakened by the presence of extensive surface flaws. The normalized slope of the Weibull plot reflects the spread of strengths at failure. The low m-values indicate a large range of surface flaw sizes, while high m-values indicate high surface quality and a close approach to the tensile strength of a flaw-free fiber under normal conditions [86]. The results summarized in Table 3.2 indicate that jacketed fibers have an absence of surface flaws, while stripped samples have a degraded surface. It is argued by Askins et al. [85] that the loss in fiber strength arises from surface abrasion and corrosion. For example, a carefully handled bare fiber may fail below 4 GPa in the presence of moisture, particularly after hours at elevated temperatures. That unirradiated, bare fiber had a breaking strength <1 GPa suggests that the degraded surface leads to a lower optical damage threshold. It is also recognized that the threshold for optical damage can be drastically reduced by a slight degradation of the glass surface, and given that light at 193 and 248 nm is absorbed differently, this could account for the wavelength-dependent optical damage presented in [82]. One can also exclude internal compaction as a source of degradation; compaction should instead

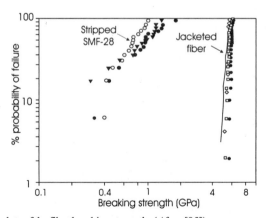

Figure 3.37   Weibull plots of the fiber breaking strengths (*After:* [85]).

Table 3.2 Proof-Testing Statistics for New Results

Sample		Mean breaking strength GPa	Weibull m-value
Jacketed	A, 1 J/cm², 248 nm	6.0	93
	B, unirradiated	5.6	40
	B, 1 J/cm², 248 nm	5.8	89
	SMF-28	5.4	45
	Flaw-free fiber	5.4–5.7	>110
Stripped SMF-28	Unirradiated	0.92	2.4
	1 J/cm², 248 nm	0.69	3.9
	1 J/cm², 193 nm	0.84	2.0

After: [85].

improve low breaking strengths, since at stresses well below pristine strength values, internal tension generally inhibits failure by placing surface flaws in compression (e.g., tempered glass).

Veleras et al. [87] have used a dynamic fatigue model to predict the probability of failure of a Bragg grating written in a section of optical fiber having its buffer coating removed and under a constant applied stress. Bragg gratings were fabricated in hydrogenated and untreated fibers using a frequency-doubled Ar⁺ laser operating at 244 nm in 22 mol% Ge-doped fiber having an acrylate coating. Three different grating groups were examined. First were gratings having 99% reflectivity in hydrogenated fiber with a total irradiation dose of 75 J/cm², producing a total refractive index change of 3.2×10^{-4}. A second group of gratings was produced with the same irradiation dose in non-hydrogenated fiber, producing a reflectivity of ~20% and a total index change of 0.29×10^{-4}. Finally, a third group consisted of 50% reflectivity gratings fabricated in non-hydrogenated fiber with a dose of 680 J/cm² producing a refractive index change of 0.7×10^{-4}. The Weibull plot for this data reveals that CW irradiation does not influence the median breaking strength compared with pristine fiber samples, and in all cases a median breaking strength of 5 GPa is noted, which is comparable to that measured in [85]. Grating inscription, however, does reduce the m-value of the fiber from 112 for the pristine fiber to a value of ~60 after grating inscription. These results agree with their earlier reported data [83] that indicate negligible mechanical degradation when the fiber is homogeneously side exposed to a CW laser. Hydrogenation appears to have no bearing on mechanical degradation. Several models for the calculation of optical fiber lifetime under tension exist. The basic assumption is that stress concentration at crack tips created on the fiber structure during the fiber fabrication process causes fracturing at stress levels lower than the stress that results from SiO tetrahedral bonds. The stress corrosion of silica glass is described by a power law $da/dt = DK_I^n$, where da/dt represents the crack growth, K_I^n is the stress intensity factor, D is the material parameter for the speed of the crack growth, and n is the corrosion susceptibility. The lifetime equation in terms of parameters that can be experimentally extracted by dynamic fatigue tests is given by

$$t_f(L,F,\sigma_\alpha) = \frac{\sigma_f^{n+1}(1)}{n+1}\left(\frac{L_0 F}{L}\right)^{\left(\frac{n+1}{m}\right)}\sigma_\alpha^{-n} \tag{3.26}$$

L_0 is the length at which the Weibull parameters are defined (i.e., the gauge length) and L is the fiber service length. σ_α is a constant applied stress. The n-value and $\sigma_f(1)$ are experimentally determined from tension Weibull plots for different stress rates. Figure 3.38 shows the calculated failure time of a grating with service stress compared with a pristine fiber. To obtain a lifetime of 50 years with a failure probability of 0.001 for $m = 57$ and $n = 14$, one must apply a constant service stress to the grating of no more than 0.96 GPa. This is a value that is 6% less than the constant stress applied to the pristine fiber for the same lifetime.

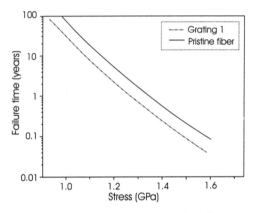

Figure 3.38 Calculated lifetime based on experimental dynamic fatigue parameters of a Bragg grating compared with the lifetime of the pristine fiber (*After:* [87]).

3.10 Incoherent Scattering Loss From Fiber Bragg Gratings

Janos et al. [88] have measured the scattering loss from UV-processed optical fibers and Bragg gratings for wavelengths in the vicinity of the Bragg wavelength. Samples of hydrogenated and untreated optical fibers were single-side exposed to a uniform dose of 193 nm radiation (18 mJ/cm^2/pulse at 10 Hz) in 10 minutes increments from 0 to 100 minutes, in order to determine the effect that UV processing has on the radially scattered pattern from the fiber. In all cases there was no observable change in the level of scattering, indicating that UV irradiation did not change in intrinsic inhomogeneity of the glass matrix that results in Rayleigh scattering. Gratings were written in B–co-doped, germanosilicate fiber using 244-nm CW radiation from a frequency-doubled Ar$^+$ ion (fluence 500 mW/mm^2). Gratings were also written in the same fiber type using pulsed 193-nm radiation (fluence 18 mJ/cm^2/pulse at 10 Hz) and finally in highly doped Er^{3+}/Yb^{3+} phosphosilicate fiber with 193-nm radiation. In all three cases gratings were written in hydrogenated fibers with reflectivity greater than 97%. Incoherent scattering losses from gratings were found to vary in the range of 5×10^{-5} dB/cm to 3×10^{-4} dB/cm for the B–co-doped germanosilicate fiber with grating inscription via CW light. Losses of 5×10^{-3} to 5×10^{-2} dB/cm for the same fiber type exposed to pulsed light and finally 0.2 dB/cm for gratings in Er^{3+}/Yb^{3+} phosphosilicate fiber. This high loss would affect the performance of any resonant device formed in this fiber. Polar plots comparing the radially scattered light for the B–co-doped germanosilicate glass exposed to pulsed UV light is shown in Figure 3.39(a); that for the Er^{3+}/Yb^{3+} phosphosilicate fiber is shown in Figure 3.39(b) (both data show best fits). The measurements reveal two significant differences between scattered light measured from gratings and from unprocessed fibers. First, in Figure 3.39(a), there is an 8- to 20-dB increase in the amount of side-scattered light detected compared with an unprocessed or uniformly processed fiber. Second, there is significant UV-induced asymmetry with clear lobe structures on opposite sides of the radial profile. Figure 3.39(b) shows that the profile of the scattered light can display significant differences depending on fiber type and processing conditions, however, the rotational symmetry is observed in all cases. This indicates that even though the grating elements are aligned symmetrically with respect to the plane of the incident light, they are not uniform across the fiber core. The spectral dependence of the scattered light exhibits a large background loss, in comparison with the loss associated with coupling to cladding modes at wavelengths shorter than the Bragg resonance. This suggests that the measured light is incoherent scattering from the ensemble of periodic grating elements, rather than collective diffraction from the grating elements that would appear to be more wavelength selective. The intensity of the scattered light is attributed to the finite size of the grating elements with respect to the wavelength of the incident light. The data in Figure 3.39 are consistent with the findings of Innis et al. [89], which suggest that the UV-affected area is confined to a crescent-shaped region at the core/cladding interface. The radial profile of the scattered light may be modeled as a superposition of scattered light from an infinite number of cylindrical scatterers. The cylinders are modeled by a Gaussian distribution with mean radius a and mean length L in order to account for random variations in the dimensions of the scatterers. The same distribution also accounts for the possible angular positions of the axis of the cylinder. The model gives a qualitative estimate of the mean dimensions of the scatterers and, therefore, the uniformity of the UV processing across the core. The amplitude of the scattered wave in the plane containing the axis of the cylinder is approximately [88]

$$S(\theta) = \frac{1}{\pi} \frac{x \sin[x \sin(\theta)]}{x \sin(\theta)} \frac{xR \sin[xR \cos(\theta)]}{xR \cos(\theta)} \tag{3.27}$$

θ is the angle of scattering, with zero being in the direction of then incident UV irradiation. $R = L/2a$ is the ratio of length to diameter of the scatterer, and $x = (2\pi n/\lambda)a$, where λ is the wavelength of the incident wave. The scatterer length L equals the core diameter, and a and θ are chosen to give a best fit to the scattering profile. It is found that scattering center dimensions are much smaller than the core diameter. The elongated scatterer shape accounts for the observed preferential scattering if light is perpendicular to the direction of UV irradiation. In general, the anisotropy observed in the radially scattered light is consistent with the scattering elements being confined to within a few microns of the core/cladding interface, as per the results of Innis et al. [89].

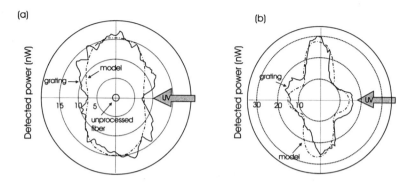

Figure 3.39 (a) Polar plot comparing the radially scattered light from a grating (written in boron-co-doped germanosilicate fiber by pulsed 193-nm irradiation) with that from an unprocessed fiber; (b) as i n (a) for a grating written in highly doped Er^{3+}/Yb^{3+} phosphosilicate fiber. The insets show the mean shape of the scatterer's distribution, giving the best fit to the scattered light profile (*After:* [88]).

3.11 Long Period Gratings

Long period gratings are special forms of the conventional Bragg grating that were first demonstrated by Vengsarkar et al. [90]. They can be used as in-fiber, low-loss, band-rejection filters. Their nature is similar to the grating mode converters (LP_{01} to LP_{11} [91] and LP_{01} to LP_{02} [92]) and rocking filters that have been written in single-mode fibers for polarization mode conversion [22, 93]. In this case, the periodicity of a long period grating is chosen to couple light from the guided fundamental mode of the fiber into the forward propagating cladding modes, where it is lost due to absorption and scattering. Therefore, the transmission spectrum is measured. Coupling from the guided to unguided modes is wavelength dependent; thus, one obtains a spectrally selective loss. The phase-matching condition that dictates the coupling of one mode to another determines the exact periodicity to be used. For two forward-propagating modes, the difference between the propagation constants is very small, leading to the requirement that the grating periodicity be long.

Fiber gratings satisfy the Bragg phase-matching condition between the guided and cladding or radiation modes, or another guided mode. This wavelength-dependent phase-matching condition is given by

$$\beta_{01} - \beta = \Delta\beta = \frac{2\pi}{\Lambda} \tag{3.28}$$

where Λ is the periodicity of the grating, β_{01} and β are the propagation constant of the fundamental guided mode and the mode to which coupling occurs, respectively. For conventional fiber Bragg gratings the coupling of the forward-propagating LP_{01} mode occurs to the reverse propagating mode ($\beta = -\beta_{01}$). In this case $\Delta\beta$ is large, thus the grating periodicity is small, typically of the order of 1 μm. Unblazed long period gratings couple the fundamental mode to the discrete and circularly symmetric forward-propagating cladding modes, resulting in smaller values of $\Delta\beta$ and hence periodicities ranging in hundreds of micrometers. The cladding modes attenuate rapidly as they propagate along the length of the fiber due to the lossy cladding coating interface and bends in the fiber. Since $\Delta\beta$ is discrete and a function of the wavelength, this coupling to the cladding modes is highly selective, leading to a wavelength-dependent loss (Figure 3.40(a)). As a result, any modulation of the core and cladding guiding properties modifies the spectral response of long period gratings. This phenomenon can be utilized for sensing purposes. Furthermore, since the cladding modes interact with the fiber jacket or any other material surrounding the cladding, changes in the properties of this ambient material can also be detected.

Long period fiber gratings are affected by external perturbations such as strain and temperature. The effect is primarily due to a differential change induced in the two modes. More specifically, since the propagation constants β_{01} and β undergo dissimilar changes owing to a change in the external conditions, the difference between the two modes $\Delta\beta$ is altered, implying a shift in the wavelength of resonant coupling between the two modes. The variations in peak wavelengths with strain for two different fibers are plotted in Figure 3.40(b). Grating A exhibits a gradient of -0.7 nm/mε, whereas grating B has a gradient of $+1.5$ nm/mε. It is apparent that with the appropriate choice of core-cladding parameters one can reduce the strain sensitivity to zero. Applications of long period gratings will be discussed in Chapters 6 and 7.

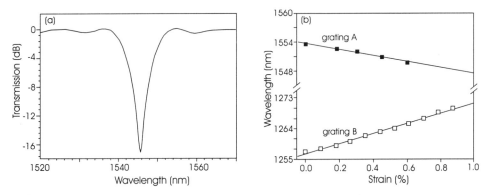

Figure 3.40   (a) Transmission spectrum of a long period grating, (b) Effect of strain on the peak wavelength of two different long period gratings (*After*:[90]).

References

[1]   Lam, D. K. W., and B. K. Garside, "Characterization of single-mode optical fiber filters," *Applied Optics*, Vol. 20, 1981, pp. 440–445.

[2]   Russell St. J., P., J. L. Archambault, and L. Reekie, "Fibre gratings," *Physics World*, October 1993, pp. 41–46.

[3]   Meltz, G., and W. W. Morey, "Bragg grating formation and germanosilicate fiber photosensitivity," International Workshop on Photoinduced Self–Organization Effects in Optical Fiber, Quebec City, Quebec, May 10–11, *Proceedings SPIE*, Vol. 1516, 1991, pp. 185–199.

[4]   Hill, K. O., and G. Meltz, "Fiber Bragg grating technology fundamentals and overview," *IEEE Journal of Lightwave Technology*, Vol. 15, 1997, pp. 1263–1276.

[5]   Delevaque, E., et al. "Optical fiber design for strong gratings photoimprinting with radiation mode suppression," Conference on Optical Fiber Communication, 1995, Postdeadline paper PD5.

[6]   Komukai, T., and M. Nakazawa, "Efficient fiber gratings formed on high NA dispersion-shifted fiber and dispersion-flattened fiber," *Japanese Journal of Applied Physics,* Vol. 34, 1995, pp. L1286–L1287.

[7]   Dong, L., et al. "Optical fibers with depressed claddings for suppression of coupling into cladding modes in fiber Bragg gratings," *IEEE Photonics Technology Letters*, Vol. 9, 1997, pp. 64–66.

[8]   Haggans, C. W., et al. "Narrow-depressed cladding fiber design for minimization of cladding mode losses in azimuthally asymmetric fiber Bragg gratings," *IEEE Journal of Lightwave Technology*, Vol. 16, 1998, pp. 902–909.

[9]   Albert, J., et al. "Apodisation of the spectral response of fibre Bragg gratings using a phase mask with variable diffraction efficiency," *Electronics Letters*, Vol. 31, 1995, pp. 222–223.

[10]   Malo, B., et al. "Apodised in-fibre Bragg grating reflectors photoimprinted using a phase mask," *Electronics Letters*, Vol. 31, 1995, pp. 223–225.

[11]   Kashyap, R., A. Swanton, and D. J. Armes, "Simple technique for apodising chirped and unchirped fibre Bragg gratings," *Electronics Letters*, Vol. 32, 1996, pp. 1226–1228.

[12]   Albert, J., et al. "Moire phase masks for automatic pure apodisation of fibre Bragg gratings," *Electronics Letters*, Vol. 32, 1996, pp. 2260–2261.

[13]   Kersey, A. D., et al. "Fiber grating sensors," *IEEE Journal of Lightwave Technology*, Vol. 15, 1997, pp. 1442–1463.

[14]   Ball, G. A., W. W. Morey, and W. H. Glenn, "Standing-wave monomode erbium fiber laser," *IEEE Photonics Technology Letters*, Vol. 3, 1991, pp. 613–615.

[15]   Hillmer, H., et al. "Novel tunable semiconductor lasers using continuously chirped distributed feedbackgratings with ultrahigh spatial precision," *Applied Physics Letters*, Vol. 65, 1994, pp. 2130–2132.

[16]   Ball, B. A., and W. W. Morey, "Continuously tunable single-mode erbium fiber laser," *Optics Letters*, Vol. 17, 1992, pp. 420–422.

[17]   Othonos, A., et al. "Fiber Bragg grating laser sensor," *Optical Engineering*, Vol. 32, 1993, pp. 2841–2846.

[18]   Ball, G. A., W. W. Morey, and P. K. Cheo, "Fiber laser source/analyzer for Bragg grating sensor array interrogation," *IEEE Journal of Lightwave Technology,* Vol. 12, 1994, pp. 700–703.

[19]   Alavie, A. T., et al. "A multiplexed Bragg grating fiber laser sensor system," *IEEE Photonics Technology Letters*, Vol. 5, 1993, pp. 1112–1114.

[20]   Meltz, G., and W. W. Morey, Conference on Optical Fiber Communication, 1991, Paper TuM2.

[21]   Kashyap, R., R. Wyatt, and P. F. McKee, "Wavelength flattened saturated erbium amplifier using multiple side-tap Bragg gratings," *Electronics Letters*, Vol. 29, 1993, pp. 1025–1026.

[22]   Hill, K. O., et al. "Birefringent photosensitivity in monomode optical fiber: Application to external writing of rocking filters," *Electronics Letters*, Vol. 27, 1991, pp. 1548–1550.

[23]   Ouellette, F., "Dispersion cancellation using linearly chirped Bragg grating filters in optical waveguides," *Optics Letters*, Vol. 12, 1987, pp. 847–849.

[24]   Brady, G. P., et al. "Extended range, coherence tuned, dual wavelength interferometry using a superfluorescent fibre source and chirped fibre Bragg gratings," *Optics Communications*, Vol. 134, 1997, pp. 341–346.

[25]   Byron, K. C., et al. "Fabrication of chirped Bragg gratings in photosensitive fibre," *Electronics Letters*, Vol. 29, 1993, pp. 1659–1660.

[26]   Sugden, K., et al. "Chirped gratings produced in photosensitive optical fibers by fibre deformation

during exposure," *Electronics Letters*, Vol. 30, 1994, pp.440–442.

[27]   Kashyap, R., et al. "Novel method of producing all fibre photoinduced chirped gratings," *Electronics Letters*, Vol. 30, 1994, pp. 996–997.

[28]   Sugden, K., et al. "Fabrication and characterization of bandpass filters based on concatenated chirped fiber gratings," *IEEE Journal of Lightwave Technology*, Vol. 15, 1997, pp. 1424–1432.

[29]   Kashyap, R., et al. "Super-step-chirped fibre Bragg gratings," *Electronics Letters*, Vol. 32, 1996, pp. 1394–1396.

[30]   Kashyap, R., et al. "1.3 m long super-step-chirped fibre Bragg grating with a continuous delay of 13.5 ns and bandwidth 10 nm for broadband dispersion compensation," *Electronics Letters*, Vol. 32, 1996, pp. 1807–1809.

[31]   Farries, M. C, C. M. Ragdale, and D. C. J. Reid, "Broadband chirped fibre Bragg filters for pump rejection and recycling in erbium doped fibre amplifiers," *Electronics Letters*, Vol. 28, 1992, pp. 487–489.

[32]   Patrick, H., and S. L. Gilbert, "Growth of Bragg gratings produced by continuous-wave ultraviolet light in optical fiber," *Optics Letters*, Vol. 18, 1993, pp. 1484–1486.

[33]   Riant, I., and F. Haller, "Study of the photosensitivity at 193 nm and comparison with photosensitivity at 240nm influence of fiber tension: Type IIA aging," *IEEE Journal of Lightwave Technology*, Vol. 15, 1997, pp. 1464–1469.

[34]   Xie, W. X., et al. "Experimental evidence of two types of photorefractive effects occurring during photoinscriptions of Bragg gratings written within germanosilicate fibres," *Optics Communications,* Vol. 104, 1993, pp. 185–195.

[35]   Archambault, J. L., L. Reekie, and P. St. J. Russell, "100% reflectivity Bragg reflectors produced in optical fibres by single excimer laser pulses," *Electronics Letters*, Vol. 29, 1993, pp. 453–455.

[36]   St. J. Russell, P., and R. Ulrich, "Grating fiber-coupler as a high-resolution spectrometer," *Optics Letter*, Vol. 10, 1985, pp. 291–293.

[37]   Askins, C. G., et al. "Fiber Bragg reflectors prepared by a single excimer pulse," *Optics Letters*, Vol. 17, 1992, pp. 833–835.

[38]   Dong, L., et al. "Single pulse Bragg gratings written during fibre drawing," *Electronics Letters*, Vol. 29, 1993, pp. 1577–1578.

[39]   Othonos, A., X. Lee, and R. M. Measures, "Superimposed multiple Bragg gratings," *Electronics Letters*, Vol. 30, 1994, pp. 1972–1973.

[40]   Eggleton, B. J., et al. "Long periodic superstructure Bragg gratings in optical fibres," *Electronics Letters*, Vol. 30, 1994, pp. 1620–1622.

[41]   Morey, W. W., G. A. Ball, and G. Meltz, "Photoinduced Bragg gratings in optical fibers," *Optics and Photonics News*, Optical Society of America, February, 1994, pp. 814.

[42]   Agrawal, G. P, and N. K. Dutta, *Semiconductor Lasers,* New York: Van Nostrand Reinhold, 1993, Chapter 7.

[43]   Haus, H. A., and C. V. Shank, *IEEE Journal of Quantum Electronics*, Vol. QE-12, 1976, p. 352.

[44]   Alferness, R. C., et al. "Narrowband grating resonator filters in InGaAsP/InP waveguides," *Applied Physics Letters*, Vol. 49, 1986, pp. 125–127.

[45]   Kashyap, R., P. F. McKee, and D. Armes, "UV written reflection grating structures in photosensitive optical fibres using phase-shifted phase masks," *Electronics Letters*, Vol. 30, 1994, pp. 1977–1978.

[46]   Canning, J., and M. G. Sceats, "Pi-phase-shifted periodic distributed structures in optical fibres by UV post-processing," *Electronics Letters*, Vol. 30, 1994, pp. 1344–1345.

[47]   Uttamchandani, D., and A. Othonos, "Phase-shifted Bragg gratings formed in optical fibres by post-fabrication thermal processing," *Optics Communications*, Vol. 127, 1996, pp. 200–204.

[48]   Taverner, D., et al. "Experimental investigation of picosecond pulse reflection from fiber gratings," *Optics Letters*, Vol. 20, 1995, pp. 282–284.

[49]   Chen, L.R., et al. "Ultrashort pulse reflection from fiber gratings: A numerical investigation," *IEEE Journal of Lightwave Technology*, Vol. 15, 1997, pp. 1503–1512.

[50]   Sipe, J. E., L. Poladian, and C. Martijn de Sterke, "Propagation through nonuniform grating

structures," *Journal of the Optical Society of America A*, Vol. 11,1994, pp. 1307–1320.

[51]   Chen, L. R., et al. "Ultrashort pulse propagation through fiber gratings: Theory and experiments," Conference on Bragg Gratings, Photosensitivity, and Poling in Glass Fibers and Waveguides: Applications and Fundamentals, Technical Digest Series (Optical Society of America, Washington, DC), Vol. 17, 1997, pp. 117–119.

[52]   Bloembergen, N., *Nonlinear Optics*, Reading, MA: Benjamin, 1977, Chapter 1.

[53]   Boyd, R. W., *Nonlinear Optics*, New York: Academic Press, 1992.

[54]   Agrawal, G. P., *Nonlinear Fiber Optics*, New York: Academic Press, 1989.

[55]   Hasagawa, A., and F. Tappert, "Optical solitons," *Applied Physics Letters*, Vol. 23, 1973, pp. 142–144.

[56]   Mollenauer, L. F., R. H. Stolen, and J. P. Gordon, "Experimental observation of picosecond pulse narrowing and solitons in optical fibers," *Physical Review Letters*, Vol. 45, 1980, pp. 1095–1098.

[57]   Aitchison, J. S., et al. "Observation of spatial optical solitons in a nonlinear glass waveguide," *Optics Letters*, Vol. 15, 1990, pp. 471–473.

[58]   Winful, H. G., J. H. Marburger, and E. Garmire, "Theory of bistability in nonlinear distributed feedback structures," *Applied Physics Letters*, Vol. 35, 1979, pp. 379–381.

[59]   St. J. Russell, P., "Bloch wave analysis of dispersion and pulse propagation in pure distributed feedback structures," *Journal of Modern Optics*, Vol. 38, 1991, pp. 1599–1619.

[60]   Eggleton, B. J., et al. "Dispersion compensation over 100 km at 10 Gbit/s using a fiber grating in transmission," *Electronics Letters*, Vol. 32, 1996, pp. 1610–1611.

[61]   Martijn de Sterke, C., B. J. Eggleton, and P. A. Krug, "High-intensity pulse propagation in uniform gratings and grating superstructures," *IEEE Journal of Lightwave Technology*, Vol. 15, 1997, pp. 1494–1502.

[62]   Yablonovitch, E., "Inhibited spontaneous emission in solid state physics and electronics," *Physical Review Letters*, Vol. 58, 1987, pp. 2059–2062.

[63]   Winful, H. G., "Pulse compression in optical fiber filters," *Applied Physics Letters*, Vol. 46, 1985, pp. 527–529.

[64]   Eggleton, B. J., et al. "Distributed feedback pulse generator based on nonlinear fiber grating," *Electronics Letters*, Vol. 32, 1996, pp. 2341–2342.

[65]   Winful, H. G., and G. D. Cooperman, "Self-pulsing and chaos on distributed feedback bistable optical devices," *Applied Physics Letters*, Vol. 40, 1982, pp. 298–300.

[66]   Martijn de Sterke, C., and J. E. Sipe, "Switching dynamics of finite periodic nonlinear media: A numerical study," *Physical Review A*, Vol. 42, 1990, pp. 2858–2869.

[67]   Martijn de Sterke, C., "Stability analysis of nonlinear periodic media," *Physical Review A*, Vol. 45, 1992, pp. 8252–8258.

[68]   Erdogan, T., et al., "Decay of ultraviolet-induced fiber Bragg gratings," *Journal of Applied Physics*, Vol. 76, 1994, pp. 73–80.

[69]   Kannan, S., J. Z. Y. Guo, and P. J. Lemaire, "Thermal stability analysis of UV-induced fiber Bragg gratings," *IEEE Journal of Lightwave Technology*, Vol. 15, 1997, pp. 1478–1483.

[70]   Poirier, M., et al. "Dynamic and orientational behavior of UV-induced luminescence bleaching in Ge-doped silica optical fiber," *Optics Letters*, Vol. 18, 1993, pp. 870–872.

[71]   Patrick, H., et al. "Annealing of Bragg gratings in hydrogen-loaded optical fiber," *Journal of Applied Physics*, Vol. 78, 1995, pp. 2940–2945.

[72]   Baker, S. R., et al. "Thermal decay of fiber Bragg gratings written in boron and germanium co-doped silica fiber," *IEEE Journal of Lightwave Technology*, Vol. 15, 1997, pp. 1470–1477.

[73]   Egan, R. J., et al. "Effects of hydrogen loading and grating strength on the thermal stability of fiber Bragg gratings," Conference on Optical Fiber Communication, 1996, Paper TuO3.

[74]   Atkins, R. M., and V. Mizrahi, "Observations of changes in UV absorption bands of single mode germanosilicate core optical fibres on writing and thermally erasing refractive index gratings," *Electronics Letters*, Vol. 28, 1992, pp. 1743–1744.

[75]   Starodubov, D. S., et al. "Bragg grating fabrication in germanosilicate fibers by use of near-UV light:

A new pathway for refractive-index changes," *Optics Letters*, Vol. 22, 1997, pp. 1086–1088.

[76] Atkins, G. R., et al. "Control of defects in optical fibers: A study using cathodoluminescence spectroscopy," *IEEE Journal of Lightwave Technology*, Vol. 11, 1993, pp. 1793–1801.

[77] Tsai, T. E., G. M. Williams, and E. J. Friebele, "Index structure of fiber Bragg gratings in Ge-SiO$_2$ fibers," *Optics Letters*, Vol. 22, 1997, pp. 224–226.

[78] Dong, L., and W. F. Liu, "Thermal decay of fiber Bragg gratings of positive and negative index changes formed at 193nm in a boron–co-doped germanosilicate fiber," *Applied Optics*, Vol. 36, 1997, pp. 8222–8226.

[79] Camibel, I., D. A. Pinnow, and F. W. Dabby, "Optical aging characteristics of borosilicate clad fused silica core fiber optical waveguides," *Applied Physics Letters*, Vol. 26, 1975, pp. 185–187.

[80] Olshansky, R., and D. R. Maurer, "Tensile strength and fatigue of optical fibers," *Journal of Applied Physics*, Vol. 47, 1976, pp. 4497–4499.

[81] Mitsunaga, Y., Y. Katsuyama, and Y. Ishida, "Reliability assurance for long-length optical fibre based on proof testing," *Electronics Letters*, Vol. 17, 1981, pp. 567–568.

[82] Feced, R., et al. "Mechanical strength degradation of UV-exposed optical fibres," *Electronics Letters*, Vol. 33, 1997, pp. 157–159.

[83] Varelas, D., H. G. Limberger, and R. P. Salathe, "Enhanced mechanical performance of single-mode optical fibres irradiated by a CW UV laser," *Electronics Letters*, Vol. 33, 1997, pp. 704–705.

[84] Varelas, D., et al. "UV-induced mechanical degradation of optical fibres," *Electronics Letters*, Vol. 33, 1997, pp. 804–806.

[85] Askins, C. G., et al. "Fibre strength unaffected by on-line writing of single-pulse Bragg gratings," *Electronics Letters*, Vol. 33, 1997, pp. 1333–1334.

[86] Kurkjian, C. R., and U. C. Paek, "Single valued strength of 'perfect' silica fibres," *Applied Physics Letters*, Vol. 42, 1983, pp. 251–253.

[87] Varelas, D., et al. "Fabrication of high-mechanical-resistance Bragg gratings in single-mode optical fibers with continuous-wave ultraviolet laser side exposure," *Optics Letters*, Vol. 23, 1998, pp. 397–399.

[88] Janos, M., J. Canning, and M. G. Sceats, "Incoherent scattering losses in optical fiber Bragg gratings," *Optics Letters*, Vol. 21, 1996, pp. 1827–1829.

[89] Inniss, D., et al. "Atomic force microscopy study of UV-induced anisotropy in hydrogen-loaded germanosilicate fibers," *Applied Physics Letters*, Vol. 65, 1994, pp. 1528–1530.

[90] Vengsarkar, A.M., et al. "Long-period fiber gratings as band-rejection filters," *IEEE Journal of Lightwave Technology*, Vol. 14, 1996, pp. 58–65.

[91] Hill, K. O., et al. "Efficient mode-conversion in telecommunication fiber using externally written gratings," *Electronics Letters*, Vol. 26, 1990, pp. 1270–1272.

[92] Bilodeau, F., et al. "Efficient narrowband LP$_{01}$ LP$_{02}$ mode convertors fabricated in photosensitive fiber: Spectral response," *Electronics Letters*, Vol. 27, 1991, pp. 682–684.

[93] Johnson, D. C. et al. "Long-length long-period rocking filters fabricated from conventional monomode telecommunications optical fiber," *Optics Letters*, Vol. 17, 1992, pp. 1635–1637.

Chapter 4

INSCRIBING BRAGG GRATINGS IN OPTICAL FIBERS

This chapter describes the various techniques used in fabricating standard and complex Bragg grating structures in optical fibers. The objective here is to give the reader a detailed outlook on the technology for inscribing Bragg grating structures. Bragg gratings may be classified as internally or externally written, depending on the fabrication technique employed. Internally inscribed Bragg gratings are not very useful, nevertheless it is important to consider them, if only to provide a complete historical perspective. Far more useful Bragg gratings are inscribed using external techniques such as the interferometric, point-by-point, and phase-mask techniques which overcome the fundamental limitation of internally written gratings. Although these processes were initially considered difficult due to the requirements of submicron resolution and thus stability, today they are well controlled and the inscription of Bragg gratings using these techniques is considered routine. Note that several of the approaches considered here are used to fabricate long period gratings.

4.1 Internal Inscription of Bragg Gratings

The internal writing technique, first demonstrated by Hill and co-workers in 1978 [1, 2], requires the use of single-frequency laser light whose two-photon absorption lies in the UV photosensitivity region of the fiber which, initiates the change in the index of refraction. This technique is simple and the experimental requirements are minimal. These gratings, however, are limited to operating at a Bragg wavelength coinciding with the excitation laser wavelength. A typical experimental setup is shown in Figure 4.1. An argon ion laser is used as the source, oscillating on a single longitudinal mode at 514.5 nm (or 488 nm) and exposing the photosensitive fiber by coupling light into its core. Isolation of the argon ion laser from the back-reflected beam is necessary to avoid instability in the pump laser, and the fiber is usually placed in a tube for thermal isolation. The incident laser light interferes with the Fresnel reflection (approximately 4% from the cleaved end of the fiber) to initially form a weak standing wave intensity pattern within the core of the fiber. The high-intensity points alter the index of refraction in the photosensitive fiber permanently. Thus, a refractive index perturbation having the same spatial periodicity as the interference pattern is formed. Due to the small index of refraction change induced via this method, useful reflectivity may only be achieved for gratings having a long length (a few tens of centimeters).

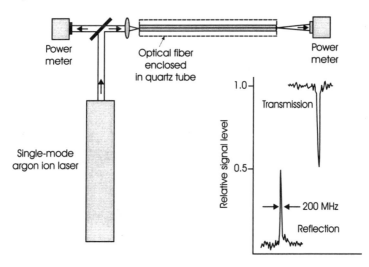

Figure 4.1   A schematic of typical apparatus used for generating self-induced Bragg gratings. The graph shows typical reflection and transmission characteristics of the self-induced gratings.

4.2 Interferometric Fabrication Technique

Meltz and co-workers [3] were the first to demonstrate the interferometric fabrication technique, which is an external writing approach for inscribing Bragg gratings in photosensitive fibers. They utilized an interferometer that split the incoming UV light into two beams that were subsequently recombined to form an interference pattern that side-exposed a photosensitive fiber, inducing a permanent refractive index modulation in the core. Bragg gratings in optical fibers have been fabricated using two types of interferometers, the amplitude-splitting and the wavefront-splitting interferometers.

4.2.1 Amplitude-Splitting Interferometer

Meltz et al. [3] used an amplitude-splitting interferometer to fabricate fiber Bragg gratings, the experimental arrangement of which is shown in Figure 4.2. A tunable excimer-pumped dye laser operating at a wavelength in the range of 486–500 nm was frequency-doubled using a nonlinear crystal and provided a UV source in the 244-nm band with adequate coherence length. The UV radiation was split into two beams of equal intensity that were recombined to produce an interference pattern, normal to the fiber axis. A pair of cylindrical lenses focused the light onto the fiber and the resulting focal line was approximately 4 mm long by 124 μm wide. A broadband source was also used in conjunction with a high-resolution monochromator to monitor the reflection and transmission spectra of the grating. The graph in Figure 4.2 shows the reflection and complementary transmission spectra of the grating formed in a 2.6-μm diameter core, 6.6 mol % GeO_2-doped fiber after 5 minutes of exposure to a 244-nm interference pattern with an average power of 18.5 mW. The length of the exposed region was estimated to be between 4.2 and 4.6 mm.

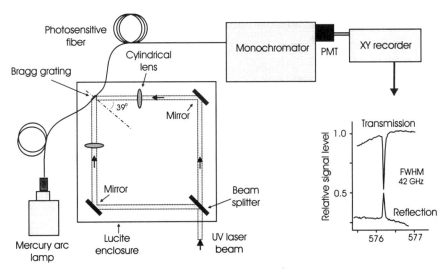

Figure 4.2   A schematic of an amplitude-splitting interferometer used by Meltz et al. [3], which demonstrated the first externally fabricated Bragg grating. Transmission and reflection spectra for a 4.4 mm long Bragg grating fabricated with this apparatus is also shown.

A further improvement to the amplitude-splitting interferometer is shown in Figure 4.3(b). In the conventional interferometer of Figures 4.2 and 4.3(a) the UV writing laser light is split into equal intensity beams that subsequently recombine after having undergone a different number of reflections in each optical path. Therefore, the interfering beams (wavefronts) acquire different (lateral) orientations. This results in a low-quality fringe pattern for laser beams having low spatial coherence. This problem is eliminated with the interferometer shown in Figure 4.3(b), which compensates for the beam splitter reflection by including a second mirror in one of the optical paths. The total number of reflections is now the same in both optical arms, thus ensuring that the two interfering beams at the fiber are identical. A cylindrical lens is normally placed outside the interferometer to focus the interfering beams to a fine line matching the fiber core; this results in higher intensities at the core, thereby improving the grating inscription. The Bragg grating period, Λ, which is identical to the period of the interference fringe pattern, depends on both the irradiation wavelength, λ_w, and the half angle between the intersecting UV beams, φ (Figure 4.3). The period of the grating is given by

$$\Lambda = \frac{\lambda_{\mathrm{w}}}{2 \sin \varphi} \tag{4.1}$$

where λ_{w} is the UV wavelength and φ is the half angle between the intersection UV beams (Figure 4.3). Given the Bragg condition $\lambda_{\mathrm{B}} = 2 n_{\mathrm{eff}} \Lambda$, the Bragg resonance wavelength λ_{B} can be represented in terms of the UV writing wavelength and the half angle between intersecting UV beams as

$$\lambda_{\mathrm{B}} = \frac{n_{\mathrm{eff}} \lambda_{\mathrm{w}}}{\sin \varphi} \tag{4.2}$$

where n_{eff} is the effective core index. From (4.2) one can easily see that the Bragg grating wavelength can be varied either by changing λ_{w} [4] and/or φ. The choice of λ_{w} is limited to the UV photosensitivity region of the fiber, however there is no restriction for the choice of the angle φ. One of the advantages of the interferometric method is the ability to introduce optical components within the arms of the interferometer, allowing for the wavefronts of the interfering beams to be modified. In practice, incorporating one or more cylindrical lenses into one or both arms of the interferometer produces chirped gratings with a wide parameter range [5] (Section 4.7.4). The most important advantage offered by the amplitude-splitting interferometric technique is the ability to inscribe Bragg gratings at any wavelength desired. This is accomplished by changing the intersecting angle between the UV beams. This method also offers complete flexibility for producing gratings of various lengths, which allows the fabrication of wavelength narrowed or broadened gratings.

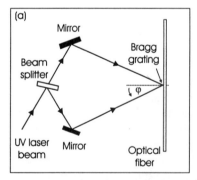 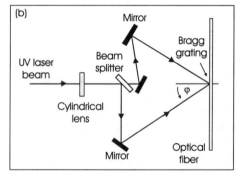

Figure 4.3 (a) A schematic of a general amplitude-splitting interferometer utilized in the fabrication of Bragg gratings in optical fibers. (b) An improved version of the amplitude-splitting interferometer, where an additional extra mirror is used to achieve an equal number of reflections, thus eliminating the different lateral orientations of the interfering beams. This type of interferometer is applicable to fabrication systems where the source spatial coherence is low, such as with excimer lasers.

The main disadvantage of this approach is a susceptibility to mechanical vibrations. Submicron displacements in the position of mirrors, beam splitter, or other optical mounts in the interferometer during UV irradiation will cause the fringe pattern to drift, washing out the grating from the fiber. Furthermore, because the laser light travels long optical path lengths, air currents, which affect the refractive index locally, can become problematic and degrade the formation of a stable fringe pattern. In addition to the above shortcomings, quality gratings can only be produced with a laser source that has good spatial and temporal coherence and excellent wavelength and output power stability.

4.2.2 Wavefront-Splitting Interferometers

Wavefront-splitting interferometers are not as popular as the amplitude-splitting interferometers for grating fabrication; however, they offer some useful advantages. Two examples of wavefront-splitting interferometers used to fabricate Bragg gratings in optical fibers are the prism interferometer [6, 7] and the Lloyd interferometer [8]. A schematic of the prism interferometer is shown in Figure 4.4(a). The prism is made from high-homogeneity, ultraviolet-grade, fused silica, which allows for good transmission characteristics. In this setup the UV beam is spatially bisected by the prism edge, and half the beam is spatially reversed by total internal reflection from the prism face. The two beam halves are then recombined at the output face of the prism, giving a fringe pattern parallel to the photosensitive fiber core. A cylindrical lens placed just before the setup helps in forming the interference pattern on a line along the fiber core. The interferometer is intrinsically stable because the path difference is generated within the prism and remains unaffected by vibrations. Writing times of over 8 hours have been reported with this type of interferometer. One disadvantage of this system is the geometry of the interference. The interferogram is formed by folding the beam onto itself; hence, different parts of the beam must interfere, thus requiring a UV source with good spatial coherence.

The experimental setup for fabricating gratings with the Lloyd interferometer is shown in Figure 4.4(b). This interferometer consists of a dielectric mirror, which directs half of the UV beam to a fiber that is perpendicular to the mirror. The writing beam is centered at the intersection of the mirror surface and fiber. The overlap of the direct and deviated portions of UV beam creates interference fringes normal to the fiber axis. A cylindrical lens is usually placed in front of the system to focus the fringe pattern along the core of the fiber. As in the case of the previous interferometer, the interferogram is formed by dividing the beam and folding it onto itself, thus requiring different parts of the beam to interfere. Therefore, the UV source must have good spatial coherence.

A key advantage of the wavefront-splitting interferometer is the requirement for only one optical component, which greatly reduces its sensitivity to mechanical vibrations. In addition, the short distance for which the UV beams are separated reduces the wavefront distortion induced by air currents and temperature differences between the two interfering beams. Furthermore, this assembly can be easily rotated to vary the angle of intersection of the two beams for wavelength tuning. One disadvantage of this system is the limitation on the grating length, which is restricted to half of the beam width. Another disadvantage is the range of Bragg wavelength tunability, which is restricted by the physical arrangement of the interferometers. As the intersection angle increases, the difference between beam path lengths increases; therefore, the beam coherence length limits the Bragg wavelength tunability.

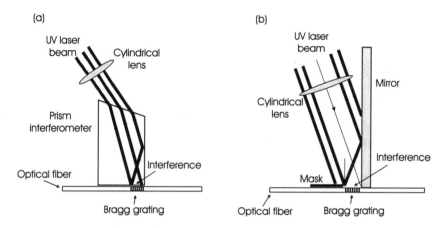

Figure 4.4   Schematics of wavefront-splitting interferometers used for Bragg grating fabrication: (a) prism interferometer, and (b) Lloyd interferometer.

4.3 Phase-Mask Technique

One of the most effective methods for inscribing Bragg gratings in photosensitive fiber is the phase-mask technique [9, 10]. This method employs a diffractive optical element (the phase mask) to spatially modulate the UV writing beam (Figure 4.5(a)). Phase masks may be formed either holographically or by electron-beam lithography. One of the advantages of the electron-beam lithography over the holographic technique is that complicated patterns can be written into the mask's structure such as quadratic chirps and Moire patterns. Lithographically induced phase masks are usually generated by stitching together small subsections (0.4 by 0.4 mm) of periodic corrugations on the mask substrate, in order to fabricate large phase structures. An error in the precise positioning of the various subsections will result in what is commonly referred to as stitching error (Section 4.3.1). Holographically induced phase masks, on the other hand, have no stitch error.

The phase mask is produced as a one-dimensional periodic surface-relief pattern, with period Λ_{pm} etched into fused silica. The profile of the periodic surface-relief gratings is chosen such that when a UV beam is incident on the phase mask, the zero-order diffracted beam is suppressed to less than a few percent (typically less than 3%) of the transmitted power. In addition, the diffracted plus and minus first orders are maximized; each typically containing more than 35% of the transmitted power. A near-field fringe pattern is produced by the interference of the plus and minus first-order diffracted beams. The period of the fringes (Λ) is one-half that of the mask ($\Lambda = \Lambda_{pm}/2$). The interference pattern photo-imprints a refractive index modulation in the core of a photosensitive optical fiber that is placed in contact with or in close proximity to the phase mask. A cylindrical lens may also be used to focus the fringe pattern along the fiber core.

One of the earliest experiments using the phase mask was performed by Hill and co-workers [9], who used a KrF excimer laser with pulse duration of 12 ns as the writing source. The period of the phase-mask corrugations was approximately 1060 nm, and the 249-nm zero-order diffracted beam was nulled below 5%. The transmitted light contained in each of the plus and minus first-order diffracted beams was measured to be 37%. Figure 4.5(b) shows the spectral response of a Bragg grating written using a phase mask into a D-type fiber. Mild focusing of the excimer laser beam was achieved using a cylindrical lens aligned with cylinder axis parallel to the fiber. This arrangement increased the fluence lever per pulse to approximately 200 mJ/cm^2. The KrF excimer laser was pulsed at 50 Hz for 20 minutes during fiber exposure. A peak reflectivity of 16% was observed for a grating approximately 0.9 mm long. From the grating reflectivity data the amplitude for the refractive index modulation is estimated to be 2.2×10^{-4}. This value is comparable to the average refractive index change, which was determined to be 6×10^{-4} from the shift in Bragg grating resonance due to photo-exposure and knowledge of the fiber's effective index dispersion at 1531 nm (the Bragg wavelength). Ideally, the induced modulation depth is expected to be the same or larger than the average index change when exposing the fiber to maximum-contrast grating-diffraction patterns. The observed difference, however, may be due to several factors such as nonlinearity in the photosensitive response of the fiber, low-quality nulling of the zero-order beam, the presence of higher order diffracted beams downstream from the mask, and fiber/phase mask misalignment during fabrication. It is worth noting that since its original demonstration in 1993, the phase-mask technique has been developed to a point where the inscription of nearly 100% reflective gratings is now routine. Furthermore, it is now possible to fabricate complex grating structures using the phase-mask technique.

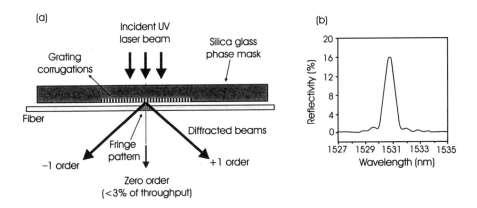

Figure 4.5   (a) A schematic of a phase-mask utilized in inscribing fiber Bragg gratings. (b) Spectral response of a Bragg grating fabricated with a KrF excimer laser using the phase-mask technique (*After:* [9]).

The phase mask greatly reduces the complexity of the fiber grating fabrication system. The simplicity of using only one optical element provides a robust and inherently stable method for reproducing fiber Bragg gratings. Since the fiber is usually placed directly behind the phase mask in the near field of the diffracting UV beams, sensitivity to mechanical vibrations and therefore stability problems are minimized. Low temporal coherence does not effect the writing capability (as opposed to the interferometric technique) due to the geometry of the setup (Figure 4.6). However, the spatial coherence plays a critical role in the fabrication of gratings. Specifically, it requires the fiber to be placed in near contact with the grating corrugations of the phase mask in order to induce a maximum modulation in the index of refraction. Clearly, the separation of the fiber from the phase mask is a critical parameter in producing quality gratings and should be minimized for UV sources that have low spatial coherence, such as excimer lasers. However, placing the fiber in contact with the phase mask (thus minimizing the separation) can cause damage to the fine grating corrugations. It should be noted that even if the fiber is placed in contact with the phase mask, the core (in most standard fibers) is several tens of microns away from the grating corrugations.

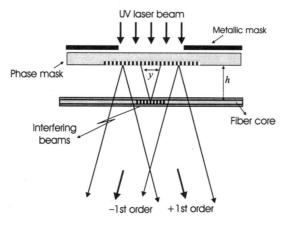

Figure 4.6   Schematic of the phase-mask geometry for inscribing Bragg gratings in optical fibers. The plus and minus first-order diffracted beam interfere at the fiber core, placed at a distance h from the mask.

Othonos and Lee [11] demonstrated the importance of the UV source's spatial coherence using the phase-mask technique. Similarly, Dyer et al. [12] also demonstrated the importance of spatial coherence in the phase-mask technique when writing gratings in polyimide films. To understand the significance of spatial coherence it is helpful to consider the schematic diagram in Figure 4.6. Consider the fiber core to be at a distance h from the phase mask. The transmitted *plus* and *minus* first orders that interfere to form the fringe pattern on the fiber emanate from different parts of the mask (separated by a distance y). Since the distance of the fiber from the phase mask is identical for the two interfering beams, the requirement for temporal coherence is not critical for the formation of a high-contrast fringe pattern. On the other hand, as the distance h increases, the separation y between the two interfering beams emerging from the mask increases. In this case, the requirement for good spatial coherence is critical for the formation of a high-contrast fringe pattern. As the distance h extends beyond the spatial coherence of the incident UV beam, the interference fringe contrast will deteriorate, eventually resulting in no interference at all.

One advantage of not having to position the fiber against the phase mask is the freedom to angle the fiber, forming blazed gratings. Placing one end of the exposed fiber section against the mask and the other end at some distance from the mask, it is possible to change the induced Bragg grating center wavelength. From simple geometry (Figure 4.7) one can derive a general expression for the tunability of the Bragg grating center wavelength, given by

$$\lambda_B = 2n_{eff}\Lambda\sqrt{1+\left(\frac{r}{l}\right)^2} \qquad (4.3)$$

where Λ is the period of the fiber grating, r is the distance from one end of the exposed fiber section to the phase mask, and l is the length of the phase grating. Clearly, due to the geometry of the fringe formation on the fiber core, the fiber Bragg gratings inscribed will be blazed. For a fixed phase-mask period, changing R will result in blazed gratings with changing center Bragg wavelength. In the experiments described in Othonos and Lee [11] a phase mask with $\Lambda = 0.531\,\mu m$ ($l = 10,000\,\mu m$) was used, resulting in $\lambda_B = 1558.0\,nm$ a $r = 0$ (the fiber placed parallel to the phase mask). Figure 4.7 shows experimental results for the tunability of the inscribed Bragg grating resonance as a function of distance r, with a corresponding theoretical curve fit showing very good agreement.

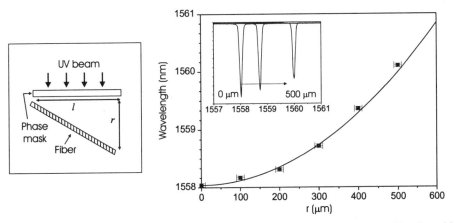

Figure 4.7   Experimental and calculated results of tuning a Bragg grating resonance by tilting the writing fiber with respect to the phase mask. This tuning technique results in blazed gratings.

A variation to the phase mask scheme with the fiber in near contact with the mask (as described above) has been demonstrated [13]. This technique is based on a UV-transmitting silica prism. The plus and minus first orders are internally reflected within a rectangular prism, as shown in Figure 4.8(a), and interfere at the fiber. This noncontact technique is flexible and allows quick changes of the inscribed Bragg wavelength. Another very useful noncontact technique is the Talbot interferometer arrangement (Figure 4.8(b)) [14]. This interferometer consists of two plane-parallel mirrors and a diffractive element (a phase mask) with its surface aligned perpendicular to the mirrors. The illumination is arranged so as to recombine the first-order beams (or other orders such as the plus first and zero orders) to form the UV interference pattern.

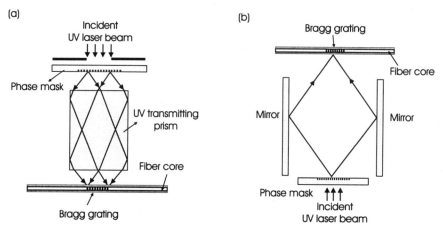

Figure 4.8   (a) Noncontact, interferometric phase-mask technique for generating fiber Bragg gratings. (b) Schematic diagram of the Talbot interferometer in which the first-order diffracted beams from a grating recombined to produce interference in the fiber core.

4.3.1 Stitch Error in Phase Masks

In view of the importance of the phase-mask technique for fabricating Bragg gratings, there has been a great deal of research directed to the use of electron-beam lithography for direct production of the phase masks. This type of phase-mask fabrication allows for very precise structures (grating period <100 nm) to be defined over large areas up to 140 by 140 mm^2. Phase masks are usually generated by stitching small 400 by 400 μm^2 electron-beam fields together. It is therefore important to understand the nature of the possible stitching errors that occur and to minimize them. It is also important from a practical point of view to determine their effect on the corresponding grating spectral response. Liu et al. [15] have reported on the spectroscopic effect of one-sign stitching errors, which have the characteristic of shifting the grating in a field to one direction only along the grating axis. For uniform phase masks, the stitching errors are random in both directions and lie within a range of <40 nm [16]. Fiber gratings produced using these phase masks show no evidence of ghost peaks or asymmetry. For chirped gratings, however, or for gratings that require phase shifts, stitching errors can play a significant role. These errors can occur in regions where there is either a discrete change in period or phase. Due to the fact that field distortion produces a regular trapezoid of stitching field rather than an ideal square (or a completely random shape), stitching errors produced by this kind of systematic fault have a random value but occur in a single direction. An example of a large, one-sign abnormal stitching error is shown in Figure 4.9. In this extreme case the period of the phase mask is 726 nm; the axial stitching error causes the grating in the field on the right-hand side of the stitching to shift by ~20–25 % of a period in a single direction. Such a large stitching error in a phase mask results in a Bragg grating that is characterized by ghost peaks on both sides of the main reflection peak (Figure 4.9).

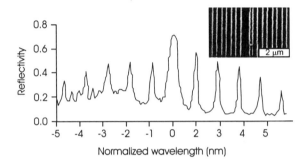

Figure 4.9 Spectral response from a Bragg grating written using a phase mask with large stitching error, shown in the upper corner (Scanning Electron Microscope picture of the mask with abnormal stitching errors). Ghost peaks are clearly visible due to the stitching error (*After:* [15]).

Many stitching errors caused by systematic fault of the electron-beam lithography have the typical one-sign characteristic shown above. Liu et al. [15] have prepared a series of phase masks with varying stitching errors of known size to study the effects on Bragg gratings. The stitch size varied from 50 nm up to half a period of the phase mask. Gratings were written into hydrogenated fibers. From Figure 4.10 it can be seen that using phase masks with relatively small stitching errors have no noticeable effect on the fiber grating response. The fundamental shape, particularly the symmetry of the spectra, remains unchanged. Larger stitching errors, however, begin to introduce an asymmetry into the spectra, starting from one-eighth of the period and ultimately changing the spectra dramatically, as seen in Figure 4.10(f).

4.3.2   Changing the Phase-Mask Periodicity

The phase-mask technique for inscribing Bragg gratings can become more flexible by allowing for the possibility of shifting the Bragg wavelength for a fixed mask periodicity. By using a simple lens to introduce a curvature on the wavefront of the diffracting light, it is possible to magnify the original periodicity formed by the mask. Prohaska et al. [17] have shown experimentally that the Bragg wavelength is shifted to shorter wavelengths by the

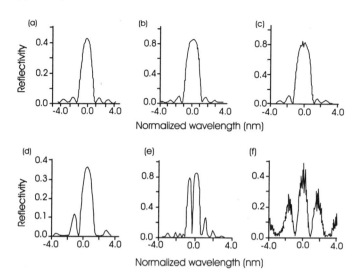

Figure 4.10   Bragg grating spectral response from fiber grating produced using phase mask with varying stitching errors: (a) no phase shift, (b) 50 nm, (c) 107 nm (1/10 period), (d) 134 nm (1/8 period), (e) 269 nm (1/4 period) and (f) 537.5 nm (1/2 period) (*After:* [15]).

addition of a converging lens before the mask. Consider a plane wave incident on a positive lens of focal length f. The exiting wavefront is a converging sphere. A simple geometric construction (seen in Figure 4.11) provides an equation for a given distance between the lens and the mask l_1 and between the mask and the fiber axis l_2. The magnification factor M is defined as the ratio of the period of the image to the period of the original mask. The expression for magnification is then given by the simple expression

$$M = \frac{f - l_1 - l_2}{f - l_1} \tag{4.4}$$

In their experiments Prohaska et al. [17] used a 180-mm focal length biconvex, fused silica lens to produce the necessary curved wavefront, and fiber Bragg reflectors were

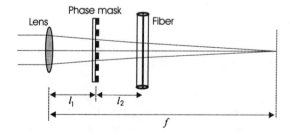

Figure 4.11   Magnifying geometry for changing the phase-mask periodicity.

made with and without the converging lens. In both cases the same type of fiber and the same exposure levels were used. It was found that the peak reflection of the Bragg grating was shifted to a shorter wavelength with the lens. The unshifted peak occurred at 1300.38 nm, whereas the shifted peak moved to 1298.67 nm. This shift in wavelength corresponds to a demagnification of the period of the phase mask by 0.9987. In this experiment $l_1 = 3.305$ mm and $l_2 = 0.245$ mm; thus, the predicted magnification was approximately $M = 0.9986$, which is in very good agreement with the measured value. To increase the demagnification of the grating period, the spacing l_2 has to increase; however, there is a limit to the value of l_2 because the fiber is required to be in the Fresnel near field of the light passing through the mask. A greater dimensional change in the grating period may be obtained using a shorter focal length lens. In view of this, the magnification tuning technique of the phase-mask period may achieve tunability on the order of a few percent.

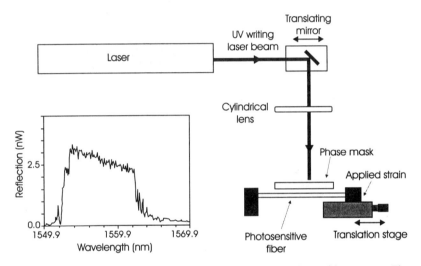

Figure 4.12   A schematic representing the controlled strain Bragg grating writing system. The graph corresponds to the spectral response from a chirped Bragg grating inscribed using strain control (*After:* [19]).

Tunability in the inscribed Bragg wavelength on a fiber using the phase-mask technique can also be achieved by applying strain to the fiber during UV illumination. Zhang et al. [18] demonstrated this technique, whereby they were able to write two gratings having different wavelengths located at the same point. In their experiment, a KrF excimer laser was used as the source of UV light along with a phase mask with dimensions 10 by 0.45 mm and period 1060 nm. The first grating was made on an unstrained fiber with Bragg wavelength centered at 1534.84 nm. The fiber was then strained by approximately 0.2%, and was exposed to the same UV light to produce the second grating at 1532.04 nm (unstrained wavelength). The tuning range of this method is limited by the mechanical strength of the fiber. Typically, a pristine fiber cannot survive more than 5% strain. A further advantage of the above technique is the fabrication of complex Bragg grating structures, such as chirped gratings [19]. A schematic of an experimental setup utilized to write Bragg gratings with the ability to apply strain to the fiber, thus altering the inscribed Bragg wavelength, is shown in Figure 4.12. The UV laser beam is directed onto a phase mask, which is focused onto the fiber with the help of a cylindrical lens. The fiber is clamped between towers, one of which is mounted on a computer-controlled, motorized translational stage. To produce a complex structure such as a chirped grating, a grating section is written at a given fiber strain, then the strain is either increased or decreased. The writing beam can be translated placing the second grating section exactly adjacent to the first and this can be repeated. The length of the phase mask and the maximum strain to which the fiber can be subjected limits the width of grating response that can be created by this method. Care must be taken to choose the rate at which the strain is decreased, to ensure that the response of each grating overlaps properly, leaving no gaps in the spectrum. Clearly the degree and rate of chirp can be easily tailored by selecting the length of each grating section and the rate of change of strain in the fiber. Controlling the exposure time can vary the profile of the grating, thus achieving apodization.

4.4 Point-by-Point Fabrication of Bragg Gratings

The point-by-point technique [20] for fabricating Bragg gratings is accomplished by inducing a change in the index of refraction corresponding to a grating plane one step at a time along the core of the fiber. In a typical experimental arrangement a single pulse of UV light from an excimer laser passes through a mask containing a slit. A focusing lens images the slit onto the core of the optical fiber from the side, as shown in Figure 4.13. As a result, the refractive index of the core in the irradiated fiber section is changed locally. The fiber is then translated through a distance Λ, corresponding to the grating pitch, in a direction parallel to the fiber axis. This process is repeated to form the grating structure in the fiber core. Essential to the point-by-point fabrication technique is a very stable and precise sub-micron translational system.

In their experiments, Malo and co-workers [20] used the point-by-point technique to fabricate Bragg reflectors via a KrF excimer laser that was passed through a 15-μm slit. A 15 mm focal length lens imaged the slit on the core of the optical fiber, with the long dimension of the image oriented perpendicular to the axis of the optical fiber. In the image plane it was estimated that the image size was 500 by 1.5 μm^2 and that the fluence level produced by a single 248-nm UV light pulse was 5 J/cm^2. With these irradiation conditions the width of the photoinduced perturbation in the fiber core was estimated to be 0.7 μm. Bragg gratings were then fabricated by translating the fiber between each irradiation step with the aid of an interferometrically controlled translation stage. Fiber Bragg reflectors were fabricated at 1550 nm by irradiating the flat cladding side of a D-type polarization-maintaining fiber. As a result of the large width of the photoinduced index perturbation, it was not possible to fabricate first-order 1500-nm Bragg gratings using this technique. Instead Bragg gratings that reflect light in the second or third order at 1500 nm were fabricated. A plot at the upper right corner of Figure 4.13 illustrates the spectral response of a third-order Bragg grating reflector in a wavelength region coinciding with the resonator wavelength. In that case, the grating period was $\Lambda = 1.59$ µm and the grating contained 225 index perturbations, resulting in a device length of 360 µm. The grating had a peak reflectivity of 70% at 1536 nm and a full width half maximum of 2.7 nm. The calculated refractive index change for the third-order Bragg reflector at 1536 nm was estimated to be 6×10^{-3}. These micro-Bragg reflectors also act as taps for light at wavelengths shorter than the first-order resonance wavelength. The spectral transmission response of the taps is similar to the response obtained for Bragg gratings fabricated using a phase mask and single-pulse excimer laser irradiation. At wavelengths shorter than the resonance wavelength, the Bragg grating couples the light out of the core into the radiation modes of the waveguide. The effect is manifested as transmission loss through the Bragg reflector.

The main advantage of the point-by-point writing technique lies in its flexibility to alter the Bragg grating parameters. Because the grating structure is built up a point at a time, variations in grating length, pitch, and spectral response can easily be incorporated. Chirped gratings can be accurately produced simply by increasing the amount of fiber translation each time the fiber is irradiated. The point-by-point method allows for the fabrication of spatial mode converters [21], polarization mode converters, or rocking filters [22] that have grating periods, Λ, ranging from tens of micrometers to tens of millimeters. Clearly, long period (transmission) gratings can also be fabricated using this technique. Furthermore, since the UV pulse energy can be varied between points of induced index change, the refractive index profile of the grating can be tailored to provide any desired apodization.

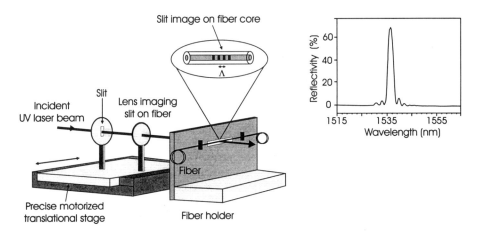

Figure 4.13 A schematic of a setup for fabricating Bragg gratings using the point-by-point technique. The plot at the upper right corner illustrates a reflection spectrum of a third-order Bragg grating fabricated using the point-by-point method (*After:* [20]).

One disadvantage of the point-by-point technique is that it is a tedious process requiring a relatively long process time. Errors in the grating spacing due to thermal effects and/or small variations in the fiber's strain can occur. This limits the gratings to very a short length. Furthermore, due to the submicron translation and requirement for tight focusing, first-order 1550-nm Bragg gratings are difficult to demonstrate. Malo et al. [20] have only been able to fabricate Bragg gratings which reflect light in the second and third order that have a grating pitch of approximately 1 and 1.5 µm, respectively.

4.5 Mask Image Projection

In addition to the above, well-known techniques for fabricating fiber Bragg gratings, high-resolution mask projection has also been demonstrated [23] as a means of inscribing Bragg gratings, as seen in Figure 4.14. In Mihailov and Gower's experiment [23] a standard KrF excimer laser, generating 20-ns duration pulses with energy approximately of 1 J/pulse, was employed. The transmission mask consisted of a series of UV opaque line-spaces of periods ranging between 5 and 120 µm on a fused silica substrate. The opaque lines on the mask were generated by vacuum-deposited, multilayer-stacked, high-reflectivity dielectric material. The transmitted beam was imaged onto the fiber core by a multicomponent, fused silica, high-resolution, 0.35 NA, lens having a demagnification of 10:1. Image demagnification resulted in irradiation fluences at the fiber 100 times greater than incident on the mask. A Ge-doped BNR (CA2114) fiber with core diameter of 3.6 µm and with a core-cladding refractive index difference of 0.028 was used as the host material. The fiber was placed on a submicrometer precision computer-controlled translation stage at the image plane (Figure 4.14). Using the imaging system, grating periods of 1, 2, 3, 4, and 6 µm were recorded onto the fiber by single pulse exposures. Due to the large grating periods recorded in the fiber, observation of their spectral response was made possible only through higher order reflections. These higher order reflections from gratings follow the

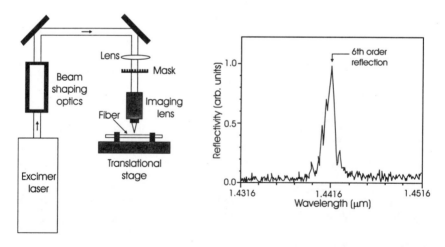

Figure 4.14 A schematic diagram of the mask projection system used to record Bragg gratings in photosensitive fiber. The plot to the right shows the spectrum for the sixth-order reflection from a 3-µm period Bragg grating. The grating was recorded using a single excimer laser pulse (*After:* [23]).

Bragg condition $m_o\lambda_B = 2n_{eff}\Lambda$ where m_o corresponds to the order of the reflection. The plot in Figure 4.14 illustrates the sixth-order reflection ($m_o = 6$), having a 1.5-nm FWHM bandwidth and ~72% reflectivity, which was observed at 1.441 μm from a 3-μm period grating recorded by mask projection. Grating periods as large as 6 μm were successfully recorded.

With the formation of the gratings an overall decrease in the Fresnel reflection across the entire spectrum was observed. A sharp fluence threshold for grating formation at 0.8 J/cm^2 was observed, which was approximately 30% of the power at which catastrophic damage in the fiber occurred. Optimal grating formation occurred at 1.4 J/cm^2. It is noteworthy that with single excimer laser pulse inscription of gratings, an associated grating structure recorded on the surface of the cladding was always observed. Because of the simplicity of the source and setup, the recording of coarse period gratings by mask imaging exposures may, in some cases, be more flexible than with other techniques. With a demagnifying imaging system, the amplitude mask is exposed to far lower irradiation fluences than is the case for proximity exposures using the phase-mask technique. Complicated grating structures (blazed, chirped, etc.) can be readily fabricated with this method by implementing a simple change of mask.

Long period or transmission gratings can also be fabricated using this mask-projection technique. Due to the long grating periodicity, this is much easier to accomplish than inscribing Bragg gratings. Furthermore, the amplitude mask may be fabricated at the exact grating period required; thus, no demagnification imaging will be needed, although imaging optics (magnification or demagnification) may be used as a means to vary the grating period inscribed on the fiber.

4.6 Laser Sources

Thus far we have described various techniques for writing Bragg gratings in optical fibers. It is apparent that each of the techniques has different requirements on the UV laser source. In the interferometric technique, which is the most demanding approach, the sources must have very good temporal and spatial coherence. In the phase-mask technique the only source requirement is good spatial coherence. The point-by-point and mask projection techniques are the least demanding. These techniques require only that enough energy pass through the slit or the mask (mask projection) in order to change the index of refraction. Although in principle this may be possible with a source other than a laser, in practice, due to the large energy requirement and the need for good focusing characteristics, this is accomplished with UV laser sources.

4.6.1 Laser Sources for Interferometric Techniques

Laser sources used for inscribing Bragg gratings via the interferometric techniques must have good temporal and spatial coherence. The spatial coherence requirements can be relaxed in the case of the amplitude-splitting interferometer by ensuring that the total number of reflections are the same in both arms. This is particularly critical in the case when a laser with low spatial coherence (e.g., an excimer laser) is used as the UV light source. The coherence length (temporal) must be at least equal to the length of the grating (assuming the two optical paths are identical) in order for the interfering beams to have a good visibility, thus resulting in high-quality Bragg gratings. This coherence requirement, together with the UV wavelength range required (240–250 nm, ~190 nm), initially forced researchers to use very complicated laser systems, such as the excimer-pumped, frequency-doubled dye laser.

4.6.1.1 Frequency-Doubled Dye Laser

A typical system consists of a main laser that is used as the pump source for a dye laser whose output frequency is subsequently doubled with the help of a nonlinear crystal. The dye laser cavity is designed to generate a narrow linewidth output corresponding to long temporal coherence. Furthermore, the appropriate dye solution must be used in order for the system to operate at a suitable wavelength that will lie within the photosensitive band of the fiber when frequency doubled. Typically excimer and Nd:YAG laser systems are used as the main pump source. Excimer lasers incorporated in a pumping scheme usually work at either 351 nm (XeF) or 308 nm (XeCl) (although other deep UV radiation is available from excimer lasers it is not recommended for pumping due to the short life of the dye when excited with highly energetic photons). Q-switched Nd:YAG lasers (1.064 μm) are also used as pump sources for a dye laser. In these systems the fundamental operating wavelength is first frequency doubled and then tripled to 355 nm prior to pumping a dye laser. For comparative purposes the fundamental advantages and disadvantages to both systems are summarized as follows. Excimer lasers can operate at high repetition rates and

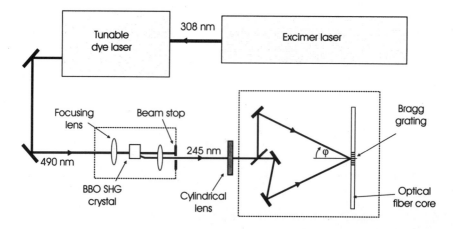

Figure 4.15 Experimental configuration of an excimer-pumped dye laser incorporating a frequency-doubled Beta-Barium Borate (BBO) crystal. The BBO crystal is used to generate UV light at 245 nm for inscribing Bragg gratings in optical fibers with an interferometer setup. A simple adjustment of the angle (2φ) between the interference beams results in fabricating gratings with different Bragg resonance wavelengths.

most often with large energy per pulse. In addition, the excimer beam is quite large and relaxes the tolerance required for alignment of the dye laser. Q-switched laser systems, with frequency-tripling capability at comparable energies, are usually more expensive than "basic" excimer lasers. The pulse-to-pulse stability may be the same in both systems assuming they operate under optimum conditions. However, the Nd:YAG laser is more sensitive to various laser parameters given that its output depends on a second and third harmonic generation (two nonlinear processes). Dye laser technology is well advanced and, thus, most dye laser systems (even with minimum specification) will meet the requirements for Bragg grating fabrication using the interferometric technique. Dye lasers typically offer a 10% conversion efficiency to the set wavelength (for example, at 490 nm) and a coherence length of approximately 20 cm. The output from the dye laser is focused on a nonlinear crystal to double the frequency of the fundamental light (Figure 4.15). Typically, this arrangement provides 5- to 20-ns pulses (depending on the pump laser) and up to 15 mJ of energy with excellent temporal and spatial coherence. One of the major drawbacks to such a system is its complicated arrangement. Maintaining two lasers and a frequency-doubling system to give optimal performance is not a simple task. Furthermore, for optimal performance the dye solution must be changed often, which is undesirable.

4.6.1.2 Frequency-Doubled Optical Parametric Oscillator

Recent advancements in crystal technology and nonlinear optics have made possible a new, innovative, tunable laser system, known as the optical parametric oscillator (OPO). In this particular system, the pump source is usually a Q-switched Nd:YAG laser equipped with amplifiers and a diode seeding system to improve temporal and spatial stability. The fundamental laser wavelength is tripled to 355 nm and used as the pump source to an OPO. The OPO can generate a beam that may be wavelength tuned from 400 nm to the infrared, a range that is system dependent. After setting the output of the parametric oscillator to the desired wavelength (i.e., at 490 nm), it may then be frequency-doubled to obtain the UV writing wavelength for Bragg gratings. There are several commercially available systems with excellent characteristics for inscribing Bragg gratings interferometrically. The major drawback of this system is its complexity, making it a very expensive approach to producing UV light for fabricating gratings.

4.6.1.3 Narrow-Linewidth Excimer Lasers

Since the introduction of excimer lasers in 1976 [24] there has been a great deal of interest in narrowing the spectral linewidth [25]. This is relevant for a number of applications such as photolithography, light-induced detection and ranging, laser fluorescence, and more recently, fiber Bragg grating fabrication. A normal excimer laser, due to its extremely short coherence length, cannot be used in an interferometric setup to fabricate Bragg gratings unless it undergoes some form of spectral line narrowing.

Commercially available narrow linewidth excimer systems are oscillator amplifier configurations with excellent characteristics for writing Bragg gratings; however, they are extremely expensive. Othonos and Lee [26] demonstrated a low cost and simple technique, whereby existing KrF excimer lasers may be retrofitted with a spectral narrowing system for inscribing Bragg gratings in a side-written interferometric configuration. In that work a commercially available KrF excimer laser (Lumonics Ex-600) was modified to produce a spectrally narrowed laser beam with a linewidth of approximately 4 pm (Figure 4.16). Typical KrF (248.5 nm) excimer lasers under normal operation have very broad spectral profiles; for the laser used in that experiment this was approximately 80 cm^{-1} (0.5 nm). There are a number of techniques for narrowing the linewidth of an excimer laser [26], including the utilization of diffractive gratings, intracavity prisms, and intracavity etalons. Intracavity etalons where chosen over the other elements due to advantages in efficiency, simplicity, and stability when used in excimer laser oscillators [25]. The system consisted of two air-spaced etalons mounted at the end of the cavity between a Brewster window attached to the gas chamber and the high reflector (Figure 4.16). In order to maintain one narrow transmission peak within the gain bandwidth of the KrF laser, an etalon with a free spectral range of 80 cm^{-1} and finesse of 16 was used. A second etalon with a free spectral range of 6 cm^{-1} and finesse of 10 was also used to further reduce the spectral linewidth. An aperture was inserted between the Brewster window and the etalon to prevent reflection from the highly reflective faces of the etalon feeding back into the gain medium. More importantly, this intracavity aperture spatially filtered the laser beam resulting in output with a nearly Gaussian intensity profile. This output was approximately 30 mJ/pulse at a repetition rate of 50 Hz. The pulse width was approximately 12 ns and the output wavelength was centered close to 248.3 nm with tunability of 0.5 nm. Although a Brewster window was incorporated at one end of the excimer chamber, due to the short pulse

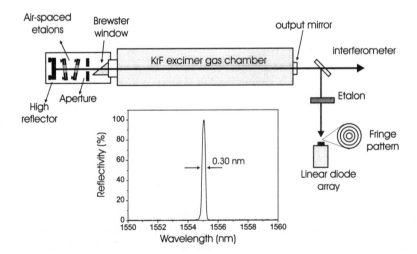

Figure 4.16   A schematic of a narrow linewidth, excimer laser system (SLN-KrF), consisting of two air-spaced etalons and an intracavity aperture placed between the KrF excimer gas chamber and the high reflector. The inserted graph corresponds to the reflection spectra from a Bragg grating inscribed in hydrogen-loaded, telecommunications fiber (Phillips fiber, 3 mol% Ge) using the SLN-KrF (*After:* [26]).

duration there were not enough cavity round trips to selectively polarize the beam in one direction. For the application of inscribing Bragg gratings, where the UV excimer beam was used in an interferometer, s-polarization (vertical) was necessary for maximum fringe contrast and higher quality Bragg gratings. This requirement made an external polarizer necessary before directing the output laser beam into the interferometer. Part of the beam was directed through an etalon, and the circular interference fringes were monitored with a linear diode array camera to keep track of the laser beams spectral characteristics, which are critical in fabrication of Bragg gratings. This system was used to successfully inscribe Bragg gratings in photosensitive optical fibers. Figure 4.16 shows a typical Bragg grating inscribed in normal telecommunications fiber (Phillips fiber, 3 mol% Ge), using this SNL-KrF laser in a interferometer setup (similar to the one in Figure 4.3(b)).

4.6.1.4 Intracavity, Frequency-Doubled, Argon Ion Laser

An alternative laser source, which is becoming very popular, is the intracavity, frequency doubled, argon ion laser [27] that uses Beta-Barium Borate (BBO) crystals. This system efficiently converts high-power visible laser wavelengths into deep ultraviolet (244 and 248 nm). The characteristics of these lasers include unmatched spatial coherence, narrow linewidth, and excellent beam pointing stability, thus making them very successful for Bragg gratings inscription. This system is highly recommended for fabricating Type I gratings that do not require high intensity. When high intensity is required, however, (such as when writing Type II gratings) it is necessary to use pulsed lasers, such as the excimer systems.

4.6.2 Laser Source Requirements for Noninterferometric Techniques

Here the laser source requirements are not as stringent as those needed for the interferometric technique. In the case of the phase-mask technique it is clear from the geometry of the situation seen in Figure 4.6 that temporal coherence is not important in inscribing high quality gratings. Thus, in addition to the laser systems listed in Table 4.1, a standard excimer laser may be used to inscribe Bragg gratings. The point-by-point and mask-projection techniques do not require any temporal or spatial coherence. The only requirement is that enough energy enters the slits to achieve a suitable change in the index of refraction.

In conclusion, the versatility of an excimer laser, in particular the ability to change the operating wavelengths at 193 and 248 nm (wavelengths for which the optical fiber exhibits photosensitivity), along with the recent availability of phase masks, has made this combination one of the most efficient and inexpensive ways of writing Bragg gratings.

Table 4.1 UV Sources for Inscribing Bragg Gratings Using Interferometric Techniques

UV laser source	UV wavelength (nm)	Energy/Power	Pulse	Coherence length (cm)
Excimer pump dye laser frequency doubling	240–250	1–10 mJ/pulse	10–20 ns	10
Optical parametric oscillator frequency doubling	240–250	1–10 mJ/pulse	8–12 ns	5
Excimer-amplifier narrow linewidth	248	200 mJ/pulse	10–20 ns	3
Excimer intracavity etalons	248	10–40 mJ/pulse	10–20 ns	3
Argon ion intracavity doubling	244	up to 0.5 W	CW	2

4.7 Special Fabrication Processing of Gratings

This section will consider in detail several techniques used in inscribing special Bragg grating structures. For example, the fabrication of fiber gratings using a single excimer laser pulse, the inscription of gratings during fiber drawing, and the writing of long and complex fiber Bragg gratings will be discussed. In addition, some of the techniques for inscribing chirped and phase-shifted gratings are described. Furthermore, various apodization techniques are discussed.

4.7.1 Single Pulse Inscription of Fiber Gratings

Fiber Bragg gratings typically require exposure periods of seconds to minutes (depending on the photosensitivity of the fiber) and the precise overlap of the intensity patterns of hundreds or thousands of individual exposures. Successful multishot interferometric exposures require the spatial stability of the interference pattern relative to points along the fiber axis to be less than the Bragg period ($\lambda_B 2n_{eff}$). This strenuous stability requirement demands the suppression of air currents, vibration, thermal excursions, and laser pointing instability over a period corresponding to the exposure time. This holds whether a pulsed or a CW laser is used, and, as a result, fabrication of a useable grating is quite difficult. Clearly, the fabrication of fiber Bragg gratings by a short laser pulse (such as a few nanosecond pulses from an excimer laser) will satisfy many of the stability requirements. Askins et al. [28] first demonstrated the inscription of Bragg reflection gratings in a Ge-doped silica core optical fibers by interfering two beams of a single 20-ns pulse from a KrF excimer laser. A commercial line-narrowed KrF excimer laser that delivered 0.1 J of energy per pulse, in a 5 by 20 mm rectangular beam, was used to write the gratings. This initial demonstration of fabricating Bragg gratings with a single laser pulse showed a maximum reflectivity up to 2%. Although such weak gratings can be sufficient for sensor applications, most practical telecommunications applications require fiber gratings with high reflectivities. Archambault et al. [29, 30] have demonstrated stronger single pulse gratings using different interferometer configurations and a highly photosensitive optical fiber. Figure 4.17 shows a schematic of the interferometer used in inscribing single pulse fiber Bragg gratings. The interferometer was used in two configurations: A and B. In configuration A, two cylindrical lenses were placed before the beam splitter. In configuration B, a single lens was used close to the fiber. To obtain good interference despite the nonuniformities in the laser beam, the number of reflections in each path was the same so to avoid beam reversal though parity violation. A single-mode silica fiber was used and had a core doped with 9% germania and 7% boron, making it highly photosensitive. In configuration A, the lens system affected the two interfering beams equally because of their placement outside the interferometer and before the beam splitter. Consequently, any phase distortions introduced by the lenses were cancelled in the interference pattern. This allowed the fabrication of relatively long (15 mm) single pulse gratings with bandwidths smaller than 0.15 nm. A typical reflection spectrum is shown in the plot of Figure 4.17. The reflectivity of the grating was ~7% and the bandwidth approximately 0.05 nm. To reduce the size of the focused beams and so increase the amount of energy absorbed by the fiber core, the two lenses of configuration A were replaced by a single cylindrical lens (40 mm), as seen in configuration B. It is important to shorten the focal distance as much as possible as this minimizes the ratio between the UV intensity at the fiber surface and that in the core, thus avoiding damage to the fiber surface. In this configuration, the fiber core was exposed to fluences of up to 25 J/cm^2 per 20-ns pulses, as opposed to 0.5 J/cm^2 for configuration A. This resulted in very large index changes (in excess of 10^{-3}) and grating reflectivities of up to 65%. The reflection spectra of these

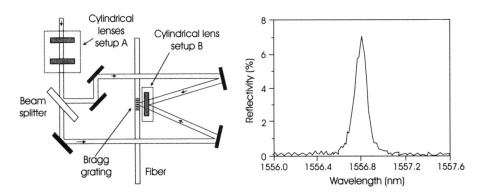

Figure 4.17 Schematic of the interferometer used to inscribe single pulse gratings. The plot to the right of the figure shows a typical reflection spectrum of a grating fabricated by a single laser pulse (*After:* [29]).

single pulse gratings were broad (between 0.3 and 2 nm) indicative of nonuniform gratings. This nonuniformity was probably caused by lens aberrations unevenly distorting the phase fronts of the two beams. Further experiments by the same group have demonstrated fiber gratings with reflectivities in excess of 99.8% and bandwidths of 7.5 nm at 1550 nm, in single pulse exposure.

Following the work by Askins et al. [28] and Archambault et al. [29, 30] on the single pulse fiber Bragg grating fabrication using the interferometric technique, Malo and co-workers [31] demonstrated the inscription of single pulse fiber Bragg gratings using the phase-mask technique. The experimental setup, used for writing fiber gratings in Malo's work, was the same as the one used by the same group in previous work [9]. The phase mask was made from high quality, fused silica glass and had a square-wave surface corrugation with a period of 1060 nm. Ultraviolet light passing through the phase mask was diffracted into several different orders. The mask was designed to suppress the amount of light that was diffracted into the zeroth-order beam (<5% at 248.5 nm). Most of the diffracted light (~80%) was contained in the first orders. In this work Andrew Corporation D-type polarization-maintaining fiber was used. The phase mask was placed such that the corrugations in the silica slab were adjacent to the curved side of the D-fiber cladding, with the long axis of the corrugations oriented perpendicular to the axis of the optical fiber. Light from the excimer laser was directed normal to the flat surface of the phase grating and passed through with the diffracted beams interfering to form the fringe pattern. The single pulse Bragg gratings were inscribed using an irradiation fluence of ~1 J/cm^2 on the cladding of the optical fiber. Figure 4.18(a) shows the transmission response of a 4-mm long Bragg reflector written in the D-type fiber. As seen in this figure, for wavelengths longer than 1535 nm the fiber has 100% transmission. However, at 1535 nm a sharp dip in the transmission occurs as a result of light reflection by the Bragg reflector (measured peak reflectivity 99.5% at 1535 nm). The FWHM of the reflection peak is approximately 3.5 nm, which is much broader than the spectral width of 0.5 nm expected for a uniform Bragg grating of length 4 mm. For wavelengths shorter than 1535 nm, the single pulse Bragg

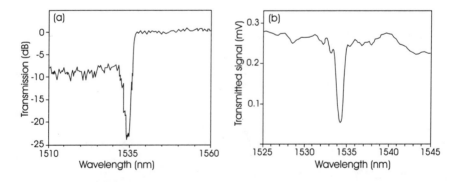

Figure 4.18 (a) Transmission spectrum of a Bragg reflector with a peak reflectivity of 99.5% fabricated with irradiation fluence of ~1 J/cm^2. (b) Transmission spectrum of a Bragg reflector with a peak reflectivity of 80% induced with a fluence level that was slightly lower (amount was not measured accurately) than that used in (a) (*After:* [31]).

reflector transmits only 20% of the light in the measured range. Therefore, the Bragg grating acts as an efficient tap (80%) for coupling light out of the optical fiber at wavelengths shorter than the Bragg wavelength. The reflection and transmission spectra obtained for this photoinduced grating is typical of Bragg gratings that do not extend uniformly across the core of the optical fiber. Similar spectra [32] have been obtained for Bragg gratings that are formed by etching a surface-relief grating on the core cladding boundary of the optical fiber. The light lost for wavelengths shorter than the Bragg resonance was attributed to the grating providing phase-matched coupling of the reflected light into the cladding and radiation modes of the fiber. The transmission response of the photo-imprinted Bragg reflector is sensitive to the fluence level of the UV light incidence upon the optical fiber. Photo-imprinted gratings were not detected (they had less than 5% reflectivity) if the fibers were irradiated with a single KrF excimer pulse having a fluence level 20% below 1 J/cm^2. Furthermore, Figure 4.18(b) shows the spectral response of a Bragg reflector that was photoinduced with a fluence level that was slightly lower than the fluence level used in fabricating the Bragg grating whose spectrum is shown in Figure 4.18(a). The reduction in fluence level decreased both the Bragg peak reflectivity (~80%) and the spectral width (FWHM ~1 nm). The light loss at shorter wavelengths is not present in the spectrum shown in Figure 4.18(b) but was observed over a larger spectral range. Malo and co-workers [31] used an optical microscope to verify directly the existence of the grating period by observing the image of the photoinduced perturbation produced by the single pulse of high-intensity KrF excimer laser light in the Andrew D-fiber. The image appeared to be a fringe pattern with the 1060-nm period of the phase mask [33]. Irregularities observed in the borders of the perturbation pattern indicated that diffusion or melting occurred, which suggested that a high temperature was achieved during the photo-imprinting process.

4.7.2 Bragg Grating Inscription During Fiber Drawing

One of the major drawbacks in inscribing a Bragg grating is that a section of the fiber must be stripped of its UV absorbing polymer coating in order for the fiber to be exposed. This process weakens the fiber at the site of the grating, due to surface contamination, even if the fiber is subsequently recoated. It is also not possible to mass produce fiber gratings in the same section of fiber within a reasonable time scale (e.g., gratings to be used in a long quasi-distributed fiber sensor). These problems can be overcome by writing the gratings during the fiber-drawing process, just before the fiber is coated.

Dong et al. [34] demonstrated Bragg grating inscription during fiber drawing using a line-narrowed KrF excimer laser with beam size of 205 mm^2 and pulse width of 20 ns. An interferometer was mounted directly on the fiber drawing tower between the fiber diameter monitor and the coating cup (Figure 4.19). The coating fluid, being a few centimeters below the point of writing, served to damp out lateral vibration in the fiber. The beam was focused to a line 70.5 mm^2 using a cylindrical lens. The position of the fiber did not alter visibly during the writing process. The drawing capstant was modified so that the free end of the fiber could be placed into the diagnostics, which consisted of 1.55-μm pigtailed 50:50 fiber coupler and an optical spectrum analyzer. This arrangement allowed online

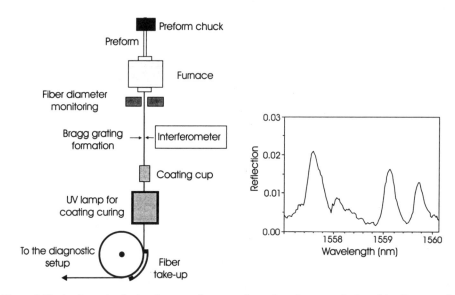

Figure 4.19   A schematic of a drawing tower incorporating an interferometer for inscribing Bragg gratings. The plot on the right shows the reflection spectra of three consecutively written Type I gratings. Strain was applied to distinguish the spectral response of each grating (*After:* [34]).

monitoring of the grating production. The fiber was drawn at speeds ranging from 2 to 23 m/min to a diameter of 110 μm, resulting in an axial displacement of only 1 nm during the 20-ns pulse. Type II and conventional Type I photorefractive gratings were written into the fiber as it was drawn. The reflection spectra from three consecutively written Type I gratings (pulse energy 10 mJ) are shown in Figure 4.19. A reflectivity of approximately 2% was obtained, limited by the refractive index change that could be induced by the intensity of the single excimer laser pulse in the above experiment. The center wavelengths of the gratings varied by approximately 0.2–0.3 nm, which was attributed to possible fluctuations in pulling tension and/or core temperature (variation of only 30°C would result in the observed wavelength shift). Applying strain to the gratings offline made it possible to separate the gratings, as shown in Figure 4.19. The pulse-to-pulse energy fluctuations in the excimer laser beam resulted in a small variation in the peak reflectivity of the written Bragg gratings. Furthermore, increasing the pulse energy of the excimer laser to 40 mJ resulted in Type II grating formation.

A further improvement has been demonstrated by Askins et al. [35] whereby an interferometer was specifically built to write Bragg gratings during the fiber draw (Figure 4.20). As shown in Figure 4.20 the laser beam arrives at the first mirror M1 in a direction precisely normal to the vertical fiber axis. It is then deflected 45 degrees upward, passing through the focusing cylindrical lens and the horizontal beam splitter. The transmitted beam reflects off the upper horizontal mirror M2 and the final upper mirror M3, before being directed to the fiber. The reflected beam is turned by the final lower mirror M4 to intersect the upper beam at the fiber. If mirrors M3 and M4 are placed to intercept their respective beams and are precisely aligned vertically, then the equilateral triangle formed

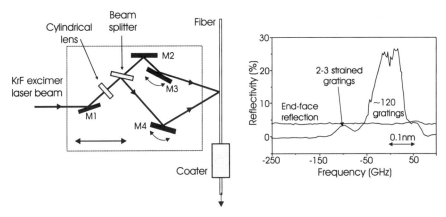

Figure 4.20   Schematic diagram of the excimer laser interferometer used to write gratings during the fiber draw. The long axis of the excimer laser beam is orientated vertically. The plot to the right of the figure corresponds to the reflectance spectrum of a number of gratings written at the same wavelength (772 nm) and at 3 Hz (*After:* [35]).

by M3, M4, and the intersection point of the beams guarantees equal distances from the beam splitter along both interferometer arms. The apparatus used motorized micrometers, which appropriately rotated M3 and M4 to define the angle of beam intersection. When this angle is varied, the entire optical assembly translates strictly parallel to the incident laser beam to maintain the beam intersection at the fiber core position, as shown by the long horizontal arrow in Figure 4.20. More than 450 single pulse gratings were generated online. One fiber section that has been partially evaluated involved the writing of more than 100 gratings within 40 seconds, all at 772 nm and spaced 10 cm apart along the fiber. Figure 4.20 is a composite spectrum obtained by aligning five overlapping data sets with the monitoring Fabry-Perot interferometer as a reference. As shown in this figure, the reflectivity of the single pulse grating ensemble is seen to fall within a 0.4-nm band, with the vast majority falling within a FWHM of 0.14 nm. All features were lowered and broadened somewhat by the averaging necessary to smooth the rapidly varying Fabry-Perot interference that arises between the gratings. The ~25% reflectivity of the ensemble represents the incoherent addition of many gratings. The small 4% reflectivity peak at 100 GHz results from two or three gratings strain-tuned to a longer wavelength by intentional loading of the fiber. The bandwidth of all the gratings was consistent with a grating length equal to the exposure region of 5 mm.

Examples of two serial wavelength-stepped arrays are given in Figure 4.21. In the first example 50 gratings were written at 0.1-nm intervals every 0.5 seconds over a total range of 5 nm. The reflectivity was measured only with a tungsten source and monochromator with the fiber still on a spool; therefore, the actual uniformity and reflectivity are unknown. Although the reflected signal was believed to be insufficient for device use, this serves to illustrate the ability to tune the writing wavelength continuously and rapidly. In the second example, shown in Figure 4.21(b), tuning was stepped by 5 nm per grating at a rate of 1 Hz.

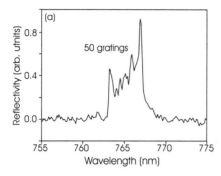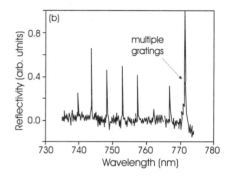

Figure 4.21   (a) Reflectance spectrum of a dense array of 50 gratings written with a 0.1-nm Bragg wavelength spacing at 2 Hz. (b) Reflectance spectrum of series of gratings written with 5-nm spacing (*After:* [35]).

4.7.3 Long Fiber Bragg Gratings with Complex Reflectivity Profiles

Many of the applications of fiber Bragg gratings require complex structures that often depend on having a relatively long length (typically a few centimeters). A method of writing very long gratings has been described by Martin et al. [36] where a UV beam is scanned over a long phase mask in a fixed position relative to the fiber. The idea behind this approach is to keep the mask and fiber held together by placing the fiber directly behind the mask and to move the writing beam along the mask and fiber assembly. As long as the mask and fiber do not move relative to each other, the phase of the fringes created in the fiber core remains determined by the mask, regardless of the writing beam's position. This results in gratings that can be as long as the mask itself. An advantage of this technique is that the writing intensity and, therefore, the grating strength can be modified at will along the grating length. Thus, gratings with a complex coupling coefficient profile can be written to suit specific applications. Complex structures can be added to the long gratings by simply varying the exposure time or via post-processing the grating by changing the dc-level of the refractive index [37]. Another method is to use specially designed long phase masks where the complex structure already has been implemented in the mask itself [38]. Complex grating structures have also been demonstrated by Loh et al. [39] where a UV beam was scanned over a phase mask while moving the fiber slightly with a piezoelectric transducer, and by Kashyap et al. [40] for a number of concatenated apodized and unapodized step-chirped gratings.

Figure 4.22 shows a schematic diagram of a typical experimental setup for producing long complex grating fiber structures. The idea is similar to the point-by-point writing method demonstrated by Malo et al. [20]; there are some differences, however. For example, instead of writing a single grating element at a time, a subgrating structure is inscribed per irradiation step. The UV interference creating the index modulation per step is accomplished either by an interferometer or a phase mask. The fiber is translated with constant speed relative to the UV fringes with an interferometer-controlled translation

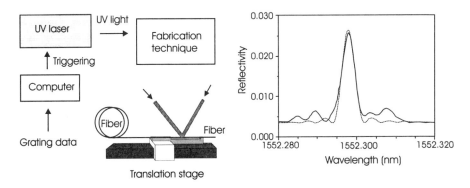

Figure 4.22   A schematic of a typical setup for inscribing long, complex Bragg grating structures. The grating parameters are entered into a computer, which controls and then synchronizes the various instruments involved in writing the grating. An interferometer is used to keep track of the position of the fiber holder relative to the UV interferometer setup. The plot on the right shows the reflectivity spectra of a weak 20-cm long uniform grating. The solid line corresponds to the actual grating, while the dotted line corresponds to a numerical simulation of an ideal 20-cm long grating with index modulation 3.75×10^{-7} and a FWHM linewidth of 3.8 pm (*After:* [42]).

stage. The position is very accurately tracked during the motion, and these data are used to trigger the UV laser when the fiber reaches the desired position for the next irradiation. The low relative speed between the fiber and the UV fringes makes their relative position virtually constant during the short 10-ns exposure time defined by the UV laser pulse. A natural way to generate a complex grating structure, using this method (Figure 4.22), is to keep the grating period of each subgrating constant while varying the phase between different sets of subgratings [41]. Clearly, by increasing the number of subgratings in each set but keeping the total length of the subgrating constant, a higher degree of flexibility can be achieved.

In the setup of Figure 4.22 the fiber subject to exposure is placed in a holder. The length of the holder and the maximum attainable length of a single grating can be varied depending on the translational stage's movement capability. In the experiment of Asseh et al. [42] the fiber holder was mounted on an air-bearing borne carriage, which was translated by a feedback-controlled linear drive. The carriage could be translated over 50 cm; thus, in principle it would be possible to write a 50-cm long grating. The UV interference pattern in their experiment was generated with a Mach-Zehnder interferometer. A second interferometer utilized a stabilized He-Ne as light source and continuously tracked the relative position between the fiber holder and the Mach-Zehnder. A Pockels cell was used for modulation of the tracking interferometer in order to achieve fringe counting and phase interpolation and to determine the direction of motion. This offers a resolution in position of about 0.3 nm. The tracking signal was used in part for feedback to the translation stage, keeping the stage at constant speed and compensating for vibrations, while also triggering the UV laser when the position reached a predefined value corresponding to the calculated grating profile data. The fiber could move back and forth with sustained control of its position, and thus, multiple exposures proved to be possible. One way of achieving apodization would be to utilize multiple exposures with different UV laser power for different grating sections. In their experiments Asseh et al. [42] demonstrated 20-cm long gratings (Figure 4.22). In the inscription of these long gratings no sign of systematic phase errors, such as stitching errors, was observed. Furthermore, the measured linewidths were approximately 4 pm, which was very close to what can be expected from an ideal grating of such length. In the plot of Figure 4.22, the measured spectrum of a weak grating is shown together with a numerical simulation of an ideal grating with a refractive index shift of 3.75×10^{-7}, peak reflectivity of 2.6% at 1552.298 nm, and FWHM linewidth of 3.8 pm. Although the grating presented is weak, strong gratings are easily obtained, considering that a long grating requires smaller change of index of refraction for the same peak reflectivity as a short grating. For example, a 10-cm long grating will need approximately one-tenth the change of index of refraction of a 1-cm long grating possessing the same peak reflectivity. The modulation of the tracking He-Ne interferometer allows phase interpolation between two He-Ne fringes. Using this information, Asseh et al. [42, 43] were able to accomplish arbitrary phase shifts φ_n between any consecutive subgrating pair n and $(n + 1)$. These phase shifts φ_n can be used to generate extremely narrow transmission resonance peaks within the stopband of the grating (Chapter 5). The phase shifts φ_n can also be used for controlled mismatch between subgratings to accomplish apodization by phase dither. In this way, almost any filter function can be implemented.

4.7.4 Chirped Gratings

There has been a great deal of effort in producing chirped gratings. The motivation behind this drive has been twofold. First, although strong, uniform-period gratings can provide a wide bandwidth, this is accompanied by substantial and unavoidable losses on the short-wavelength side of the Bragg resonance. Ideally, one would fabricate gratings with the desired bandwidth and minimal losses. The second motivation is to exploit applications in telecommunications for dispersion compensation in high-bit-rate transmission systems and in laser cavities. Chirped gratings may be fabricated using all the techniques referred to thus far for inscribing conventional Bragg gratings. The first report of chirped grating fabrication was by Byron et al. [44] who used a conventional two-beam, UV interferometer to produce a uniform-period fringe pattern in a tapered photosensitive fiber, as seen in Figure 4.23(a). In this method the chirp is achieved by the approximately linear variation of the fiber effective index along the tapered section. In another method with the same uniform-period exposure arrangement, chirp can be introduced by bending the fiber with respect to the interference fringes, as seen in Figure 4.23(b) [45]. This in effect results in a fringe separation that varies continuously along the exposed fiber length. This method is capable of producing >99% reflectivity over a 7.5-nm reflection bandwidth in hydrogen-loaded, high-germania fiber without incurring short-wavelength loss.

A far more flexible and controllable approach to chirped grating fabrication relies on two-beam interference and is based on the use of dissimilar curvatures in the interfering wavefronts. This is accomplished by placing two lenses in the two arms of the

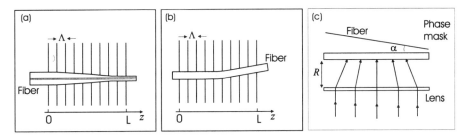

Figure 4.23 Formation of chirped gratings (a) in a tapered fiber, (b) by bending the fiber during exposure, and (c) using a lens and tilting the fiber with respect to a uniform-period phase mask.

interferometer [5]. Figure 4.24(a) shows the general arrangement, where the interfering beams 1 and 2 pass through lenses of focal lengths f_1 and f_2, respectively. The coordinate z is the distance along the grating from the origin, which is defined by the point where the two beam axes intersect on the fiber. The parameters d_1 and d_2 are the distances from the lens' focal point to the point $z = 0$, and θ_1, θ_2 are the angles that the respective beams make with the optical fiber. The fringe spacing $\Lambda(z)$ along the fiber axis may be obtained from the geometry of the interfering beams and has the following form [5]:

$$\Lambda(z) = \frac{\lambda_w}{\dfrac{d_1 \cos(\theta_1) + z}{\sqrt{d_1^2 + 2d_1 z \cos(\theta_1) + z^2}} + \dfrac{d_2 \cos(\theta_2) - z}{\sqrt{d_2^2 - 2d_2 z \cos(\theta_2) + z^2}}} \qquad (4.5)$$

Here, diffraction is not taken into account, and the refractive index modulation is assumed to be sinusoidal. Given (4.5), the Bragg wavelength of the grating is clearly a function of the position z, given by $\lambda_B = 2n_{eff}\Lambda(z)$. A single cylindrical lens used in just one arm of the interferometer results in large-bandwidth, linearly chirped gratings. Figure 4.24(b) shows a plot of $\lambda_B(z)$ against z for a grating with a center wavelength at 1550 nm, where $d_1 = 100$

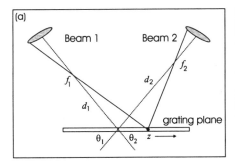
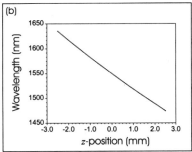

Figure 4.24 (a) Arrangement for the formation of chirped Bragg gratings by interfering dissimilar wavefronts using cylindrical lenses in each arm of the interferometer. (b) Variation of the peak reflected wavelength along the axis of the chirped grating calculated using a single cylindrical lens ($d_1 = 100$ mm) in one arm of the interferometer used in (a).

mm and $d_2 \to \infty$, which corresponds to having no lens in the second arm of the interfero-meter. It clearly indicates an almost linear variation of the reflected wavelength with distance along the grating over a bandwidth exceeding 150 nm. Therefore, the bandwidth of the chirped grating may be selected by the appropriate combination of lens focal length and position. For smaller bandwidths, the single cylindrical lens is substituted by a two-lens arrangement with a magnification of unity. Slight defocusing of the telescope produces either a slowly diverging or converging wavefront. This method offers a fine control over chirp rate and bandwidth. With cylindrical lenses placed in both arms of the interferometer, as seen in Figure 4.24(a), nonlinear chirp profiles may be accomplished. As shown in Figure 4.25, this situation is modeled by a series of plots of $\lambda_B(z)$ against z calculated from (4.5). In all the cases in Figure 4.25 the center wavelength $\lambda_B(z = 0)$ was designed to be ~1544 nm and d_2 was held constant at 100 mm. In each case, the chirp profile is quadratic in shape over the range of z illustrated. Clearly, a greatly varied chirped grating parameter may be obtained by selecting the appropriate d_1 and d_2. The parameter d_1 was varied as follows: (a) $d_1 = 100$ mm, (b) $d_1 = 102$ mm, (c) $d_1 = 105$ mm, and (d) $d_1 = 110$ mm.

The phase mask is probably one of the most precise and controlled techniques for inscribing chirped gratings. In this technique the mask may be divided into subsections, each having its own period that is progressively changing in a linear or other functional form depending on the type of chirped grating required (Section 3.5.1). The number of sections depends on the chirp value and the length of the grating (see Chapter 5). Several other methods have been described in which a uniform phase mask may be used for (e.g., the "stretch-and-write" method mentioned earlier), producing a piecewise chirped grating.

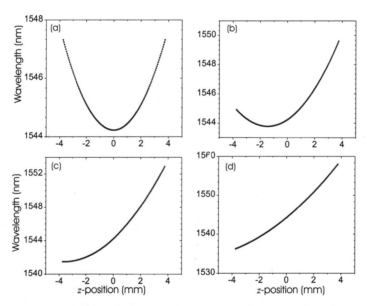

Figure 4.25 Chirped Bragg grating profiles calculated using (4.5) for a 1544-nm center wavelength with $d_2 = 100$ mm and d_1 values of (a) 100 mm, (b) 102 mm, (c) 105 mm, and (d) 110 mm.

In another technique first introduced by Hill et al. [46] a double exposure was used to produce a continuously chirped grating. In the first exposure, an opaque mask was positioned between the fiber and the single UV exposing beam. The mask was translated through the beam at constant velocity, thereby continuously increasing the length of fiber exposed and so photo-inducing a linear variation of the index of refraction $(n_{eff}(z))$ in the irradiated section of the fiber. In the second exposure, a grating with uniform fringe spacing was written into the same section of the fiber using a phase mask. As a result of the linear variation of the index of refraction along the fiber, the resultant grating was chirped. The important parameter here is the first exposure of the fiber and in particular the linear dependence of the index of refraction along the grating. Another technique for inscribing chirped gratings with a uniform phase mask uses a lens (focal length f) at distance r from the mask, with the fiber tilting at angle α with respect to the mask [47]. Using the arrangement shown in Figure 4.23(c), one can show that [47] a grating with period $\Lambda(z)$ may be given by the following expression:

$$\Lambda(z) = \frac{\Lambda_{pm}}{2}\left[1 - z\left(\frac{\alpha}{f-r}\right)\left(1 - \frac{\lambda_w^2}{\Lambda_{pm}^2}\right)^{-\frac{1}{2}}\right] \tag{4.6}$$

where Λ_{pm} is the phase period, λ_w is the writing beam wavelength, and z is the distance along the fiber. Equation 4.6 shows a linear variation of the grating period with distance z. Using a tilting angle of $\alpha = 1.6$ degrees, gratings with peak reflectivity of 96% and spectral bandwidth of 6 nm have been produced in hydrogen-loaded standard telecommunications fiber. The point-by-point technique may also be used to fabricate chirped gratings; however, it is a tedious and time-consuming process. A number of other methods have been employed using strain [48] or temperature [49] gradients to introduce chirp into the already written uniform-period grating. These types of chirped gratings are intended to be used for dispersion compensation in telecommunications systems where a carefully controlled chirp profile is more important than a strong reflected signal.

4.7.5 Phase-Shifted Gratings

The phase-shifted gratings (Section 3.7.3) may be fabricated either directly or via post-processing after the grating has been written. In the direct approach the phase mask may be constructed with a phase shift incorporated into the corrugated structure of the mask [50]. The phase-mask approach is probably the most controlled and precise way of fabricating phase-shifted gratings in optical fibers. Another approach is to inscribe the first section of the grating on the fiber, move the fiber by an amount corresponding to the required phase shift, and then inscribe the second part of the grating structure. This technique requires an interferometrically controlled translation stage for accurate displacement of the fiber. Other post-processing techniques include: exposing a grating region to pulses of UV laser radiation to alter the index of refraction, which in effect inserts a phase shift at that particular area [51], and or localized heat treatment to achieve a similar result [52]. Unfortunately, neither of these post-processing methods is well controlled.

4.7.6 Apodization of Gratings

Apodization [53–55], in practice, corresponds to the modulation of the index of refraction of the fiber grating with a much larger period than the period of the grating. This may be accomplished using various techniques, such as, exposing the optical fiber with the interference pattern from two nonuniform beams. An alternative approach uses a phase mask with apodization achieved by varying the exposure time along the length of the grating, either from a double exposure, or by scanning a small writing beam or using a variable diffraction efficiency phase mask.

4.8 Hydrogenation

Regardless of which technique is used to fabricate Bragg gratings in optical fibers, one requires a fiber that is photosensitive to the irradiating wavelength. As discussed in Chapter 2, fibers that are doped with high concentrations of germanium have been shown to be photosensitive to ultraviolet irradiation in the range between 240–250 nm and in the 193-nm band. Unfortunately, standard telecommunications fibers, where Bragg gratings are expected to have a strong applications impact, are not photosensitive. Research carried out to photosensitize standard fibers led Lemaire et al. [56] in 1993 to report on a new technique for achieving high photosensitivity in normal GeO_2-doped fiber. This technique involved optical fibers being soaked in high-pressure hydrogen for several weeks to allow hydrogen molecules to diffuse into the core of the optical fibers. These optical fibers were hydrogen loaded at high pressure between 500–1000 psi (Chapter 2).

In view of the fact that fibers have to be hydrogen loaded for periods of up to several weeks, it becomes evident that having only one hydrogen-loading fiber chamber is not very efficient. In the design of a hydrogen-loading system, safety is of primary concern given that one is dealing with an explosive gas under high pressure. The fiber chambers are best made from copper tubing of small diameter (i.e., ¼ inch). The ¼ -inch copper tubing can hold approximately 20 optical fibers in a small volume, which minimizes the amount of hydrogen required to fill the chamber. A typical system for hydrogen loading may consist of several chambers, each connected to the gas system via a two-way valve that maintains the pressure in each fiber chamber and isolates them from the rest of the system. The fiber chambers are pumped out with the vacuum pump each time before being filled with high-pressure hydrogen. Heating elements can be placed at various positions along the fiber chamber heating up the hydrogen gas and accelerating the diffusion of hydrogen molecules into the core of the optical fiber.

The concentration of hydrogen molecules and the rate at which these molecules diffuse into the core of the optical fiber depend on the pressure and temperature of the hydrogen gas. The equilibrium solubility of the hydrogen (the saturated hydrogen concentration in the fiber core), κ_{sat} is expressed as [56]

$$\kappa_{sat} = 3.3481 p \exp\left[\frac{8.67 \text{ kJ/mol}}{RT}\right] \quad \text{(ppm)} \qquad (4.7)$$

where p is the pressure of the hydrogen gas in units of atmospheres, T is the absolute temperature, and R is the gas constant. The saturated hydrogen concentration increases linearly with pressure and decreases as the temperature increases. Here, 1 ppm is defined as 10^{-6} moles of H_2 per mole of SiO_2. The diffusivity of hydrogen molecules in silica is expressed by [56]

$$d_{H_2} = 2.83 \times 10^{-4} p \exp\left[\frac{-40.19 \text{ kJ/mol}}{RT}\right] \quad (\text{cm}^2/\text{s}) \tag{4.8}$$

and increases with temperature. Hence, at higher temperatures the saturation level is reached considerably faster. Using the above expression in conjunction with classical diffusion solutions for a cylindrical geometry, the diffusion rate of hydrogen into optical fibers can be calculated from the following equation:

$$\frac{\kappa}{\kappa_{\text{sat}}} = 1 - 2\sum_{n=1}^{\infty} \frac{1}{\beta_n J_1(\beta_n)} \exp\left[-\beta_n^2 \frac{d_{H_2} t}{r^2}\right] \tag{4.9}$$

where κ is the hydrogen concentration in the fiber core, β_n is the nth zero of the zeroth-order Bessel function $J_0(\beta)$, t is the time, r is the fiber radius (62.5 mm for most single-mode fibers), and $J_1(\beta)$, is the first-order Bessel function. Figure 4.26 shows two graphs that illustrate the effect of pressure and temperature on the hydrogen concentration saturation level and the diffusion rate of H_2 into the core of the optical fiber. These plots were evaluated using the above equations. A comparison of the plots in Figure 4.26(a) illustrates the linear relation between pressure and hydrogen concentration saturation level. It also shows that a change in pressure does not affect the diffusion rate of H_2 into the core of the optical fiber. A comparison of graphs (a) and (b) illustrates that a decrease in temperature results in an increase in the hydrogen concentration saturation level and a significant decrease in the diffusion rate of H_2 into the core of the optical fiber.

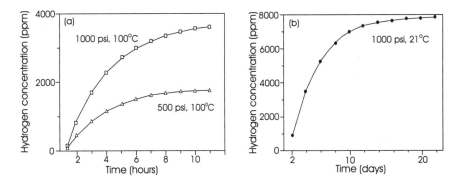

Figure 4.26 Build-up of hydrogen concentration in the fiber core as a function of time for various pressures and temperatures.

4.9 Fabrication of Bragg Gratings Through Polymer Jacket

An expensive and time consuming step for Bragg grating fabrication is the removal of the protective polymer coating of the fiber, decreasing its mechanical strength, in order to expose the fiber core with UV radiation. One method to avoid fiber stripping and recoating is to fabricate Bragg gratings directly through the polymer coating. Unfortunately, standard polymer coatings are not sufficiently transparent at the usual writing wavelengths at 240 and 193 nm. An alternative approach is to use near-UV light around 334 nm [57], a wavelength that is far less destructive to the polymer coating. Moreover, most of the standard polymer coatings are transparent to near-UV light. Recently, Starodubov et al.[58] have reported the fabrication of strong Bragg gratings with a reflectivity of >15 dB through a common polymer in hydrogen-loaded fibers using near-UV light. The polymer coating thickness was approximately 40 μm and the concentration of germanium in the core of the fiber was estimated to be 20 mol%. Figure 4.27(a) shows the transmission spectrum of 150 μm silicone resin film, which was used to coat the fiber, demonstrating a negligible absorption above 300 nm. Figure 4.27(b) shows the spectrum of a 7 mm long grating fabricated with a phase mask using ~200 mW of Ar laser light at 302 nm and 4 minutes exposure time. No mechanical degradation or visible changes in the polymer layer after near-UV exposure is observed. Detailed measurements of the fiber strength using bending techniques demonstrate no strength degradation in exposed fibers.

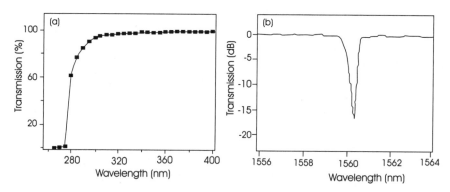

Figure 4.27   (a) Transmission spectrum of 150-μm thick film of silicone resin used for fiber coating. (b) Spectral response from a Bragg grating written through a ~40-μm thick polymer fiber coating using near UV light (*After:* [58]).

References

[1]   Hill, K. O., et al. "Photosensitivity in optical fiber waveguides: Application to reflection filter fabrication," *Applied Physics Letters,* Vol. 32, 1978, pp. 647–649.

[2]   Kawasaki, B. S., et al. "Narrow-band Bragg reflectors in optical fibers," *Optics Letters*, Vol. 3, 1978, pp. 66–68.

[3]   Meltz, G., W. W. Morey, and W. H. Glenn, "Formation of Bragg gratings in optical fibers by a transverse holographic method," *Optics Letters*, Vol. 14, 1989, pp. 823–825.

[4]   Dockney, M. L., J. W. James and R.P. Tatam, "Fiber Bragg grating fabricated using a wavelength tunable source and a phase-mask based interferometer," *Measurements Science and Technology*, Vol. 7, 1996, pp. 445.

[5]   Bennion, I., et al. "UV-written in-fiber Bragg gratings," *Optical and Quantum Electronics*, Vol. 28, 1996, pp. 93–135.

[6]   Kashyap, R., et al. "All-fiber narrow band reflection grating at 1500 nm," *Electronics Letters*, Vol. 26, 1990, pp. 730–732.

[7]   Eggleton, B. J., P. A. Krug, and L. Poladian, "Experimental demonstration of compression of dispersed optical pulses by reflection from self-chirped optical fiber Bragg gratings," *Optics Letters*, Vol. 19, 1994, pp. 877–880.

[8]   Limberger, H. G., et al. "Photosensitivity and Self-Organization in Optical Fibers and Waveguides," *SPIE*, Vol. 2044, 1993, pp. 272.

[9]   Hill, K. O., et al. "Bragg gratings fabricated in monomode photosensitive optical fiber by UV exposure thorough a phase mask," *Applied Physics Letters*, Vol. 62, 1993, pp. 1035–1037.

[10]   Anderson, D. Z., et al. "Production of in-fiber gratings using a diffractive optical element," *Electronics Letters*, Vol. 29, 1993, pp. 566–568.

[11]   Othonos, A., and X. Lee, "Novel and improved methods of writing Bragg gratings with phase-masks," *IEEE Photonics Technology Letters*, Vol. 7, 1995, pp. 1183–1185.

[12]   Dyer, P. E., R. J. Farley, and R. Giedl, "Analysis of grating formation with excimer laser irradiated phase masks," *Optics Communications*, Vol. 115, 1995, pp. 327.

[13]   Kashyap, R., et al. "Light-sensitive optical fibers and planar waveguides," *BT Technol. J.*, Vol. 11, 1993.

[14]   Dyer, P. E., R. J. Farley, and R. Giedl, "Analysis and application of a 0/1 order Talbot interferometer for 193 nm laser grating formation," *Optics Communications*, Vol. 129, 1996, pp. 98–108.

[15]   Liu, X., et al. "The influence of phase mask stitch errors on the performance of UV-written Bragg gratings," *Bragg gratings, photosensitivity, and poling in glass Fibers and waveguides: Applications and fundamentals*, Vol. 17, 1997, BMG9–1.

[16]   Liu, X., et al. "Electron-beam lithography of phase-mask gratings for near-field holographic production of optical fiber gratings," *Microelectronics Engineering*, Vol. 35, 1997, pp. 345.

[17]   Prohaska, J. D., et al. "Magnification of mask fabricated fiber Bragg gratings," *Electronics Letters*, Vol. 29, 1993, pp. 1614–1616.

[18]   Zhang, Q., et al. "Tuning Bragg wavelength by writing gratings on prestrained fibers," *IEEE Photonics Technology Letters*, Vol. 6, 1994, pp. 839–842.

[19]   Byron, K. C., and H. N. Rouke, "Fabrication of chirped fiber gratings by novel stretch and write technique," *Electronics Letters*, Vol. 31, 1995, pp. 60–61.

[20]   Malo, B., et al. "Point-by-point fabrication of micro-Bragg gratings in photosensitive fiber using single excimer pulse refractive index modification techniques," *Electronics Letters*, Vol. 29, 1993, pp. 1668–1669.

[21]   Hill, K. O., et al. "Efficient mode conversion in telecommunication fiber using externally written gratings," *Electronics Letters*, Vol. 26, 1990, pp. 1270–1272.

[22]   Hill, K. O., et al. "Birefringent photosensitivity in monomode optical fiber: Application to the external writing of rocking filters," *Electronics Letters*, Vol. 27, 1991, pp. 1548–1550.

[23]   Mihailov, S. and M. Gower, "Recording of efficient high-order Bragg reflectors in optical fibers by mask image projection and single pulse exposure with an excimer laser," *Electronics Letters*, Vol. 30, 1994, pp. 707–708.

[24]   Burnham, R., N. W. Harris, and N. Djeu, *Applied Physics Letters*, Vol. 28, 1976, pp. 86.

[25]   Gower M. C., et al. "Efficient line-narrowing and wavelength stabilization of excimer lasers," OE/Technology'92 Excimer Lasers, Boston MA, 1992.

[26]   Othonos, A., and X. Lee, "Narrow linewidth excimer laser for inscribing Bragg gratings in optical fibers," *Review of Scientific Instruments*, Vol. 66, 1995, pp. 3112–3115.

[27]   Cannon, J. and S. Lee "Fiberoptic Product News," *Laser Focus World*, Vol. 2, 1994, pp.5051.

[28]   Askins, C. G., et al. "Fiber Bragg reflectors prepared by a single excimer pulse," *Optics Letters,* Vol. 17, 1992, pp.833–835.

[29]   Archambault, J. L., L. Reekie and P. St. J. Russel, "High reflectivity and narrow bandwidth fiber gratings written by single excimer pulse," *Electronics Letters*, Vol. 29, 1993, pp.28–29.

[30]   Archambault, J. L., L. Reekie and P. St. J. Russel, "100% reflectivity Bragg reflectors produced in optical fibers by single excimer laser pulses," *Electronics Letters*, Vol. 29, 1993, pp.453–454.

[31]   Malo, B., et al. "Single-excimer–pulse writing of fiber gratings by use of a zero-order mulled phase mask: Grating spectral response and visualization of index perturbations," *Optics Letters*, Vol. 18, 1993, pp.1277–1279.

[32]   Bennion, I., et al. *Electronics Letters*, Vol. 22, 1986, pp.341.

[33]   Dyer, P. E., et al. "High reflectivity fiber gratings produced by incubated damage using a 193 nm ArF laser," *Electronics Letters,* Vol. 30, 1994, pp.860–861.

[34]   Dong, L., et al. "Single pulse Bragg gratings written during fiber drawing," *Electronics Letters*, Vol.29, 1993, pp. 1577–1578.

[35]   Askins, C. G., et al. "Stepped-wavelength optical-fiber Bragg grating array fabricated in line on a draw tower," *Optics Letters,* Vol. 19, 1994, pp.147–149.

[36]   Martin, J. and F. Ouellette, "Novel writing technique of long and highly reflective in-fiber gratings," *Electronics Letters*, Vol. 30, 1994, pp.811–812.

[37]   Kashyap, R., et al. "Novel method of producing all fiber photoinduced chirped gratings," *Electronics Letters*, Vol. 30, 1994, pp. 996–997.

[38]   Hill, K. O., et al. "Aperiodic in-fiber Bragg gratings for optical fiber dispersion compensation," *Proc. OFC'04*, PF–77 postdeadline paper, 1994.

[39]   Loh, H.W., et al. "Complex grating structures with uniform phase masks based on the moving fiber-scanning beam technique," *Optics Letters*, Vol. 20, 1995, pp. 2051.

[40]   Kashyap, R., et al. "Superstep-chirped fiber Bragg gratings," *Electronics Letters*, Vol. 32, 1996, pp.1394–1395.

[41]   Eriksson, U., P. Blixt, and J. A. Tellefsen Jr., "Design of fiber gratings for total dispersion compensation," *Optics Letters*, Vol. 19, 1994, pp. 1028.

[42]   Asseh, A., et al. "A writing technique for long fiber Bragg gratings with complex reflectivity profiles," *Journal of Lightwave Technology,* Vol. 15, 1997, pp. 1419–1423.

[43]   Asseh, A., et al. "10 cm Yb^{+3} DFB fiber laser with permanent phase shifted grating," *Electronics Letters*, Vol. 31, 1995, pp. 969–970.

[44]   Byron, K. C., et al. "Fabrication of chirped Bragg gratings in photosensitive fiber," *Electronics Letters*, Vol. 29, 1993, pp.1659–1660.

[45]   Sugden, K., et al. "Chirped gratings produced in photosensitive optical fiber deformation during exposure," *Electronics Letters*, Vol. 30, 1994, pp.440–441.

[46]   Hill, K. O., et al. "Chirped in- fiber Bragg grating for compensating of optical-fiber dispersion," *Optics Letters*, Vol. 19, 1994, pp.1314–1316.

[47]   Painchaud Y., A. Chandonnet and J. Lauzon, "Chirped fiber gratings produced by tilting the fiber," *Electronics Letters*, Vol. 31, 1995, pp.171–172.

[48]   Hill, P. C. and B. J. Eggleton, "Strain gradient chirp of Bragg gratings," *Electronics Letters*, Vol. 30, 1994, pp.1172–1173.

[49]   Lauzon, J., et al. "Implementation and characterization of fiber Bragg grating linearly chirped by a temperature gradient," *Optics Letters*, Vol. 19, 1994, pp.2027–2030.

[50]   Kashyap, R., P. F. McKee, and D. Armes, "UV written reflection grating structures in photosensitive optical fibres using phase-shifted phase-masks," *Electronics Letters*, Vol. 30, 1994, pp. 1977–1978.

[51]   Canning, J., and M. G. Sceats, "Π-phase-shifted periodic distributed structures in optical fibres by UV post-processing," *Electronics Letters*, Vol. 30, 1994, pp. 1344–1345.

[52]   Uttamchandani, D., and A. Othonos, "Phase shifted Bragg gratings formed in optical fibres by post-fabrication thermal processing," *Optics Communications*, Vol. 127, 1996, pp. 200–204.

[53]   Albert, J., et al. 'Apodisation of the spectral response of fibre Bragg gratings using a phase-mask with

variable diffraction efficiency," *Electronics Letters*, Vol. 31, 1995, pp. 222–223.

[54]   Malo, B., et al. "Apodised in-fibre Bragg grating reflectors photoimprinted using a phase mask," *Electronics Letters*, Vol. 31, 1995, pp. 223–225.

[55]   Kashyap, R., A. Swanton, and D. J. Armes, "Simple technique for apodising chirped and unchirped fibre Bragg gratings," *Electronics Letters*, Vol. 32, 1996, pp. 1226–1228.

[56]   Lemaire, P. J., "Reliability of optical fibers exposed to hydrogen: prediction of long-term loss increases," *Optical Engineering*, Vol. 30, 1991, pp. 780–781.

[57]   Starodubov, et al. "Bragg grating fabrication in germanosilicate fibers by use of near-UV light; a new pathway for refractive-index changes," *Optics Letters*, Vol. 22, 1997, pp.1086.

[58]   Starodubov, D.S., et al. "Fiber Bragg gratings with reflectivity >97% fabricated thorough polymer jacket using near-UV light," Bragg gratings, photosensitivity and poling in glass fibers and waveguides: Application and Fundamentals, OSA 1997 Technical Digest Series Volume 17, 328/PDP1–2 (1997).

Chapter 5

FIBER BRAGG GRATING
THEORY

5.1 Introduction

Thus far we have shown that an understanding of fiber Bragg gratings encompasses a variety of different fields in the physical sciences, from the detailed description of fiber photosensitivity in Chapter 2, to the fabrication of fiber gratings and the multitude of important physical properties of gratings discussed in the chapters that followed. Now we will focus on the optical properties of fiber Bragg gratings, providing the reader with an intuitive understanding and the necessary tools for designing fiber gratings. It is also important to look at the different theoretical treatments of the Bragg grating. In this chapter we will investigate key models used in simulating Bragg grating behavior, as each approach often offers a unique insight into the physical mechanism of the grating-electric field interaction. Recent developments in fiber and integrated optics fabrication methods have resulted in the capability of producing a myriad of periodic and aperiodic structures, and it has become necessary to carefully model this behavior. Of course, our interest is focussed on fiber Bragg gratings, for which almost any chosen grating period (uniform or chirped) and apodization are possible. Examples of the applications of such structures include distributed feedback lasers [1] and all fiber lasers, optical filters for WDM systems [2], pulse compression, dispersion compensation in both digital [3, 4] and analog [5] optical communications, and optical sensor systems [6]. In parallel with this interest in such structures several methods have been developed for the analysis of the field propagation in corrugated structures. The most widely used of these techniques has been the coupled-mode theory [7] where the counter-propagating fields inside the grating structure, obtained by convenient perturbation of the fields in the unperturbed waveguide, are related by coupled differential equations. The coupled-mode theory was initially developed for uniform gratings; however, Kogelnik [8] extended the model to cover aperiodic structures. The coupled-mode approach in the most general case, and for complicated grating structures, involves the numerical solution of two coupled differential equations, since analytic solutions are only possible for the uniform gratings. Matrix methods such as the effective index method (EIM) [9] and the transfer matrix method (TMM) [10] have also been developed for the purpose of grating analysis. In the effective index method the grating is divided into sections, with the length of each one being much smaller than the smallest value of the corrugation period. The fields are computed inside each section, under the hypothesis that the refractive index remains constant, using the effective index method of integrated optics. The electric fields in each section are impedance matched to those of its preceding and succeeding sections, yielding a matrix relationship between the fields at the left and right part of each section. A global matrix, obtained from the multiplication of the individual matrices of its sections, characterizes the overall structure. This approach is especially suitable for integrated optic gratings, where the maximum structure lengths are of the order of a few millimeters. This technique, however, may require excessive computation time for fiber gratings where the structure lengths are typically in the centimeter range.

In the transfer matrix method the grating is divided into sections, with the length of each one being much bigger than the biggest period of the corrugation. Furthermore, the index variation inside each one of these sections is such that they can be considered uniform gratings. Each of these sections is described by a transfer matrix corresponding to a uniform grating, and an overall structure is characterized by a global matrix obtained as the product of the individual matrices. This approach is suitable for periodic and aperiodic structures, as well as for long gratings. Several other less-known techniques have also been developed for analyzing grating structures. For example, Weller-Brophy and Hall [11] presented an extension of the Rouard's method in thin-film design for the characterization of waveguide gratings, and Frolik and Yagle [12] developed an elegant discrete-time approach, based on digital signal processing (DSP) formulation for the analysis of periodic gratings. The advantage of the DSP formulation is that fast algorithms may be employed in analyzing both the determination of the grating response for an arbitrary structure and the inverse problem of the determination of the grating structure from its response. This formulation also includes all multiple reflections, scattering and absorption losses, and, as such, constitutes an exact grating description. Other methods based on more fundamental approaches such as the WKB (phase-integral) [13], Hamiltonian [14], and variational [15] principles have been developed and are available in the literature.

An alternative approach to the above techniques for grating analyzing is the Bloch wave approach [16]. Optical Bloch waves are the eigenmodes of periodic media in the same way as plane waves are the natural modes of free space propagation. It is interesting to compare the philosophies behind the coupled-mode and Bloch wave approaches. The latter involves first solving the dispersion relation to obtain the full set of permitted normal modes for the periodic medium. The modes that are excited by matching to the incident plane wave spectrum along the boundaries are then identified. The coupled-mode approach attempts a "one-stage" system solution, the interpretation of which can be difficult. We shall begin with a description of the two-mode coupling in uniform and non-uniform gratings, followed by an examination of tilted gratings and coupling to cladding and radiation modes. Some of the more tractable modeling approaches will be discussed and compared to the coupled-mode theory. We close by presenting an outline of the nonlinear optical properties of fiber Bragg gratings.

Exposing photosensitive fiber to a spatially varying pattern of ultraviolet light (see previous chapters) produces the necessary refractive index perturbation that produces the fiber gratings. We assume that the resultant perturbation to the index of refraction n_{eff} of the guided modes of interest is given by the following expression [17]:

$$\delta n_{eff}(z) = \overline{\delta n}_{eff}(z)\left[1 + s\cos\left(\frac{2\pi}{\Lambda}z + \varphi(z)\right)\right] \qquad (5.1)$$

where s is the fringe visibility associated with the index change, Λ is the grating period, $\varphi(z)$ accounts for the grating chirp, and $\overline{\delta n}_{eff}$ is the "dc" index change spatially averaged over the grating period, or the slowly varying envelope of the grating. For a fiber with a step-index profile and a uniform induced index change across the core $\delta n_{co}(z)$, one finds that $\delta n_{eff} \sim \Phi\delta n_{co}$ where Φ is the core power confinement factor for the mode of interest. Typically, the fundamental LP_{01} propagation mode for a single-mode fiber, for which the normalized frequency $V < 2.4$, gives a confinement factor of 0.8 and an effective index parameter b of ~0.5. $V = (2\pi/\lambda)a(n_{co}^2 - n_{cl}^2)^{1/2}$ where a is the core radius, and n_{co} and n_{cl} the core and cladding indices, respectively. The effective index is related to the parameter b by the simple expression $b = (n_{eff}^2 - n_{cl}^2)/(n_{co}^2 - n_{cl}^2)$. Knowledge of V and b results in determination of the confinement factor through

$$\Phi = \frac{b^2}{V^2}\left[1 - \frac{J_l^2(V\sqrt{1-b})}{J_{l+1}(V\sqrt{1-b})J_{l-1}(V\sqrt{1-b})}\right] \qquad (5.2)$$

where l is the azimuthal order of the mode and J is the Bessel function of the first kind. We have made the assumption that the grating is confined to the core of the fiber. Therefore, the mode confinement factor approaching unity tells us that the grating is effective in scattering to the extent that the fiber mode is confined to the core, as one may have anticipated given the core's inherent photosensitivity.

5.2 Coupled-Mode Theory

The coupled-mode theory is often use as a technique for obtaining quantitative information about the diffraction efficiency and spectral dependence of fiber gratings. It is one of the most popular techniques utilized in describing the behavior of Bragg gratings, mainly due to its simplicity and accuracy in modeling the optical properties of most fiber gratings of interest. We will not provide a derivation of the coupled-mode theory since it is described in detail in numerous articles and texts [1, 7]. The derivation in this section closely follows the work by Erdogan [18]. We begin by writing the transverse component of the electric field in the ideal-mode approximation to coupled-mode theory as a superposition of the ideal modes (the modes in an ideal waveguide where no grating perturbation exist). Given that the modes are labeled with index m, we then have

$$\mathbf{E}^T(x,y,z,t) =$$
$$\sum_m\left[A_m(z)\exp(i\beta_m z) + B_m(z)\exp(-i\beta_m z)\right]\mathbf{e}_m^T(x,y)\exp(-i\omega t) \qquad (5.3)$$

where the coefficients $A_m(z)$ and $B_m(z)$ are slowly varying amplitudes of the mth mode traveling in the $+z$ and $-z$ directions, respectively, and the propagation constant β is simply $\beta = (2\pi/\lambda)n_{\text{eff}}$. The transverse mode field $e_m^T(x,y)$ might describe the bound-core or radiation LP modes, or they might describe cladding modes. In an ideal waveguide situation, the modes are orthogonal and there is no mechanism available for the exchange of energy between them. However, the presence of a dielectric perturbation associated with the fiber grating forces coupling between the modes. In that case the amplitudes A_m and B_m of the mth mode evolve along the z direction according to

$$\frac{dA_m}{dz} = i\sum_q A_q(C_{qm}^T + C_{qm}^L)\exp[i(\beta_q - \beta_m)z]$$
$$+ i\sum_q B_q(C_{qm}^T - C_{qm}^L)\exp[-i(\beta_q + \beta_m)z] \tag{5.4}$$

$$\frac{dB_m}{dz} = -i\sum_q A_q(C_{qm}^T - C_{qm}^L)\exp[i(\beta_q + \beta_m)z]$$
$$- i\sum_q B_q(C_{qm}^T + C_{qm}^L)\exp[-i(\beta_q - \beta_m)z] \tag{5.5}$$

The transverse coupling coefficient between the m and q modes in the above equations is given by the following integral:

$$C_{qm}^T(z) = \frac{\omega}{4}\iint_\infty \Delta\varepsilon(x,y,z)e_q^T(x,y)\cdot e_m^{T*}(x,y)dx\,dy \tag{5.6}$$

where $\Delta\varepsilon(x,y,z)$ is the permittivity perturbation, which is approximately $2n\delta n$ for δn much smaller than n. The longitudinal coupling coefficient $C_{qm}^L(z)$ is defined in a similar way to the above transverse coupling coefficient $C_{qm}^T(z)$; however, for fiber modes $C_{qm}^L(z)$ is usually neglected since $C_{qm}^L \ll C_{qm}^T$.

In most fibers considered, the UV-induced index changes $\delta n(x,y,z)$ are approximately uniform across the fiber core and negligible outside the core. Following this assumption we may describe the core index change by an expression similar to that of (5.1), where $\overline{\delta n}_{\text{eff}}(z)$ is replaced by $\overline{\delta n}_{\text{co}}(z)$. Now let us define two new coefficients, the self- and cross-coupling coefficients, in the following way [18]:

$$\zeta_{qm}(z) = \omega\frac{n_{\text{co}}}{2}\overline{\delta n}_{\text{co}}(z)\iint_{\text{core}} e_q^T(x,y)\cdot e_m^{T*}(x,y)dx\,dy \tag{5.7}$$

$$\kappa_{qm}(z) = \frac{s}{2}\zeta_{qm}(z) \tag{5.8}$$

where $\zeta_{qm}(z)$ is a "dc" coupling coefficient and $\kappa_{qm}(z)$ is an "ac" coupling coefficient. Thus, the general coupling coefficient may now be written as

$$C_{qm}^T(z) = \zeta_{qm}(z) + 2\kappa_{qm}(z)\cos\left[\frac{2\pi}{\Lambda}z + \varphi(z)\right] \tag{5.9}$$

Equations 5.4 through 5.9 are the coupled-mode equations that can be used to describe the spectral response of fiber gratings.

5.2.1 Bragg Gratings

For a Bragg grating the dominant interaction lies near the wavelength for which reflection occurs from a mode of amplitude $A(z)$ into an identical counter-propagating mode of amplitude $B(z)$. Under such conditions (5.4) and (5.5) may be simplified [18, 19] to the following equations:

$$\frac{dA^+}{dz} = i\zeta^+ A^+(z) + i\kappa B^+(z) \tag{5.10}$$

$$\frac{dB^+}{dz} = -i\zeta^+ B^+(z) - i\kappa^* A^+(z) \tag{5.11}$$

where $A^+(z) = A(z)\exp(i\delta_d z - \varphi/2)$, $B^+(z) = B(z)\exp(-i\delta_d z + \varphi/2)$, and ζ^+ is the general "dc" self-coupling coefficient defined as

$$\zeta^+ = \delta_d + \zeta - \frac{1}{2}\frac{d\varphi}{dz} \tag{5.12}$$

with δ_d being the detuning, which is independent of z and is defined in the following way:

$$\begin{aligned} \delta_d &= \beta - \frac{\pi}{\Lambda} \\ &= 2\pi n_{\text{eff}}\left[\frac{1}{\lambda} - \frac{1}{\lambda_d}\right] \end{aligned} \tag{5.13}$$

here $\lambda_d = 2n_{\text{eff}}\Lambda$ is the *design peak reflection wavelength* for an infinitesimally weak index of refraction change grating ($\delta n_{\text{eff}} \to 0$). A complex coefficient ζ can describe the absorption loss in the grating where the power loss coefficient will be given by $a = 2\text{Im}(\zeta)$. For a single-mode Bragg reflection we find the following simplified relations [18]:

$$\zeta = \frac{2\pi}{\lambda}\overline{\delta n}_{\text{eff}} \tag{5.14}$$

$$\kappa = \kappa^* = \frac{\pi}{\lambda}s\overline{\delta n}_{\text{eff}} \tag{5.15}$$

If the grating is uniform along the z direction, then δn_{eff} is constant and $d\varphi/dz = 0$ (i.e., there is no grating chirp). Thus, κ, ζ, and ζ^+ are constants. This simplifies (5.10) and (5.11) into coupled first-order ordinary differential equations with constant coefficients. One may arrive at a closed-form solution to these equations given the appropriate boundary conditions. For a uniform fiber grating of length L the reflectivity can be found assuming a forward-propagating wave incident from $z = -\infty$, while requiring that no backward-propagating wave exist for $z \geq L/2$. The amplitude $\rho = B^*(-L/2) / A^*(-L/2)$ and the power reflection coefficients $R = |\rho|^2$ can be shown to be

$$\rho = \frac{-\kappa \sinh \sqrt{(\kappa L)^2 - (\zeta^+ L)^2}}{\zeta^+ \sinh \sqrt{(\kappa L)^2 - (\zeta^+ L)^2} + i\sqrt{\kappa^2 - \zeta^{+2}} \cosh \sqrt{(\kappa L)^2 - (\zeta^+ L)^2}} \tag{5.16}$$

and

$$R = \frac{\sinh^2 \sqrt{(\kappa L)^2 - (\zeta^+ L)^2}}{-\frac{\zeta^{+2}}{\kappa^2} + \cosh^2 \sqrt{(\kappa L)^2 - (\zeta^+ L)^2}} \tag{5.17}$$

Figure 5.1 shows the reflectivity of a uniform Bragg grating calculated from (5.17) for $\kappa L = 2$ and $\kappa L = 10$. The two curves are plotted against the normalized wavelength

$$\frac{\lambda}{\lambda_{max}} = \frac{1}{1 + \zeta^+ L/\pi N} \tag{5.18}$$

where N is the total number of grating periods ($N = L/\Lambda$), and λ_{max} is the wavelength at which maximum reflectivity occurs. For this calculation it was assumed that $N = 10,000$. It is interesting to note that for a given κL with increasing N the reflectivity bandwidth becomes narrower (i.e., longer gratings produce narrower spectral linewidths), as expected.

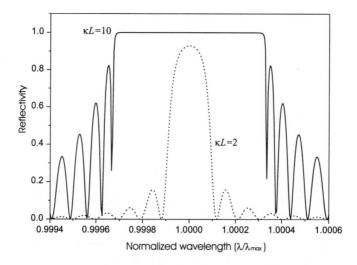

Figure 5.1   Reflection spectral response versus normalized wavelength for a uniform Bragg gratings with $\kappa L = 2$ and $\kappa L = 10$.

Using (5.17) we find that the maximum reflectivity for the Bragg grating is

$$R_{max} = \tanh^2(\kappa L) \qquad (5.19)$$

This maximum occurs when $\zeta^+ = 0$, or at the wavelength of

$$\lambda_{max} = \left(1 + \frac{\overline{\delta n}_{eff}}{n_{eff}}\right)\lambda_d \qquad (5.20)$$

A bandwidth for the uniform Bragg grating may be defined as the width between the first zeros on either side of the maximum reflectivity [18]. Thus, from (5.17) we find

$$\frac{\Delta\lambda_0}{\lambda} = \frac{s\overline{\delta n}_{eff}}{n_{eff}}\sqrt{1 + \left(\frac{\lambda_d}{s\overline{\delta n}_{eff}L}\right)^2} \qquad (5.21)$$

For the case where the index of refraction change is weak (weak-grating limit) $s\overline{\delta n}_{eff}$ is very small; thus, $s\overline{\delta n}_{eff} \ll \lambda_d/L$ and

$$\frac{\Delta\lambda_0}{\lambda} \rightarrow \frac{\lambda_d}{n_{eff}L} = \frac{2}{N} \qquad (5.22)$$

which implies that the bandwidth of weak gratings is limited by their length (length limited). However, in the case of strong gratings where $s\overline{\delta n}_{eff} \gg \lambda_d/L$,

$$\frac{\Delta\lambda_0}{\lambda} \rightarrow \frac{s\overline{\delta n}_{eff}}{n_{eff}} \qquad (5.23)$$

In the strong grating limit, since light does not penetrate the full length of the grating, the bandwidth is independent of the length; however, it is directly proportional to the index of refraction change. This means that the bandwidth is similar whether measured at the band edges, at the first zeros, or as the full width half maximum.

Figure 5.2 shows an excellent agreement of the theoretical calculation (solid line) with the experimental data (square points) from a 2-mm uniform Bragg grating. The design wavelength of the grating was approximately 1550 nm with an induced index change estimated to be approximately 3×10^{-4}.

The dispersive properties of Bragg gratings are of interest because of their applications in dispersion compensation, pulse shaping, and fiber and semiconductor laser components. The group delay and dispersion of the reflected light from the grating can be determined from the phase of the amplitude reflection coefficient ρ given in (5.16). Let this phase be denoted by ψ_ρ. Since the first derivative $d\psi_\rho/d\omega$, from a Taylor series expansion of ψ_ρ about a local frequency ω_0, is directly proportional to the frequency ω, this quantity can be identified as a time delay. Therefore, the time delay τ for the light reflected back from the grating is

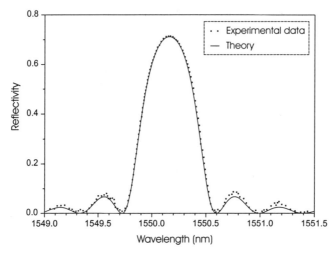

Figure 5.2   Calculated reflection data (solid line) and experimental result (square points) from a 2-mm long uniform Bragg grating, with a design wavelength of 1550 nm. The index of refraction change was estimated to be approximately 3×10^{-4}.

$$\tau = \frac{d\psi_\rho}{d\omega} = \left(-\frac{\lambda^2}{2\pi c}\right)\left(\frac{d\psi_\rho}{d\lambda}\right) \tag{5.24}$$

The dispersion D, with units of picoseconds per nanometer is the rate of change of delay with wavelength; thus, we find

$$\begin{aligned}
D &= \frac{d\tau}{d\lambda} \\
&= \frac{2\tau}{\lambda} - \frac{\lambda^2}{2\pi c}\frac{d^2\psi_\rho}{d\lambda^2} \\
&= -\frac{2\pi c}{\lambda^2}\frac{d^2\psi_\rho}{d\omega^2}
\end{aligned} \tag{5.25}$$

Figure 5.3 shows the delay τ and dispersion D calculated from the above equations for a uniform grating of length ~5.3 mm. The design wavelength for this grating was 1550 nm and the index of refraction of the fiber was set at $n_{eff} = 1.45$. Figure 5.3 also shows the reflectivity spectral response of the same Bragg grating. Clearly, both reflectivity and delay are symmetric about the wavelength λ_{max}. The dispersion is zero near λ_{max} for uniform gratings and becomes appreciable near the band edges and sidelobes of the reflections spectrum where it tends to vary rapidly with wavelength. A useful qualitative understanding of this delay and dispersion behavior is provided using the effective medium approach developed by Sipe et al. [20]. For wavelengths beyond the band gap, the boundaries of the uniform grating at $z = \pm L/2$ act at abrupt interfaces, resulting in the formation of a Fabry-Perot-like cavity. Analogous to the Fabry-Perot, nulls in the reflection spectra are observed, with the trapped light inside the cavity experiencing enhanced delay.

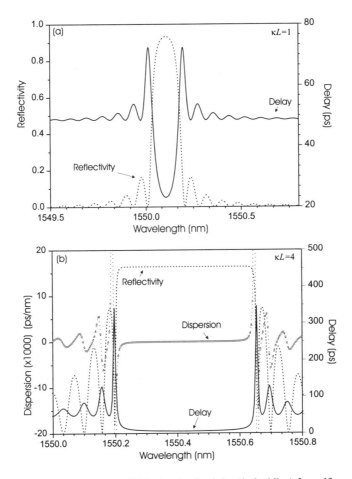

Figure 5.3   (a) Calculated group delay (solid line) and reflectivity (dashed line) for uniform weak Bragg grating $\overline{\delta n}_{eff} = 1 \times 10^{-4}$ with $L=5$ mm long. (b) Calculated group delay (solid line) and dispersion (squares) for a strong $\overline{\delta n}_{eff} = 4 \times 10^{-4}$, 5-mm long Bragg grating. The reflectivity of the grating is also shown for comparison, where its maximum value corresponds to ~1 and minimum 0 reflectivity. For both gratings the designed wavelength was 1550 nm and the fringe visibility was $s = 1$.

5.3 Two-Mode Coupling in Nonuniform Gratings

Many of the practical applications for Bragg gratings require that the grating pitch be nonuniform. One advantage of taking this approach is the reduction of undesirable side-lobes that are so prevalent in uniform grating spectra. We have already seen in earlier chapters how apodization has proven extremely effective in reducing these features. The applications demanding immediate attention are in dense wavelength division multiplexing (WDM) communication systems. Similarly, chirping allows for controlling the dispersive properties of the grating and is utilized for pulse compression and shaping in short fiber lasers [21], and for creating stable CW and tunable mode-locked external cavity semiconductor lasers [22, 23].

One may consider two standard approaches for arriving at the reflection and transmission spectra resulting from two-mode coupling in nonuniform gratings. First, one may numerically integrate the coupled-mode equations. Second, one may apply a piecewise-uniform approach, whereby the grating is divided into discrete uniform sections. The closed form solution to each of these uniform sections are combined by multiplying an array of matrices, each of which are the associated uniform sections. This has proven to be a powerful approach. A similar approach is to use Rouard's method, where each grating half-period is treated as a layer in a thin-film stack. Like in the piecewise approach, this requires the multiplication of a large number of matrices; however, the matrix number scales with the number of grating periods, and therefore, this approach can quickly become computationally intensive for fiber gratings that are centimeters in length, with 10^5 periods or more.

For the calculation of the spectral response of apodized gratings using direct numerical integration, one may simply use the coefficients $\zeta_{qm}(z)$ and $\kappa_{qm}(z)$ of the coupled-mode equations giving rise to $\zeta^+(z)$. In the apodization of gratings one may incorporate different shape functions for the index of refraction change (e.g., the Gaussian function). Fiber gratings are often written with a Gaussian laser beam, and as a result, the index of refraction change may be approximated with a Gaussian shape profile of the form

$$\overline{\delta n}_{\text{eff}}(z) = \overline{\delta n}_{\text{eff}} \exp\left[-\frac{1}{2}\left(\frac{z - a_0}{a_1} \right)^2 \right] \tag{5.26}$$

where $\overline{\delta n}_{\text{eff}}$ is the peak value of the "dc" effective index of refraction change. The coefficient corresponds to the center position of the Gaussian function, and $a_1 = \text{FWHM} / (8\ln 2)^{1/2}$ where FWHM is the full-width-at-half-maximum of the grating profile. Another very common index of refraction change profile is the "raised-cosine" shape which has the following form:

$$\overline{\delta n}_{\text{eff}}(z) = \frac{1}{2}\overline{\delta n}_{\text{eff}}\left[1 + \cos\left(\frac{\pi z}{L} \right) \right] \tag{5.27}$$

where L is the length of the grating, thus setting the raised cosine function identically to zero at each end of the grating structure.

Chirped gratings can also be modeled using the direct integration technique. This can be achieved by simply including a nonzero z-dependent phase term in the self-coupling coefficient ζ^+ of (5.12). The phase term of a linear chirp may be expressed as

$$\frac{1}{2}\frac{d\varphi}{dz} = -\frac{4\pi n_{\text{eff}}}{\lambda_d^2}\frac{d\lambda_d}{dz} \tag{5.28}$$

where $d\lambda_d/dz$ is a measure of the rate of change of the design wavelength with position along the grating structure, usually given in units of nanometers/centimeters. The chirp can also be specified in terms of a dimensionless parameter χ given by

$$\chi = -4\pi n_{eff} \left(\frac{FWHM}{\lambda_d} \right)^2 \frac{d\lambda_d}{dz} \tag{5.29}$$

where χ is the fractional change in the grating period over the whole length of the grating. Because chirp is incorporated into the coupled-mode equations as a z-dependent term in the self-coupling coefficient ζ^+, its effect is identical to the "dc" index variation ζ with the same z dependence. This has been used to modify the dispersion of gratings without varying the physical length of the grating pitch [24].

The spectral response from phase-shifted gratings may be calculated using the direct integration approach by simply adding a constant phase shift to the expression (5.1) as the integration proceeds along the z direction. Thus, one multiplies the value of the current (in the coupled-mode equations (5.10) and (5.11)) by phase-shift term $\exp(\iota\varphi)$ (where φ is the given phase shift). In general, complicated structures like superstructures may be implemented though the z dependence in $\zeta(z)$ and $\kappa(z)$.

The piecewise-uniform approach to modeling nonuniform grating structures is based on identifying 2 by 2 matrices for each uniform section of the grating (see Figure 5.4) and then multiplying all of them together to obtain a single 2 by 2 matrix that describes the whole grating [10]. The compound grating structure can be divided into M uniform matrix components, with A_k^+ and B_k^+ being the field amplitudes after traversing the section k. Therefore, in the case of Bragg gratings one starts with the boundary conditions $A_0^+ = A^+(L/2) = 1, B_0^+ = B^+(L/2) = 0$ and calculates the final matrix component $A_M^+ = A^+(-L/2)$ and $B_M^+ = B^+(-L/2)$. The propagation through each of the uniform sections k is described by a matrix $\mathbf{T}_k$, which is defined such that

$$\begin{pmatrix} A_k^+ \\ B_k^+ \end{pmatrix} = \mathbf{T}_k \begin{pmatrix} A_{k-1}^+ \\ B_{k-1}^+ \end{pmatrix} \tag{5.30}$$

where the matrix $\mathbf{T}_k$ for Bragg gratings is given by

$$\mathbf{T}_k = \begin{pmatrix} \cosh(\Omega dz) - i\dfrac{\zeta^+}{\Omega}\sinh(\Omega dz) & -i\dfrac{\kappa}{\Omega}\sinh(\Omega dz) \\ i\dfrac{\kappa}{\Omega}\sinh(\Omega dz) & \cosh(\Omega dz) + i\dfrac{\zeta^+}{\Omega}\sinh(\Omega dz) \end{pmatrix} \tag{5.31}$$

dz is the length of the kth uniform section, ζ^+ and κ are the local coupling coefficients for the kth section, and

$$\Omega = \sqrt{\kappa^2 - \zeta^{+2}} \tag{5.32}$$

The total grating structure may be expressed as

$$\begin{pmatrix} A_M^+ \\ B_M^+ \end{pmatrix} = \mathbf{T}_M \cdot \mathbf{T}_{M-1} \cdot \ldots \cdot \mathbf{T}_k \cdot \ldots \cdot \mathbf{T}_1 \begin{pmatrix} A_0^+ \\ B_0^+ \end{pmatrix} \tag{5.33}$$

This piecewise-uniform approach is ideal for analyzing chirped gratings. For example, in the case of linearly chirped gratings the total structure will be made up of M subgratings each with its own period Λ_k, which increases linearly (Figure 5.4). Simulation for linearly chirped Bragg gratings with a chirp value of $\overline{\chi}_L$ (over the length of the structure) indicates that sufficient accuracy may be achieved if it is constructed from $10\overline{\chi}_L$ sections (subgratings). However, the number of sections M cannot be made arbitrarily large, since the coupled-mode theory approximations that lead to (5.10) and (5.11) are no longer valid when a uniform grating section is only a few grating periods long [10]. The above approximation is valid for $dz \gg \Lambda$, which means that it must be maintained:

$$M \ll \frac{2n_{\text{eff}}L}{\lambda_d} \tag{5.34}$$

The number of sections determines the accuracy of the piecewise-uniform calculation. For most cases $M > 50$ sections gives sufficient accuracy.

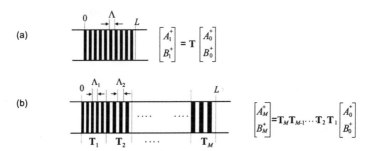

Figure 5.4   Illustration of the piecewise-uniform simulation: (a) a single uniform Bragg grating, (b) a series of gratings with different periods (chirped grating structure) represented by M, $\mathbf{T}_k$ matrices (see (5.33)), $k = 1...M$.

In general, to implement the piecewise-uniform method for apodized and chirped gratings, we simply assign constant values ζ, κ, and $0.5d\varphi/dz$, which are evaluated at the center of each of the sections. In the case of a phase-shifted grating, a phase-shift matrix $\overline{\mathbf{T}}_k$ is inserted between the matrices $\mathbf{T}_k$ and $\mathbf{T}_{k+1}$ in the product of (5.33), which represents a phase shift after the kth section. The phase-shift matrix has the following form:

$$\overline{\mathbf{T}}_k = \begin{bmatrix} \exp\left(-\dfrac{i\varphi_k}{2}\right) & 0 \\ 0 & \exp\left(\dfrac{i\varphi_k}{2}\right) \end{bmatrix} \tag{5.35}$$

In the case of discrete phase shifts, φ_k is the shift in the phase of the grating where for sample gratings (gratings placed some finite distance apart dz_0)

$$\frac{\varphi_k}{2} = \frac{2\pi n_{\text{eff}}}{\lambda} dz_0 \qquad (5.36)$$

where dz_0 is the separation between two grating sections.

Thus far we have described two basic techniques for calculating the spectral response of nonuniform grating structures, which are based on the coupled-mode theory. We will now give several examples of some nonuniform grating structures that will demonstrate the effects of apodization, phases shift, and chirp on the optical properties of the fiber gratings. To demonstrate the effects of apodization, Figure 5.5 shows the reflection and group delay versus wavelength for two gratings that have been apodized with a Gaussian shape function (similar to the one described in (5.26)).

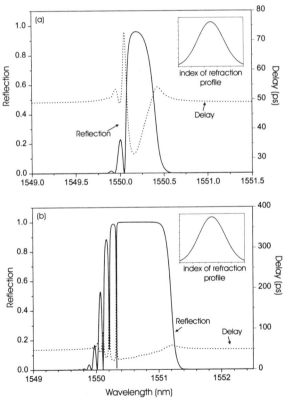

Figure 5.5 Reflection and group-delay calculations versus wavelength of Gaussian apodized Bragg gratings. The FWHM = $L/2$, $L = 10$ mm (length of grating), designed wavelength $\lambda_d = 1550$ nm, and the fringe visibility $s = 1$. The index of refraction variation along the fiber grating is shown at the upper right corner of the graphs. (a) Results for a weak grating, where the maximum index of refraction change is 2.5×10^{-4}. (b) Results for a strong grating, where the maximum index of refraction change is 7.5×10^{-4}.

The index of refraction variation along the fiber grating is shown at the upper right corner of the graphs. The maximum value of the index of refraction change of these Gaussian shape variations is 2.5×10^{-4} for the first grating and 7.5×10^{-4} for the other grating, with the FWHM in both cases being $L/2$. The designed wavelength for the Bragg gratings was 1550 nm, and its length was chosen to be 10 mm. We notice that the spectrum does not possess the typical periodic sidelobes on each side of the peak reflection that is seen in the case of uniform gratings, but rather on the short wavelength side there exists a very different periodic structure. This short wavelength structure is caused by the nonuniform "dc" index of refraction change [17, 20]. This can be understood as a kind of Fabry-Perot effect. Due to the increase in the average index of refraction, the entire grating resonance is shifted to longer wavelengths, but the region in the center of the grating, where the averaged index is increased the most, has its resonance shifted the farthest. Thus, there is a frequency region near the short wavelength side of the grating resonance where the edges of the grating are near their local Bragg resonance, but the center of the grating is not. Qualitatively, the edges of the grating behave as partially reflecting mirrors, and the center as a transparent region. The resulting Fabry-Perot resonance accounts for the high reflection peaks at the short wavelength side of the Bragg grating wavelength in Figure 5.5.

Figure 5.6(a) shows the reflection spectral response of uniform and nonuniform Bragg gratings, where for each the "dc" index change is assumed to be small (approximately zero). In the simulations the "dc" index change is set to a very small value (i.e., $\overline{\delta n}_{\text{eff}} \rightarrow 0$) while the "ac" index change was maintained to $s\overline{\delta n}_{\text{eff}} = 7.5 \times 10^{-4}$ by assuming a large fringe visibility (s). The gratings are assumed to be 20 mm long with the design wavelength set at 1550 nm. The index of refraction change profiles include Gaussian and a raised-cosine with the FWHM = 10 mm ($L/2$).

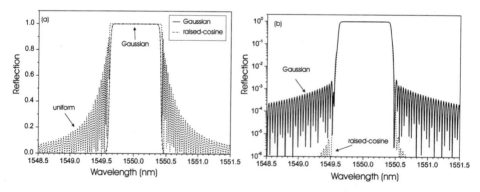

Figure 5.6   Reflectivity as a function of wavelength calculated for a uniform and nonuniform grating with "ac" index change $s\overline{\delta n}_{\text{eff}} = 7.5 \times 10^{-4}$ and a very small value of "dc" index change. (a) The dashed line corresponds to the spectral response from a uniform grating (sidelobes evident), where the solid and dash-point-dash curves correspond to Gaussian and raised-cosine profiles (these two curves are indistinguishable on this scale). (b) Reflection spectra for the two nonuniform gratings plotted on a logarithmic scale. It is quite evident that both the Gaussian and the raised-cosine shape gratings possess sidelobes. The sidelobe structure for the raised-cosine shape grating appears to be two orders of magnitude smaller than that of the Gaussian shape grating.

Figure 5.6(a) shows these two nonuniform gratings along with the uniform grating (dashed line) for comparative purposes. The sidelobes are apparent for the uniform grating; however, the Gaussian and the raised-cosine show no sidelobes (at this scale) and appear to be indistinguishable. Figure 5.6(b) shows the two nonuniform gratings on a logarithmic scale. The solid and dashed lines represent the Gaussian and raised-cosine gratings, respectively. It is interesting to note that both gratings have sidelobes; however, they appear to be reduced by a factor of 1000, in comparison to the case of the uniform grating. More specifically, the sidelobes of the Gaussian grating appear at 10^{-3}, whereas those for the raised-cosine appear at below 10^{-5}, making the differences between the two types of gratings evident.

Two examples of chirped gratings are shown in Figure 5.7. In both examples the gratings are 20 mm long and have a raised-cosine profile with a positive chirp of 3 nm over the length of the grating. The "dc" index of refraction change is set to be a very small value with $s\overline{\delta n}_{eff} = 7.5 \times 10^{-4}$. Figure 5.7(a) shows the reflection spectra and group delay as a function of wavelength assuming a linear chirp. Similarly, Figure 5.7(b) shows the reflection and group delay as a function of wavelength for the same parameters, but assumes a chirp with quadratic functional dependence. The nonlinear dependence of the chirp becomes evident from the functional dependence of the group-delay that is clearly different from that in Figure 5.7(a). This controlled functional dependence of the group delay can be very useful in applications such as dispersion compensation in telecommunications. The implementation of this nonlinear chirp or any other functional chirp dependence becomes quite straightforward using the phase-mask technique for inscribing Bragg gratings.

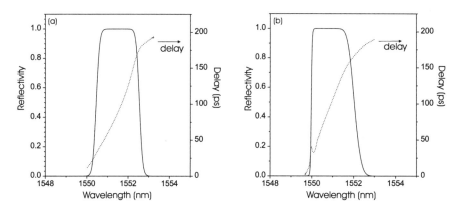

Figure 5.7 Calculated group delay and reflection as a function of wavelength for two differently chirped gratings. In both examples the gratings are 20 mm long and have a raised-cosine profile with a positive chirp value of 3 nm over the length of the grating. (a) In this grating the chirp variation is assumed linear; (b) the grating dependence here is assumed quadratic.

Examples of phase-shifted Bragg gratings are shown in Figure 5.8. In this figure the calculated reflectivity is shown as a function of wavelength for a 10 mm long grating with a raised-cosine shape whose maximum "ac" index change is $s\overline{\delta n}_{eff} = 5 \times 10^{-4}$ and a very small value for a "dc" index change. For comparison purposes, the spectral response from the zero phase-shift grating is also shown. The reflection responses from the same grating with the added phase shifts placed in the middle of the structure, are shown in Figure 5.8. Specifically a phase shift in the grating results in a narrow transmission resonance at the center of the spectral band and a broadening of the reflection spectrum. $\pi/2$ and $3\pi/2$ phase shifts result in a narrow transmission resonance at either side of the center of the band, thus providing evidence of tunability of the transmission resonance.

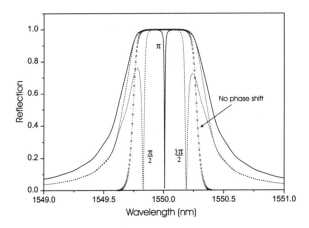

Figure 5.8   Calculated reflectivity as a function of wavelength for a Bragg grating with several phase-shift values. The grating used in the simulations is 10 mm long with a raised-cosine profile whose maximum "ac" index change is $s\overline{\delta n}_{eff} = 7.5\times10^{-4}$ and "dc" index change $\rightarrow 0$. The discrete phase shifts spectra shown correspond to π, $\pi/2$, and $3\pi/2$ phase shifts placed exactly in the middle of the grating structure. For comparison purposes the zero phase-shift spectra is also shown (square symbols joined by dotted line).

Figures 5.9(a) and (b) illustrate some of these interesting properties of the phase-shifted transmission resonance. Tuning the transmission resonance peak is achieved by simply changing the phase-shift angle, as seen in Figure 5.9(a). The resonance peak moves away from the center of the stop band by changing the phase shift from π to 0.5π (i.e., π, $0.9\pi,...,0.5\pi$). It is interesting to note that the linewidth of the transmission resonance increases as it moves away from the center. In particular, for the case shown in Figure 5.9(a) the linewidth is approximately 6.8 pm at a π phase shift, and broadens to 24 pm for a 0.5π phase shift. Similar results are also observed on the other side of the center of the stop band. Increasing the length of the grating also reduces the linewidth of the transmission resonance. As an example we have calculated for the same grating parameters described above for Figure 5.8 the spectral response of the transmission resonance (π phase shift) with the grating length doubled to 20 mm. The resultant transmission profile is shown in Figure 5.9(b). Clearly, the linewidth of the resonance is extremely narrow, estimated to be 0.043 pm from a Lorentzian fit to the data (shown on the graph with the square symbols), which is approximately 160 times narrower than that observed for the 10-mm long grating.

Undoubtedly, the tunability of the transmission resonance within the stop band of the grating structure, along with the control bandwidth of the resonance, finds immediate applications in telecommunications, as will be seen in the next chapter.

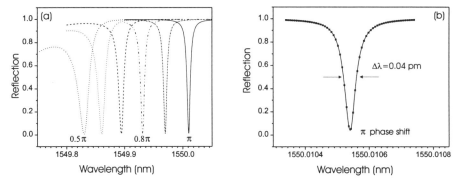

Figure 5.9   (a) Tunability of the transmission resonance accomplished via the change of the phase shift in the gratings structure (with phase-shift values at π, 0.9π, 0.8π, 0.7π, 0.6π, and 0.5π). The linewidth of the phase-shift grating was 6.8 pm and broadens to 24 pm for the 0.5π phase shift. Note the parameters of the grating are exactly the same as the one described in Figure 5.8. (b) Transmission resonance of a π phase-shift grating (double the length of the grating in 5.8 (i.e., 20 mm long). The linewidth of this transmission resonance was estimated to 0.043 pm.

5.4 Tilted Gratings

From the first reported discussions of radiation-mode coupling in fiber gratings, it has been recognized that this coupling can be enhanced and even controlled by incorporating a suitable degree of tilt in the fringes of the phase grating. If one assumes the ray picture description of the grating, then the tilt can be interpreted as blaze. From a mode description it is also clear how adding a degree of tilt to the grating can enhance radiation-mode coupling [25]. Using symmetry arguments one can show that an LP_{01} bound mode in a standard grating can couple only to LP radiation modes with azimuthal mode orders of 0 and 2. The introduction of tilt allows for coupling to all odd and even radiation modes. In addition to enhancing radiation-mode coupling, variation of the tilt affects the separation of the wavelength region at which maximum radiation-mode coupling occurs from that at which Bragg reflection occurs, and also affects the Bragg reflection spectrum. Modification of the Bragg reflection spectrum can prove useful for filter applications that cannot tolerate any reflections and for which the ability of grating tilt to negate this reflection while also enhancing radiation-mode coupling is advantageous [26]. Grating tilt also enables coupling between forward- and back-propagating bound modes, an example of which is the coupling of bound modes having dissimilar azimuthal symmetry that would otherwise remain disallowed. Examples of this interaction, for Bragg and transmission gratings and for coupling the LP_{01} mode to the LP_{11} mode of a dual-mode fiber, may be found in [27] and [28], respectively. In much the same way, coupling can occur between the bound core mode and bound cladding modes of the fiber. The most immediate effect of introducing tilt to a single-mode Bragg grating is a reduction in the fringe visibility s.

Let us consider the induced index change in the core of a single-mode optical fiber, δn_{co}, rotated by an angle θ and defined along an axis z', as shown in Figure 5.10, with

$$\delta n_{co}(x,z) = \overline{\delta n}_{co}(z')\left[1 + s\cos\left(\frac{2\pi}{\Lambda_g}z' + \varphi(z')\right)\right] \tag{5.37}$$

and $z' = x\sin(\theta) + z\cos(\theta)$. The grating period along the fiber axis that determines the resonant wavelengths for coupling is simply $\Lambda = \Lambda_g/\cos(\theta)$. By taking the projection of $\overline{\delta n}_{co}(z')$ and $\varphi(z')$ along the fiber axis, whereby $z' \cong z\cos(\theta)$, the general coupling coefficient (5.9) becomes

$$C_{\mp\pm}^T(z) = \zeta(z) + 2\kappa_{\mp\pm}(z)\cos\left[\frac{2\pi}{\Lambda}z + \varphi z\cos(\theta)\right] \tag{5.38}$$

where the forward-propagating mode (+) is associated with subscript m and the backward propagating mode (−) with q. This modifies the self- and cross-coupling coefficients as follows:

$$\zeta(z) = \omega\frac{n_{co}}{2}\overline{\delta n}_{co}z\cos(\theta)\iint\limits_{core}\mathbf{e}_\mp^T(x,y)\cdot\mathbf{e}_\pm^{T^*}(x,y)dx\,dy \tag{5.39}$$

and

$$\kappa_{\mp\pm}(z,\theta) = \frac{s}{2}\omega\frac{n_{co}}{2}\overline{\delta n}_{co}z\cos(\theta)$$
$$\times \iint\limits_{core}\exp\left(\pm i\frac{2\pi}{\Lambda}x\tan(\theta)\right)\mathbf{e}_\mp^T(x,y)\cdot\mathbf{e}_\pm^{T^*}(x,y)dx\,dy \tag{5.40}$$

with $\kappa+-=(\kappa-+)^*$. Finally, the effect of the tilt can be described by an effective fringe visibility [18], $s_{\mp\pm}(\theta)$ defined such that

$$\frac{s_{\mp\pm}(\theta)}{s} = \frac{\iint\limits_{core}\exp\left(\pm i\frac{2\pi}{\Lambda}x\tan(\theta)\right)\mathbf{e}_\mp^T(x,y)\cdot\mathbf{e}_\pm^{T^*}(x,y)dx\,dy}{\iint\limits_{core}\mathbf{e}_\mp^T(x,y)\cdot\mathbf{e}_\pm^{T^*}(x,y)dx\,dy} \tag{5.41}$$

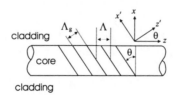

cladding

core

cladding

Figure 5.10   Schematic diagram of the core of a step-index optical fiber showing a tilted fiber grating.

This allows us to produce an equation in direct analogy to (5.8):

$$\kappa_{\mp\pm}(z,\theta) = \frac{s_{\mp\pm}(\theta)}{2}\zeta(z)$$

(5.42)

This result supports our earlier statement that the introduction of tilt reduces the effective fringe visibility according to (5.41). This factor describes how effective the grating perturbation is in backward-scattering and is essentially the backward-scattering parameter g_b in [25], which gives the overlap integral between the forward- and backward-propagating modes in the core for the LP_{01} mode profiles with the presence of grating tilt. A plot of the normalized effective fringe visibility associated with the single-mode Bragg reflection in a tilted grating is shown in Figure 5.11 for a cladding index of 1.44, a core-cladding $\Delta = 0.0055$, and a core radius $a = 2.635$ μm for a wavelength of 1550 nm.

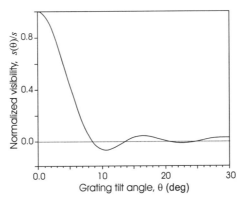

Figure 5.11   Normalized effective fringe visibility of a single-mode Bragg grating as a function of grating tilt (*After:* [18]).

Figure 5.12 shows a calculation of the grating reflectivity over a range of tilt angles. Note that the reflectivity nulls are independent of grating amplitude. When this curve is plotted with a third axis that accounts for wavelength detuning [18, 25], it is observed that the modulation on the short wavelength side of the reflectivity spectra occurs even for zero tilt angles and for a zero tilt visibility of unity. The modulation results from the effective Fabry-Perot cavity formed by the wings of the gratings (Gaussian profile). It is possible for short wavelengths to lie within the band gap associated with the wings, but not within the band gap associated with the center of the grating because of the higher average refractive index [17]. This is because the increase in the average refractive index shifts the entire grating resonance to longer wavelengths, with the region in the center of the grating shifting farthest, as this is where the averaged index has increased the most. Thus, there is a frequency region near the short wavelength side of the grating resonance where the edges of the grating are near local Bragg resonances, whereas the center of the grating is not. Therefore, the edges of the grating behave as partially reflecting mirrors, and the center is a transparent region. Similarly, this effect can be dominant for high tilt values, where the grating amplitude is significantly reduced, without a concomitant reduction in the background index.

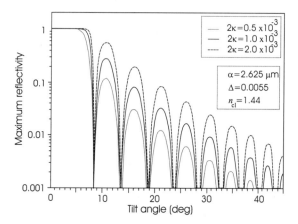

Figure 5.12 Maximum grating reflectivity versus grating tilt angle for three different modulation indices.

5.5 Cladding-Mode Coupling

In Section 3.4.2 we discussed the impact of cladding-mode coupling on the short wavelength properties of the Bragg grating wavelength spectrum. In fact, there are many applications of fiber gratings that can involve coupling light to and from the bounded core modes of the fiber, for example in spectral filtering [26] and sensing [29]. Figure 5.13 shows the transmission spectra of an LP_{01} core mode through a Gaussian fiber grating (zero tilt). In Figure 5.13(a) a bare fiber section containing a grating $(\overline{\delta n}_{eff} = 2 \times 10^{-3})$ is immersed in index-matching liquid to simulate a cladding of infinite extent. The featureless transmission profile below 1540 nm demonstrates loss through coupling to a continuum of radiation modes (see the following section). When the fiber grating is immersed in glycerin, for which the refractive index exceeds that of the cladding, the transmission spectrum now exhibits fringes that are caused by Fabry-Perot-like interference resulting from partial reflection of the radiation modes off the cladding-glycerin interface (Figure 5.13(b)). Figure 5.13(c) shows the bare fiber surrounded by air. Here the LP_{01} mode couples to a distinct number of cladding modes, producing the well-defined resonances.

Using the coupled-mode theory outlined earlier, Erdogan [30] has produced a detailed description of the coupling between the LP_{01} core and cladding modes. One may treat the core-cladding modal interaction in a fiber grating by treating the coupling of the core mode with multiple cladding modes simultaneously at a given wavelength. Limiting the analysis to untilted gratings (circularly symmetric index perturbation), one finds that the only nonzero coupling coefficients between the core and cladding modes involve cladding modes having azimuthal order $l = 1$; all other orders result in zero modal coupling from the orthogonality relations. We quote the salient equations as described in [30]. These

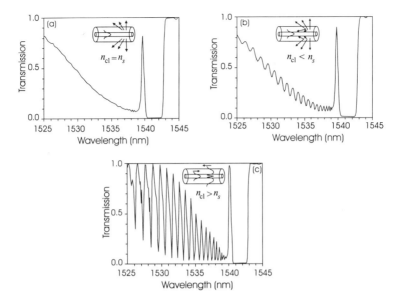

Figure 5.13 Measured transmission through a Bragg grating where (a) the uncoated fiber is immersed in index-matching liquid to simulate an infinite cladding, (b) the fiber is immersed in glycerin, and (c) the bare fiber is surrounded by air and thus supports cladding modes (*After:* [30]).

equations take into account the coupling of the LP_{01} core mode to itself (i.e., counter-propagating Bragg reflection) and with both counter- and co-propagating $l = 1$ cladding modes (i.e., with the exact cladding modes of order 1μ):

$$\frac{dA^{co}}{dz} = i\zeta_{01-01}^{co-co} A^{co} + i\frac{s}{2}\zeta_{01-01}^{co-co} B^{co} \exp(-2i\delta_{01-01}^{co-co}z)$$
$$+ i\sum_{\mu}\kappa_{1\mu-01}^{cl-co} B_{\mu}^{cl} \exp(-2i\delta_{1\mu-01}^{cl-co}z) \tag{5.43}$$

$$\frac{dB^{co}}{dz} = -i\zeta_{01-01}^{co-co} B^{co} - i\frac{s}{2}\zeta_{01-01}^{co-co} A^{co} \exp(2i\delta_{01-01}^{co-co}z) \tag{5.44}$$

$$\sum_{\mu}\left[\frac{dB_{\mu}^{cl}}{dz} = -i\kappa_{1\mu-01}^{cl-co} A^{co} \exp(2i\delta_{1\mu-01}^{cl-co}z)\right] \tag{5.45}$$

where A^{co} and B^{co} are the amplitudes for the core mode, and B_{μ}^{cl} is the amplitude for the μth cladding mode. The corresponding detuning parameters are given by

$$\delta_{01-01}^{co-co} \equiv \frac{1}{2}\left(2\beta_{01}^{co} - \frac{2\pi}{\Lambda}\right) \tag{5.46}$$

$$\delta_{1\mu-01}^{cl-co} \equiv \frac{1}{2}\left(\beta_{01}^{co} + \beta_{1\mu}^{cl} - \frac{2\pi}{\Lambda}\right) \tag{5.47}$$

Here, ζ_{01-01}^{co-co} is the self-coupling coefficient for the LP_{01} mode given by (5.14), $\kappa_{1\mu-01}^{cl-co}$ is the cross-coupling coefficient defined using (5.7) and (5.8). Terms that involve self- and cross-coupling between the cladding modes have been neglected as they are either very small or $\zeta_{1\mu-1\mu}^{cl-cl}$, $\kappa_{1\mu-1\nu}^{cl-cl} \ll \kappa_{1\mu-01}^{cl-co} \ll \zeta_{01-01}^{co-co}$. The above equations are appropriate in the wavelength range for which the detuning parameters are nearly zero for a given Λ. Therefore, the wavelength at which $\delta_{01-01}^{co-co} = 0$ is the resonant wavelength, or Bragg wavelength, for core-mode to core-mode coupling, whereas the wavelength at which $\delta_{1\mu-01}^{cl-co} = 0$ is the resonant wavelength for the core mode to μth cladding-mode coupling. When the Bragg reflection of the LP_{01} mode is not included in the above equations ($B^{co}=0$), they are only valid at wavelengths far from the Bragg resonance and reduce to the form presented in [18]. If the cladding mode resonances do not overlap, then each resonance may be calculated separately retaining only the core mode and appropriate cladding mode. In this case the problem reduces to the two-mode coupling as outlined earlier. However, if it is found that there is significant overlap between the cladding mode resonances, an accurate fit can be obtained by simply including more resonances of multiple cladding modes at each wavelength [18]. This theory is only pertinent to gratings with a circularly symmetric index perturbation; this necessarily excludes tilted gratings.

5.6 Radiation-Mode Coupling

Sipe and co-workers [17, 25] have studied radiation-mode coupling as a loss mechanism on core-mode transmission. As discussed in Section 5.3, the coupling to these higher order modes is very small unless significant grating tilt is introduced. However, we have also seen from Figure 5.13(a) that if the fiber is immersed in a suitable index-matching liquid, bound cladding modes are removed and substantial coupling to low order radiation modes can occur near the Bragg wavelength resonance even for zero-tilt gratings. One may arrive at a useful quantity in the form of a radiation mode cut-off wavelength λ_{cut}. It is assumed that the core confinement factor for the radiation mode is much smaller than that for the bound mode for which

$$\lambda_{cut} \cong \frac{1}{2}\left(1 \pm \frac{n}{n_{eff}}\right)\left(1 + \frac{\overline{\delta n_{eff}}}{n_{eff}}\right)\lambda_d \tag{5.48}$$

for the case of an infinitely clad fiber, $n = n_{cl}$ (Figure 5.13(a)), or $n = 1$ when the fiber is surrounded by air such that bound cladding modes can propagate. The + sign in the first factor applies to reflection, while the − sign corresponds to forward coupling. The second factor describes the shift of λ_{cut} increasing dc index change.

Consider the coupling of an LP_{01} core mode to backward-propagating radiation modes labeled $LP_{j\rho}$, where the discrete index j identifies the polarization and azimuthal order, while the continuous label $\rho = \sqrt{(2\pi/\lambda)^2 n_{cl}^2 - \beta_{jp}^2}$ denotes the transverse wavenumber of the radiation mode (β_{jp} is the axial propagation constant). The fiber is assumed to have a cladding index n_{cl} with an infinite radius. Thus, the coupled-mode equations are now [18]

$$\frac{dA}{dz} = i\zeta_{01-01}^{co-co} A + i\sum_j \int \kappa_{jp-01}^{ra-co} B_{jp} \exp(-2i\delta_{jp-01}^{ra-co} z) d\rho \qquad (5.49)$$

and

$$\frac{dB_{jp}}{dz} = -i\kappa_{01-jp}^{co-ra} A \exp(2i\delta_{jp-01}^{ra-co} z) \qquad (5.50)$$

where

$$\delta_{jp-01}^{ra-co} \equiv \frac{1}{2}\left(\beta_{01} + \beta_{jp} - \frac{2\pi}{\Lambda}\right) \qquad (5.51)$$

where A is the amplitude of the core mode, and B_{jp} are the amplitudes of the continuous spectrum of radiation modes. Additionally, ζ_{01-01}^{co-co} is the LP_{01} self-coupling coefficient given by (5.14) and $\kappa_{01-jp}^{co-ra} = (\kappa_{jp-01}^{ra-co})^*$ is the cross-coupling coefficient defined by (5.7) and (5.8). The above equations can be used to show that the core mode amplitude approximately obeys an equation of the form [18]

$$\frac{dA}{dz} = i\zeta_{01-01}^{co-co} A - \left[\sum_j \frac{\pi\beta_{jp}}{\rho}\left|\kappa_{jp-01}^{ra-co}\right|^2\right] A \qquad (5.52)$$

where the term in square brackets is evaluated at $\beta_{jp} = 2\pi/\Lambda - \beta_{01}$. This real term gives rise to exponential loss in the amplitude of the core mode. Furthermore, the loss coefficient is proportional to the square of the cross-coupling coefficient and therefore to the square of the induced index change. This simplified analysis can nevertheless be used to predict the radiation-mode coupling loss spectrum, even when the grating is tilted. As the tilt angle is increased, many radiation mode azimuthal orders must be included in the summation in (5.52) to accurately model radiation modes propagating more normal to the fiber axis. Examples of this may be found in [25]. Figure 5.14 shows experimentally measured transmission loss spectra versus the grating tilt angle for bound-mode to radiation-mode coupling, including a comparison with the corresponding theoretical calculation.

5.7 Long Period Gratings

We now briefly discuss the behavior of long period gratings, also known as transmission gratings, in which coupling occurs between modes traveling in the same direction [29, 31]. Although they are not Bragg gratings, they nevertheless merit discussion, as many of the applications that we will discuss in later chapters regarding the Bragg grating are pertinent

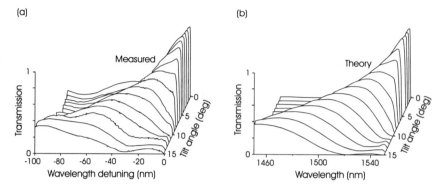

Figure 5.14 (a) Experimentally measured transmission loss spectra versus grating tilt angle for bound-mode to radiation-mode coupling; (b) the corresponding calculated transmission loss spectra. Loss that results from Bragg scattering has been excluded (*After:* [25]).

to the long period grating. It is also of interest to examine the physical properties of this grating type for comparative purposes. The fundamental coupled-mode equations are found to be pertinent to the long period grating description, particularly if the application is limited to coupling between two isolated modes. Therefore, close to the wavelength for which a forward-propagating mode of amplitude $A_1(z)$ is strongly coupled into a co-propagating mode with amplitude $A_2(z)$ (where the subscripts denote the different modes), (5.3) and (5.4) may be modified. We retain terms that involve the amplitudes of these two modes and make the synchronous approximation to similarly give as before

$$\frac{dA^+}{dz} = i\zeta^+ A^+(z) + i\kappa B^+(z)$$

$$(5.53)$$

$$\frac{dB^+}{dz} = -i\zeta^+ B^+(z) + i\kappa^* A^+(z)$$

However, the amplitudes are now defined as

$$A^+(z) = A_1 \exp\left[-i(\zeta_{11} + \zeta_{22})\frac{z}{2}\right]\exp\left(i\delta_d z - \frac{\varphi}{2}\right) \qquad (5.54)$$

and

$$B^+(z) = A_2 \exp\left[-i(\zeta_{11} + \zeta_{22})\frac{z}{2}\right]\exp\left(-i\delta_d z + \frac{\varphi}{2}\right) \qquad (5.55)$$

where ζ_{11} and ζ_{22} are the "dc" coupling coefficients defined in (5.7), and $\kappa = \kappa_{21} = \kappa_{12}^*$ is the "ac" cross-coupling coefficient from (5.8) with the general "dc" self-coupling coefficient now defined by [18]

$$\zeta^+ \equiv \delta_d + \frac{\zeta_{11} - \zeta_{22}}{2} - \frac{1}{2}\frac{d\varphi}{dz} \qquad (5.56)$$

for a uniform grating ζ^+ and κ are constants. When the detuning is assumed to be constant along the z axis, one finds

$$\delta_d \equiv \frac{1}{2}(\beta_1 - \beta_2) - \frac{\pi}{\Lambda}$$

$$= \pi\Delta n_{eff}\left(\frac{1}{\lambda} - \frac{1}{\lambda_d}\right) \qquad (5.57)$$

where again $\lambda_d = \Delta n_{eff}\Lambda$ is the design wavelength for a grating approaching zero index modulation. As was highlighted earlier for Bragg gratings, $\delta_d = 0$ corresponds to the grating condition predicted by the qualitative picture of grating diffraction, or $\lambda = \Delta n_{eff}\Lambda$. When the appropriate boundary conditions are applied, closed form solutions are possible. By assuming that only one mode is incident from $z = -\infty$ ($A^+(0) = 1$ and $B^+(0) = 0$), the transmission may be evaluated [7]:

$$\frac{\left|A^+(z)\right|^2}{\left|A^+(0)\right|^2} = \cos^2\sqrt{\kappa^2 + \zeta^{+2}}z + \frac{\zeta^{+2}}{\zeta^{+2} + \kappa^2}\sin^2\sqrt{\kappa^2 + \zeta^{+2}}z \qquad (5.58)$$

$$\frac{\left|B^+(z)\right|^2}{\left|A^+(0)\right|^2} = \frac{\kappa^2}{\zeta^{+2} + \kappa^2}\sin^2\sqrt{\kappa^2 + \zeta^{+2}}z \qquad (5.59)$$

$|A^+(z)|^2/|A^+(0)|^2$ refers to the measured transmission response of the bound core mode, with an induced loss in transmission through the interaction with, for example, the lowest order cladding mode. $|B^+(z)|^2/|A^+(0)|^2$ represents the ratio of power coupled into the relevant cladding mode to the initial power contained in the fundamental LP_{01} core mode. This analysis excludes tilted gratings, and the modal overlap considered is that between the fundamental core mode with coupling to azimuthally symmetric cladding modes that peak in intensity within the core diameter. The cross-coupling coefficient κ is increased to maximize the power transfer to the cladding mode; it is increased until the condition $\kappa L = \pi/2$ is met. In Figure 5.15, we show examples of typical transmission grating spectra for (5.58) and (5.59) for which the grating length is L with $\kappa L = \pi/2$ and $5\pi/2$, respectively. The grating period is set to $\Lambda = 500$ μm, which is typical of transmission gratings, and the total number of grating periods N is 100. The total number of grating periods for a given value of κL may either narrow or broaden the spectral response, depending on whether N increases or decreases in value.

The maximum cladding- to core-mode transmission occurs when $\zeta^+ = 0$, that is

$$\frac{\left|B^+(z)\right|^2}{\left|A^+(0)\right|^2} = \sin^2(\kappa L) \qquad (5.60)$$

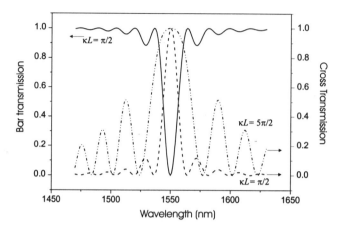

Figure 5.15   Calculated transmission through a uniform long period grating, for bound core-mode to core-mode and cladding-mode to core-mode resonances, from (5.58) and (5.59).

at a corresponding wavelength approximately given by

$$\lambda_{max} \cong \left(1 + \frac{\overline{\delta n}_{eff}}{\Delta n_{eff}}\right)\lambda_{d} \qquad (5.61)$$

for which $\overline{\delta n}_{eff} \ll \Delta n_{eff}$. We may compare (5.61) and (5.20) and observe that the wavelength of maximum coupling in a long period cladding-mode coupler grating shifts towards longer wavelengths during inscription $n_{eff}/\Delta n_{eff}$ times more rapidly than the shift that occurs in a Bragg grating. To estimate the bandwidth of the long period grating, one may consider the separation between the first zeros on either side of the spectral peak for which we find the normalized bandwidth to be

$$\frac{\Delta\lambda_{0}}{\lambda} = \frac{2\lambda}{\Delta n_{eff}L}\sqrt{1 - \left(\frac{\kappa L}{\pi}\right)^{2}} \qquad (5.62)$$

for a grating in which there is at most one complete exchange of power between the two modes; that is, $\kappa L \le \pi$, as is inferred graphically in Figure 5.15. In the case of a very weak grating, one arrives at a normalized bandwidth $\Delta\lambda_{0}/\lambda = 2/N$, which is identical to a weak Bragg grating. For the case of strong overcoupling (i.e., $\kappa L \gg \pi$ as in Figure 5.15), we observe that the sidelobes become significant relative to the central peak, and therefore, we must take the envelope of the complete spectrum, including the sidelobes, to be the FWHM [18]

$$\frac{\Delta\lambda_{env}}{\lambda} = \frac{2\lambda\kappa}{\pi\Delta n_{eff}} \qquad (5.63)$$

The above estimates for the grating bandwidth must be treated with caution since the effects of dispersion, which are of concern for long period gratings given their increased bandwidths (compared with Bragg gratings) under dispersion-free conditions, have been ignored. If $\Delta n_{\text{eff}} \propto \lambda$, then the detuning δ_d could potentially remain small over the wavelength span considered, and this would necessarily result in strong coupling over the relevant wavelengths, ultimately leading to a very broadened grating.

For the case of nonuniform transmission gratings one may again apply the piecewise-uniform approach, as implemented in Section 5.3 for the modeling of nonuniform Bragg gratings. The grating matrix, however, must be modified as follows [18]:

$$
\mathbf{L}_k =
\begin{bmatrix}
\cos(\Omega_L \, dz) + i \dfrac{\zeta^+}{\Omega_L} \sin(\Omega_L \, dz) & i \dfrac{\kappa}{\Omega_L} \sin(\Omega_L \, dz) \\[3mm]
i \dfrac{\kappa}{\Omega_L} \sin(\Omega_L \, dz) & \cos(\Omega_L \, dz) - i \dfrac{\zeta^+}{\Omega_L} \sin(\Omega_L \, dz)
\end{bmatrix}
\tag{5.64}
$$

with

$$
\Omega_L \equiv \sqrt{\kappa^2 + \zeta^{+2}}
\tag{5.65}
$$

Finally, when coupling to cladding modes, the equations that describe coupling of the LP_{01} core mode and the cladding modes of order 1μ by a transmission grating are given by [30]

$$
\frac{dA^{\text{co}}}{dz} = i\zeta^{\text{co-co}}_{01-01} A^{\text{co}} + i\sum_{\mu} \kappa^{\text{cl-co}}_{1\mu-01} A^{\text{cl}}_{\mu} \exp(-2i\delta^{\text{cl-co}}_{1\mu-01} z)
\tag{5.66}
$$

$$
\sum_{\mu} \left[\frac{dA^{\text{cl}}_{\mu}}{dz} = i\kappa^{\text{cl-co}}_{1\mu-01} A^{\text{co}} \exp(2i\delta^{\text{cl-co}}_{1\mu-01} z) \right]
\tag{5.67}
$$

and

$$
\delta^{\text{cl-co}}_{1\mu-01} \equiv \frac{1}{2}\left(\beta^{\text{co}}_{01} - \beta^{\text{cl}}_{1\mu} - \frac{2\pi}{\Lambda} \right)
\tag{5.68}
$$

using the synchronous approximation. Figure 5.16 gives an example of cladding-mode coupling, for a strong, uniform transmission grating (index change of 10^{-3}, period of 500 μm). The grating is designed to couple the LP_{01} core mode to the $\mu = 7$ cladding mode at 1550 nm. The five main dips in this spectrum correspond to the $\mu = 1, 3, 5, 7, 9$ cladding modes. The notch associated with the $\mu = 9$ cladding mode at 1800 nm is marginally weaker than the $\mu = 7$ dip at 1550 nm, despite the fact that the higher order mode has a stronger coupling coefficient [30]. This behavior results from the overcoupling of the incident core mode light at 1800 nm into the cladding mode and back again to the core mode. We close this section by highlighting that for the long period or transmission grating the co-propagating coupling can take place between a core mode and radiation modes, a core mode and a cladding mode, and two different core modes (i.e., LP_{01} to LP_{11} modes). The aforementioned coupling occurs from the longest to shortest wavelength of excitation, respectively. The papers of Erdogan give numerous, detailed examples of the possible resonances for long period gratings [30].

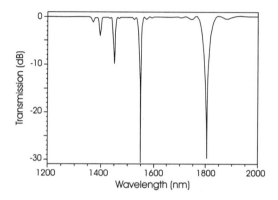

Figure 5.16   Calculated transmission spectrum for a strong long period grating designed to couple the LP_{01} core mode to the $\mu = 7$ cladding mode at 1550 nm (*After:* [30]).

5.8 Rouard's Method

Here we discuss in some detail one of the many computational techniques for arriving at the spectral characteristics of the Bragg grating. Rouard's method is a recursive computational technique that is generally used in the design of thin-film coatings, but which has been modified by Weller-Brophy and Hall for the analysis of diffraction gratings in waveguides [11]. This technique is applicable to periodic and aperiodic gratings and is a particularly tractable approach. It does not require the solution of a set of coupled-mode equations and yet has proven to give the same results as presented by Kogelnik, through finding the amplitude reflectivity of each grating period individually. Therefore, one is only required to know the fundamental characteristics of a single grating element, and the recursive procedure that follows allows for the evaluation of the total grating response. This method is suitable for all grating structure types and has been shown to be applicable to linearly and quadratically chirped gratings. The fundamental step in this technique is the replacement of an effective grating interface, characterized by an effective complex reflectivity, with a single interface having the same properties, with each grating half-period treated as a layer in a thin-film stack. Rouard's method starts by considering the layer at the bottom of the "stack" and working up through the stack, replacing the layer considered during each iteration by an interface having the same reflectivity and phase shift. There is one disadvantage that is immediately obvious, the computation time is directly related to the number of grating elements; this scales with the grating length and can therefore become very large indeed.

The solution for the single grating period is found by replacing L (the grating length) with Λ (the grating period) in the Bragg grating amplitude reflectivity term $\rho = \rho_{max}$ (i.e., for $\zeta^* = 0$). A specific example is taken from [11] for the case of a grating having three grating periods, each of length Λ, Bragg-matched to the incident and reflected modes such that the detuning δ_d from the Bragg condition defined as

$$2\delta_d = \beta_i \cos(\theta_i) + \beta_r \cos(\theta_r) - \frac{2m\pi}{\Lambda}, \quad m = 1, 2, 3, \cdots \quad (5.69)$$

is given by $2\delta_d = 0$ for $m = 1$. The effective complex reflectivity is found using the Airy equation

$$\rho_2 = \frac{r_1 + r_2 \exp(-2i\Delta_2)}{1 + r_1 r_2 \exp(-2i\Delta_2)} = \frac{r_1 + r_2}{1 + r_1 r_2} \quad (5.70)$$

where r_1 and r_2 are the Fresnel coefficients of the isolated interfaces (Figure 5.17), and Δ_2 is the phase shift occurring on traversing layer 2, with

$$\Delta_2 = \frac{2\pi}{\lambda} n_2 d_2 \cos(\theta_2) \quad (5.71)$$

for which the phase shift between two consecutive layers is given by

$$2\Delta_n = d_n [\beta_i \cos(\theta_i) + \beta_r \cos(\theta_r)] \quad (5.72)$$

and $2\Delta_2 = 2m\pi$ for $\delta_d = 0$. We may therefore write ρ_2 in the form

$$\text{arctanh}(\rho_2) = \text{arctanh}(r_1) + \text{arctanh}(r_2) \quad (5.73)$$

Using the Airy equation again, the complex reflectivity of the entire grating becomes, via Rouard's method,

$$\rho = \frac{r_0 + \rho_2 \exp(-2i\Delta_1)}{1 + r_0 \rho_2 \exp(-2i\Delta_1)} = \frac{r_0 + \rho_2}{1 + r_0 \rho_2} \quad (5.74)$$

This may again be rewritten as

$$\text{arctanh}(\rho) = \text{arctanh}(r_0) + \text{arctanh}(r_1) + \text{arctanh}(r_2) \quad (5.75)$$

The reflectivity of each effective interface of the grating is given by $-\tanh(\kappa L)$, and the total grating reflectivity takes the form $\rho = -\tanh(\kappa L)$, where $L = 3\Lambda$ is the grating length and is identical to the form presented in (5.16). This technique only requires the amplitude reflectivity of an isolated grating period and the phase shift to occur between successive periods. Any variation in the grating period can be handled by recognizing that each period of the grating has a different r_n and Λ_n that results from the difference in the period along the length of the grating. Therefore, it becomes clear that Rouard's method can be extended to any straight line grating, including periodic, linearly or quadratically chirped, and other more complex structures.

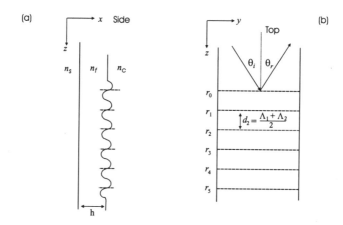

Figure 5.17   Side and top views of a periodic waveguide grating illustrating the thin-film waveguide-grating analogy. Each period of the grating depicted in the side view is represented by an effective interface, as shown by the dashed lines in the top view. The effective interfaces have reflectivities $r_0 - r_s$.

5.9 Bloch Waves

The Floquet-Bloch waves are the simplest electromagnetic disturbances, forming the eigenmodes in a lossless stratified structure, whose effective propagation constant is periodically modulated, such as a distributed feedback structure (i.e., a Bragg grating). In the most rigorous description, a Bloch wave consists of a stable infinite superposition of plane waves (called "partial" waves), with wave vectors having constant amplitudes. In the limiting case of two such partial waves, the Flocquet-Bloch wave is represented by the superposition of two spectral waves, chosen such that they interfere to produce fringes of spacing Λ. The Bloch wave may be represented by an electric field consisting of a superposition of forward- and backward-propagating waves [16]:

$$E(z,t) = \frac{1}{2}\left\{ A_f \exp(-ik_f z) + A_b \exp[i(K - k_f)z] \right\} \exp(i\omega t) + c.c. \qquad (5.76)$$

where the subscripts f and b refer to forward and backward directions, V_b and V_f are the constant field amplitudes, and $K = 2\pi/L$ is the grating vector. The grating perturbation may again be described by the sum of dc and ac contributions

$$\xi^1 = \xi_0^1 + \xi_m^1 \cos(Kz), \quad z > 0$$
$$= \xi_0^1, \quad z \leq 0 \tag{5.77}$$

with $\xi_0^1 > 0$. A further parameter P is defined by

$$P = \frac{\xi_m^1}{1 + \xi_0^1} \tag{5.78}$$

which is the modulation depth of the relative dielectric permittivity. When used in conjunction with Maxwell's equations one arrives at a homogeneous matrix equation for the eigenvalues k_f and eigenvectors $\{V_f, V_b\}$ of the system. The mean wavevector $k_0 = \omega n_0/c$, where $n_0 = (1 + \xi_0^1)^{1/2}$. The approximation $k_f^2 - k_0^2 \approx 2k_0(k_f - k_0)$, with the perturbation to k_0 in the vicinity of the Bragg condition, $\Gamma = k_f - k_0$, leads to a matrix equation

$$\begin{pmatrix} -\Gamma & \varpi \\ \varpi & \Gamma + \varepsilon \end{pmatrix} \begin{pmatrix} V_f \\ V_b \end{pmatrix} = \begin{pmatrix} 0 \\ 0 \end{pmatrix} \tag{5.79}$$

where the coupling coefficients are given by

$$\varpi = Pk_0/4 \tag{5.80}$$

and the dephasing parameter by

$$\vartheta = 2k_0 - K \approx \frac{4\pi n_0}{\lambda_B^2}(\lambda_B - \lambda) \tag{5.81}$$

Two independent Bloch waves exist for every value of ϑ, having eigenvalues

$$k_f(\pm) = k_0 + \Gamma = \left(\frac{K}{2}\right) \mp \left(\frac{\vartheta}{2}\right)\sqrt{1 - \left(\frac{1}{\Delta}\right)^2} \tag{5.82}$$

for traveling waves, with

$$k_f(\pm) = \left(\frac{K}{2}\right) \pm j\varpi\sqrt{1 - \Delta^2} \tag{5.83}$$

for evanescent waves. The parameter $\Delta = \vartheta/2\varpi$ distinguishes between the two regimes. For $\Delta^2 > 1$ traveling Bloch waves appear; for $\Delta^2 < 1$ the Bloch waves are evanescent [16]. The traveling Bloch wave may be divided into two additional categories, slow waves for $\Delta < -1$ ($k_f > k_0$) and fast waves for $\Delta > 1$ ($k_f < k_0$). Further details pertaining to the fundamentals of Bloch waves may be found in [32–34].

5.10 Nonlinear Grating Effects

The propagation of short intense pulses in Bragg gratings introduces nonlinear effects that may be described by an intensity-dependent index of refraction. Thus, in its simplest form the general index of refraction may be written as

$$n(I) = n + n_{nl}I \tag{5.84}$$

where n is the refractive index at low intensities, and n_{nl} is the nonlinear refractive index. Using the coupled-mode theory to describe pulse propagation through fiber gratings at wavelengths close to the Bragg wavelengths we write the electric field $\mathbf{E}$ as

$$\mathbf{E}(z,t) = [A(z,t)\exp(ik_0 z) + B(z,t)\exp(-ik_0 z)]\exp(-i\omega_0 t) + c.c. \tag{5.85}$$

where k_0 and ω_0 are the wave-number and angular frequency associated with the Bragg wavelength λ_B. Furthermore, $A(z,t)$ and $B(z,t)$ are envelope functions, which, if the grating is weak and the spectrum of the field is not too far from the Bragg wavelength, are slowly varying in space and time. It can be shown by substituting the above equations into the wave equation for the electric field, that the envelopes $A(z,t)$ and $B(z,t)$ satisfy the nonlinear coupled-mode equations (NLCME) [35]

$$i\frac{\partial A}{\partial z} + i\frac{n}{c}\frac{\partial A}{\partial t} + \zeta A + \kappa B + G|A|^2 A + 2G|B|^2 A = 0 \tag{5.86}$$

$$-i\frac{\partial B}{\partial z} + i\frac{n}{c}\frac{\partial B}{\partial t} + \zeta B + \kappa A + G|B|^2 B + 2G|A|^2 B = 0 \tag{5.87}$$

where κ and ζ are defined in a similar way as in (5.14) and (5.15). Thus, κ represents the strength of the grating per unit length and is proportional to the refractive index modulation; it leads to coupling between the forward- and backward-propagating modes. The coefficient ζ represents a small detuning, due to the contribution from the "dc" index change to the background refractive index. Finally, the nonlinearity leads to the term proportional to G, which is given by

$$G = \frac{4\pi Z_0}{\lambda_B} n_{nl} \tag{5.88}$$

where Z_0 is the vacuum impedance. In the limit of $G \to 0$ and taking harmonic time dependence, NLCME (5.87) and (5.88) reduce to a set of coupled, ordinary, linear differential equations. It can be shown that in this limit, these equations lead to the reflection spectrum of Chapter 3, Figure 3.26. It has been shown that at higher intensities the NLCME reduce to the nonlinear Schrodinger equation [36]. More generally, however, it was pointed out by Aceves and Wabnitz [37] that the NLCME allow for a two-parameter set of pulse-like solutions. The first of these parameters determines the pulse's velocity, which can be anywhere between zero and the speed of light in the bare medium, while the second determines the spectral content, peak power, and width of the pulse.

5.11 Discussion

We have presented many of the important principles and tools for understanding the spectra exhibited by Bragg gratings, and have also included a brief overview of the spectral characteristics of transmission gratings for the purpose of general interest. Finally, we close with some comments regarding the techniques available. The coupled-mode theory is used for its relative simplicity and flexibility. The key parameter is the coupling coefficient between the forward and backward modes due to the presence of the index perturbation. The transfer matrix method, on the other hand, can be used in two different ways. First, as a convenient tool to represent the solutions to the conventional coupled-mode theory; and second through an approximate mode-matching technique where the waveguide structures with large modulation depth and index difference can be accurately treated [38]. The transfer matrix also has the advantage of being able to model arbitrary structures considering strong coupling, refractive index and gain without having to solve the coupled-mode equations.

References

[1]   Kogelnik, H., and C. W. Shank, "Coupled wave theory of distributed feedback lasers," *Journal of Applied Physics*, Vol. 43, 1972, pp. 2327–2335.

[2]   Bilodeau, F., et al. "An all-fiber dense-wavelength-division multiplexer/demultiplexer using photoimprinted Bragg gratings," *IEEE Photonics Technology Letters*, Vol. 7, 1995, pp. 388–390.

[3]   Ouellette, F., "Dispersion cancellation using linearly chirped Bragg grating filters in optical waveguides," *Optics Letters*, Vol. 16, 1987, pp. 847–849.

[4]   Loh, W. H., et al. "10cm chirped fiber Bragg grating for dispersion compensation at 10Gb/s over 400km of nondispersion shifted fibre," *Electronics Letters*, Vol. 31, 1995, pp. 2203–2205.

[5]   Marti, J., et al. "Optical equalization of dispersion-induced nonlinear distortion in subcarrier systems by employing tapered linearly chirped gratings," *Electronics Letters*, Vol. 32, 1996, pp. 236–237.

[6]   Davis, M. A., and A. D. Kersey, "Matched-filter interrogation technique for fiber Bragg grating arrays," *Electronics Letters*, Vol. 31, 1995, pp. 822–825.

[7]   Yariv, A., "Coupled-mode theory for guided-wave optics," *IEEE Journal of Quantum Electronics* Vol. QE-9, 1973, pp. 919–933.

[8]   Kogelnik, H., "Filter response of nonuniform almost-periodic structures," *Bell System Technical Journal*, Vol. 55, 1976, pp. 109–126.

[9]   Winick, K. A., "Effective-index method and coupled-mode theory for almost periodic waveguide gratings: A comparison," *Applied Optics*, Vol. 31, 1992, pp. 757–764.

[10] Yamada, M., and K. Sakuda, "Analysis of almost-periodic distributed feedback slab waveguide via a fundamental matrix approach," *Applied Optics*, Vol. 26, 1987, pp. 3474–3478.

[11] Weller-Brophy, L. A., and D. G. Hall, "Analysis of waveguide gratings: Application of Rouard's method," *Journal of the Optical Society of America A*, Vol. 2, 1985, pp. 864–871.

[12] Frolik, J. L., and A. E. Yagle, "An asymmetric discrete-time approach for the design and analysis of periodic waveguide gratings," *IEEE Journal of Lightwave Technology*, Vol. 13, 1995, pp. 175–185.

[13] Poladian, L., "Graphical and WKB analysis of nonuniform Bragg gratings," *Physical Reviews E*, Vol. 48, 1993, pp. 4758–4767.

[14] Hirono, T., and Y. Yoshikuni, "A Hamiltonian formulation for coupled wave equations," *IEEE Journal of Quantum Electronics,* Vol. QE-8, 1974, pp. 1751–1755.

[15] Poladian, L., "Variational technique for nonuniform gratings and distributed feedback lasers," *Journal of the Optical Society of America A*, Vol. 11, 1974, pp. 1846–1853.

[16]   Russell St. J., P., "Bloch wave analysis of dispersion and pulse propagation in pure distributed feedback structures," *Journal of Modern Optics*, Vol. 38, 1991, pp. 1599–1619.

[17]   Mizrahi, V., and J. E. Sipe, "Optical properties of photosensitive fiber phase gratings," *IEEE Journal of Lightwave Technology*, Vol. 11, 1993, pp. 1513–1517.

[18]   Erdogan, T., "Fiber grating spectra," *IEEE Journal of Lightwave Technology*, Vol. 15, 1997, pp. 1277–1294.

[19]   Kogelnik, H., "Theory of optical waveguides," in *Guided-Wave Opto-Eletctronics*, T. Tamir, editor New York: Springer-Verlag, 1990.

[20]   Sipe, J.E., L. Poladian, and C. M. de Sterke, "Propagation through nonuniform grating structures," *Journal of the Optical Society of America A*, Vol. 11, 1994, pp. 1307–1320.

[21]   Fermann, M. E., K. Sugden, and I. Bennion, "High-power soliton fiber laser based on pulse width control with chirped fiber Bragg gratings," *Optics Letters*, Vol. 20, 1995, pp. 172–174.

[22]   Morton, P. A., et al. "Stable single-mode hybrid laser with high power and narrow linewidth," *Applied Physics Letters*, Vol. 64, 1994, pp. 2634–2636.

[23]   Morton, P. A., et al. "Mode-locked hybrid soliton pulse source with extremely wide operating frequency range," *IEEE Photonics Technology Letters*, Vol. 5, 1993, pp. 28–31.

[24]   Malo, B., et al. "Apodised in-fibre Bragg grating reflectors photoimprinted using a phase mask," *Electronics Letters*, Vol. 31, 1995, pp. 223–225.

[25]   Erdogan, T., and J. E. Sipe, "Tilted fiber phase gratings," *Journal of the Optical Society of America A*, Vol. 13, 1996, pp. 296–313.

[26]   Kashyap, R., R. Wyatt, and R. J. Campbell, "Wideband gain flattened erbium fibre amplifier using a photosensitive fibre blazed grating," *Electronics Letters*, Vol. 29, 1996, pp. 154–156.

[27]   Strasser, T. A., J. R. Pedrazzani, and M. J. Andrejco, "Reflective mode conversion with UV-induced phase gratings in two-mode fiber," Conference on Optical Fiber Communication, OFC 1997, Dallas, TX, Feb. 16-21, 1997, paper FB3.

[28]   Hill, K. O., et al. "Efficient mode conversion in telecommunication fiber using externally written gratings," *Electronics Letters*, Vol. 26, 1990, pp. 1270–1272.

[29]   Bhatai, V., and A. M. Vengsarkar, "Optical fiber long-period grating sensors," *Optics Letters*, Vol. 21, 1996, pp. 692–694.

[30]   Erdogan, T., "Cladding-mode resonances in short- and long-period fiber grating filters," *Journal of the Optical Society of America A*, Vol. 14, 1997, pp. 1760–1773.

[31]   Vengsarkar, A. M., et al. "Long-period fiber gratings as band-rejection filters," *Journal of Lightwave Technology*, Vol. 14, 1996, pp. 58–65.

[32]   Peral. E., and J. Capmany, "Generalized Bloch wave analysis for fiber and waveguide gratings," *IEEE Journal of Lightwave Technology*, Vol. 15, 1997, pp. 1295–1302.

[33]   Elachi, C., "Waves in active and passive periodic structures: a review," *Proceedings of the IEEE*, Vol. 64, 1976, pp. 1666–1698.

[34]   St. J. Russell, P., "Optical superlattices for modulation and deflection of light," *Journal of Applied Physics*, Vol. 59, 1986, pp. 3344–3355.

[35]   De Sterke, C. M., and J. E. Sipe, "Gap solitons," in *Progr. Optics XXXIII*, E. Wolf, Editor. Amsterdam, The Netherlands, 1994, Chapter III.

[36]   De Sterke, C. M., and J. E. Sipe, "Coupled modes and the nonlinear Schrödinger equation," *Physical Review A*, Vol. 42, 1990, pp. 550–555.

[37]   Aceves, A. B., and S. Wabnitz, "Self-induced transparency solitons in nonlinear refractive periodic media," *Physical Review A*, Vol. 41, 1989, pp. 37–42.

[38]   Boon-Gyoun, K., and E. Garmire, "Comparison between the matrix method and coupled-wave method in the analysis of Bragg reflector structures," *Journal of the Optical Society of America A*, Vol. 9, 1992, pp. 132–136.

Chapter 6

APPLICATIONS OF BRAGG GRATINGS
IN
COMMUNICATIONS

6.1 Introduction

Lightwave communication networks enable high-capacity interconnection between producers and consumers of information to most locations around the world. There is increased demand for bandwidth resulting from new services and users, which has stimulated many innovations in the lightwave industry. This includes such intriguing prospects as direct connection of all network users to gigabit-per-second optical links. Allocating communication channels to partitions in the optical frequency domain has established wavelength division multiplexed (WDM) transmission as a conventional method for increasing capacity and has enabled optical methods for implementing network functions. Existing commercial point-to-point WDM lightwave communication systems operating in the 1550-nm wavelength region have 8, 16, or more channels, each carrying 2.5- or 10-Gbps traffic, and there are already proposals of finer channel spacing appearing. Furthermore, the traditional point-to-point link will be supplanted by networks that include optical add/drop or cross-connect capability. Accompanying this evolution of the optical networks in communications, with its emphasis on WDM transmission, is the proliferation of new optical devices, many of which will probably be based on fiber Bragg gratings.

Fiber Bragg gratings have emerged as important components in a variety of lightwave applications. Their unique filtering properties and versatility as in-fiber devices is illustrated by their use in wavelength stabilized lasers, fiber lasers, remote pump amplifiers, Raman amplifiers, phase conjugators, wavelength converters, passive optical networks, wavelength division multiplexers and demultiplexers, add/drop multiplexers, dispersion compensators, and gain equalizers. This chapter will describe in detail those applications of fiber gratings that are of interest to lightwave communications as either auxiliary components (i.e., gratings to stabilize pump laser wavelengths) or as network elements performing critical functions, such as fiber grating add/drop multiplexers. Table 6.1 shows a list of some of these applications and includes brief descriptions of the principal optical features and grating parameters.

Table 6.1 A List of Some Applications of Fiber Bragg Gratings in Lightwave Communications

Applications	Description	Parameters
Fiber laser	Narrowband reflector	$\Delta\lambda = 0.1-1$ nm R = 1-100 %
Laser wavelength stabilization (980 nm, 1480 nm)	Narrowband reflector	$\Delta\lambda = 0.2-3$ nm R = 1-10 %
Pump reflector in fiber amplifiers (1480 nm)	Highly reflective mirror	$\Delta\lambda = 2-25$ nm R = 100 %
Raman amplifiers (1300 nm, 1550 nm)	Several highly reflective mirror pairs	$\Delta\lambda = 1$ nm R = 100 %
Isolation filters in bidirectional WDM transmission (1550 nm)	Matched sets of WDM gratings	$\Delta\lambda = 0.2-1$ nm R = 100 %
Pump reflector in phase conjugator (1550 nm) and isolation filter in wavelength converter	Highly reflective mirror	$\Delta\lambda = 1$ nm R = 100 %
WDM Demultiplexer (1550 nm)	Multiple high-isolation reflectors	$\Delta\lambda = 0.2-1$ nm Isolation > 30 dB
WDM add/drop filter (1550 nm)	High-isolation reflector	$\Delta\lambda = 0.1-1$ nm Isolation > 50 dB
Optical amplifier gain equalizer (1530-1560 nm)	Blazed Bragg gratings or long period grating	$\Delta\lambda = 30$ nm loss = 0-10 dB
Dispersion compensation for long-haul transmission (1550 nm)	Chirped grating	$\Delta\lambda = 0.1-15$ nm 1600 ps/nm

6.2 Fiber Lasers

Advancements in material science have made the doping of optical fiber core possible with rare-earth ions that posses both a low propagation loss and interesting laser properties. These properties make possible laser oscillation at low thresholds and in low-gain materials with appropriate lasing wavelengths for telecommunications applications. Furthermore, these novel fiber laser systems are targeted to meet the increasing demand of high-speed, high-capacity transmission lines in communications with the added advantage of simple integrability with the basic fiber systems. The development of fiber Bragg gratings has further enhanced the functionality of fiber lasers. The ability to incorporate gratings within the doped fiber with low loss wavelength selectivity and insensitivity to outside perturbations has revolutionized fiber laser technology. This section will include a detailed description on applications of fiber Bragg gratings in fiber lasers and in particular on the erbium-doped fiber laser. It will also cover the various modes of laser operation, such as single-mode lasing, cladding pumped fiber lasers, and pulse generation in fiber lasers.

6.2.1 Erbium-Doped Fiber Laser

The majority of Bragg grating fiber laser research has been carried out on erbium-doped fiber due to its potential for communications applications. The characteristic broadband gain profile of the erbium-doped fiber around the 1550-nm region makes it an extremely useful tunable light source. Employing this doped fiber in an optical cavity as the lasing medium, along with some from of tuning element, results in a continuously tunable laser source over its broad gain profile ranging from 1530–1565 nm. In fact, a tunable erbium-doped fiber with an external grating was reported by Reekie et al. [1] in 1986. Since then several laser configurations have been demonstrated with two or more intracavity gratings [2–5].

6.2.1.1 Normal Operation

An example of a simple erbium-doped fiber laser configuration is shown in Figure 6.1 [6]. This laser cavity consists of a 2-m long erbium-doped fiber with Bragg gratings at each end (broadband and narrowband) providing feedback. The output coupler was the narrowband grating with approximately 80% reflectivity and 0.12-nm linewidth. The broadband mirror was constructed from a series of Bragg gratings resulting in the broadband reflector having a bandwidth of approximately 4 nm. It should be noted that with today's advancements in

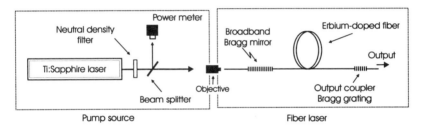

Figure 6.1 A schematic of a simple erbium-doped fiber laser with Bragg grating at each end providing feedback to the laser cavity.

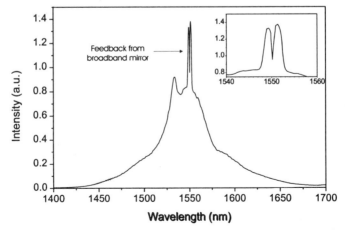

Figure 6.2 Broadband fluorescence obtained from an erbium-doped fiber laser. Superimposed on the gain profile is the broadband mirror at 1550 nm. Within this peak there is a notch at 1550 nm corresponding to the Bragg grating

photosensitivity and writing techniques, such a broadband mirror may have any shape and bandwidth desired. Figure 6.2 shows the broadband fluorescence obtained from the erbium- doped fiber laser system before lasing threshold is reached. The spectrum is the typical characteristic broadband gain profile from an erbium-doped fiber spanning a range of several tens of nanometers, namely between 1.45 and 1.65 μm. Superimposed on the gain profile is a broadband peak (4 nm) at 1550 nm corresponding to the reflection of the fluorescence from the broadband Bragg mirror, and within this peak there is a notch at 1550 nm corresponding to the narrow Bragg grating. With increasing incident pump power, the losses in the fiber laser cavity are overcome and lasing begins. At pump powers just above threshold value, the notch due to the Bragg grating begins to grow in the positive direction, and as the pump power increases further, the laser grows even stronger by depleting the broadband fluorescence (Figure 6.3). In Figure 6.3 the output spectrum from the erbium-doped fiber laser is shown for various coupled pump powers into the doped fiber, starting below lasing threshold at 0.50 to 1.0 mW and 1.5 mW where the laser line at 1550 nm begins to grow. At 3.0 mW of coupled pump power lasing is dominant. This is shown in the inset at the upper left-hand corner where a vertical line at 1550 nm represents the lasing wavelength.

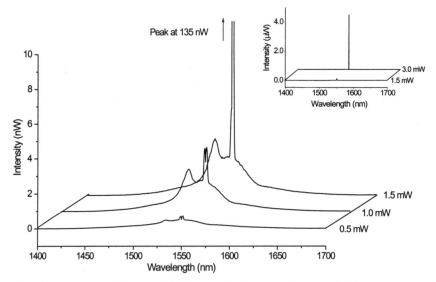

Figure 6.3 Output spectrum from an erbium-doped fiber laser at various coupled-input powers varying from below the lasing threshold of 0.5 mW to nearly lasing at 1.0 mW and fully lasing at 1.5 mW. The inset at the upper-right corner shows the lasing spectrum as a function of coupled pump power of 1.5 and 3.0 mW (note that the laser power at 3 mW is approximately 40 times stronger than that of 1 mW).

6.2.1.2 Multiwavelength Operation

Erbium-doped fiber lasers may also be configured to simultaneously emit several wavelengths. This makes them potentially useful as multichannel sources in wavelength division multiplexed optical fiber telecommunication systems. Various approaches have been attempted to achieve simultaneous multiple wavelength lasing not only in fiber lasers but also in semiconductors. However, they all have two characteristics in common: (1) a means of obtaining gain at several wavelengths simultaneously, and (2) one or more filters to define the lasing wavelengths. A simple approach is to produce this multiple wavelength source using a single gain medium, provided that it is possible to overcome or eliminate cross-saturation between different wavelengths. One technique uses a compound cavity with the net gain at each wavelength controlled in separate arms of the cavity, which are then combined for feedback to the gain medium [7, 8]. Due to the relatively large homogenous linewidth of doped fiber amplifiers at room temperature, this technique requires careful balancing of the cavity losses at each wavelength, particularly if a large number of wavelengths are to be generated. One useful approach is to use a comb filter to promote multiwavelength lasing [9]. Figure 6.4 shows a schematic of a ring laser configuration used to produce multiwavelength lasing. This unidirectional ring cavity is used to avoid spatial hole burning in the gain medium. The fiber cavity is approximately 30 m long, with round trip loss estimated to be 3 dB, not including the intracavity filter. The doped fiber was cooled in liquid nitrogen (77K) to reduce the homogeneous broadening of the fiber, and hence reduce the gain cross-saturation between the lasing wavelengths [10]. It was observed that cooling the doped fiber was necessary to enable stable lasing at multiple wavelengths. An isolator was included in this configuration to ensure the laser operated unidirectionally, while a 25:75 coupler acted as the output for the laser. As illustrated in Figure 6.4, two-laser configurations were examined by Chow et al. [9], namely, a wideband Fabry-Perot transmissive comb filter, in series with a 6-nm tunable filter and a reflective sampled Bragg grating coupled into the cavity using a 50:50 fused fiber coupler. The 6-nm filter in the first configuration is used to prevent lasing outside the bandwidth of the chirped grating used in the filter, while the lasing wavelengths in the second configuration are determined by the sampled gratings. The output from the laser, which incorporates the Fabry-Perot transmissive comb filter, is shown in Figure 6.5(a), whereas the one incorporating the reflective samples Bragg grating filter is shown in

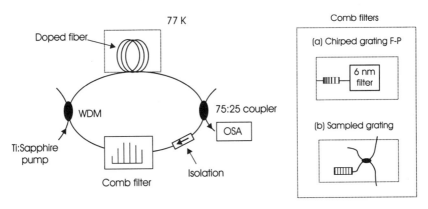

Figure 6.4 Multiwavelength ring laser configuration showing how the different Bragg grating filters were incorporated (*After:* [9]).

Figure 6.5(b). In both cases the multiwavelength capability is demonstrated. The unevenness evident in the output powers at different wavelengths in the lasing combs for both configurations was due to the combined effect of the uneven gain-spectrum of the doped fiber and the difference in transmission or reflection between fringes in the comb filter. Improved filters with higher finesse and more fringes over a wider range of wavelengths are expected to provide further improvements to the characteristics of the multiwavelength laser systems.

Simultaneous multiple wavelength lasing in fiber optics has been demonstrated using several techniques with various degrees of complexity. With the development of fiber Bragg grating technology, however, such sources are becoming simpler and easier to manufacture. This revolution in multi-wavelength fiber optic sources is providing a simple means by which optical fiber communication networks may achieve a multichannel capability for wavelength division multiplexing.

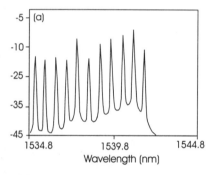 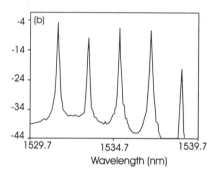

Figure 6.5   Optical spectrum of the laser output using (a) the wideband in-fiber transmissive comb filter and (b) the sampled Bragg grating reflective comb filter (*After:* [9]).

6.2.1.3 Single-Frequency Operation

Single-frequency erbium-doped Fabry-Perot fiber lasers using fiber Bragg gratings as the end mirrors [11, 12] are emerging as an interesting alternative to DFB diode lasers for use in future optical community antenna television (CATV) networks and high capacity WDM communications systems [13]. They are fiber-compatible, simple, scaleable to high output powers, and have low noise and kilohertz linewidth. In addition, the lasing wavelength can be determined to an accuracy of better than 0.1 nm, which is very difficult to achieve in DFB diode lasers. Fiber lasers can operate in a single frequency mode provided that the grating bandwidth is kept below the separation between the axial mode spacings. Furthermore, it is necessary to keep the erbium concentration low enough (a few 100 ppm) to reduce ion-pair quenching, which causes a reduction in the quantum efficiency and may also lead to strong self-pulsation of the laser [12–14]. The combination of these practical limits implies that the pump absorption of an erbium-doped fiber system can be as low as a few percent resulting in low output lasing power. One solution to this problem is to use the residual pump power to pump an erbium-doped fiber amplifier following the fiber laser [15]. In such cases, however, the amplified spontaneous emission from the amplifier increases the output noise. Another approach is by co-doping the erbium-doped fiber with ytterbium [16]. This increases the absorption at the pump wavelength by more than two orders of magnitude and enables highly efficient operation of centimeter-long lasers with relatively low erbium concentration. Kringlebotn et al. [16] reported a highly efficient, short, robust single-frequency and linearly polarized erbium-ytterbium co-doped fiber laser with fiber grating Bragg reflectors, an output power of 19 mW and a linewidth of 300 kHz for 100 mW of 980- nm diode pump power.

6.2.2 Single-Mode/Single Frequency Fiber Lasers

Stable, spectrally narrow wavelength, sources are a key requirement for wavelength division multiplexed optical communication systems and for externally modulated high-data-rate systems. Fiber lasers are attractive for such applications because of their fiber compatibility, compact size, and simplicity. The introduction of fiber gratings has further simplified the design of single longitudinal-mode fiber lasers first by reducing the required components to a minimum and second by minimizing intracavity loss. A single-mode fiber laser may be constructed using a small diode pumped Fabry-Perot cavity consisting of a few centimeters long rare-earth-doped fiber as the lasing medium with fiber Bragg gratings at each end [17–22]. Another approach is to use spectral filtering to reduce the number of longitudinal modes that will have significant gain, allowing the oscillation mode of interest to be locked to a transmission maximum of this filter, thereby preventing other modes from reaching threshold. One such technique uses a single phase-shifted grating written over the full length of the rare-earth-doped fiber, a configuration similar to that of a DFB laser diode [23–29]. These single-mode fiber lasers have been shown to achieve high output powers, low-noise, narrow linewidths, and excellent wavelength stability and accuracy. Furthermore, stretching the fiber cavity provides continuous tunability in wavelength [16, 20]. This laser system does not mode-hop when it is strain-tuned, simply because the cavity modes shift in wavelength at the exact same rate as the reflection spectrum of the fiber Bragg gratings. Overall, single-frequency fiber lasers possess some unique characteristics that make them very useful in the field of communications. One such example is the consideration of single-mode fiber lasers as an alternative to the high-power DFB laser diodes used in CATV transmission systems.

6.2.2.1 Single-Mode Fiber Laser Design

The fiber laser in its most elementary form can be described as a length of doped fiber (lasing material) enclosed by two reflectors. When intracore Bragg gratings are used for cavity feedback in an appropriate length of fiber, single-mode operation can be achieved. Although it appears simple, parameters such as the length of the cavity, reflectivity, and bandwidth of the Bragg reflectors are some of the important design factors that have to be considered for proper lasing operation. Thus, in this section a detailed model of a single-mode fiber laser (erbium-doped) will be described. This will allow to a first approximation the determination of some of these important design constraints for single-mode operation within a linear-cavity fiber laser. This model deals only with erbium-doped fiber laser due to its importance in telecommunications; however, it can be extended to other fiber laser dopants.

Consider a fiber laser cavity of length l incorporating two Bragg gratings of desired reflectivity as the end-reflector and output coupler (Figure 6.6). Each Bragg grating is characterized by its reflectivity maximum and associated bandwidth. Since lasing will only take place within this bandwidth and the grating bandwidth is much less than the homogeneous gain linewidth in rare-earth-doped fibers such as erbium, spatial hole burning is the primary source of multilongitudinal-mode operation in the linear fiber laser. Single longitudinal-mode operation will therefore be maintained when the residual gain from spatial hole burning is insufficient for any mode other than the dominant mode to lase. The gain in an erbium fiber may be obtained by solving the appropriate atomic rate equations. The three-level laser transition in an erbium-doped germanosilicate fiber may be modeled according to the simplified three-level system shown in Figure 6.6. A pump photon is absorbed by an electron resting in energy level 1 (ground state), which is sent to level 3 (pump band). This electron rapidly relaxes to level 2. From level 2 the electron decays down to level 1, with the concomitant emission of either a spontaneous or a stimulated photon. The spontaneous emission is characterized by the spontaneous lifetime τ_2, and the stimulated emission by the effective cross section σ_E. Signal photon ground state absorption also takes place between levels 1 and 2, with an absorption cross section σ_A. In this analysis, pump excited state absorption has been ignored since it does not pertain to the commonly used 980- and 1480-nm pump bands [3]. Finally, an ion may be excited from the ground sate to level 2 through the absorption of a signal photon.

Figure 6.6 shows a schematic of an erbium-doped fiber laser and its corresponding simplified three-level energy diagram with the various transition rates (W_P, W_A, and W_E).

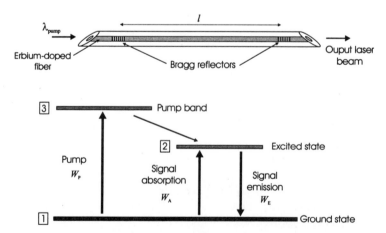

Figure 6.6   A schematic of a linear fiber laser configuration that utilizes Bragg grating reflectors for cavity feedback. Also seen is a simplified three-level energy diagram for erbium showing the various transitions.

In the case where the spontaneous lifetime τ_3 is much shorter than τ_2, the ion population densities (N_i) are given by

$$\frac{N_1}{N_0} = \frac{W_E + 1/\tau_2}{W_A + W_E + 1/\tau_2 + W_P} \tag{6.1}$$

and

$$\frac{N_2}{N_0} = \frac{W_A + W_P}{W_A + W_E + 1/\tau_2 + W_P} \tag{6.2}$$

where the total population satisfies $N_0 \cong N_1 + N_2$ since N_3 is negligible. In this rate equation the transition rates are given by

$$W_P = \frac{\sigma_P I_P}{h\nu_P}, \quad W_A = \frac{\sigma_A I_S}{h\nu_S}, \quad \text{and} \quad W_E = \frac{\sigma_E I_S}{h\nu_S}$$

where σ_P is the stimulated pump absorption cross section for ground state absorption, I_P and I_S are the pump and signal photon intensities in the fiber, and $h\nu_P$ and $h\nu_S$ are the pump and signal photon energies. The gain is defined in terms of the steady state ion population densities as

$$\alpha = \sigma_E N_2 - \sigma_A N_1 \tag{6.3}$$

This may be rewritten as

$$\alpha = \frac{\alpha_0}{1 + \dfrac{\sum\limits_j I_j}{I_{sat}}} \tag{6.4}$$

where

$$\alpha_0 = N_0 \frac{\sigma_E \tau_2 W_P - \sigma_A}{1 + \tau_2 W_P} \tag{6.5}$$

and

$$I_{sat} = \frac{h\nu_S \left(W_P + 1/\tau_2 \right)}{\sigma_A - \sigma_E} \tag{6.6}$$

Given the gain and loss mechanisms within the resonator, the design constraints required to achieve single-mode operation within a linear-cavity fiber laser, which utilizes Bragg gratings for the feedback, can be found. In modeling the laser, the transient energy density per mode within the resonator will be considered. In general this can be written as

$$\frac{dE_j}{dt} = \alpha I_j(z) - \gamma_{\text{loss}} I_j(z) \tag{6.7}$$

where α is the modal gain, γ_{loss} is the modal loss, and j is the index of the lasing mode. For fiber lasers in which propagation in the fiber and Bragg grating losses are negligible, γ_{loss} is given by

$$\gamma_{\text{loss}} = -\ln[R_1(\lambda) R_2(\lambda)] \tag{6.8}$$

where $R_1(\lambda)$ and $R_2(\lambda)$ are the reflectivity spectra response of the Bragg gratings. To determine the conditions for single-mode operation, two potential longitudinal modes, E_1 and E_2, will be considered.

When considering the potential lasing of two modes within the resonator, the gain (6.4) may be written as follows:

$$\alpha = \frac{\alpha_0}{1 + \dfrac{I_1^+(z) + I_2^+(z)}{I_{\text{sat}}}} \approx \alpha_0 \left[1 - \frac{I_1^+(z) + I_2^+(z)}{I_{\text{sat}}} \right] \tag{6.9}$$

where

$$I_1^+(z) = 2I_1 + 2I_1 \sin\left[\frac{2\pi m_1 z}{2l} \right]$$

$$\tag{6.10}$$

$$I_2^+(z) = 2I_2 + 2I_2 \sin\left[\frac{2\pi m_2 z}{2l} \right]$$

l is the laser cavity length, and m_1 and m_2 are integers. In expanding (6.9) we are considering the strong pump regime that will result in a high saturation intensity. In this regime the lasing power within the resonator will be much less than the saturation intensity. Substitution of the linearized gain (6.9) into (6.7) and integrating over the length of the fiber result in the rate equation of the energy buildup for each mode:

$$\frac{dE_{1,2}}{dt} = (\alpha_0 - \gamma_{\text{loss1,2}}) I_{1,2} - 3 \frac{\alpha_0}{I_{\text{sat}}} I_{1,2}^2 - 2 \frac{\alpha_0}{I_{\text{sat}}} I_1 I_2 \tag{6.11}$$

At steady state the intensity of each mode is found to be

$$I_{1,2} = \frac{I_{\text{sat}}}{3} \left(1 - \frac{\gamma_{\text{loss1,2}}}{\alpha_0} \right) - \frac{2}{3} I_{2,1} \tag{6.12}$$

Thus, when the laser is operating in a single mode, the slope efficiency, as given by (6.12), with respect to lasing power versus pump power is determined to be

$$\eta_s = \frac{1}{3}\frac{\lambda_P}{\lambda_1}\frac{\sigma_P}{\sigma_A + \sigma_E}\left[1 - \frac{\gamma_{loss1}}{\alpha_0}\right]\frac{A_L}{A_P} \qquad (6.13)$$

where A_L and A_P are the laser and pump mode areas, respectively. For computational purposes the mode areas will be computed as a function of the mode field diameter, which is defined as $2.6\lambda/\pi NA$. Stable single-mode operation will occur when the laser is operated above threshold so that mode 1 is operating at steady state with intensity I_1 and the loss incurred by mode 2 is greater than the gain so that the rate of energy buildup in mode 2 is negative. In this case

$$\frac{dE_2}{dt} < 0 \qquad (6.14)$$

and when the steady state value of I_1 (6.12) is substituted into the energy density rate equation for mode 2 (6.11), we find

$$\gamma_{loss2} > \frac{1}{3}\alpha_0 + \frac{2}{3}\gamma_{loss1} \qquad (6.15)$$

which is the condition for single-mode operation.

The condition for stable single-mode operation is dependent on both the gain and the loss for the two lowest loss cavity modes. Since the gain medium is considered to be homogeneous and constant over the narrow grating reflector bandwidth, the gain is assumed to be identical for both modes. Mode discrimination must therefore result from the cavity loss mechanism. When narrowband intracore Bragg gratings are used for cavity feedback, a differential loss as a function of wavelength is naturally imposed on the cavity. Two neighboring longitudinal modes will therefore see different cavity losses.

Under the model described, the bounds of single-frequency operation can be determined as a function of grating reflectivity and resonator length. These bounds of single-mode operation will range between a lower limit, which is defined by the lasing threshold, and an upper limit in which the laser will change to multimode operation due to spatial hole burning, as given by (6.15). In the calculations that are described in [3] it is assumed that the fiber laser has two identical gratings of peak reflectivity R_{max}. The single-frequency domain, defined between lasing threshold and the solution to (6.15), is shown in Figure 6.7 as a function of grating reflectivity and resonator length. In the calculation it is assumed that the laser wavelength is 1530 nm and the index of refraction of the fiber is 1.46. In Figure 6.7(a) the dark shaded lower region represents the use of 1.25-cm length grating, and the light shaded regions represent the use of 2.5-cm length grating. Figure 6.7(b) illustrates similar results; however, the fiber small signal gain is assumed to be 4.0 dB/m. The laser slope efficiency is found to be maximum at the single/multimode boundary for the design parameters given in Figure 6.7. The normalized slope efficiency, given by (6.13), is illustrated in Figure 6.8(a) as a function of grating reflectivity for fiber gains of 2.5 and 4.0 dB/m.

When considering single-frequency fiber lasers which utilize intracore Bragg gratings

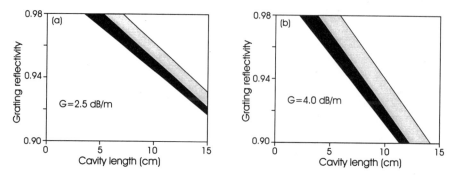

Figure 6.7   Single-frequency operation domain for a linear fiber laser utilizing identical grating pairs: (a) gain 2.5 dB/m, (b) gain 4.0 dB/m. The dark shaded lower region represents lasers with 1.25-cm gratings, and the combined dark and light shaded regions represent lasers with 2.5-cm gratings (*After:* [3]).

as reflectors, an important factor is the stability of the single-mode condition. Therefore, we may ask how far can the single lasing mode be tuned off the grating resonance peak before the cavity loss of the neighboring mode decreases to the point that multimode operation exists or (6.15) is not satisfied. The stability of the single-mode operation is important because the fiber's optical path length is strongly affected by thermal, acoustic, and strain variation perturbations that make the cavity length difficult to fix, and because small optical path length variations (i.e., $\sim\lambda/2n$) will shift the Bragg resonance frequency by one free spectral range. Thus far it was assumed that the dominant lasing mode coincides with the resonant Bragg frequency. Under small resonator optical path length perturbations, the longitudinal modes of the resonator will shift under the Bragg grating spectrum, however, the free spectral range will not vary significantly. This shift can therefore be normalized with respect to the laser free spectral range. Let q represent the fraction of a free spectral range by which the dominant mode is shifted from the grating

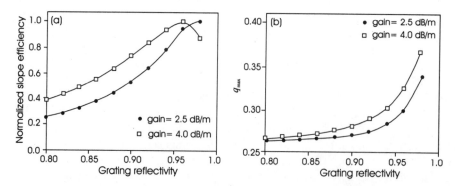

Figure 6.8   Plot (a) showing the normalized slope efficiency versus gating reflectivity for fiber having gains of 2.5 and 4.0 dB/m. Plot (b) showing q_{max} versus grating reflectivity. Both plots (a) and (b) are calculated for 1.25-cm grating length (*After:* [3]).

centerline, where $q = 0$ indicates that the dominant lasing mode is on the center of the grating resonance. Then the maximum resonance shift q_{max}, at which point the laser will undergo transition from single to multimode operation, can be determined. This maximum resonance shift q_{max} can then be plotted as a function of the grating pair reflectivity. The single-mode stability for a laser length, which lies half way between laser threshold and the single-mode/multimode boundary, is shown in Figure 6.8(b). It should be noted that $q = 0$ when the reflectivity and length are chosen to lie on either the single-frequency operation boundary or the threshold boundary. Although the slope efficiency of the fiber laser, when operating half way between the single-mode/multimode boundaries, is approximately 60% of that when the laser is designed to operate on the single-mode/multimode boundary, the solution is significantly more stable.

Using the above model and criterion required to achieve single-mode operation, a fiber laser was designed and constructed [3]. This laser used erbium-doped germanosilicate fiber doped at approximately 450 ppm, with a $NA = 0.24$ and cutoff wavelength of 1.02 μm. The small signal gain was approximately 3 dB/m in the strong pump regime when pumped at 980 nm. This particular laser consisted of a pair of Bragg gratings whose peak reflectivities were 94% separated by approximately 9.5 cm. The length of the gratings was 1.25 cm, whereas the linewidth was approximately 0.2 nm. Single-mode operation was confirmed with a scanning Fabry-Perot interferometer. All experimentally obtained results from this laser system were in good agreement with the predicted values from the described model [3].

6.2.2.2 Examples of Single-Mode Fiber Lasers

In view of the importance of single-mode fiber lasers in telecommunications and the plethora of systems that have been demonstrated over the past few years, it is useful to examine in detail several interesting examples and their characteristics.

One of the earlier demonstrations of a single-mode linear-cavity fiber laser that utilized intracore Bragg gratings as reflectors for cavity feedback was continuously tuned without mode hopping [17]. This was accomplished when both the gratings and enclosed fiber were stretched uniformly. The laser was fabricated with erbium-germanosilcate fiber that was doped at approximately 550 ppm, with a numerical aperture of 0.24 and a cutoff wavelength at 1.02 μm. The laser cavity was formed in the fiber core by inscribing two Bragg gratings with reflectivity of ~95% separated by 10 cm. The fiber ends were polished to angles of 10 degrees to ensure that reflections from the fiber ends did not contribute to the fiber laser's single-mode operation. The fiber laser was pumped at 980 nm generated by a Ti:Sapphire laser. This system was able to operate in single mode with a linewidth of less than or equal to 15 MHz. The fiber laser was mounted at both ends onto a piezoelectric translator (PZT), which allowed tuning over a range of 0.72 nm. Continuous tuning was verified with a high-resolution scanning Fabry-Perot interferometer with no mode hopping observed.

Following this initial work in single-frequency erbium-doped fiber lasers, a much shorter cavity length system has been demonstrated [12]. This fiber laser was fabricated in

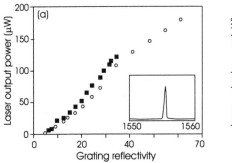 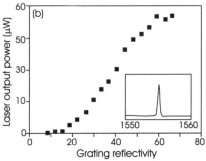

Figure 6.9   Fiber laser output as a function of pump power (solid squares correspond to pumping with 1480 nm, open circles correspond to pumping with 980 nm). Plot (a) correspond to the 2-cm long laser cavity. Plot (b) illustrates the results for the 1-cm long cavity. The insets correspond to the fiber laser output spectra (*After:* [12]).

erbium-doped fiber by inscribing intracore Bragg gratings using side irradiation with UV interfering beams at 240 nm. The doped fiber was fabricated with an alumino-germano-silicate core 2.5 μm in diameter, an erbium concentration of 2500 ppm, and NA of 0.33. Standard AT&T single-mode fiber was fusion-spliced to a short piece of the doped fiber where the Bragg gratings were inscribed. The grating centers were 1 or 2 cm apart. The fiber was pumped through the output coupler by a diode laser using a wavelength division multiplexer. The pump laser was protected from reflection by two optical isolators, whereas the second fiber end was terminated by small diameter turns and angle polishing. Figure 6.9(a) shows the fiber laser output power as a function of pump power for a 2 cm resonator both for pumping at 980 and 1480 nm. The threshold appears to be at 5 mW for both pump wavelengths (980 and 1480 nm), and the emission wavelength was 1540 nm. The output power reached 122 μW for a pump power of 34 mW at 1480 nm, and 182 μW for 61 mW of pump power at 980 nm. Figure 6.9(b) shows the output power as a function of pump power for the 1-cm long fiber laser. The pump wavelength was set at 980 nm and the lasing wavelength was 1556 nm. The output power exceeded 57 μW for a pump power of 66 mW. For such short cavities the longitudinal-mode spacing is comparable to the bandwidth of the distributed Bragg reflector assuring robust single-mode operation.

6.2.2.3 Master-Oscillator Power-Amplifier Configuration

Although single-mode operation may be relatively simple to achieve from a short cavity fiber laser, one problem remaining is the low output power generated from such systems. Master-oscillator power-amplifier (MOPA) configurations can be used to increase this output power. In such configurations an erbium-doped fiber amplifier is used to absorb some of the pump power to boost the weak single-mode output signal from the laser [5, 19]. An optical isolator is placed between the oscillator and amplifier to prevent backward-amplified spontaneous emission from the amplifier coupling into the oscillator and destabilizing the system. Diode-pumped MOPA fiber lasers have achieved single-mode operation in excess of 10 mW and have been implemented successfully in high-bit-rate transmission experiments [5]. In fact, a single-frequency MOPA erbium-doped distributed Bragg reflector fiber laser for WDM digital telecommunications systems has been demonstrated successfully [30]. A schematic of the device configuration is shown in Figure 6.10(a). Two Bragg reflectors were written 2 cm apart, close to one end of a 3-m long piece elliptical core erbium-doped fiber. The elliptical core fiber allowed for high birefringence and high Ge concentration in the core and thus increased photosensitivity so that peak reflectivities above 95% were easily achievable for the Bragg reflectors. The center wavelength and bandwidth for the gratings used were 1556.8 and 0.08 nm, respectively. The fiber birefringence led to the existence of separate reflectivity peaks for the two orthogonal polarization states 0.15 nm apart. The peak reflectivities were 96% and 88% for the rear and output Bragg reflectors, respectively. The rear part of the fiber was spliced to a WDM for 980-nm pumping, while the remaining part acted as power amplifier in a MOPA configuration. A WDM was incorporated to provide the optical feedback signal used in an electronic circuit to suppress the intensity ripple. The amplifier section length was optimized for pumping at 65 mW in order to obtain high output power with good optical signal-to-noise ratio. The measured output power was approximately 13 mW at 66-mW pump power with a slope efficiency of ~24%. In tests of this single-mode laser in a long transmission telecommunications line, it became necessary to phase modulate the signal due to the extremely narrow linewidth of the MOPA in order to avoid detrimental effects due to Brillouin scattering. Even with the phase modulation, however, a bit error rate (BER) test in a complete 475 km long 2.5-Gbps wavelength division multiplexed transmission line showed only a 0.5-dB penalty (Figure 6.10(b)). This indicates that simple MOPA configurations represent an attractive alternative to semiconductor laser in WDM transmitters, especially for 16- or 32-channel systems, where the wavelength control becomes a critical factor.

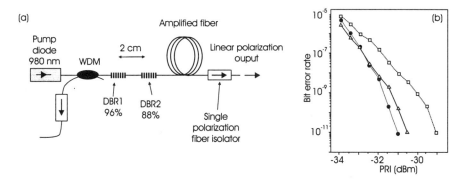

Figure 6.10   (a) A schematic of a single-frequency MOPA erbium-doped distributed Bragg reflector fiber laser. (b) Bit error reading measurement results. The curves clearly show that when the phase modulation is applied a penalty of only 0.5 dB is introduced with respect to the back-to-back configuration. The open squares correspond to the 475-km line, the open triangles to the 475-km line with the phase modulation, and the solid circles correspond to the back-to-back configuration.

6.2.2.4 Erbium-Ytterbium Configuration

Erbium-doped fiber lasers have received a great deal of attention because of their compatibility with telecommunications systems; however, these systems are not well suited for fabricating single-mode lasers. This is due to the their relatively low pump absorption per unit length. Compared to other rare-earth elements, erbium ions have a relatively low absorption cross section at the usual pump wavelengths and they can only be incorporated in silica fibers at low concentrations to avoid problems due to clustering and ion pair interactions. For example, in the length of fiber required for single-mode operation (approximately less than 10 cm) a typical erbium-doped fiber is only able to absorb a few milliwatts of pump power, which results in a fraction of a milliwatt at 1.55 μm. Higher efficiency 1.55-μm single-mode lasers are possible using phosphosilicate fiber co-doped with erbium and ytterbium [22, 23]. These erbium-ytterbium fibers are designed so that the ytterbium ions absorb most of the pump light and resonantly transfer this absorbed energy to erbium. Since ytterbium does not suffer from the concentration-quenching problems of erbium and has a larger abortion cross section, it is possible to construct a fiber laser that absorbs tens of milliwatts of pump power in just a few centimeters. Furthermore, ytterbium gives access to a much wider range of pump wavelengths, starting from 900 nm to 1070 nm, with peak absorption occurring around 975 nm. Initially, erbium-ytterbium single-mode fiber lasers were constructed by splicing fiber gratings at each end of a short section (3–10 cm) of doped fiber [22, 16]. Efficiencies as high as 55% with respect to the launched pump power have been demonstrated for these simple Fabry-Perot cavities [16]. The ability to inscribe Bragg gratings in the erbium-ytterbium fibers has lead to the demonstration of a distributed feedback (DFB) fiber laser [23]. An erbium-ytterbium DFB fiber laser with an output of 10 mW and slope efficiency of 11% has been constructed [31]. In this configuration a single grating inscribed in the doped fiber covers the entire gain region, whereas the single-mode operation is realized by the introduction of a $\pi/2$ phase shift in the middle of the grating [24–26]. DFB fiber lasers offer better stability and stronger side-mode suppression than the conventional Fabry-Perot laser cavities. DFB fiber lasers have also been demonstrated in erbium-doped [26–29] and in ytterbium-doped [25] fiber with grating lengths of up to 10 cm.

A major problem with phosphosilicate erbium-ytterbium fiber is the lack of photosensitivity. Phosphorous doping is known to reduce fiber photosensitivity even though germanium can be incorporated into the material. One can make erbium-ytterbium distributed Bragg reflector single-frequency lasers by splicing the doped fiber to fiber Bragg gratings; however, intracavity splice losses are an obvious concern. Erbium-ytterbium lasers have also been made by hydrogenation [23] and Sn co-doping [31]. A significant loss at the pump wavelength, however, arises from a broad UV absorption band induced when hydrogenation is used, lowering the laser efficiency. In particular, UV exposure of hydrogenated fibers results in substantial OH formation, with associated losses near 0.95 μm, which is resistant to annealing (Chapters 2 and 3). On the other hand, Sn co-doping has permitted the implementation of a 10-cm long DFB laser with 11% slope efficiency, without the need to resort to hydrogenation. Shorter devices, which are desirable from a practical point of view, are, however, still difficult to realize without further improvement in the photosensitivity of such fibers. Dong et al. have demonstrated a simple and efficient method to achieve highly photosensitive erbium-ytterbium fiber that requires no modification of the host glass [32]. In this approach a highly photosensitive B/Ge-doped silica cladding was used to surround a standard erbium-ytterbium–doped phosphosilicate core. The germanium doping increased the refractive index, even though the boron doping lowers it, so the same refractive index as that of pure silica can be achieved for the cladding. The erbium-ytterbium–doped core was not affected by the B/Ge-doped silica cladding. The highly photosensitive cladding allows strong gratings (>99%) ~1 cm long to be easily achieved in these fibers. A fiber laser was constructed from this erbium-ytterbium fiber with a 5-mm long grating as the output coupler (~90%) and a 10-mm long Bragg grating as the high reflector (>99%) and a lasing section of only 2 cm long (Figure 6.11). The threshold was 4 mW and the slope efficiency approximately 25%. The laser operated in a single frequency and a single polarization for output power of as much as 8 mW, beyond which a second polarization mode started lasing as well.

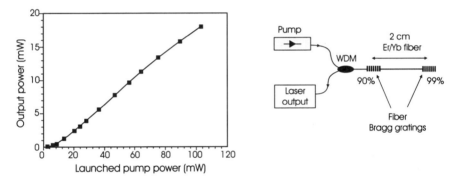

Figure 6.11 Fiber laser constructed with the erbium-ytterbium fiber. A plot of the output power as a function of the launch pump power is shown (*After:* [32]).

6.2.2.5 Polarization Properties

Polarization properties of single-mode fiber lasers are an important issue because of the birefringence present in fibers and fiber gratings themselves. As a result of birefringence, fiber lasers tend to operate over two distinct sets of modes with orthogonal linear polarization [5, 23]. The difference in optical frequencies between the two modes is given by

$$\Delta v(\text{pol}) = \frac{B_v v}{n} \tag{6.16}$$

where B_v is the fiber birefringence, v is the nominal optical frequency, and n is the mode index. The fiber birefringence is usually around 10^{-6}–10^{-5}, thus the polarization mode splitting is typically a few hundred megahertzes [23, 24, 27]. Observations indicate that close to threshold short cavity fiber lasers operate mostly in a single polarization [5]. At higher powers, however, the lasers tend to operate in both polarizations, with unequal distribution of power between the two modes [23, 24]. In this situation the fiber laser has a dual-frequency output, with a frequency separation $\Delta v(pol)$. This is a serious problem since polarization mixing after the output of the laser causes beat noise to appear at the frequency $\Delta v(pol)$. A simple solution to this problem is to operate the laser just above its threshold; since each polarization state experiences different net gain, the laser will be forced to operate in a single polarization [5]. Another solution consists of twisting the fiber [27], thus introducing circular birefringence in the laser. It has been demonstrated experimentally that with increasing twist number the polarization beat frequency reduces and approaches zero. At the same time, one of the polarization modes is suppressed until it is eliminated and the laser operates in a single polarization. A third possible solution arises from the fact that gratings written in certain high-birefringence fiber exhibit anisotropic reflection efficiency [33]. This effect can be used to discriminate between the two orthogonal polarizations, thus forcing the laser into one of the polarization states.

6.2.2.6 Noise Spectrum

One other important characterization property of single-mode fiber lasers is their noise spectrum. This is particularly important in their applications in communications network and high-data-rate transmissions. The noise spectrum of single-mode fiber lasers is dominated by a relaxation-oscillation peak, which usually occurs at a few hundred kilohertzes. This noise can be eliminated with a feedback circuit to the pump laser [18, 5, 21]. At very high frequencies (megahertz to gigahertz range) the noise level can be less than 150 dB/Hz [16], which is comparable to the noise from a DFB laser diode.

6.2.2.7 Nonlinear Cavity Single-Mode Fiber Lasers

In addition to the above linear cavities, single-mode operation has been demonstrated successfully in other types of architectures with some clear advantages. It is well known that spatial hole burning can be eliminated using a traveling wave cavity. This allows the use of much longer intracavity fiber lengths and hence low dopant levels. Several different traveling wave cavity designs have been studied including unidirectional rings and Sagnac-type geometries. The Sagnac-type geometry is of interest as it allows the use of reflective rather than transmissive filters for wavelength control. A single-frequency fiber laser based on a Sagnac-type loop has been demonstrated to generate resolution-limited linewidths of 35 kHz at 1530 nm [34]. The laser cavity is shown in Figure 6.12 where a 1-m long erbium-doped fiber was excited by two InGaAsP laser diodes lasing at 1.48 μm. The pump signal from each laser diode was coupled onto the gain fiber via two 1480/1550-nm wavelength division multiplexers, and an average power of 35 mW was measured in the common arm of each WDM. A 3-dB fused fiber coupler was then fusion-spliced to the gain section in a Sagnac configuration. Unidirectional operation of this loop was achieved by including a polarization-independent optical isolator in the fiber loop. A fiber grating and a 50% reflective mirror were attached to the free port of the 3-dB coupler to provide wavelength selective feedback. The peak reflectivity of the grating was approximately 55% at 1530 nm with a full width at half maximum of 0.15 nm. The laser output was taken from the transmitted signal at either coupler arm. The 50% reflective mirror was necessary to ensure that the laser reached threshold at the Bragg wavelength, and a set of standard polarization controllers was used to manipulate the polarization state of the laser resonator. The loop optical path length was 7.25 m and the optical path lengths of the fiber arms were 0.87 and 1.015 m, which implies a cavity mode spacing of approximately 27 MHz. The optical path difference between the two fiber arms resulted in the formation of an effective intracavity etalon which had a free spectral range of ~1.05 GHz. Optimization of the polarization controllers resulted in a threshold for laser oscillation of 35 mW and a total output power of 1.5 mW. This corresponded to a slope efficiency of 5%. Measurements of the linewidth of the laser placed an upper limit on the value to be approximately 37 kHz. The long-term frequency stability of such lasers will make their output suitable for coherent communications and high-resolution spectroscopy applications.

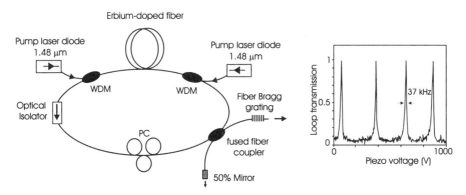

Figure 6.12 Basic configuration for traveling wave cavity erbium-doped fiber laser with intracore fiber gratings. The plot shows the output from a loop resonator (fringes of 37 kHz) used to measure the linewidth of the traveling wave laser (*After:* [34]).

6.2.3 Cladding-Pumped Fiber Lasers

Laser output powers achievable through single-mode fiber either via pigtailed laser diodes or from fiber lasers are limited and are often insufficient for many applications. The optical powers that are currently achievable from single-mode fiber pigtailed laser diodes are on the order of 100 mW. These powers are limited by intrinsic material properties of the laser diodes themselves, and it is unlikely that there will be significant improvement over the next few years. In the case of fiber lasers the major limitation in the output is the amount of coupled pump power available into the core of the doped fiber. An alternative scheme that increases the amount of power available from single-mode fibers by nearly two orders of magnitude is the cladding-pumped technique.

Cladding-pumped fiber lasers are designed to have two distinct waveguiding regions, a large multimode guiding region for the diode pump light and a rare-earth-doped single-mode core from which the diffraction-limited laser output is extracted. Figure 6.13 shows a schematic diagram of a high-power ytterbium cladding-pumped fiber laser. A silica ($n = 1.46$) rectangular waveguiding region of dimensions 360 by120 μm, which is referred to as the pump cladding, confines the diode laser pump. This pump cladding region is typically surrounded by a low-index polymer ($n = 1.39$) giving a high numerical aperture pump region ($NA = 0.48$) into which to couple diode laser power. Furthermore, a protective polymer coating surrounds this low-index polymer. The ytterbium- or neodymium-doped single-mode core is located at the center of the pump cladding. In this configuration the pump light is guided along the cladding and absorbed when the rays occasionally cross the rare-earth-doped single-mode core. Placing Bragg gratings within the fiber core results in extracting wavelength-selective and high laser power from the single-mode core. The most important advantage of the cladding-pumped fiber laser is an ability to couple far more power into the pump cladding, thus resulting in much higher fiber laser powers. A diode-pumped neodymium cladding-pumped fiber laser has produced output powers of 5W at slope efficiencies of 51% [35]. Recently an output power of 9.2W has been obtained from a diode-pumped neodymium cladding-pumped fiber laser [36]. The smaller slope efficiency (25%) observed was a direct consequence of the circular geometry for the cladding-pumped structure and the resultant inefficiency of pump light absorption by the single-mode core.

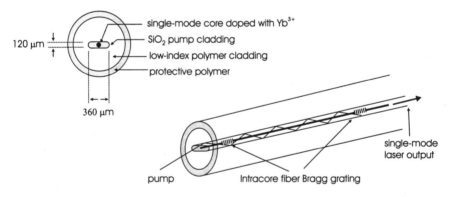

Figure 6.13   Schematic diagram of cladding-pumped fiber laser with intracore fiber Bragg gratings.

As in the case of core-pumping the output spectra of cladding-pumped fiber lasers is broad, with emission extending over several tens of nanometers for ytterbium and neodymium fiber lasers. The use of fiber Bragg gratings as the feedback mirrors has become necessary to narrow and stabilize the wavelength of these high-power fiber lasers. This is especially true when the fiber lasers are used as pump sources for high-power erbium-ytterbium amplifiers, where the wavelengths of the pumps must be maintained within the absorption of the erbium-ytterbium fiber spectral bandwidth. There are now commercially available single-mode fiber lasers than generate output powers of 9W [37]. The feedback elements in this system are fiber Bragg gratings, which were inscribed in the innermost single-mode germania-doped core. No power scaling limitation of cladding-pumped fiber lasers has become apparent, although a power limit of several tens of watts has been estimated [38]. Probably the ultimate limiting mechanism will be the nonlinear effects in these types of high-power lasers rather than thermal effect. Stimulated Raman scattering will be probably the ultimate limiting nonlinearity with an estimated threshold of around 40W [39]. Clearly the future of cladding-pumped lasers will depend on new and more efficient methods of coupling of high-power diode laser power into cladding-pumped fibers and the development of a higher brightness laser source that would increase the efficiency and wavelength range over which cladding-pumped fiber laser operates. The use of all fiber Bragg feedback structures in cladding-pumped fiber lasers will increase both the wavelength range and efficiency available from these high-power fiber laser systems.

6.2.4 Fiber Raman Lasers

Stimulated Raman scattering corresponds to the frequency downshifted Stokes signal that results from the scattering of light by optical vibrational modes (optical phonons) of the host material. In germanosilicate fiber this shift occurs at approximately 450 cm^{-1} (13.2 THz) [39]. Although the nonlinear cross section for this process in germanosilicate fiber is relatively weak, the long lengths and low loss of optical fiber compensate for that. Stolen, Lin, and co-workers were the first to demonstrate the potential of both fiber amplifiers and lasers based on Raman scattering in the 1980s, when Raman lasers operating between 0.3 and 2.0 μm were constructed [40]. At the time it was not apparent that practical laser sources could be constructed to pump Raman fiber lasers or that an efficient continuous-wave pumped, Raman fiber laser would be possible. The recent availability of high-power single-mode fiber coupled output powers from cladding-pumped fiber laser and the ability to construct ultralow-loss fiber cavities through the use of fiber Bragg gratings have dramatically changed this situation.

The development of fiber Bragg gratings has made it possible to fabricate numerous highly reflecting elements directly in the core of germanosilicate fibers with reflection widths of several nanometers and out of band insertion losses of a few hundredths of a decibel. This technology, coupled with that of cladding-pumped fiber lasers, has made Raman fiber lasers possible (Figure 6.14). The pump light is introduced through one set of

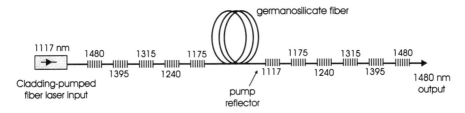

Figure 6.14   Schematic diagram of fifth-order 1480-nm cascaded Raman fiber laser.

highly reflecting fiber Bragg gratings. The cavity consists of several hundred meters to a kilometer of germanosilicate fiber. The output consists of a set of highly reflecting gratings through Raman order $n-1$ and the output wavelength of Raman order n is coupled out, by means of a partially reflecting fiber grating ($R \sim 20\%$). The intermediate Raman Stokes orders are contained by sets of highly reflecting fiber Bragg gratings, and this power is circulated until it is nearly entirely converted to the next successive Raman Stokes order. It has been shown that it is possible to efficiently convert the output from ytterbium-doped cladding-pumped fiber laser at 1117 nm by five Raman Stokes orders to 1480 nm, with a cascaded Raman laser resonator, as shown schematically in Figure 6.14. An output power of 1.7 W at a slope conversion efficiency of 46% has been demonstrated [41]. The typical spectral output of a Raman fiber laser ranges between 1 and 2 nm wide and is controlled by the widths of the fiber Bragg gratings. These types of high-power single-mode fiber lasers at 1480 nm are designed as the pump source for erbium-doped post-amplifiers as well as for remote pumping of in-line erbium-doped fibers [42]. Figure 6.15 shows a repeaterless transmission experiment at 2.5 Gbps over a distance of 529 km using a high-power erbium-ytterbium post-amplifier and three high-power Raman lasers at 1480 nm [43]. Clearly, high-power fiber Raman lasers are becoming very important sources with many possible applications in the area of telecommunications.

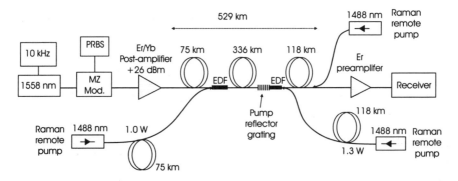

Figure 6.15   Schematic of 529-km repeaterless transmission experiment at 2.5 Gbps using three high-power 1488-nm Raman lasers and a +26-dBm booster amplifier (*After:* [43]).

6.2.5 Mode-Locked Fiber Lasers

For telecommunications applications there is a need for pulses having a duration of a few picoseconds and sometimes even femtoseconds. These short pulses, generally referred to as ultrafast pulses, can be generated in a laser source through the coherent process known as mode locking. This involves locking together the axial modes of the laser across the oscillation bandwidth into some fixed linear relation. This is normally accomplished by modulating a cavity parameter at the fundamental, or a harmonic, of the cavity round-trip frequency.

In the temporal domain, pulses are formed that have a duration inversely proportional to the bandwidth over which locking is achieved. Actively mode-locked erbium-doped fiber lasers have been demonstrated, using both uniform [44] and chirped gratings [45] as the end reflector to a laser cavity. The chirped grating configuration is shown in Figure 6.16, where the fiber dispersion is used to control the properties of the cavity in which mode locking is carried out. This was achieved using a dual-wavelength, nonlinear, optical loop mirror modulator switched by 90-ps pulses generated by a model-locked Nd:YAG laser whose repetition rate was 76 MHz. The chirped grating was approximately 4 mm long and had ~100% reflectivity centered at 1556 nm and a 3.5-nm bandwidth, which gave a dispersion of 11 ps/nm: this was the dominant contribution to the overall cavity dispersion. The cavity had a negative, non-soliton-supporting group delay dispersion for one orientation of the grating. In this configuration stable mode-locked operation was achieved for output pulses of 25 ps. The chirped grating provided stabilization against cavity length fluctuations. For the reversed orientation of the grating, the cavity operated with positive group delay dispersion, which is the soliton-supporting regime. Under this condition the mode-locked laser produced 8-ps stable soliton pulses [45]. The use of chirped gratings in mode-locked fiber lasers gives the added advantage of allowing the cavity dispersion to be selected independently of its length.

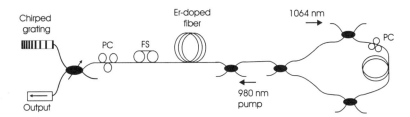

Figure 6.16 Actively mode-locked Er-doped fiber laser using a chirped fiber grating as the reflector. Mode locking is accomplished by switching the nonlinear optical loop mirror using 90-ps pulses at 1064 nm (*After:* [45]).

Passive mode locking has also been achieved in optical fibers using both uniform [46] and chirped gratings [47]. Figure 6.17 shows the experimental setup where a chirped grating was used to passively mode lock a fiber laser. The chirped grating had a bandwidth of 13 nm and was centered near 1555 nm with maximum reflectivity approaching 100%. The grating length was 5 mm, and the group velocity dispersion at 1580 nm was externally measured to be 3.40 ± 0.05 ps^2. The active fiber was doped with 1200 parts-in-10^6 Er^{3+} and had a numerical aperture of 0.16. In these experiments a 2-m long section of fiber was used. The system was pumped through a wavelength division multiplexing coupler with a launched power as high as 400 mW at 980 nm. The chirped grating was oriented to produce a negative group delay dispersion (GDD), with an intracavity polarizer serving as an output coupler. As a result, a wavelength-tunable Kerr-type mode-locked erbium-doped fiber laser has produced an average mode-locked output power of 170 mW while generating near-bandwidth limited 4-ps pulses with up to 10 nJ pulse energy. The advantage of such a system comes from the fact that the large negative GDD contributed by the chirped grating increases the pulse width while the cavity nonlinearity remains effectively constant, resulting in greatly increased pulsed energies compared to conventional soliton fiber lasers.

Another interesting and simple self-mode-locking, all-fiber system has been demonstrated using three intracore Bragg gratings [48]. The main fiber laser cavity consisted of an erbium-doped fiber having a core diameter of 3 μm and a gain of 3.7 dB/m at 1530 nm. Two intracore Bragg gratings having reflectivities of 82% and 74%, respectively, at 1531.5 nm were fusion-spliced on this fiber. An auxiliary cavity was made by fusion-splicing a standard single-mode silica fiber to the main cavity and was terminated by another matched Bragg grating having a reflectivity of 14% at 1531.5 nm. The fiber laser was pumped by a Ti-Al$_2$O$_3$ laser operating at 980 nm. Mode-locked pulses at a repetition rate of 100 MHz have been obtained with an input pump threshold as low as 10 mW. This suggests that mode locking may be achieved in this simple configuration by using a relatively low-powered semiconductor AlGaAs diode laser. The key advantages of this self mode-locked all fiber configuration are its simplicity, low loss and scalability. Such simple self-mode-locked systems can easily be incorporated in communication networks where ultrashort optical pulses are necessary.

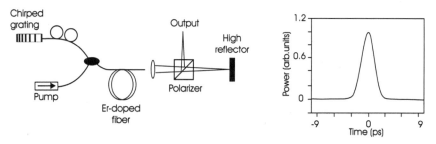

Figure 6.17   A schematic of a passively mode-locked fiber laser incorporating a chirped grating for pulse width control. The plot corresponds to an autocorrelation of the mode-locked pulses produced by the system (*After:* [47]).

6.3 Fiber Amplifiers

No single subject has received more attention in modern optical communication systems than erbium-doped fiber amplifiers. This optical amplifier, which operates in the important third low-loss communication window around 1.5 μm, exhibits a high gain, low pump power requirement, a high saturation power and a low noise and low inter-channel cross talk. These properties, which stem to a large extent from the long lifetime of the metastable level of erbium in silica, make erbium-doped fiber amplifiers well suited for in-line optical amplification, so much so that in just a few years they have significantly altered the future of communication systems. Furthermore, the incorporation of fiber Bragg gratings as selective reflectors for the signal and pump power has improved the efficiency, functionality, and integrability of these devices with fiber optic systems. Other amplifiers providing amplification at different wavelengths have also been demonstrated; one such system is the Raman amplifier. This section will deal with some of these amplifier types and their modification using fiber Bragg gratings.

6.3.1 Fiber Bragg Gratings in Erbium-Doped Fiber Amplifiers

The development of erbium-doped fiber amplifiers has revolutionized the field of communications. The capacity of a telecommunications system has increased from the gigabit per second range to the terabit per second range as a direct consequence of the development of the erbium-doped fiber amplifier and wavelength division multiplexing technology. Fiber photosensitivity and specifically fiber Bragg grating technology has dramatically improved erbium-doped fiber amplifier performance over the past several years, in the areas of pump laser wavelength stabilization, pump reflectors, and gain wavelength flattening. Fiber Bragg gratings have now established themselves as an integral part of the erbium-doped fiber amplifier system tailoring its performance.

Numerous fiber amplifier configurations have been proposed that utilize Bragg reflectors or filters to enhance performance. Placing, for example, a single broadband reflector at the output of erbium-doped fiber amplifiers double-passes the input signal, thus increasing the small-signal gain and recycling the remnant pump light. Reflecting only the pump light may increase the amplifier saturated output power, in the case of amplifiers having marginal pump power. Figure 6.18 shows a variety of reflector configurations used to enhance amplifier performance [49]. Clearly, fiber Bragg gratings can be used as efficient wavelength selective reflectors that discriminate between pump and signal light. A simple analysis under the assumptions of negligible amplified spontaneous emission and perfect fiber Bragg reflector grating (100% reflection) may be used to provide an insight to the various amplifier configurations [50]. Typically, reflecting the signal doubles the small-signal gain, while reflecting the pump may yield a 1–3-dB improvement in small-signal gain. Figure 6.18 shows one calculation of amplifier performance modified by various reflector configurations. Calculations of amplifier noise show marginal improvements (< 0.2 dB) in comparing the case with and without pump reflection. It is interesting to note that pump Bragg reflectors have been used to enhance performance of remotely pumped amplifiers in repeaterless systems.

6.3.2 Gain Equalizers and Gain Control

The useful optical bandwidth of amplified lightwave systems is limited because of gain-narrowing through optical amplifiers. Erbium-doped silica fiber amplifiers show gain peaking at 1530 and 1560 nm and the useful gain bandwidth may be reduced to less than 10 nm. Exotic glasses may be used instead of silica to improve the gain flatness, but handling and structural integrity are issues, as is the inability to effectively pump these glass systems at 980 nm [51]. Cryogenic cooling of erbium-doped fibers to exploit inhomogeneous gain

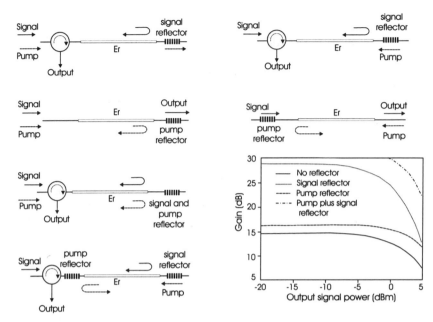

Figure 6.18   Several configurations of reflected pump and signal in an erbium-doped amplifier. Signal reflection nearly doubles the small-signal gain and reflecting the pump increases saturated output power. The plot shows the calculated gain of an erbium-doped fiber amplifier with pump and signal reflectors (*After:* [49]).

has also been reported [52], but it does not lend itself to practical implementation. Optical filters appear to be the best candidates for gain flattening (equalizing), thus increasing the gain bandwidth of an amplifier lightwave system [53, 54]. The loss spectrum of the gain-equalizing filter must match the erbium-gain spectrum at the nominal operating condition of the amplifier. Placing such a filter inside the optical amplifier flattens the gain spectrum and, with appropriate design, may have minimal effect on the amplifier noise figure or saturated output power. Kashyap et al. [55, 56] first reported the use of photorefractive in-fiber gratings to improve erbium-doped fiber amplifier performance. Slanted-fringe, out-coupling gratings were used to produce equalizing fibers to flatten the spectral profile of the amplifier gain. Incident guided modes satisfying the phase-matching condition are coupled to radiation modes by the slanted-fringe grating, thereby introducing a degree of wavelength-dependent transmission loss in the fiber. Measurements show the amplified spontaneous emission (ASE) spectrum in the 1550-nm window to be flat to within ± 0.5 dB over a bandwidth of 35 nm in a 3-m long fiber amplifier when pumped with 15 mW of pump power at 980 nm from a diode laser. In another scheme [56] by writing a series of gratings, which together provide an appropriate spectral loss profile, a filter was constructed that reduced the gain variation form ± 1.6 to ± 0.3 dB across a 33-nm bandwidth in a saturated erbium-doped fiber amplifier, incurring an overall loss penalty of only 0.3 dB. The gain spectrum of the amplifier before and after flattening is shown in Figure 6.19.

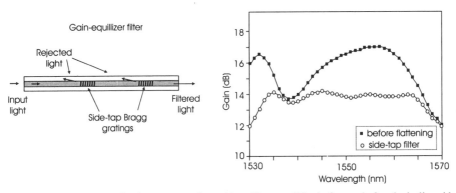

Figure 6.19 The saturated gain spectrum of an erbium fiber amplifier before and after the in-line side-tap filter (*After:* [56]).

Significant progress in gain flattening has been achieved using long period fiber gratings as filters in erbium silica fiber amplifiers. Unlike the short period fiber Bragg gratings, long period gratings work by phase-matching guided modes to cladding modes in the fiber, inducing nonreflective loss. A 22-dB gain with 1-dB flatness from 1528 to 1568 nm has been demonstrated using long period grating equalizers [57]. In another experiment, 0.2-dB gain ripple over 30 nm was achieved, resulting in 5-dB signal variation after traversing 500 dB of system loss [58]. These results greatly enhance the WDM capacity of amplified lightwave systems and may in the future enable practical terabit per second transmission through a single-mode fiber [59].

In addition to flattening the gain spectral profile, gain control in erbium-doped fiber amplifiers is desirable for improved system performance. An elegant approach to achieve all-optical automatic gain control has been demonstrated by Delevaque [60]. Later on, Massicott et al. [61] achieved gain control by simply causing the erbium-doped fiber amplifier to lase at a control wavelength λ_c, which had the effect of clamping the population inversion and hence the gain. In their experimental arrangement a laser cavity was formed by two Bragg gratings at each end of the erbium-doped fiber. The control wavelength λ_c was set at 1520 nm. A slanted-fringe grating at the signal output end served to reject the laser output at 1520 nm. With 80-mW pump power, 15.2-dB gain at 1550 nm with a gain variation of less than 0.2 dB for input signal powers up to 0.32 mW was demonstrated. Gain control in an erbium-doped fiber amplifier has also been demonstrated using two fiber Bragg grating and a fiber-stretcher [62]. The principle of operation of this device is simple. At a steady lasing state, the gain of an erbium-doped fiber amplifier with two fiber Bragg gratings at both its ends is clamped to the total cavity loss due to homogeneous line broadening. Assuming that the reflectivity of each grating is the only factor contributing to a total cavity loss, then

$$g_a = 5\log\left(\frac{1}{R_t}\right) \tag{6.17}$$

where g_a is the gain of the erbium-doped fiber amplifier in the lasing state, and R_t is the total cavity reflectance at lasing wavelength, which is given by

$$R_t = \max[R_1(\lambda)R_2(\lambda)] \qquad (6.18)$$

where $R_1(\lambda)$ and $R_2(\lambda)$ are the reflection spectra of the two gratings. Since the bandwidth of the grating reflection is much narrower than that of the erbium-doped fiber amplifier's gain spectrum, lasing is expected to occur at the wavelength where the product of $R_1(\lambda)$ and $R_2(\lambda)$ is maximum. Due to the shape of the reflection spectra, as the peak center wavelength detuning increases, R_t decreases, resulting in an increase in gain. Thus, gain control of the erbium-doped fiber amplifier can be accomplished by tuning the center wavelength of one of the fiber Bragg gratings. Using this gain control technique [62], a stabilized gain level has been controlled from 4.6 to 22.6 dB by stretching one of two gratings, with initially identical center wavelengths.

Gain equalizers and gain control of erbium-doped fiber amplifiers will clearly play a crucial role in their direct application in telecommunications and in particular enhancing the wavelength division multiplexing capacity. The development of simple innovating techniques for gain equalization and gain control may enable in the future very high transmission through single-mode optical fiber.

6.3.3 Fiber Raman Amplifiers

Fiber Raman amplifiers were considered to be the main technology for optical amplification prior to the revolutionary development of the erbium-doped fiber. While these types of amplifiers possessed many attractive features such as low noise, polarization insensitive gain, and the ability to achieve amplification in ordinary germanosilicate transmission fiber, it was primarily the unavailability of high-power diode laser pump sources that prevented their acceptance. This has now changed with the development of cladding-pumped fiber lasers.

Raman fiber amplification may be obtained from a simple cascaded Raman resonator [63]. In fact, a cascaded Raman resonator amplifier system was the first silica fiber–based optical fiber amplifier demonstrated at 1310 nm with gains of 40 dB and output powers of +24 dBm. Figure 6.20 shows a schematic of such a system, where laser light at 1060 nm generated from a high-power cladding-pumped fiber laser is injected into a long length of germanosilicate fiber. At both ends of the fiber there are three highly reflective Bragg gratings, with center wavelengths at the first three Stokes-lines starting at 1060 nm. This configuration converts the fundamental pump light (1060 nm) efficiently to light at 1240 nm. When a signal at 1310 nm is injected through this structure, it will experience amplification since it is designed to be at the next Stokes Raman shift from the 1240-nm pump light. The efficiency of the Raman amplification process is controlled both by the amount of germania dopant in the fiber core and the cross-sectional area of the core. Gain of 25 dB for only 350 mW of pump power has been obtained in highly germania-doped, small core fibers [64]. In the cascaded Raman approach, lasing or amplification should be

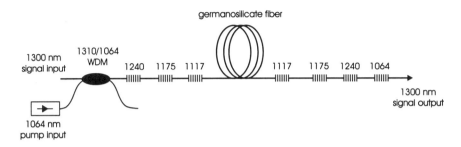

Figure 6.20 Schematic of 1.3 μm Raman amplifier utilizing a series of Bragg gratings as the high reflectors to the various Stokes-lines.

possible from 1.1 to 2.0 μm. Since the bandwidth of the Raman process is broad and the pump wavelength obtained from a neodymium- or ytterbium-doped cladding-pumped laser can be varied by nearly 100 nm, one can efficiently down-convert to virtually any arbitrary wavelength with the use of the appropriate fiber Bragg gratings. Clearly, Bragg gratings are playing an important role in Raman amplifier technology and are having a strong impact in communications lightwave applications.

High-speed transmission in the 1.3-μm wavelength window is attractive since it would eliminate the need for dispersion compensation of nondispersion-shifted fiber, such as standard fiber of which a vast amount has already been installed. The 1.3-μm window also provides an upgrade path to higher capacity of existing 1.5-μm systems by adding it as a second band. Such an upgrade would require 1.3-μm amplifiers that can facilitate the power budget as dictated by the amplifiers spacing in the 1.5-μm systems, which typically is 80–120 km corresponding to a power budget of up to 40 dB for the 1.3-μm channels. Recently, a WDM transmission experiment in the 1.3-μm wavelength band utilizing Raman post- and pre-amplifiers has been demonstrated [65]. The system consisted of eight DFB lasers, which were combined in fiber couplers such that odd- and even-numbered channels were encoded with a $2^{31} - 1$ pseudorandom bit sequence. The two signals were then delayed complementary, in two Mach-Zehnder modulators. The combined signal from the two modulators was amplified in a Raman post-amplifier to a level of 17.8 dBm corresponding to a launch power of 8.8 dBm per channel. The fiber span consisted of a low-loss silica-core fiber with a zero-dispersion wavelength. A 141-km span of fiber was connected directly to the pre-amplifier input. The receiver included a Raman pre-amplifier, a 37-GHz bandwidth tracking Fabry-Perot filter, and a photodiode. The Raman post- and pre-amplifiers, which were identical in topology, are shown in Figure 6.21. The upper half of the ring is a cascaded Raman laser that converts 1117-nm pump light to a wavelength of 1240 nm. The 1240-nm light is lasing clockwise around the ring and provides Raman gain for the 1.3-μm input signal, which propagates in the lower part of the ring. The counter-propagating pump scheme minimizes the noise transfer from the 117-nm Ytterbium dual cladding-pumped pump source to the signal. Furthermore, both amplifiers are divided into two stages separated by a midstage isolator to reduce the length of continuous fiber, as amplified Rayleigh back-scattering noise will increase the noise

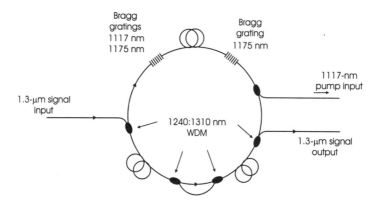

Figure 6.21   Schematic of the Raman post- and pre-amplifier topology.

figure at high gain. Measurements on the system indicate no degradation due to polarization sensitivity or other multichannel effects such as cross-phase modulation. The back-to-back performance of the transmitted (excluding the post-amplifier) and the pre-amplified receiver is −36.7 dBm resulting in penalties for eight-channel transmission between 0.4 and −0.2 dB. The BER performance of 10^{-10} for transmission through a 141-km long fiber has been observed. These results demonstrate the versatility of Raman amplifiers and their potential for WDM transmission at a wavelength outside the conventional erbium window such as the 1.3-µm wavelength band.

6.4 Fiber Bragg Grating Laser Diodes

Bragg gratings have been used with laser diodes for many years: initially with the grating etched into the surface of side-polished fibers [66–68] and now with UV-induced Bragg gratings [69–81]. A simple but effective means of controlling the laser wavelength is by incorporating a Bragg grating in the pigtail of the diode laser. This technique compares favorably to other types of feedback techniques such as distributed feedback (DFB) or distributed back reflector (DBR) due to the low cost and the simple manufacturing procedure. One problem in DBR laser manufacture is the precise control of the laser wavelength. Routine production of DBR lasers with the wavelength specified to better than 1 nm is difficult; however, fiber Bragg gratings can be manufactured precisely to the wavelength required. With antireflection coating on the semiconductor chip, the lasing wavelength may be selected from anywhere in the gain bandwidth by choosing the appropriate Bragg resonant wavelength of the grating. Clearly, such an approach will increase the yield from the semiconductor wafers. In addition, since each laser has to be coupled to a fiber, the Bragg grating may be written after the packaging process has proved to be successful, thus reducing the time spent on unsuccessful products.

   To date there have been three main types of fiber grating laser diodes reported: first, the single-mode laser [66–75], which has a grating located close to the output facet of the diode to ensure stable single-frequency operation; second, the mode-locked laser [68, 76] in which the diode is modulated at a multiple of the characteristic external cavity frequency; and finally, the coherence-collapsed laser [77–79] where the distance between the grating and the diode is greater than the coherence length of the diode modes. These grating-stabilized laser techniques have unique characteristics and advantages that make them suitable for specific applications. Each of these techniques will be discussed in detailed in the following sections.

6.4.1 Single-Mode Operation

A fiber Bragg grating may be coupled to a semiconductor laser chip to obtain what is commonly referred to as a "hybrid" or a fiber Bragg diode laser. In this process a semiconductor laser chip is antireflection coated on the output facet and coupled to a fiber incorporating a Bragg grating, as illustrated in Figure 6.22. With the Bragg grating set to reflect within the gain bandwidth of the semiconductor material, it is possible to obtain lasing at the Bragg grating wavelength. The grating bandwidth can be narrow enough to force single-frequency operation with a linewidth of much less than a gigahertz (GHz). This type of hybrid laser has been studied for many years [66, 67] and has recently seen renewed interest because of its potential as a low-cost WDM source [69–75]. One of the main advantages of hybrid lasers over conventional DFB or DBR lasers is their relative insensitivity to the changes in temperature or drive current. In a typical semiconductor laser, temperature or current changes result in a variation of the optical pathlength (δL_{sl}) of the laser cavity, which leads to a corresponding change in the emission wavelength

$$\delta\lambda = \frac{\lambda \delta L_{sl}}{L_{sl}} \tag{6.19}$$

where L_{sl} is the optical pathlength of the semiconductor laser. In the case of the hybrid laser, the same change in optical pathlength results in a wavelength shift given

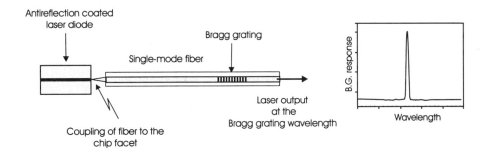

Figure 6.22 A schematic illustration of the semiconductor laser coupled to a fiber with a fiber Bragg grating.

$$\delta\lambda = \frac{\lambda \delta L_{sl}}{L_{tot}}$$

(6.20)

where L_{tot} is the total optical cavity pathlength covering the length between the end facet of the semiconductor and the fiber Bragg grating (the output reflector). Since L_{tot} is much greater than L_{sl} and because of the low sensitivity of optical fiber to environmental changes, the temperature and current dependence of the hybrid laser is usually much less than that of a DFB or DBR laser. A consequence of this improvement on the behavior of the hybrid lasers is apparent when used as a WDM source, which may be implemented without an external wavelength reference and even without a thermoelectric cooler [74]. A second important advantage of the hybrid fiber laser is the ability to perform direct high-speed modulation with a very low level of chirp [74, 75]. Furthermore, the wavelength stability of the hybrid laser can be improved by shortening the semiconductor chip, as was demonstrated using diodes as short as 150 μm [73]. In that experiment the wavelength change with temperature was only 0.007 nm/°C as opposed to 0.01 nm/°C for longer devices.

Figures 6.23(a) and (b) show optical output power versus drive current and spectral characteristic of a 500-μm antireflection coated InGaAsP/InP diode coupled to 30% reflectivity fiber grating [75]. It is evident from these two graphs that high power can be achieved without mode hopping and with good linearity and high side-mode suppression. This device had a linewidth of 100–250 kHz and a relative intensity noise level below 135 dB/Hz in the range from 20 MHz to 10 GHz. One of the important parameters in the performance of the hybrid laser is the quality of the AR coating. Even weak residual reflections ($\sim 10^{-3}$) can result in instabilities in the output power and mode hops with current ramping. This problem may be eliminated by simply reducing the front facet reflectivity to less than 10^{-4} using a laser diode with a curved waveguide [74]. Perhaps a more important issue is the packaging of the device, which will undoubtedly be the main contributing factor in determining the long-term reliability of hybrid lasers. As noted from the above

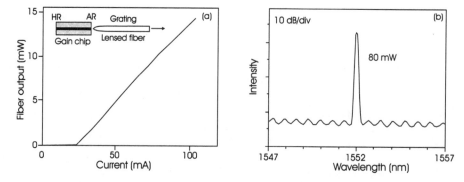

Figure 6.23   Plot (a) illustrates a pigtailed fiber output from a 1.55-μm hybrid laser versus current at 20°C. The inset shows a schematic of a hybrid laser. Plot (b) shows the spectral profile of this hybrid laser (*After:* [74]).

properties of hybrid lasers, they clearly have the potential of not only replacing DFB lasers, but also creating a whole system comprised of an external modulator and an external wavelength reference with feedback circuitry. Clearly, the hybrid laser could be the preferred approach for a low-cost WDM source.

Single-mode operation at 1.55 μm of a fiber grating external cavity diode has been demonstrated using an interesting and effective approach (Figure 6.24) [72]. In this configuration a length of unpumped erbium-doped fiber was placed between the diode and the fiber grating. During laser operation, a standing-wave intensity pattern is formed in the fiber, which, at high powers, periodically saturates the erbium absorption. This spatial hole burning causes a reduction of the cavity loss for this particular longitudinal mode. On the other hand, the loss reduction is less important for other modes as they have a different intensity distribution and therefore interact with unbleached sections of the fiber. Therefore, this spatial hole burning in the doped fiber produces a self-narrow band-pass filter that increases the mode discrimination of the laser and forces single-frequency operation. One interesting aspect of this laser configuration is that the single-mode stability is fairly independent of the external cavity length as the optical bandwidth of the self-written filter is inversely proportional to the fiber length, which in effect compensates for the denser mode-spacing. Stable single-frequency operation has thus been observed for cavities up to 3m long with subkilohertz bandwidths.

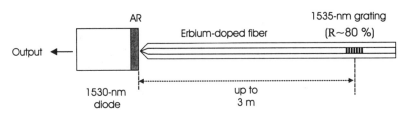

Figure 6.24 Schematic of single-mode grating-stabilized laser diode with erbium fiber external cavity (*After:* [72]).

6.4.2 Mode-Locked Operation

Fiber Bragg gratings have also been used to provide feedback for active mode-locked semiconductor laser diodes. Mode locking in these systems is accomplished by modulating the injection current of the laser at the characteristic frequency of the optical cavity, which is determined primarily by the position of the fiber grating. The reflection profile of the grating limits the number of modes that can oscillate and thus determines the duration of the mode-locked pulses. This type of a hybrid soliton pulse source utilizing a uniform Bragg grating into a fiber external cavity has been shown to operate close to 1.5 μm with transform-limited pulses (~19 ps) and very high output powers [76]. This device, however, worked well only under specific operating conditions and showed spectral instabilities when those operating parameters were changed. The use of a linearly chirped fiber grating has overcome this spectral instability problem [80]. It also resulted in a tunable system over the range of mode-locked frequencies as the cavity length was wavelength-dependent. High-quality pulses (~50 ps) with repetition rates up to 2.4 GHz have been demonstrated using this type of configuration, making them a very attractive source for high-speed soliton transmission in communication networks [80].

6.4.3 Coherence-Collapsed Operation

One of the modes of operation of a laser diode with an external fiber Bragg grating as a feedback is known as "coherence-collapse" [81]. The coherence-collapsed regime can be observed by providing a sufficiently strong source of feedback (e.g., a Bragg grating reflector) and positioning it beyond the coherence length of the diode laser (Figure 6.25) [77, 78]. In the presence of "incoherent" feedback from the Bragg grating, the laser enters the coherence-collapsed regime, which is characterized by a substantial broadening of the laser diode modes and a strong reduction of the coherence length of the laser. The output from such a system consists of a large number of external cavity modes whose amplitudes are modulated by the mode structure of the diode and the reflection spectrum of the grating (Figure 6.26(a)). The coherence-collapse operation forces the diode to operate within the spectral bandwidth of the fiber grating and thus stabilizes the laser wavelength. Despite the high-frequency noise observed in the laser output, mainly due to mode beating, the large number of modes and their lack of coherence averages out the low-frequency power fluctuations normally associated with mode hopping. This makes this device an excellent source for low-speed applications (Figure 6.26(b) and (c)). Stabilized 980-nm laser diodes that operate in the coherence-collapsed regime are now widely used as the pump sources in commercial erbium-doped fiber amplifiers [78, 79]. In this application, the high-frequency noise of the coherence-collapse pump laser is of no consequence due to the slow response time of the erbium ions. The absence of low-frequency noise greatly improves the power stability of the amplifier. Grating-stabilized 980-nm pump lasers typically consist of the standard high-power 980-nm Fabry-Perot laser coupled to a fiber Bragg reflector with a reflectivity of a few percent (comparable to the output facet reflectivity). Since the fiber grating has very low loss and a small reflectivity, the output power from a stabilized laser is approximately the same as the nonstabilized system. An additional benefit of the coherence-collapsed mode of operation is that it is much less sensitive to feedback compared to nonstabilized counterparts. Measurements have shown that a stabilized 980-nm laser remain, unaffected by optical feedback up to about 20 dB, whereas non-

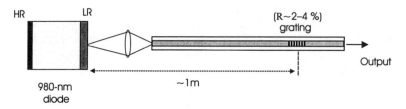

Figure 6.25 Schematic of a coherence-collapse 980 nm grating-stabilized laser diode.

stabilized systems can become unstable at the 40-dB feedback level [78]. Furthermore, a stabilized 980-nm laser has been demonstrated without a thermoelectric cooler [79]. This is possible due to the low temperature dependence of the grating.

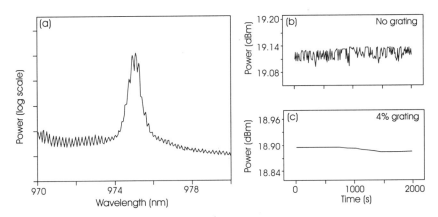

Figure 6.26 Coherence-collapse stabilization. Plot (a) illustrates the spectrum output of a commercial grating-stabilized 980-nm pump laser. Plots (b) and (c) show power monitoring of a 980-nm laser diode: (b) without grating stabilization, mode hop causes noticeable power fluctuations; (c) with grating stabilization, coherence-collapse eliminates sudden power changes (*After:* [78]).

6.5 Basic Band-Pass and Other Types of Fiber Bragg Filters

Band-pass filters are considered one of the most fundamental devices in multiwavelength optical networking and in most communication systems where wavelength demultiplexing is required. There are several techniques for fabricating these band-pass filters utilizing Bragg gratings. One approach is based on the interferometric principle, where gratings are incorporated into Saganac [82], Michelson [83, 84] or Mach-Zehnder [85, 86] configurations. Another approach, based on the principle of the moiré grating resonator [87], has also been applied with uniform period [88] and chirped [89] grating types. Furthermore, resonant filter structures have been formed by introducing a phase shift into the grating by an additional UV exposure [90], or by using a phase-shifted phase mask [91]. In general the resonant-type transmission filters are capable of large wavelength selectivity and are, in principle, simple to fabricate and do not require carefully balanced arms or identical gratings, as in the case of interferometric filters. Typically, these types of filters tend to produce a narrow band-pass within a relatively narrow stop-band outside of which the filter returns to essentially total transmission.

The application of multiple phase shifts to produce multiple band-passes has also been investigated [92]. Coarser wavelength discrimination has been achieved using simple methods, including amplitude masking the fiber to remove a central portion of a broadband chirped grating during exposure [93] using two broadband reflection gratings in series that are chosen such that their reflection spectra do not overlap [94] and comb filters. In the following sections various important band pass and other filter types will be examined.

6.5.1 Basic Bragg Grating Filter

One of the most basic filters may be constructed by simply splicing or inscribing a fiber grating to a 50:50 (3 dB) fiber coupler (Figure 6.27(a)). This is one of the simplest methods of accessing the narrowband signal reflected by a fiber Bragg grating. The disadvantage is obviously that the reflected signal suffers a 6-dB loss by passing through the coupler twice. Moreover, half of the out-of-band signal is also lost and the return loss is only 6 dB. For most systems applications these drawbacks are unacceptable, and more elaborate solutions must therefore be employed.

6.5.2 Circulator-Based Filters

Optical circulators are nonreciprocal bulk optics components with multiple input and output ports. Figure 6.27(b) shows how a high-performance add/drop filter can be obtained simply by attaching a fiber grating to the second port of a 3-port circulator. With port 1 as the input port, the wavelength channel reflected by the grating is transmitted through port 3 while the out-of-band signals are transmitted through the grating in port 2. Commercially available circulators have very high port-to-port isolation (typically 50–60 dB) and thus give the filter excellent wavelength isolation and return loss. However, circulators are very expensive and have high insertion losses (1–2 dB port to port), which would make an all-fiber solution more suitable in many applications.

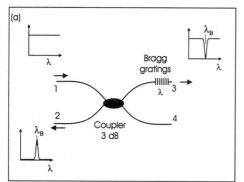 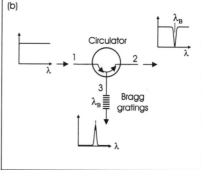

Figure 6.27   Bragg grating filters showing input (port 1) and output (ports 2 and 3) spectra: (a) basic Bragg grating filter; (b) circulator-based filter.

6.5.3 Interferometric Band-Pass Filters

A single Bragg grating in a single-mode fiber acts as a wavelength selective distributed reflector or a band rejection filter by reflecting wavelengths around the Bragg resonance. By placing identical gratings in two ports of a fiber coupler, however, as in a Michelson arrangement, one can make a band-pass filter [95]. Figure 6.28 schematically shows such a filter. The principle of operation is as follows: Consider a multiwavelength signal $S(\lambda_1,\lambda_2,...,\lambda_B,...,\lambda_N)$ incident at port 1, where λ_B is the Bragg wavelength of the identical gratings. This signal is divided into two equal-power portions by a 3-dB coupler, which are incident on the respective gratings where only the light at λ_B is reflected. The reflected light is incident on the coupler with a phase difference of π between them, which results in the emergence of all of the signal $S(\lambda_B)$ from port 2. The remaining power at all other wavelengths is output in equal portions from ports 3 and 4. In principle, this is a low-loss filter; however, there is a 3-dB loss penalty for the wavelengths that are not reflected. An efficient band-pass filter was demonstrated by Bilodeau et al. [96–98]. The device had a back reflection of -30dB. However, all wavelengths out of the passband suffered from the 3-dB loss associated with the Michelson interferometer.

When a second 3-dB coupler is added to the Michelson arrangement, the Mach-Zehnder interferometric arrangement is formed (Figure 6.29). In this case, the relative phase of the two transmitted signals is adjusted to maximize the output through either port 3 or 4, thus avoiding the 3-dB loss inherent in the simpler Michelson configuration. This arrangement offers added functionality to the filter by converting it to a wavelength add/drop multiplexer (Section 6.5.2.1). The fabrication of these interferometric filters is technically difficult as the performance of each filter is strongly affected by any imbalance in the power-splitting of the couplers, the spectral characteristics of the two gratings or the relative phase of the reflected and transmitted signals. In practice, these couplers can be made by writing the gratings simultaneously, as close as possible to the fused coupler. The relative phase of the reflected and transmitted signals can be accurately adjusted by UV-trimming, which means exposing one side of the interferometer to a uniform UV beam to permanently change the refractive index.

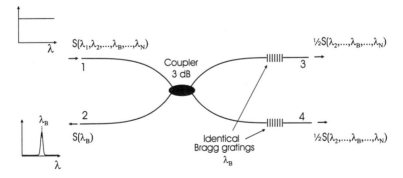

Figure 6.28 A fiber Bragg grating Michelson interferometric band-pass filter.

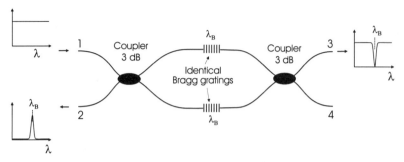

Figure 6.29   A fiber Mach-Zehnder interferometric band-pass filter.

6.5.4 Moiré Grating Filters

An example of a resonant band-pass filter is a moiré grating produced by the sequential exposure of the same length of fiber to two slightly different grating fringe patterns. Assuming uniform period gratings of equal refractive index, of amplitude δn, with periods Λ_1 and Λ_2, the resulting index of refraction perturbation will have the following form:

$$\Delta n(z) = 2\delta n \left[1 + \cos\left(\frac{2\pi z}{\Lambda_a}\right) \cos\left(\frac{2\pi z}{\Lambda_b}\right) \right] \tag{6.21}$$

where z is measured along the fiber axis, $\Lambda_a = 2\Lambda_1\Lambda_2 / (\Lambda_1 + \Lambda_2)$ and $\Lambda_b = 2\Lambda_1\Lambda_2 / (\Lambda_1 - \Lambda_2)$. The above equation represents a spatial amplitude-modulated waveform with a rapid variation of period Λ_1, and a slow varying envelope whose period is Λ_2. In the fabrication of the moiré grating, the small change in period between the constituent gratings is accomplished either by changing the angle of the writing beams [89] or by changing the writing wavelength between the exposures [88]. Typically, the band-pass widths are <0.1 nm with associated stop-band widths less than 1 nm. However, it was demonstrated by Zhang et al. [89] that the moiré concept can be extended to chirped gratings resulting in broadening of the stopband. The authors have reported several structures with subnanometer band-pass and stop-band widths up to 12 nm. They have also demonstrated a further increase in the stop-band width by concatenating broadband chirped reflectors with a chirped moiré filter [89].

6.5.5 Phase-Shifted Grating Filters

Phase-shifted gratings may also be considered as resonant band-pass filters and, as in the case of moiré gratings, they correspond to narrowband transmission filters within a narrow stopband. The narrow stopband renders these devices inadequate for most practical WDM applications. One approach to overcome this problem is to apply the phase shift on broad chirped fiber gratings as opposed to uniform gratings [99]. As a result, a highly efficient in-fiber band-pass filter with arbitrary band-pass/stop-band combination can be successfully produced. This may be achieved either using UV post-fabrication techniques [99] or with a phase mask. Agrawal et al. [92] presented theoretical results on the insertion of multiple phase shifts (three π-phase-shift) equidistant along a fiber grating, resulting in three transmission peaks inside the stopband. As a further improvement to the operation of such devices as a band-pass filter, theoretical and experimental results on the introduction to a fiber Bragg grating of two π-phase-shifts located at optimized positions have been reported [100]. Although giving a wider and flattened band-pass peak, compared with the singly phase-shifted grating, the stop-band depth was not high enough for band-pass filter. The insertion of a third phase shift has been reported [101] giving a more rectangular band-pass shape while the increased phase-shift number allowed tailoring this rectangular spectral shape. Band-pass peaks with negligible ripples at the top (<0.01 dB) have been achieved through the optimization of distances between the phase shifts along the grating. These band-pass filters should find useful applications as noise filters or channel selectors in WDM systems.

6.5.6 Fabry-Perot Etalon Filters

Placing two identical Bragg gratings in series on a single-mode fiber results in a Fabry-Perot etalon within the fiber core. With the advancements in the inscription of Bragg gratings in optical fiber it is now possible to obtain etalons with finesse as high as several thousands. A simple filter application of the Fabry-Perot consists of an optical circulator and another fiber grating [102]. The input signal is filtered with a Fabry-Perot (grating pair) and directed forward to the fiber grating by an optical circulator. The reflected signal from the fiber Bragg grating is then redirected to the output port by the circulator. Although narrowband Bragg grating Fabry-Perot filters have been reported with very high finesse, for applications in short-pulse lasers and wideband communication systems, a response over several nanometers or more may be required with a wide variety of free spectral ranges needed. One technique to accomplish this is to use linear chirped gratings instead of constant period Bragg gratings [103]. Town et al. demonstrated this approach using a resonator formed with two linearly chirped gratings having reflectivities exceeding 50% over a 150-nm spectral width. The gratings in each pair were chirped in the same direction along the fiber axis. For lower vales of the free spectral range, the gratings were spatially separated; for higher values they were partially overlapped. This arrangement produced a resonator operating over a wavelength span exceeding 150 nm with a free spectral range value in the range 0.09–11.27 nm. These types of structures have been used to demonstrate CW multiwavelength operation of erbium-doped fiber lasers [9].

6.5.7 Comb and Superstructure Filters

The ability to permanently change the index of refraction in an optical fiber has proven to be extremely useful in the area of telecommunications and, in particular, in constructing fiber optic filters. One type of such filter is the comb filter, which provides a way to multiplex/demultiplex telecommunication signals. A comb filter, as the name suggests, is a device that has a large number of transmission or reflection windows. Othonos et al. [104] have demonstrated a comb filter by inscribing up to seven gratings each with a different Bragg wavelength within a 60-nm span superimposed at the same location in the fiber. Superstructure fiber gratings also exhibit a comb-like response in reflection [105]. Gratings of this type, have been produced by the phase mask scanning technique where the writing beam is modulated by slowly varying the periodic envelope.

6.5.8 Blazed Filters

By tilting or blazing the Bragg grating at angles to the fiber axis, light can be coupled out of the fiber core into loosely bound cladding modes or to radiation modes outside the fiber. This wavelength selective tap occurs over a rather broad range of wavelengths that can be controlled by the grating and waveguide design. One of the advantages is that the signals are not reflected in, thus the tap forms an absorption type filter. An application, already mentioned in this chapter, utilizes this grating tap as a gain flattening filter for an erbium fiber amplifier. Furthermore, with a small tilt of the grating planes to the fiber axis (~1 degree), one can make a reflecting spatial mode coupler such that the grating reflects one guided mode into another.

6.5.9 Grating-Frustrated Coupler Filter

The grating-frustrated coupler is quite different from other types of filters as it is based on the dispersion, rather than the reflection, characteristics of a fiber Bragg grating. This filter was proposed and demonstrated by Archambault et al. [106]. In a grating-frustrated coupler, a grating is written in one side of a perfectly phase-matched coupler designed for 100% power transfer in the absence of the grating. The role of the Bragg grating is to introduce a phase mismatch over a narrow wavelength band due to its strong dispersion near the Bragg wavelength. Thus, the transfer of power by the coupler can be frustrated over the grating bandwidth. The main advantage of the grating-frustrated coupler over the Mach-Zehnder design is that it is noninterferometic, which makes its fabrication easier and should give it better stability.

6.5.10 Long Period Grating Filters

Long period gratings have become increasingly popular in a multitude of applications owing to their ease of fabrication, low insertion losses, low back-reflection, low polarization sensitivities, and compact sizes. Long period fiber gratings that couple light from guided core modes to nonguided cladding modes result in wavelength-dependent losses in optical fibers [107]. As a consequence, these in-fiber devices can be used as band-rejection filters in high-power cascaded Raman fiber lasers and amplifiers and as gain equalizers in optical fiber amplifiers [54, 57].

It is interesting to note that by making long period gratings, one can perturb the fiber to couple to other forward-going modes. A wavelength filter based on this effect has been demonstrated by Hill et al. [108]. The spatial mode-converting grating was written using the point-by-point technique with a period of 590 μm over a length of 60 cm. Using mode strippers before and after the grating makes a wavelength filter. In a similar manner, a polarization mode converter, or rocking filter, in polarization-maintaining fiber can be made. A rocking filter of this type, generated with the point-by-point technique, was demonstrated by Hill and co-workers [109]. In their work they demonstrated an 87-cm long, 85-step rocking filter that had a bandwidth of 7.6 nm and peak transmission of 89%.

6.6 Wavelength Division Multiplexers/Demultiplexers

As telecommunication traffic increases due to the rapid growth in use of phones, faxes, computer networks, and the Internet, fiber capacity will have to keep pace. Wavelength division multiplexing is a promising solution for increasing capacity on long-distance telecommunication links. A WDM system can be compared to a highway transportation system that has multiple lanes for traffic; it provides multiple wavelengths to carry optical signals simultaneously through a fiber.

Optical fiber communication systems employing wavelength division multiplexing/ demultiplexing (WDM/D) techniques require low loss, compact, stable and reliable components, which can be used as wavelength-selective channel dropping or inserting filters [110–118].

6.6.1 Fused Fiber Splitter Bragg Grating WDM

One of the earliest demonstrations of WDM utilizing fiber Bragg gratings was presented by Mizrahi et al. [94]. The concept of this demultiplexer is illustrated in Figure 6.30 and consists of a single commercial fused fiber 1x4 splitter followed by a fiber grating transmission filter in each of the output ports. The filters were designed to pass one signal wavelength at each port, while rejecting the three adjacent signal channels. By pairing gratings that do not overlap in wavelength, it was possible to make a simple band-pass transmission filter in which each of the two gratings rejected a broad range of wavelengths, but the desired transmission wavelength passed unaffected between the grating stopbands. The use of Gaussian spatial-profile gratings eliminated the spectral sidelobes that might otherwise interfere with the signal transmission. Although the basic concept for this device is simple, many subtleties exist, necessitating careful grating design and fiber optimization. One such problem is a pronounced fine structure within the stopband of a Gaussian profile fiber grating, which would render the filter "leaky". This structure is a consequence of the nonuniform increase in the average refractive index of the fiber core that follows from inscribing the grating with a UV laser beam having a Gaussian spatial profile. Thus, the local Bragg wavelength, which is proportional to the local index of refraction, is longer in the center of the grating than at the wings. The wings then act a Fabry-Perot resonator. To eliminate this problem the gratings were written with a strong linear chirp of 4.7 nm/cm.

To fully test the applicability of this concept, a sample device based on the 1x4 WDM was assembled and packaged with access through ST connectorised jacketed fiber cable. High-resolution measurements (0.02-nm step size) obtained from one of the channels of the packaged device is shown in Figure 6.30. The noticeable 7.4-dB insertion loss includes losses from the gratings, the splitter, the splices, and the connectors. The solid dots show the possible signal wavelength spaced at 200-GHz intervals (~1.6 nm). The peak attenuation of the filter in the stop-band region has been measured to be in excess of 80 dB. This high attenuation results in unparalleled crosstalk performance. Of particular importance is the flat transmission band of the grating filter for each desired signal wavelength. Furthermore, the small temperature sensitivity of the Bragg grating eliminates the need for temperature stabilization for this WDM device. It is also polarization independent since the gratings are highly isotropic. The only concern with this design is the splitter loss, which is unavoidable. Nevertheless, this first generation of all-fiber four-channel demultiplexers that utilizes fiber Bragg gratings is polarization insensitive, has a flat passband, has unparalleled crosstalk performance, and offers total fiber compatibility.

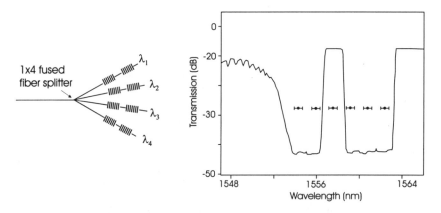

Figure 6.30   A schematic diagram of a four-channel fiber grating demutliplexer. The plot illustrates a high-resolution measurement of a complete demultiplexer system for one of the channels, including a connector-to-connector insertion loss of 7.4 dB (*After:* [94]).

6.6.2 Interferometric Wavelength Division Multiplexer

In this section wavelength division multiplexers based on the interferometric principle will be examined. Architectures such as the basic Mach-Zehnder and the twin-core Mach-Zehnder WDMs will be discussed with reference to actual application tests when used in communication networks as wavelength division add/drop multiplexers. It should be noted that fiber Bragg gratings are the ideal basic components for the interferometric WDMs since, in addition to providing high reflectivity at designer specified wavelengths with negligible transmission losses, for others they can be accurately duplicated. This allows the fabrication of identical Bragg gratings, which are needed for interferometric WDMs.

6.6.2.1 A Simple Mach-Zehnder WDM

A simple add/drop filter [110] may be obtained by simply adding a second coupler to the Michelson arrangement to close the legs containing the gratings (Mach-Zehnder arrangement; Figure 6.29). Such a device can operate as a demultiplexer (Figure 6.31) or as a multiplexer (Figure 6.32). Figure 6.31 shows the demultiplexing of a single wavelength channel from a multiwavelength transmission line. A stream of several wavelengths λ_1, λ_2, ...,λ_N is launched into the input port (port 1) of the device. Assuming the grating resonant wavelength is λ_k, light at λ_k emerges from port 2 and the remaining light emerges from the output port (port 4). Ideally, a properly balanced interferometer will have no light emerging from port 3. Because of the inherent symmetry of the interferometer, however, it is possible to use the device as a multiplexer, as shown in Figure 6.32. In that case, assuming again that the grating resonant wavelength is λ_k, light at λ_k launched in port 3 exits port 4 and is multiplexed with the other wavelengths that are launched in port 1. The placing of an additional matched pair of gratings with different resonant wavelengths would permit the multiplexer/demultiplexer device to extract or insert several different wavelength channels at once.

As a result of the symmetry of the device, multiplexing and demultiplexing can take place simultaneously in the same device. In this case crosstalk between multiplexed and demultiplexed signals occurs unless the reflectivity of the Bragg gratings is infinitely high. To minimize cross-talk in practice, reflectivities above 99% are needed. Bilodeau et al. [110] constructed such a multiplexer/demultiplexer device. A balanced Mach-Zehnder interferometer of monolithic construction and consisting of two identical 3-dB fused

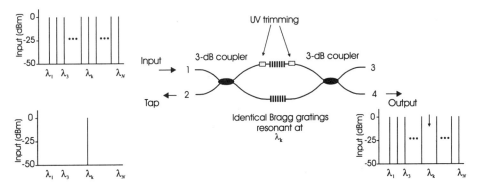

Figure 6.31 A schematic of a multiplexer/demultiplexer device using a Mach-Zehnder arrangement showing extraction of wavelength channel λ_k (*After:* [110]).

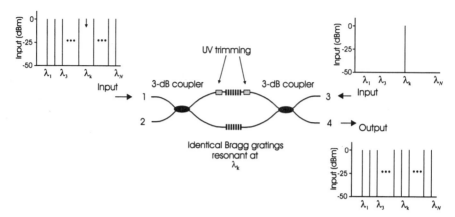

Figure 6.32 A schematic of a multiplexer/demultiplexer device using a Mach-Zehnder arrangement showing insertion of wavelength channel λ_k (*After:* [110]).

couplers was made from two continuous strands of Coring SMF-28 fiber without the use of splicing. The optical fibers were laid parallel and in contact on a fused taper coupler fabrication jig [111]. The two couplers were made in succession without moving the fibers off the jig. The couplers were kept short by stopping the elongation at the first 3-dB splitting point. Although the fabrication of the two identical 3-dB fused taper couplers was accomplished easily, there was typically a small path length difference between the two interfereomter's arms that needed to be nulled in order to balance the Mach-Zehnder interferometer. This was accomplished by exposing one arm of the interferometer to uniform UV light to photoinduce an average index change in the fiber core on each side of one of the gratings. Following this fabrication procedure, the device was used to drop/insert a single wavelength channel from/into a multiple wavelength transmission link with 100-GHz channel spacing at 1550 nm. Measurements indicated an extraction/coupling efficiency of 99.4% with an excess loss <0.5 dB, an adjacent channel isolation >20 dB, and a return loss >23 dB.

6.6.2.2 Twin-Core Mach-Zehnder

To minimize the balance requirements of a Mach-Zehnder interferometer in the add/drop multiplexers, a twin-core fiber-based system has been proposed and demonstrated, where two waveguides were embedded in the same cladding [112]. Figure 6.33(a) shows the principle of the optical add/drop multiplexer based on the twin-core fiber Mach-Zehnder interferometer. Light launched at the input port is reflected by the Bragg gratings and exits at the drop port if the optical path between the input coupler and the Bragg gratings is phase matched. From symmetry considerations, the add function must also require the same condition. The phase mismatch is proportional to the optical path difference between the two arms via $\Delta(n_{eff}L) = \Delta n_{eff}L + n_{eff}\Delta L$ where Δn_{eff} is the effective index difference between

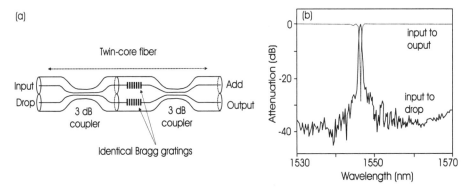

Figure 6.33 (a) Schematic of the add/drop multiplexer based on inscription of two Bragg gratings in a twin-core fiber Mach-Zehnder interferometer. (b) Measured spectrum of a drop twin-core fiber filter (*After:* [112]).

the waveguide and ΔL is the length difference between the two arms of the Mach-Zehnder interferometer. Owing to the use of a twin-core fiber for manufacturing the interferometer, the second term influence can be minimized because the 3-dB couplers are tapered on the same structure, leading to equivalent lengths of the arms. An add/drop multiplexer was achieved by fabricating two fused tapered couplers on a twin-core fiber [112]. The output and the drop response of the add/drop multiplexer is presented in Figure 6.33(b). The cladding-mode coupling losses for each passed channel was measured to be under 0.5 dB. At the Bragg wavelength, extinction was near 29 dB and 0.8-nm bandwidth was measured at 20 dB in the transmitted spectrum. Furthermore, isolation between the add and the output ports at the Bragg wavelength was measured to be at least 30 dB confirming the functionality of this new twin-core system as an add/drop multiplexer that may play an important role in communication networks.

6.6.3 Noninterferometric Wavelength Division Multiplexer

Although interferometric-type wavelength division multiplexers have demonstrated some excellent characteristics, they also have some disadvantages: for example, the need of fine tuning the interferometer arms, high cost, and difficulties in fabrication. Alternatives to such WDMs are based on the noninterferometric systems that will now be described.

6.6.3.1 Polarization Beam Splitter Add/Drop Multiplexer

A novel add/drop multiplexer that employs fiber Bragg gratings and polarization beams splitters (PBSs) has been demonstrated [114] and its performance tested in a 2.5-Gbps WDM transmission system with 0.8-nm channel separation. This add/drop multiplexer device has the advantage of being constructed with low cost elements, fabricated without any special techniques, and yet shows high stability against perturbations. Figure 6.34 shows a schematic of this highly stable add/drop multiplexer which consists of two PBSs with polarization controllers (PCs) and identical Bragg gratings in each arm. Multi-channel WDM signals with arbitrary polarizations are injected into port 1 and are split into two linearly polarized orthogonal states by a PBS. The light in each arm passes through a $\lambda/4$ polarization controller, becoming circularly polarized. Among all the WDM channels, only the signal at the Bragg wavelength is reflected, reversing its handedness of circular polarization and passing through the polarization controllers again. The light again becomes linearly polarized, but is now orthogonal to the original input polarization. This makes the signal channel at the Bragg wavelength drop through port 2, whereas the other wavelength channels are transmitted through port 4. Because of the symmetric structure of the device, a new signal at the same Bragg wavelength can be added to the transmitted port by launching the light into port 3. The device is relatively easy to fabricate without any special techniques such as UV trimming and shows very stable performance with less than 0.3 dB crosstalk power penalty in a 0.8-nm spaced, 2.5-Gbps-per-channel wavelength division multiplexing transmission system. One of the major merits of the ADM is its low sensitivity to ambient temperature change, because the signals from the two arms are directed to the output port not by interference but by simple addition of intensity of two orthogonal polarized beams.

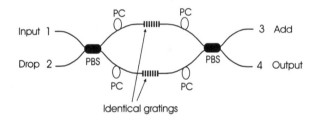

Figure 6.34   A schematic of an optical add/drop multiplexer using fiber Bragg gratings and polarization beam splitters. In this diagram PBS corresponds to polarization beam splitter, and PC is a polarization controller (*After:* [114]).

6.6.3.2 Frustrated Coupler Add/Drop Multiplexer

Another interesting add/drop filter is based on the grating-frustrated coupler demonstrated by Archambault et al. [106]. The device consists of a mismatched coupler with a Bragg grating written in one core over the coupling region (Figure 6.35). The coupler would not normally transfer power from one core to another due to the strong mismatch of the two cores. However, with the existence of the Bragg grating, power transfer of the guided fundamental mode from port 1 to port 4, R_{12}, can happen if

$$\beta_{01}(\lambda_{12},1) + \beta_{01}(\lambda_{12},2) = \frac{2\pi}{\Lambda} \tag{6.22}$$

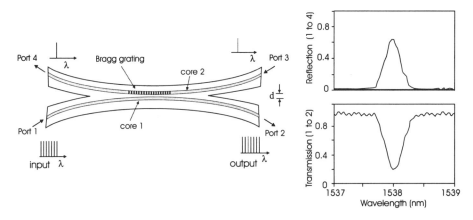

Figure 6.35 A schematic of the add/drop filter for wavelength division multiplexing using Bragg grating assisted mismatched coupler. The plot shows the outputs from ports 2 and 4 with an input to port 1 for a mismatched coupler add/drop filter (*After:* [115]).

where $\beta_{01}(\lambda_{12},1)$, $\beta_{01}(\lambda_{12},2)$, λ_{12}, and Λ are the propagation constant of the LP_{01} mode in core 1, propagation constant of the LP_{01} mode in core 2, Bragg wavelength for the coupling, and the grating pitch, respectively. This cross-coupling transfers the guided optical power at λ_{12} in core 1 into back-propagating optical power in core 2. The wavelengths that are not reflected continue to propagate in core 1. The device, therefore, performs the demultiplexing function in a WDM system. Any other optical signal at λ_{12} launched from port 3 in core 2 can then be reflected by the same grating to exit from port 2, and, therefore, a multiplexing function can be performed by the same device at the same time. Assuming the two cores are well separated, the coupling is very similar to Bragg reflection of the ordinary type, except that the effective refractive index modulation of the grating is replaced by

$$\Delta n_{12} = \int_{A_0} \psi_{01}(1)\psi_{01}(2)\Delta n dA \qquad (6.23)$$

where A_0 is the cross-sectional plane of the coupler, $\psi_{01}(1)$, $\psi_{01}(2)$ are the normalized modal field of the LP_{01} mode in cores 1 and 2, respectively, and Δn is the UV-induced refractive index modulation. The peak reflectivity can be calculated from the simple equation $R_{12} = \tanh^2(\kappa L)$, where $\kappa = \pi \Delta n_{12}/\lambda_{12}$ is the grating coupling coefficient and L is the interaction length of the grating-assisted coupler. An advantage of this type of system is that the cross-coupling is reflective as in an ordinary optical fiber Bragg grating and is not interferometric. This allows easy implementation of the device without the need for fine-tuning of the coupler. The drawback of this scheme, however, is the small modal overlap in such devices, which are typically two orders of magnitude lower than that in an ordinary fiber Bragg gratings. Therefore, to achieve the same reflectivity as in ordinary fiber Bragg gratings, it is necessary to have a large index of refraction. One other design parameter that has to be considered is the back-reflected light at the launched port. When light is launched into port 1, a partial reflection (R_{11}) is expected back at port 1 at λ_{11} when $2\beta_{01}(\lambda_{11},1) = 2\pi/\Lambda$, with the effective index modulation of the grating given by (6.23) when the two cores are well separated. If the grating is only written over core 2, Δn is nonzero only over core 2, where $\psi_{01}(1)$ is very small. Thus, Δn_{11} is much smaller than Δn_{12} when the two cores are well separated and R_{11} can be designed to be negligible. Dong et al. [115] have demonstrated this new type of add/drop filter based on the Bragg grating–assisted mismatched coupler. They were able to fabricate a device with the fiber 1 having a NA of 0.12 and a 3.98-μm core radius and fiber 2 having a NA of 0.15 and a 3.63-μm core radius. A grating with length of 15 mm and refractive index modulation $\Delta n = 0.005$ was inscribed into the core 2. Both fibers were then fixed into polishing blocks and were polished to very close proximity of the cores. The output from port 2, when light was launched into port 1, was measured to be approximately 80%; however, this can be improved easily by simply using longer interaction lengths. These devices are relatively easy to fabricate and much more stable than similar devices that use interferometric techniques.

6.6.3.3 Zero-Insertion-Loss Add/Drop Multiplexer

An all-fiber zero-insertion-loss (~0.1dB) add/drop filter for wavelength division multiplexing has also been demonstrated [116]. This new filter is based on grating-assisted mode conversion and backward coupling in the merged region of two dissimilar waveguides, formed by adiabatic tapering and strong fusing of two single-mode optical fibers. The modes of the input fibers will evolve on a one-to-one basis into the first two eigenmodes of the merged region. These eigenmodes will propagate through the waist independently of each other, transform back into their original modes, and exit the device through their respective output fibers. This adiabatic mode evolution is lossless and free of optical cross-talk. Figure 6.36(a) illustrates the one-to-one relationship between the input modes 1 and 2 in waveguides A and B and the eigenmodes into which they are transformed.

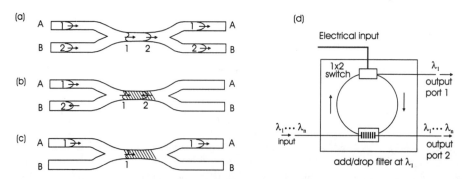

Figure 6.36 Operating principle of the grating-assisted mode converter coupler: (a) adiabatic mode evolution without a Bragg grating, (b) mode transformation when the incoming wavelength satisfies the phase-matching condition, and (c) mode transformation when the incoming wavelength does not satisfy the phase-matching condition. (d) An example of a wavelength selective optical switch (*After:* [116]).

By inscribing a transversely asymmetric Bragg grating in the waist of the device, one can convert eigenmode 1 and backward couple it into eigenmode 2, which will then exit the merged region as mode 2 via waveguide B, as shown in Figure 6.36(b). This transformation will occur only at a specific wavelength λ that satisfies the Bragg condition $\beta_1(\lambda) + \beta_2(\lambda) = 2\pi/\Lambda$, where Λ is the period of the grating and β_1 and β_2 are the modal propagation constants. Because the eigenmodes and grating fully overlap in the merged region, the coupling coefficient of this device is much larger than that of a grating-assisted directional coupler based on evanescent wave coupling. If the Bragg condition is not satisfied, the input mode 1 will simply pass through the merged region, unaffected by the presence of the grating and thus exit waveguide A unattenuated, as seen in Figure 6.36(c). Conversely, if light that satisfies the Bragg condition is launched into waveguide B it will be added onto waveguide A by the grating. Thus, this device functions as an efficient narrow-bandwidth, four-port, all-fiber, optical add/drop filter.

To test the performance of this add/drop filter a narrow-bandwidth grating was inscribed in the waist of the device from which a peak add/drop efficiency of ~98% centered at 1547 nm was measured. This add/drop filter efficiency is generally not sufficient to provide adequate channel rejection for WDM applications. To obtain 30-dB rejection, a grating efficiency of 99.9% is required. However, because the reflectivity of a standard fiber Bragg grating can easily surpass this value, much greater add/drop efficiency is expected with further development. Clearly, with improved channel rejection the device can simultaneously function as both an add and a drop filter and thus bidirectional operation is possible. Furthermore, it is possible to construct a wavelength-selective optical switch or router by combining a higher efficiency four-port add/drop filter with a standard optical switch as shown in Figure 6.36(d).

6.6.3.4 Bragg Grating Circulator Add/Drop Multiplexer

It is noteworthy to mention another simple add/drop multiplexer, which is constructed using Bragg gratings and optical circulators. The connection of a reflective fiber Bragg grating to the second port of a three-port optical circulator creates a band-pass filter from the first to the third port of the device. Adding a second circulator on the other end of the grating enables optical add/drop multiplexing of one or more channels of the WDM traffic. This type of add/drop multiplexer has low insertion loss (2 dB for add/drop channels) and one of the highest isolations of the drop and add channels (>50 dB). A four-channel transmission architecture [49] with three low-loss fiber Bragg gratings/optical circulators, including one programmable add/drop multiplexer has been constructed and tested (Figure 6.37). No impairment from cross-talk between dropped and added channels were observed at any of the add/drop multiplexers. Figure 6.37 shows a link that has three add/drop multiplexers, including one with optical switches to allow selecting two drop channels [49]. The low insertion loss of the add/drop multiplexers enabled 100-km transmission without a repeater amplifier. The progression of the optical spectra through the link are also shown in Figure 6.37. It should be noted that a four-channel programmable add/drop multiplexer has also been reported in similar [117] and other configurations [118].

6.7 Dense Wavelength Division Multiplexing

Dense wavelength division multiplexing (DWDM), as for the case of wavelength division multiplexing (WDM), is used to separate or combine lines of communication in the form of different wavelengths. The difference between these systems (as the name implies) lies in the channel spacing, with the DWDM having more channels per unit wavelength. As a result of the small channel spacing, DWDM networks impose stringent requirements on components in a system. One cost-effective device that has been developed to address such requirements is a hybrid device, which combines fiber Bragg gratings and dielectric-coated band-pass filters with a channel spacing of 0.4 nm.

To understand the advantages of a hybrid system, one must first understand the requirements of a DWDM system. A simplified point-to-point DWDM setup consists of several key functional components such as wavelength division multiplexers (to combine signals at different wavelengths) and demultiplexers (to distribute the signals to different destinations according to the wavelengths). The DWDM channel wavelengths of the networks are usually equally spaced in optical frequency based on the International Telecommunication Union industrial standard. The distance between the channel wave-

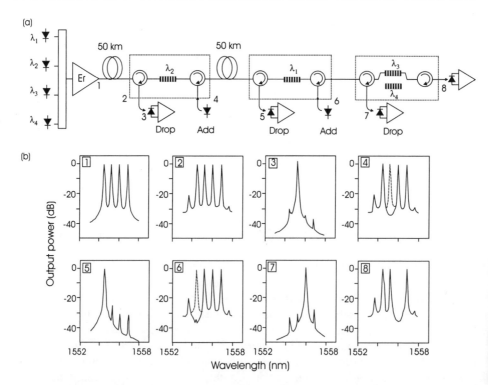

Figure 6.37   (a) A four-channel transmission experiment with three low-loss FBG/optical circulators, including one programmable add/drop multiplexer. The low insertion loss of the add/drop multiplexers enabled 100-km transmission without repeater amplifiers. (b) A progression of the optical spectra through the link (*After:* [49]).

lengths is the channel spacing–100 GHz, approximately equal to 0.8 nm. Furthermore, the devices require broad flat-top transmission bandwidth to tolerate a larger laser-source-wavelength selection range, high channel isolation, and low insertion loss.

DWDM devices incorporate one of four categories of devices: arrayed waveguide gratings, fiber Bragg gratings, dielectric-coated band-pass filters, and a hybrid of fiber Bragg gratings and dielectric band-pass filters. Arrayed waveguide-grating DWDM devices possess some shortcomings in terms of polarization-dependent wavelength, polarization-dependent loss, relatively low channel isolation, channel cross-talk to all other channels, and a low figure of merit. A fiber Bragg grating provides excellent filter spectral shapes because it can be made from thousands of refractive index modulation layers, allowing a square-like spectra-filter shape to be created with a high figure of merit. Fiber Bragg grating DWDM devices must, however, contain an expensive optical circulator to convert the band rejection of fiber Bragg grating to band-pass for multiplexing or demultiplexing, or must use the fiber Bragg gratings in an interferometric Mach-Zehnder setup, which requires accurate phase control and suffers from environmental sensitivity. The dielectric-coated band-pass filters have been widely used with channel spacing of 200 GHz and provide attractive performance in high channel isolation, low insertion loss. However, the filters have problems of low yield for channel spacing below 200 GHz.

The hybrid DWDM [119] devices combine fiber Bragg grating and dielectric-coated band-pass filters and can meet the required specification in various cost-effective structures. In an example (Figure 6.38) of a hybrid demultiplexer, an eight-wavelength optical input signal passes through an isolator to a 3-dB coupler and is split into two equal output arms. Each of the output arms has four apodized fiber Bragg gratings for a 0.4-nm channel spacing with a high reflectivity to block the optical signals at the four alternative wavelengths. The same structures can be used to extend DWDM multiplexers and demultiplexer from 8 channels to 16, 32, or 64. By proportionally adding more fiber Bragg gratings on the two arms of the 3-dB fiber coupler and on port 2 of two optical circulators,

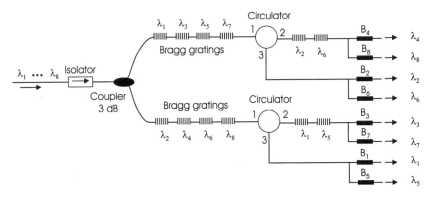

Figure 6.38 Hybrid dense wavelength division demultiplexer, each with a channel spacing of 0.4 nm. The system is constructed by incorporating fiber Bragg gratings and dielectric-coated band-pass filters.

16-, 32-, or 64-channel DWDM demultiplexers and multiplexers can be built. This may be achieved by proportionally cascading more 1.6-nm channel spacing band-pass filters at the selected wavelengths. The two optical circulators can also be replaced by two 3-dB fiber couplers with an increase of insertion loss of only 1.6 dB. This hybrid approach to DWDM devices should provide designers with many options for optical network design as the demand for increased bandwidth continues to increase.

6.8 Dispersion Compensation

One of the main problems occurring in single-mode optical fibers that inhibits ultimate performance in optical communication systems is chromatic dispersion, causing different wavelength components of a data pulse to travel at different group velocities. This causes broadening of the signal pulse and increasing bit error rates. With increasing network data rates, chromatic dispersion in standard single-mode fiber is the main limiting factor in performance. For the relatively low data rate of 2.5 Gbps, a signal can be transmitted without significant degradation for up to 1000 km. However, this distance drops to 60 km at 10 Gbps and to 15 km at 20 Gbps. In addition, a large portion of the fiber already installed worldwide is optimized for transmission at 1.31 μm. This type of fiber exhibits high chromatic dispersion of the order of 17 ps/nm km when used to transmit at the more commonly used telecommunication wavelength of 1.55 μm, which coincides with the operating wavelength of erbium-doped fiber lasers and amplifiers.

6.8.1 Dispersion Compensation from Reflective Bragg Gratings

Although large values of group velocity dispersion are obtainable from uniform Bragg gratings [120], and these Bragg gratings have been proposed for dispersion compensation, the dispersion changes rapidly within the bandwidth of short optical pulses making uniform gratings unsuitable for this application. It has been recognized, however, that dispersion compensation may be obtained from aperiodic structures, and Ouellette [121] pointed out that linearly chirped gratings can offer a large, constant dispersion over bandwidths sufficient to support such pulses. Demonstrations of the use of linearly chirped gratings to compensate for the pulse broadening arising from transmission in dispersive fiber were first given in 1994 [122, 123]. The basic principle of operation is that different wavelength components of the broadened pulse are reflected from different locations along the Bragg grating. Assuming the grating is oriented (Figure 6.39) with the shorter period facing the incident light, then the shorter wavelengths λ_{short} are reflected at the near end of grating and the longer wavelengths λ_{long} at the far end. Thus, the longer wavelengths must travel farther within the grating before they are reflected, thereby experiencing an additional time delay with respect to the shorter wavelengths. Based on this picture one may write a simple expression for the group delay dispersion of a linear chirped grating of length L

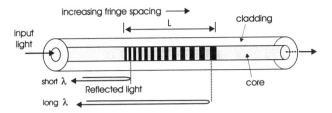

Figure 6.39 Schematic representation of the principle of dispersion in a chirped grating. Longer wavelengths travel farther within the grating than shorter wavelengths.

$$d = \frac{2n_{\text{eff}}L}{c}\left(\frac{1}{\Delta\lambda_{\text{c}}}\right) \qquad (6.24)$$

where n_{eff} is the effective refractive index, c is the speed of light in vacuum, and $\Delta\lambda_{\text{c}}$ is the difference between the wavelengths reflected at either end of the grating.

From (6.24) it follows that the dispersion for a 100-km length of standard fiber with 17 ps/nm km dispersion at 1550 nm may be compensated over a bandwidth of $\Delta\lambda_{\text{c}} = 0.2$ nm by a linearly chirped grating of approximately 3.5 cm in length. Disregarding optical nonlinearities and higher order dispersion, the chirp is selected so that grating dispersion upon reflection cancels that of the fiber. The grating must also be long enough to ensure that the entire signal spectrum is accommodated. For example, a 5-cm long grating with 0.06 nm/cm chirp may compensate chromatic dispersion of a 0.3-nm bandwidth, 1550-nm signal traversing through 100 km of conventional fiber. This analysis is oversimplified as grating apodization may be required in order to reduce group delay ripple in the compensators, thereby lowering the optical bandwidth [124–126]. Detailed calculations show the required grating length is 5.2 cm without apodization, and with a tanh apodization profile of the grating amplitude, the length increases to 5.7 cm.

The first practical demonstrations of dispersion compensation used short pulses (Figure 6.40) traveling in single-mode optical fiber and a chirped Bragg grating to compensate for the pulse broadening arising from negative group delay dispersion and nonlinear self-phase modulation in the fiber. Specifically, short pulses of 1.8 ps were sent through 200m of optical fiber, which had a measured group delay dispersion of 100 ps/nm km. These pulses suffered significant dispersive broadening in the fiber, as seen in Figure 6.40(b). A 50:50 coupler between the fiber and a linearly chirped fiber Bragg grating provided an output for the pulses directly from the fiber and those reflected off the grating. The pulses reflected by the grating were measured temporally by cross-correlation with pulses derived directly from the source. These results indicated that the chirped grating provided satisfactory dispersion compensation, which was in good agreement with numerical simulation. At approximately the same time, dispersion compensation was also demonstrated with 400-fs pulses at 1560 nm in 320m of fiber with 17 ps/nm km dispersion and an 8-mm long grating [123].

The discussion thus far has mainly addressed linear dispersion compensation, however

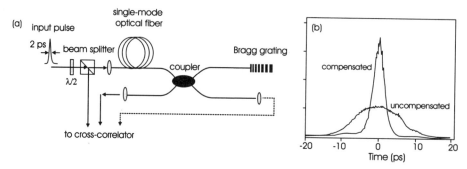

Figure 6.40 (a) Experimental configuration used to investigate dispersion compensation of a linearly chirped fiber Bragg grating. Plot (b) shows a comparison between dispersive pulse and dispersion compensated pulse from the experimental setup shown in (a) (*After:* [122]).

higher levels of compensation may be achieved using more complex grating structures or using a combination of gratings. Cubic dispersion compensation is important in long distance, high-bit-rate transmission systems and in femtosecond-pulse lasers and can be achieved using quadratically chirped grating. Using the phase-mask technique, almost any type of dispersion compensating grating may be photoimprinted in photosensitive fiber, thus making high-bit-rate transmission telecommunication networks achievable with already existing fiber lines.

6.8.2 High-Bit-Rate Long Distance Transmission

Long distance transmission at high bit rates over standard telecommunications fiber at 1.3 μm is of great interest because of the large base of such fibers already installed in the ground whose serviceable lifetime may be usefully extended. The low loss of these fibers, together with the ready availability of erbium-doped fiber amplifiers, makes the 1.55-μm window an attractive region of operation. The dispersion of these fibers, however, is relatively large within this window, severely limiting transmission distances unless compensating techniques such as chirped gratings are employed [127, 128]. Some of the interesting dispersion compensation architectures used for telecommunication purposes will be examined in detail in the following sections.

In one report [127] a 120-mm long fiber Bragg grating, which was uniformly fabricated and was subsequently chirped by mechanical means, was used to compensate for dispersion in a 10 Gbps optical communication system operating at 1.54 μm over 270 km of nondispersion-shifted fiber. The mechanical chirping was accomplished by a scheme [127] in which the core of a fiber was laterally displaced from the fiber axis, so that the fiber acted as its own cantilever. A bend applied to the fiber at axial position z then caused a shift in the resonant wavelength $\Delta\lambda$ (z) given by

$$\frac{\Delta\lambda_B(z)}{\lambda_0} = \frac{k\delta}{r(z)}\cos(\theta) \qquad (6.25)$$

where k is a dimensionless constant, δ is the lateral offset of the fiber core, $r(z)$ is the radius of curvature of the bend at position z, and θ is the angle between the plane in which the bend lies and the plane containing the axis of the fiber and the core. Any desired reasonably smooth chirp profile can be produced by imposing an axial bend profile $r(z)$ on the fiber grating according to the above equation. This offset-core, adjustable chirped grating was connected to port 2 of a three-port circulator that was positioned approximately midway along a variable span L of standard nondispersion-shifted fiber. The bit error rate was measured as a function of received optical power for a $2^{15}-1$ pseudo-random, non–return-to-zero bit sequence for $0 < L < 270$ km. Measurements demonstrated error-free transmission over 270 km in a 10-Gbps transmission system. Similar dispersion compensation measurements of up to 400-km transmission line have also been accomplished using a phase mask–inscribed, 10-cm long chirped fiber Bragg grating [128]. Loh et al. [129] have reported results of a system involving dispersion compensation chirped fiber gratings, demonstrating that transmission up to 537 km of standard telecommunications fiber is now feasible at 10 Gbps, with a pair of long Bragg gratings cascaded together. The linearly chirped fiber grating in this work were 10 cm long and fabricated from a 10-cm uniform phase mask. Apodization and chirping of the grating was accomplished during the writing process using the moving fiber-scanning beam technique, with a cosine apodization profile adopted to reduce excessive ripples in the reflection/dispersion spectra. The first grating had a bandwidth of 0.3 nm and a dispersion of 2200 ps/nm and the second grating a bandwidth of 0.12 nm and dispersion 5400 ps/nm. These gratings should thus be able to compensate for 130 and 320 km of standard single-mode fiber (17 ps/nm km), respectively. When operated in reflection mode, in combination with an optical circulator, the losses were −8 dB for the first grating and −5.5 dB for the second grating. The gratings were individually mounted on separate heatsinks so that their center wavelength could be temperature fine-tuned to match the transmitter. A system trial showed that with the 10-cm long dispersion compensating chirped fiber gratings, transmission at 10 Gbps over standard single-mode fiber of up to 400 km is possible with negligible penalty using single gratings, and up to 540 km is feasible with small (<2 dB) penalty using the cascaded gratings.

The recent development of 40-cm long broadband (4 mm) linearly chirped gratings [130] have aided in the demonstration of a 40-Gbps transmission over 109 km of nondispersion-shifted fiber at 1.55 μm [131]. Two linearly chirped gratings, whose characteristics are shown in Figure 6.41, were employed in combination with a four port circulator. Grating 1 exhibits ~98% reflectivity with < ± 0.1 dB amplitude deviation over 3.8 nm, dispersion of 837 ps/nm, and maximum delay deviation from linear fit of < ± 50 ps. Similarly for grating 2, the reflectivity was approximately 98% with < ± 0.1 dB amplitude deviation over 3.5 nm, dispersion of 870 ps/nm, and maximum delay deviation from linear fit of < ± 40 ps. A schematic diagram of the system is shown in Figure 6.42. A 10-GHz pulse train was generated by an actively mode-locked erbium ring laser. A two-stage interleaver was used to generate the 40-Gbps data. The grating compensator was inserted at the beginning of the link to reduce the launched power intensity, thus avoiding any possible nonlinear effect. In the case where gratings 1 and 2 were used in series in conjunction with the 109-km fiber, the pulse width following the reflection from the gratings was estimated

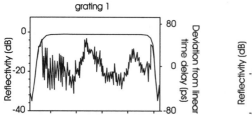
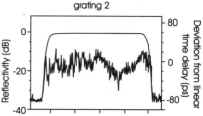

Figure 6.41   Characteristics of the two 40-cm long linearly chirped gratings used in the 40-Gbps transmission test (*After:* [131]).

to be ~720 ps corresponding to a stretching factor of ~110. Following the transmission through the fiber the pulse was recompressed to 1.4 times the original width. Error-free transmission was achieved with a penalty of 6 dB and a source wavelength tolerance of ~0.5 nm. The tunability of the transmission is mainly limited by the local dispersion deviation of the chirped gratings from linearity. This can cause local dispersion to deviate from as much as ~100 ps/nm over a fraction of a nanometer in the worst case. A variation of <5 ps in the received pulse width, while tuning, will require the fiber and grating dispersion to be matched to <13 ps/nm across the spectrum of the grating, assuming the pulse is 0.4 nm FWHM. Furthermore, the third-order dispersion will cause a dispersion slope change of ~7 ps/nm^2, which will effectively give a tuning range of ~1.8 nm for a 100-km link. Clearly, improving the fabrication of the chirped gratings will result in an improvement in the high data rate transmission in telecommunications.

A rather interesting and effective approach of dispersion compensation has been demonstrated [132] for a 10-Gbps soliton transmission over 1000 km of standard fiber. It is well known that there are two approaches to compensate fiber dispersion: (1) linear, based on dispersion compensation using compensating fibers or chirped fiber grating, and (2) nonlinear, based on the use of optical solitons. Both methods have their specific advantages

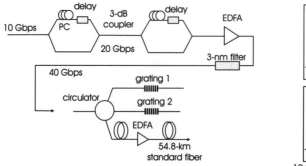
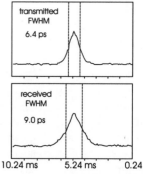

Figure 6.42   Experimental setup of the 40-Gbps transmission test. The plot shows the transmitted pulse and received pulse width after passing through the test transmission line (*After:* [131]).

and disadvantages. In particular, linear systems suffer from fiber nonlinearity, which deteriorates the system performance at distances exceeding just 500 km. The major problems with nonlinear systems are the short amplifier spacing required and interactions between the transmitting pulses. Experiments have revealed that the best performance is offered by a combination of the two methods, where the first part of the intraspan distance operates in the nonlinear regime, while the dispersion of the second part is compensated by a dispersion compensating element. Using this hybrid system [132], error-free transmission was obtained for 10-Gbps data over 1000 km of standard fiber with nominal dispersion of 17 ps/nm km. For the chirped gratings used in the above techniques to compensate for dispersion, the amplitude of the ripple in the group delay characteristics should be less than the bit period over the bandwidth of the signal, to keep the system power penalty to below 1 dB. This is accomplished by apodizing the chirped grating so that the fabrication process is well controlled to simultaneously chirp and apodize the grating. Despite the impressive results to date, the manufacture of chirped apodized Bragg gratings suitable for commercial dispersion compensation applications is not a simple task. Stephens et al. [133] have presented a grating-based dispersion compensator that employs unchirped gratings. The complexity in manufacturing the grating dispersion compensator is therefore greatly reduced. By tapering the grating strength of an unchirped grating, it is possible to achieve a smooth group delay characteristic that is suitable for dispersion compensation over a significant bandwidth.

6.8.3 Multichannel High-Bit-Rate Long Distance Transmission

A further improvement to the capacity of a high-bit-rate transmission is the addition of multiple channels. This added dimensionality to the transmission lines is achieved by simultaneously transmitting information at several wavelengths (wavelength division multiplexing/demultiplexing). Clearly the main concern in the transmission of such high-bit-rate signals in standard single-mode fiber is dispersion, which has to be compensated for, not just for one wavelength but for all the channel wavelengths used in the transmission. In this section some of the most recent results on multichannel high-bit-rate long distance transmission are presented.

An eight-channel WDM transmission over 480 km of single-mode fiber at 10 Gbps and over 315 km at 20 Gbps using chirped fiber gratings for dispersion compensation has been demonstrated [134]. This work used several broadband high-dispersion fiber gratings cascaded to allow long distance high-speed WDM transmission in conventional high dispersion fiber. The chirped fiber gratings were 1 m in length with a nominal dispersion of −1330 ps/nm over a bandwidth of 6.5 nm. The time delay ripple was measured to be less than ± 30 ps when measured with 2-pm resolution using 100-MHz signal modulation. One of the gratings has been designed with an additional parabolic shape to also compensate for 35 ps/mm^2 dispersion slope (equivalent to 480 km of single-mode fiber).

In the demonstration of the 8x20-Gbps WDM transmission over 315 km of single-mode fiber, the outputs of eight external-cavity lasers with nominal spacing of 0.8 nm were combined and optically amplified. The signals were then modulated at 20 Gbps in a chirp-free LiNbO$_3$ Mach-Zehnder modulator. The transmission path consisted of four dispersion compensated, amplified single-mode fiber spans with nominal span length of 80 km making the total transmission distance of 315 km, or more than 20 times the dispersion limit of 15 km at 20 Gbps. Measurements on the performance of this system indicated satisfactory result with very small polarization and wavelength dependence. Similarly using, 10 Gbps, six-fiber gratings were cascaded in an eight-wavelength 480-km WDM system resulting in the demonstration of a multiple channel transmission.

6.8.4 Dispersion Compensation in Transmission of Bragg Gratings

Thus far we have considered dispersion compensation using Bragg gratings in the reflection mode where different wavelengths in a dispersed pulse are reflected at different positions within the grating. This leads to different optical path lengths and thus provides the possibility of compensating for dispersion in long haul fiber links. While the results are quite impressive, the experiments typically require the use of an optical circulator or a 3-dB coupler and the design and fabrication of complex grating structures. An alternative to this would be a transmission-based system in which the gratings are placed in-line with the fiber [135]. In view of the importance of this type of compensation, we will consider the dispersion properties of finite uniform Bragg gratings. It should be noted that the dispersion characteristics of an infinitely long grating are similar to those of a finite-length apodized grating. The grating dispersion can be described by the following dispersion relation of a periodic structure, which is obtained using the coupled mode equation:

$$\gamma^2 = \delta^2 - \kappa^2 \tag{6.26}$$

where $\delta = n(\omega - \omega_B)/c$ is the detuning of the channel carrier frequency ω from the resonant Bragg frequency ω_B and $\gamma = \beta - \beta_B$ is the deviation of the propagation constant β from the Bragg wavenumber β_B, n is the node index in the single-mode silica fiber, c is the speed of light in vacuum, and κ is the coupling coefficient defined in the usual manner. The grating stop band corresponds to detuned frequencies $|\delta| < \kappa$, where the reflectivity is high. At frequencies close to the stop band ($|\delta| > \kappa$), the grating exhibits strong second and higher order dispersion, which strongly affects the light propagation. On the short wavelength side of the stop band ($\delta > 0$), where the dispersion is anomalous, Bragg solitons have been predicted and experimentally demonstrated [136]. On the long wavelength side ($\delta < 0$), the dispersion is normal and can be used for compensation of anomalous group velocity dispersion (GVD) of fiber in the 1.55-μm wavelength region [137]. Clearly, at frequencies far from the stop band the grating plays no significant role and the dispersion relation becomes identical to that of uniform medium. The effect of the grating dispersion can be accounted for mathematically by expanding β in a Taylor series about δ_0

$$\beta(\delta) = \beta_{g0} + \left(\frac{c}{n}\right)\beta_{g1}(\delta - \delta_0) + \frac{1}{2}\left(\frac{c}{n}\right)^2\beta_{g2}(\delta - \delta_0)^2 + \frac{1}{6}\left(\frac{c}{n}\right)^3\beta_{g3}(\delta - \delta_0)^3 + \cdots \tag{6.27}$$

where β_{gq} is the qth derivative of β with respect to δ evaluated at δ_0. The second-order term β_{g2} is the group velocity dispersion which is given by

$$\beta_{g2}(\delta) = -\left(\frac{n}{c}\right)^2 \frac{\kappa^2}{(\delta^2 - \kappa^2)^{3/2}} \operatorname{sign}(\delta) \qquad (6.28)$$

and $\beta_{g3}(\delta)$ is the cubic dispersion, which is given by

$$\beta_{g3}(\delta) = 3\left(\frac{n}{c}\right)^3 \frac{\kappa^2 \delta}{(\delta^2 - \kappa^2)^{5/2}} \qquad (6.29)$$

Clearly, both term β_{g2} and β_{g3} diverge at the band edge ($\delta = \kappa$). In an ideal dispersion compensator, the dispersion of the filter is matched to that of the fiber

$$L_g |\beta_{g2}| = L_f |\beta_{f2}| \qquad (6.30)$$

where β_{g2} and L_g is the quadratic dispersion and length of the grating, respectively, whereas β_{f2} and L_f are the quadratic dispersion and length of the fiber, respectively. It should be noted that the cubic dispersion of the grating acts to diminish the efficiency of compression and to distort the pulse and thus must be as small as possible to accomplish ideal compression.

The impact of dispersion is important for DWDM lightwave systems making use of grating-based add/drop filters. In such filters, fiber gratings add or drop a selected channel by reflecting it, while the neighboring channels are transmitted through the grating. Such transmitted channels are affected by the strong out-of-band dispersion possessed by the grating [138]. Given that each channel may pass through numerous fiber gratings during its transmission, the degradation of the signal from the out-of band dispersion of fiber gratings may limit the bit rate or the transmission distance achievable. Next we will give a brief overview of the dispersion effects of cascaded gratings used in WDM systems (a detailed analysis may be found in [139]).

Let us consider the simplest case where two gratings are cascaded with Bragg frequencies v_1 and v_2 such that the frequency separation $\Delta v = |v_1 - v_2|$ is larger than the width of the stop band of each grating. Then there is no spectral overlap and we can ignore interference effects between the gratings. We assume for simplicity that the coupling strength κ is the same for both gratings. The dispersion relation for the second grating may be written as

$$(\gamma + \Delta)^2 = (\delta + \Delta)^2 - \kappa^2 \qquad (6.31)$$

where $\Delta = 2\pi n \Delta v / c$. Following the approach in [137], we find the following analytical expression for the GVD coefficient β_{g2} and third-order dispersion parameter β_{g3} valid in the region between the stop bands of the two uniform fiber gratings:

$$\beta_{g2}(\delta) = \left(\frac{n}{c}\right)^2 \left[\frac{\kappa^2}{[(\delta + \Delta)^2 - \kappa^2]^{3/2}} - \frac{\kappa^2}{(\delta^2 - \kappa^2)^{3/2}} \right] \operatorname{sign}(\delta) \qquad (6.32)$$

$$\beta_{g3}(\delta) = 3\left(\frac{n}{c}\right)^3\left[\frac{\kappa^2\delta}{(\delta^2-\kappa^2)^{5/2}}+\frac{\kappa^2(\delta+\Delta)}{[(\delta+\Delta)^2-\kappa^2]^{5/2}}\right] \qquad (6.33)$$

Both expressions (6.32) and (6.33) diverge at the band edge of each grating. Figure 6.43(a) and (b) show the GVD and third-order dispersion parameters β_{g2} and β_{g3} for two cascaded gratings, as well as those of individual gratings, as a function of the detuning parameter δ for channel spacing of 100 GHz (i.e., $\Delta = 31.4$ cm^{-1}) after choosing $\kappa = 4$ cm^{-1} and $n = 1.5$. Clearly, as seen in Figure 6.43(a) (the solid curve), in the presence of another grating the GVD becomes zero at $|\delta| = \Delta/2$. Therefore, in contrast to the single grating case, there exists a zero GVD wavelength. Furthermore, this zero-GVD wavelength can be shifted by simply varying the grating design parameters κ and Δ. It is noteworthy that the third-order dispersion is always positive for each grating, and since the contribution from the two gratings is additive between the stop bands, the total dispersion does not vanish at any wavelength, as seen in Figure 6.43(b). Since third-order dispersive effects can distort short optical pulses and have the potential of becoming detrimental for devices such as dispersion compensator and optical add/drop multiplexers, the ratio between the second- and third-order dispersion terms is often used as a figure of merit . This figure of merit is used for characterizing the performance of a fiber grating and is defined as follows:

$$F(\delta) = \sqrt{2}\left|\frac{\beta_{g2}}{\beta_{g3}}\sigma_0\right| \qquad (6.34)$$

where σ_0 is the transform-limited root mean square (rms) pulse width (for a Gaussian pulse, rms pulse width σ_0 is related to the 1/e width). Clearly to minimize the detrimental effect of third-order dispersion, F should be as large as possible for a given set of design parameters.

Next let us investigate the design of a dispersion compensator for an N-channel WDM

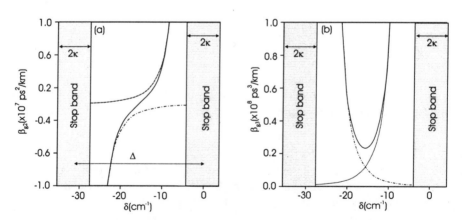

Figure 6.43   (a) Group velocity dispersion and (b) third-order dispersion as function of δ for a single grating centered at $\delta = 0$ (dashed line) and $\delta = -\Delta$ (dot-dashed line) and for two cascade gratings (solid line). The parameters used are $\kappa = 4$ cm^{-1}, $\Delta\nu = 100$ GHz (*After:* [139]).

system after the WDM signal has been transmitted through standard telecommunications fiber (Figure 6.44(a)). For simplicity we assume that the channels in the WDM system are equally spaced by Δ, that this spacing is large compared to the pulse bandwidth, and that the grating stop band satisfies $2\kappa < \Delta$. As the dispersion of each grating reduces significantly far away from its own stop band to a good approximation, we can consider only the effect of two consecutive gratings on the transmission of any particular channel and neglect the effect of all other gratings. Compensation for the anomalous GVD in optical fibers at communication wavelengths is achieved via the use of the normal dispersion side of the grating spectrum (i.e., $|\omega_0| < |\omega_B|$, where $|\omega_0|$ is the central frequency of the channel under consideration). In addition, the nearest adjacent grating centered at $|\omega_B - (c/n)\Delta| < |\omega_0|$, which exhibits anomalous dispersion at ω_0, must also be considered in the design, as it will degrade the overall performance of the compensator. In this analysis we will consider the propagation of Gaussian shape pulses of rms width σ_0 in an optical fiber of length L_f with dispersion β_{f2}. The shape of the Gaussian pulse is maintained with propagation through the fiber; however, due to the GVD its width increases resulting in the rms width [39]

$$\sigma_1 = \sigma_0 \sqrt{1 + \left(\frac{L_f}{L_D}\right)^2} \tag{6.35}$$

where $L_D = 2\sigma_0^2 / |\beta_{f2}|$ is the dispersion length. The pulse also becomes linearly chirped with the chirp parameter α given by the following expression:

$$\alpha = \frac{L_f}{\beta_{f2}\left[\dfrac{1}{L_f^2 + L_g^2}\right]} \tag{6.36}$$

It follows that an ideal dispersion compensator that recompresses the dispersion-

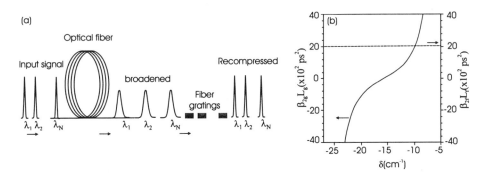

Figure 6.44 (a) A schematic illustration of a dispersion compensator designed for simultaneous compensation of GVD in a WDM communication system by using multiple cascade gratings. Plot (b) corresponds to the graphical solution of the design equation (6.37) for $L_f = 100$ km, $\beta_{f2} = -20$ ps^2/km, $\kappa = 4$ cm^{-1}, $L_g = 40$ cm, and $\Delta \nu = 100$ GHz (*After:* [139]).

broadened pulse to its original width must satisfy

$$\beta_{g2}L_g = -\beta_{f2}L_f \tag{6.37}$$

where L_g is the grating length. The solution to (6.37) may be obtained by plotting the left-(solid line) and right-hand side (dotted line) of the equation. Figure 6.44(b) shows such a plot for $\kappa = 4$ cm^{-1}, $L_g = 40$ cm, $\beta_{f2} = -20$ ps^2/km, $L_f = 100$ km. The intersection of the two lines provides the solution. Using (6.34) and (6.37) one can find the parameter of two adjacent gratings for optimum performance of the dispersion compensator. Let us describe the performance of the device in terms of input/output pulse-width ratio. The ratio of initial pulse width σ_0 and recompressed pulse width σ_2 can be written as [39]

$$\frac{\sigma_0}{\sigma_2} = \frac{\sigma_0}{\sigma_1}\left[(1+\beta_{g2}L_g\alpha)^2 + \left(\frac{L_g\beta_{g2}}{2\sigma_1^2}\right)^2 + [1+(2\alpha\sigma_1^2)^2]\frac{1}{8}\left(\frac{L_g\beta_{g3}}{2\sigma_1^3}\right)^2\right]^{-1/2} \tag{6.38}$$

This ratio is plotted in Figure 6.45(a) for the parameters $L_g = 40$ cm, $\kappa = 4$ cm^{-1}, $\Delta = 31.4$ cm^{-1} (channel spacing 100 GHz) shown as the solid line, along with a dashed line corresponding to the case when only one grating is used at $\delta = 0$. This plot shows a slight decrease in σ_0/σ_2 and a slight increase in the optimum value of δ due to the second grating at $\delta = -\Delta$. However, the performance of compensator at $\delta = \delta$ (optimum) is virtually unaffected by the second grating. The ratio of the initial pulse to the recompressed pulse (σ_0/σ_2) in a single grating case can be improved by making gratings stronger and longer [137]. However, this is not always true when gratings are cascaded because of the fixed value of the channel spacing. A plot of the figure of merit F as a function of the coupling strength κ may be used to determine the best κ value before detrimental effects of the adjacent grating is not excessive. Figure 6.45(b) shows F as a function of κ for three values of channel spacing: $\Delta\omega = 200$ GHz (solid line), 100 GHz (long dash line), 50 GHz (dot-dashed line).

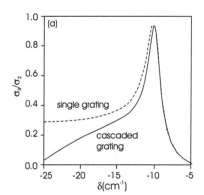 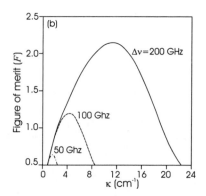

Figure 6.45 Plot (a) shows the ratio σ_0/σ_2 for two cascaded grating (solid line) and for the single grating case (dashed line). Plot (b) shows the figure of merit F versus coupling coefficient with optimized detuning for each κ for $\Delta\nu = 200$ GHz (solid line), 100 GHz (dashed line), and 50 GHz (dot-dashed line) (*After:* [139]).

As the channel spacing becomes smaller the recompressed pulse is increasingly affected by the third-order dispersion effects. In addition, the range of detuning for which the third-order dispersion is not excessive becomes narrower as Δv decreases.

It is clear from the above analysis that the design of the performance of cascaded grating-based dispersion compensators operating in the transmission mode for dispersion compensation of multiple-channel WDM lightwave systems is complicated. This is mainly because the parameters should be chosen such that adjacent gratings do not significantly affect the performance of each other, implying a channel spacing $\Delta > 2\kappa$. As the four grating parameters κ, δ, L_g and Δ are coupled, they cannot be chosen independently. It has been pointed out that the grating GVD is high only in a limited range of detunings close to the edge of the stop band, of the order of the bandwidth of the grating. This limitation becomes even more severe, especially for dense WDM systems.

Analysis of the dispersion properties of cascaded grating-based add/drop filters [139] has shown that for a channel transmitted at the zero-GVD wavelength, the performance is limited mostly by the third-order dispersion. However, the limitation on the minimum allowed pulse width (which determines the maximum allowed bit rate) is less strict when gratings are cascaded, when compared with the single grating case. One can thus conclude [138] that the grating dispersion can affect the performance of a WDM network in two ways. First, in dense WDM systems with a single grating-based add/drop filter the transmitted channel may be broadened due to GVD dispersion and the maximum allowed bit rate is inversely proportional to the quadratic root of the grating length. Second, in WDM networks incorporating cascaded gratings, the transmitted channel may be distorted due to cubic dispersion and the maximum achievable bit rate is inversely proportional to the cubic root of the grating length.

6.9 Temperature Sensor in a Passive Optical Network

Long distance and local access networks may utilize WDM signaling, which imposes special requirements on operations and maintenance. One type of access network that is studied is the passive optical network (PON), which connects the subscriber's home to the central office through fiber. This is of interest because of its potential for high bandwidth and improved outside plant reliability. There are instances, however, where the passive elements in PON may still cause difficulties. Temperature drift of a wavelength router in a WDM PON, for example, may cause misalignment between the router passbands and source wavelengths. This will cause loss of signal or cross-talk at the subscriber's optical network unit. One solution is to actively monitor changes in the router passbands by adjusting the source wavelengths. This can be done by placing a fiber Bragg grating at the remote node and reflecting back a probe beam to the central office to detect an error signal. Choosing a fiber Bragg grating with 0.1-nm bandwidth and 200-GHz channel spacing enables the probe wavelength to be placed between channels without causing interference. Both the WDM channels and the probe wavelength may be obtained from a single multifrequency laser. This arrangement allows tuning of the laser to align both probe wavelength to the fiber grating and the WDM channels to the router. Implementing such a system in a control loop enables automatic wavelength management with no impact on the subscriber channels.

Monitoring of uptstream channels in a loop-back WDM PON has been proposed by Frigo and co-workers [140] as a method of wavelength tracking to register the WDM channel wavelengths to the WDM router. Giles and Jiang [141] have proposed a tracking method that uses a narrowband reflective fiber grating at the remote node as a highly accurate temperature sensor that is interrogated with a monitor channel originating from the central office. The temperature coefficient of the fiber-grating filter is very close to that of the silica-based array waveguide grating router, enabling a simplified method of channel tracking. Figure 6.46 shows the experimental setup used to test this wavelength-tracking scheme. The WDM and monitor channels were generated from a 24-channel integrated multifrequency laser [142]. The lasers emitted in the 1555-nm wavelength region with 100-GHz channel spacing and a total fiber-coupled output power of 9.90 dBm. As many as seven WDM channels with 200-GHz channel spacing were modulated with 50-Mbps pseudo-random non-return-to-zero (NRZ) data. A temperature-tuning coefficient of 0.114 nm/°C was measured on this device. The monitor laser was biased to threshold and modulated at 505 Hz with a 20-mA peak-to-peak drive current, enabling synchronous detection of the monitor light reflected from the fiber grating. Output from the multifrequency laser was connected to a 3-dB fiber coupler. The test port from this coupler was connected to a variable optical attenuator and the second output port was used to monitor the multifrequency laser (MFL) output. The fourth coupler port was connected to an InGaAsP photodiode that was directly terminated to the input of the lock-in amplifier used to recover the monitor signal. In practice, a wavelength-tracking control loop would use the output from the lock-in amplifier as the error signal to adjust the source wavelength. An eight-channel arrayed-waveguide grating router with 200-GHz channel spacing and 8.9-dB average insertion loss was connected to the fiber-grating output and both were

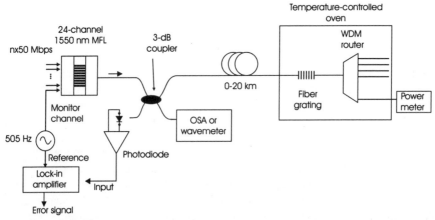

Figure 6.46 An experimental setup demonstrating that a fiber grating at a remote mode of a PON can be used as a temperature sensor to track thermally-induced changes in the WDM router filter response and to derive an eror signal to control the transmitter source wavelengths for optimum network performance (*After:* [141]).

placed inside a temperature-controlled oven to test their relative temperature sensitivity. A 6.3-km fiber span was used during the test and the array-waveguide-grating (AWG) router's output ports were left unterminated to simulate a worst-case condition. At 19.3°C (room temperature) the fiber-grating wavelength was offset from the nearest AWG-router channel by 0.6 nm. The frequency plan of the MFL was altered by changing the monitor channel so that it was spaced 100 GHz from the nearest WDM channel. Locking the monitor channel to the fiber grating wavelength placed the WDM channels in the passband of the AWG router. This arrangement takes advantage of the fiber grating's narrow reflection band, allowing full utilization of the router ports without interference from the monitor channel. With T(router) = T(grating) = 19.3°C, the MFL was temperature-tuned to obtain maximum error signal. Using the temperature-controlled oven, T(router) and T(grating) were elevated to 73°C resulting in loss of both the AWG-router output and the error signal. The maximum error signal was recovered by increasing the MFL temperature from 14.5°C to 19.3°C. This temperature tuning of the MFL realigned the WDM channels to the AWG-router, verifying the potential for wavelength-tracking applications.

6.10 Optical Fiber Phase Conjugator

Optical phase conjugation attracted considerable attention because of its application to the compensation of chromatic dispersion and nonlinearities in optical fiber communication systems, using the mid-span spectral inversion technique [143, 144] (reversing the signal optical about the middle of its span). As a result, the dispersion-induced signal distortion from the first half of the span is cancelled by dispersion in the second half. There is also interest in the phase conjugator because of its potential application in wavelength conversion future wavelength division multiplexed optical networks.

Phase conjugation may be obtained through four-wave mixing in semiconductor lasers or optical fibers. Practical use of this technique in lightwave systems requires independence of the input polarization. An example of a polarization-independent phase conjugator [145] is shown in Figure 6.47, which utilizes a fiber Bragg grating reflector to double-pass the pump beam and is used to generate the four-wave mixing products. The input signal light and the pump are injected simultaneously into dispersion-shifted fiber through an optical circulator. The efficiency of the four-wave mixing following the first pass through the dispersion-shifted fiber depends on the relative polarization of the pump and signal beams. There is no phase-conjugate signal produced when the beams have orthogonal polarizations. In Figure 6.47 the pump's input polarization is adjusted so that it is unchanged by a reflection from the fiber Bragg grating. The signal, on the other hand, is reflected by an ortho-conjugate mirror into its orthogonal polarization. In this configuration the four-wave mixing efficiency of the phase-conjugate product at the output of the optical circulator becomes independent of the signal polarization. The important features of the fiber Bragg grating used in these systems are the high reflectivity at the pump wavelength, low loss at the signal wavelength, and the narrow reflection bandwidth. Recently, a simple polarization-independent all-fiber phase conjugator using four-wave mixing has been demonstrated [146]. This configuration uses two orthogonally polarized pumps that are in-line erbium:ytterbium fiber DFB lasers pumped with 980-nm, 100-mW laser diodes. Since the DFB fiber lasers are transparent at the signal wavelength, the signal and the DFB generated four-wave mixing pumps are combined through direct injection of the signal into one end of the fiber DFB laser. This eliminates the need for a polarization combiner and a signal/pump coupler as required in a conventional polarization-independent device. After amplification by the erbium-doped fiber amplifier the signal and pumps are launched to a dispersion-shifted fiber, generating a conjugate output that is insensitive to the signal polarization owing to the two orthogonally polarized pumps. This technique features polarization-independent operation and a simple all-fiber configuration without the need for externally injected pumps.

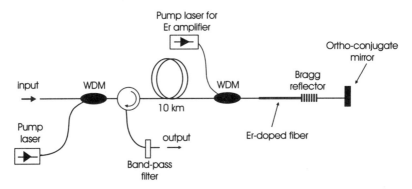

Figure 6.47   Polarization-independent reflective optical mixer for generating phase-conjugate copies of the input signal (*After:* [145]).

6.11   Phased-Array Antenna Beam-Forming Control

The use of fiber optics for implementing true time-delay (TTD) control of microwave phased-array antenna systems has been investigated for many years [147, 148] and several systems have been demonstrated [149, 150]. True time-delay is used in place of simple phase-shifting techniques in wide-bandwidth applications, in order to prevent the occurrence of beam squint at the extremes of the frequency scan. While this can be realized by replacing phase-shifting microwave waveguides with switched lengths of electrical waveguide or cable, such components sustain high loss at high RF frequencies and are susceptible to electrical cross-talk and temperature-induced time-delay changes. In contrast, optical true time-delay control networks are lightweight, compact, immune to electromagnetic interference and cross-talk, and can offer significantly lower transmission loss and higher signal bandwidth capacity. The use of optical fiber transmission-line beam-forming networks for both communications and radar antennas can therefore overcome many of the problems associated with electrical true time-delay control networks and have the potential to become a low-cost alternative to them.

The phased-array antenna beam direction is determined by the relative phases of the RF signals at each antenna element. In simple RF beam-forming systems, the phases of the individual RF signals are determined by their path lengths in their respective waveguides and consequently on the RF frequency. If this frequency is changed, the relative phasing of the signals at the antenna elements is altered, resulting in the beam shifting away from its original direction, a circumstance commonly referred to as a beam "squint". For wide-bandwidth applications, true time-delay control is used instead of the simple phase-shifting technique in order to prevent beam squint at the extremes of the scan.

To implement an optical RF beam-forming network, the RF signal is impressed on an optical carrier as an intensity modulation and the optical carrier is distributed to the antenna elements via optical fibers. While fiber optic based, true time-delay systems have been demonstrated using switched lengths of fiber, as mentioned above such systems required bulk optical elements, thus compromising the compact nature of fiber optics. Alternatively, a dispersive fiber optic element may be used and the acquired time delay varied by tuning the optical-carrier wavelength. One such dispersive element may be a high-dispersion optical fiber [151]; however, a far simpler and more compact solution is to use a series of Bragg gratings fabricated along a single length of optical fiber. Each Bragg grating has a different center wavelength; thus individual gratings may be accessed by wavelength tuning the optical carrier, thereby selecting the length of fiber traversed by the optical signal and choosing the time delay. The first variable delay line based on fiber gratings was designed with a set of six gratings spaced 1m apart. This system was tested with a 17-MHz RF signal [152] and its frequency performance has been improved in subsequent experiments [153]. With the development of chirped Bragg grating technology it became evident that this type of grating may be incorporated in beamforming networks for phased-array antenna steering, whereby wavelength tuning provides an almost continuous time delay selection.

6.11.1 Analysis Time Delay Beam Former

Let us consider an optical carrier of frequency ω, amplitude modulated by a microwave signal of frequency Q [154]. The modulated lightwave is reflected by a fiber grating of reflectivity $R(\omega) = R(\omega)\exp(i\phi(\omega))$ and detected by a fast photodiode. The output current of the photodiode has a component of frequency Q that generates a microwave signal in the output transmission line. The time dependence of the electric field in the line can be expressed as follows:

$$E(t) = \frac{1}{2} R(\omega)[A(\omega,Q)^2 + B(\omega,Q)^2]^{1/2} \exp\left[iQt + i\text{actg}\left(\frac{B(\omega,Q)}{A(\omega,Q)}\right)\right] \quad (6.39)$$

where

$$A(\omega,Q) = R(\omega+Q)\cos[\varphi(\omega+Q) - \varphi(\omega)] + R(\omega-Q)\cos[\varphi(\omega) - \varphi(\omega-Q)]$$

$$B(\omega,Q) = R(\omega+Q)\sin[\varphi(\omega+Q) - \varphi(\omega)] + R(\omega-Q)\sin[\varphi(\omega) - \varphi(\omega-Q)]$$

If the modulation frequency is much smaller than the grating bandwidth, then the above equation may be approximated as

$$E(t) = |R(\omega)|^2 \exp\left[iQt + i\frac{\varphi(\omega+Q) - \varphi(\omega-Q)}{2}\right] \tag{6.40}$$

Therefore, the phase can be written as

$$\psi(\omega,Q) = t(\omega)Q \tag{6.41}$$

where

$$t(\omega) = \frac{d\varphi(\omega)}{d\omega} \tag{6.42}$$

is the time delay of the fiber grating. Hence, provided that the modulation frequency is low enough to preserve the linewidth of the optical carrier, the phase of the modulating signal is linear with the modulation frequency Q, and the phase slope can be modified by changing the frequency ω of the optical carrier. The phase difference between two consecutive elements k and $k-1$ of a phased-array antenna radiating its main lobe in the direction θ must

$$\psi_k - \psi_{k-1} = Q\frac{d}{c}\sin(\theta) \tag{6.43}$$

where d is the distance between the array elements and c is the velocity of light. If a fiber grating is used as phase shifter, two consecutive array elements are fed with two different optical wavelengths, ω_k and ω_{k-1}, and the resultant beam-pointing angle can be written in terms of the grating time delay by substituting (6.41) in (6.43):

$$\sin(\theta) = \frac{c}{d}\left[t(\omega_k) - t(\omega_{k-1})\right] \tag{6.44}$$

Equation 6.44 states that the main beam direction of a phased array steered by a fiber grating is independent of the radio frequency under the approximations mentioned above. Hence, the grating is a true time-delay beam former suitable for wideband applications. The frequency of the optical carrier determines the scan angle of the antenna.

6.11.2   Examples of Phased-Array Antennas

A true time-delay control of microwave phased-array antenna systems has been demonstrated with a discrete Bragg grating array 3-bit delay lines and chirped Bragg grating 6-bit delay lines [155]. The 3-bit delay line was fabricated from an array of discrete and uniform period Bragg gratings, with a minimum selectable time delay of 9.09 ps, making this system suitable for phased-array antenna beam-forming control at RF frequencies of up to about 3 GHz with 10 degree phase resolution. The 6-bit delay line was fabricated using a chirped Bragg grating and was suitable for beam-forming control at RF frequencies of up to ~48 GHz with 10 degree phase resolution. This system is illustrated schematically in Figure 6.48. In the true time-delay line the gratings are used in reflection, so a 3-dB coupler, introducing >6-dB loss, is required to route the signal to the photo-detector for conversion to an electrical signal. This loss is comparable to a typical 4-bit nonwavelength-switched, true time-delay device. The maximum number of time-delay elements that can be fabricated in each delay line is determined by the tuning range of the optical source (Δf) and the optical bandwidth of each grating. The optical bandwidth of the Bragg grating determines the minimum wavelength spacing between adjacent gratings in the delay line and, hence, the maximum number of discrete time-delay elements N_{max}, which can be addressed by a single optical source as follows:

$$N_{max} \approx \frac{\Delta f}{\Delta \lambda} \tag{6.45}$$

Clearly, the minimum achievable time delay T_{min} is given by

$$T_{min} = \frac{2nd_c}{c} \tag{6.46}$$

where d_c is the center-to-center spacing between the gratings, n is the refractive index of the optical fiber, and c is the speed of light in vacuum. The 3-bit true delay line consisted of eight discrete Bragg gratings, each having a length of 2 mm, a full-width at half-maximum bandwidth of ~0.5 nm, and a peak reflectivity of ~60%. The delay line was fabricated to have 1-mm center-to-center spacing between the gratings with central wavelengths at 1494.10, 1498.75, 1501.05, 1503.75, 1515.05, 1518.05, 1521.85, and 1524.65 nm, respectively. The 1-mm spacing between adjacent gratings in the delay line corresponds to a minimum time delay of 9.6 ps. Measurements of the delay line confirmed the suitability for beam-forming control with 10 degree phase resolution at RF frequencies of up to ~3 GHz [155]. The ~9-ps delay step represents the practical lower limit on the delay step size, which can be produced through the use of delay lines comprising discrete linear Bragg gratings. In order to produce time-delay steps of significantly smaller duration and thus increase the RF frequency at which 10 degree phase resolution beam-forming control can

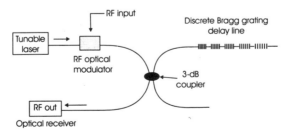

Figure 6.48 A schematic of a Bragg grating true time-delay system with a delay line accomplished using discrete Bragg gratings.

be achieved, a single chirped Bragg grating was used in place of the array of linear gratings. This enabled the production of an effective, continuously variable time delay, rather than discrete delay steps. Four chirped Bragg gratings were fabricated having bandwidths of 7, 12, 20, and 30 nm, respectively, each with a length of 4 mm and a reflectivity of ~60%. The grating amplitude profile was approximately Gaussian to improve the linearity of the group delay characteristics. The above chirped grating can therefore in principle be used to create time delays from as small as 40 fs to as large as 59 ps. In practice, the smallest time-delay step that can be realized is determined by the system characteristics and the optical linewidth broadening effect of modulating the optical carrier. In order to maintain a 10 degree phase resolution beam-pointing accuracy, the phase error incurred by modulating the optical carrier must be less than 10 degrees. For the 30-nm chirped grating, this means that the maximum RF modulation signal frequency that can be applied to the optical carrier is ~48 GHz.

In order to increase the RF-signal modulation frequency range that can be used, the chip rate for the gratings must be decreased, thus reducing the spatial displacement of the reflection points of the optical-carrier wavelength and thus reducing the time, phase, and error incurred by the signal. This can be achieved using longer length chirped Bragg gratings.

A true time-delay optical feeder for phased-array antennas utilizing a chirped fiber grating has also been demonstrated by Corral et al. [156]. They have constructed a continuously variable true time delay using a wide-bandwidth 40-cm chirped fiber Bragg grating as the dispersive element. A general, true time-delay optical feeder is shown in Figure 6.49. The output light from M narrow tuning range tunable lasers (λ_1 to λ_M) are combined and all the optical wavelengths are modulated by the same microwave signal, employing an external electro-optic modulator. The modulated lightwaves pass through an optical circulator to a long, chirped fiber grating. Following the reflection from the chirped grating, the M-delayed, modulated optical carriers λ_1 to λ_M are separated by a wavelength division multiplexed demultiplexer and then passed to each antenna element after the photo-detectors. To demonstrate the principle of the above device, a four-antenna element has been constructed [156]. The experimental setup is similar to the one shown in Figure 6.49; it actually corresponds to the single branch of the general scheme.

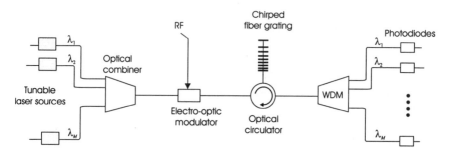

Figure 6.49 A schematic of a true time-delay beam former transmitting-mode system using a chirped grating for a dispersive element.

The grating used in this experiment was a 40-cm long linear chirped grating with a linear group-delay response within a 4-nm bandwidth between 1547 and 1551 nm. The main group-delay slope was approximately 835 ps/nm. In order to show the potential broad-bandwidth of the system, due to its true time-delay feature, group-delay measurements were made at four different microwave frequencies, namely at 2 ,5, 10, and 18 GHz. The results obtained confirm the expected good performance of the optical beam-forming scheme. These results, however may be improved by tailoring a fiber Bragg grating, which should be more focused on achieving very low group delay response ripple instead of a highly linear response as required for dispersion compensation applications. The effect of the M tunable lasers on the overall system cost is not dramatic because the lasers do not require a broad tuning bandwidth, but just tenths of a nanometer, which should reduce costs. The use of just one fiber Bragg grating for the whole system is a clear advantage compared to some previous TTD schemes proposed.

6.12 Summary

This chapter has given a detailed outline of some of the lightwave applications of fiber Bragg gratings. Some of the examples, including the use of fiber Bragg gratings in wavelength-stabilized pump lasers, fiber Raman lasers, fiber amplifiers, and add/drop multiplexers, are now commercial products. Bragg gratings have truly revolutionized applications in communication networks. Devices that were thought impossible to construct are now becoming a reality. There are competitive technologies in each application and the trade-offs and advantages need careful evaluation. For example, in the case of dispersion compensators, dispersion compensating fibers were first to be marketed, but as fiber grating manufacturing matures, the linearly chirped grating is becoming a very attractive alternative as a compact device for dispersion compensation. Add/drop multiplexers using arrayed-waveguide grating routers have been used in many demonstrations, but the low insertion loss of the fiber Bragg grating/circulator add/drop multiplexers is appealing. Furthermore, specially tailored add/drop multiplexers such as the "frustrated coupler," the "zero-insertion-loss," and the "twin-core Mach-Zehnder" are adding a new dimension and providing alternatives for designers in communication networks. As WDM networks evolve, stringent filtering needs resulting from close channel spacing and complex optical operation could be met with specially tailored fiber Bragg gratings. Reflectivity, isolation, bandwidth, edge response, and dispersion characteristics can all be adjusted to fabricate optimally designed gratings. This ability to make application-specific fiber Bragg gratings assures their long viability as components in advanced systems research and to enhance real systems in the generation, detection, and conditioning of light.

References

[1]   Reekie, L., et al. "Tunable single-mode fiber laser," *Journal of Lightwave Technology*, Vol. LT4, 1986, pp. 956–957.

[2]   Ball, G. A., W. W. Morey, and J. P. Waters, "Nd^{3+} fiber laser utilizing intra-core Bragg reflectors," *Electronics Letters*, Vol. 26, 1990, pp. 1829–1830.

[3]   Ball G. A., and W. H. Glenn, "Design of a single-mode linear-cavity erbium fiber laser utilizing Bragg reflector," *Journal of Lightwave Technology*, Vol. 10, 1992, pp. 1338–1343.

[4]   Ball G. A., et al. "Modeling of short, single-frequency fiber laser in high-gain fiber," *IEEE Photonics Technology Letters*, Vol. 5, 1993, pp. 649–651.

[5]   Mizrahi, V., et al. "Stable single-mode erbium fiber-grating laser for digital comminations," *Journal Lightwave Technology*, Vol. 11, 1993, pp. 2021–2025.

[6]   Othonos, A., X. Lee and D.P. Tsai, "Spectrally broadband Bragg grating mirror for an erbium-doped fiber laser," *Optical Engineering*, Vol. 35, 1996, pp. 1088–1092.

[7]   Okamura H. and K. Iwatsuki, "Simultaneous oscillation of wavelength-tubable, singlemode lasers using an Er-doped fiber amplifier," *Electronics Letters*, Vol. 28, 1992, pp. 461–463.

[8]   Park, N., J. W. Dawson, and K. J. Vahala, "Multiple wavelength operation of an erbium-doped fiber laser," *IEEE Photon Technology Letters*, Vol. 4, 1992, pp. 540–541.

[9]   Chow, J., et al. "Multi-wavelength generation in an erbium-doped fiber laser using in-fiber comb filters," *IEEE Photonics Technology Letters*, 1996, pp. 60–62.

[10]   Goldstein, E. L., et al. "Suppression of dynamics cross-saturation in mutli-wavelength lightwave networks with inhomogeneously broadened amplifiers," *IEEE Photonics Technology Letters*, Vol. 5, 1993, pp. 937–938.

[11]   Ball, G. A., W. W. Morey, and W. H. Glenn, "Standing-wave monomode erbium fiber laser," *IEEE Photonics Technology Letters*, Vol. 3, 1991, pp. 613–615.

[12]   Zyskind, J. L., et al. "Short single frequency erbium-doped fiber laser," *Electronics Letters*, Vol. 28, 1992, pp. 1385–1386.

[13]   Zyskind, J. L. et al. "Transmission at 2.5 Gbits/s over 654 km using an erbium-doped fiber grating laser source," *Electronics Letters*, Vol. 29, 1993, pp. 1105–1106.

[14]   Sanchez, F., et al. "Effects of ion pairs on the dynamics of erbium-doped fiber lasers," *Physics Review A*, Vol. 48, 1993, pp. 2220–2229.

[15]   Ball, G. A., and W. W. Morey, "Narrow-linewidth fiber laser with integrated master oscillator-power amplifier," in *Tech. Dig. Opt. Fiber Communication Conference*, Vol. 13, Paper WA3, 1992.

[16]   Kringlebotn, J. T., et al. "Highly efficient low-noise grating-feedback $Er^{3+}:Yb^{3+}$ co-doped fiber laser," *Electronics. Letters*, Vol. 30, 1994, pp. 972–973.

[17]   Ball, G. A., and W. W. Morey, "Continuously tunable single-mode erbium fiber laser," *Optics Letters*, Vol. 17, 1992, pp. 420–422.

[18]   Ball, G. A. et al. "Low noise single frequency linear fiber laser," *Electronics Letters*, vol. 29, 1993, pp. 1623–1625.

[19]   Ball, G. A., et al. "60 mW 1.5 im single-frequency low-noise fiber laser MOPA," *IEEE Photonics Technology Letters*, Vol. 6, 1994, pp. 192–194.

[20]   Ball, G. A. and W. W. Morey, "Compression-tuned single frequency Bragg grating fiber laser," *Optics Letters*, Vol. 19, 1994, pp. 1979–1981.

[21]   Ball, G. A., C. G. Hull-Allen, and J. Livas, "Frequency noise of a Bragg grating fiber laser," *Electronics Letters*, Vol. 30, 1994, pp. 1229–1230.

[22]   Kringlebotn, J. T., et al. "Efficient diode-pumped single-frequency erbium:ytterbium fiber laser," *IEEE Photonics Technology Letters*, Vol. 5, 1993, pp. 1162–1164.

[23]   Kringlebotn, J. T., et al. "$Er^{3+}:Yb^{3+}$ co-doped fiber DFB laser," *Optics Letters*, Vol. 19, 1994, pp. 2101–2103.

[24]   Loh, W. H., and R. I. Laming, "1.55 μm phase-shifted distributed feedback fiber laser," *Electronics Letters*, Vol. 31, 1995, pp. 1440–1442.

[25] Asseh, A., et al. "10 cm Yb^{3+} DFB fiber laser with permanent phase shifted grating," *Electronics Letters*, Vol. 31,1995, pp. 969–970.

[26] Sejka, M., et al. "Distributed feedback Er^{3+} doped fiber laser," *Electronics Letters*, Vol. 31, 1995, pp. 1445–1446.

[27] Harujunian, Z. E., et al. "Single polarization twisted distributed feedback fiber laser," *Electronics Letters*, Vol. 32, 1996, pp. 346–348.

[28] Loh, W. H., et al., "Intracavity pumping for increased output power from a distributed erbium fiber laser," *Electronics Letters*, Vol. 32, 1996, pp. 1204–1205.

[29] Loh, W. H., S.D. Butterworth, and W.A. Clarkson, "Efficient distributed feedback erbium-doped germanosilciated fiber laser pumped in 520 nm band," *Electronics Letters*, Vol. 32, 1996, pp. 2088–2089.

[30] Bonfrate, G., F. Vaninetti, and F. Negrisolo, "Single-frequency MOPA Er^{3+} DBR fiber laser for WDM digital telecommunication systems," 1998, pp. 1109–1111.

[31] Loh, W. H., L. Dong and J.E. Caplen, "Single-sided output Sn/Er/Yb distributed feedback fiber laser," *Applied Physics Letters*, Vol. 69, 1996, pp. 2151–2153.

[32] Dong, L., et al. "Efficient single-frequency fiber lasers with novel photosensitive Er/Yb optical fibers," *Optics Letters*, 1997, pp. 694–696.

[33] Douay, M., et al. "Birefringence in optical fiber laser with intracore fiber Bragg grating," *IEEE Photonics Technology Letters*, Vol. 4, 1992, pp. 844–846.

[34] Gloag, A. J., et al. "Single-frequency travelling-wave erbium doped fiber laser incorporating a fiber Bragg grating," *Optics Communications*, Vol. 123, 1996, pp. 553–557.

[35] Po, H., et al. "High power neodymium doped single transverse mode fiber laser," *Electronics Letters*, Vol. 29, 1993, pp. 1500–1501.

[36] Zellmer, H., et al. "High-power cw neodymium doped fiber laser operating at 9.2 W with high beam quality," *Optics Letters*, Vol. 20, 1995, pp. 578–580.

[37] SDL FL-10, SDL. Inc. 80 Rose Orchard Way, San Jose, CA.

[38] Zentano, L.A., "High-power double-clad fiber lasers," *Journal of Lightwave Technology*, Vol. 11, 1993, pp. 1435–1446.

[39] Agrawal, G.P., Nonlinear Fiber Optics,"2nd edition New York: Academic Press, 1995.

[40] Stolen, R. H., and C. Lin, "Fiber Raman lasers," in CRC Handbook of Laser Science and Technology, Supplement 1: Lasers, Boca Raton, FL: CRC Press, 1991.

[41] Grubb, G. S, et al. "High-power 1.48 μm cascaded Raman laser in germanosilicate fibers," *Proc. Optical Amplifiers and Their Applications*, Davow, Switzerland, Paper SaA4, 1995.

[42] Stafford, E. K., J. Mariano, and M.M. Sanders, "Undersea nonrepeatered technologies, challenges, and products," *AT&T Technologies Journal*, Vol. 74, 1995, pp. 47–59.

[43] Hansen, P. B. et al. "529 km urepeatered transmission at 2.488 Gbit/s using dispersion compensation, forward error correction, and remote post- and pre-amplifiers pumped by diode-pumped Raman lasers," *Electronics Letters*, Vol. 31, 1995, pp. 1460–1461.

[44] Davey, R. P., et al. "Mode-locked erbium fiber laser with wavelength selection by means of fiber Bragg grating reflector," *Electronics Letters*, Vol 27, 1991, pp. 2087–2088.

[45] Kean, P. N., et al. "Dispersion-modified actively mode-locked erbium fiber laser using a chirped fiber grating," *Electronics Letters*, Vol. 30, 1994, pp. 2133–2135.

[46] Noske, D. U., et al. "Dual-wavelength operation of a passively mode-locked figure-of-eight ytterbium-erbium fiber soliton laser," *Optics Communications*, Vol. 108, 1994, pp. 297.

[47] Fermann, M. E, K. Sugden, and I. Bennion, "High-power soliton fiber laser based on pulse width control with chirped fiber Bragg gratings," *Optics Letters*, Vol. 20, 1995, pp. 172–174.

[48] Cheo, P. K., V. G. Mutalik, and G. A. Ball, "Mode-locking of in-line coupled-cavity fiber lasers using intra-core Bragg gratings," *IEEE Photonics Technology Letters*, Vol. 7, 1995, pp.980–982.

[49] Giles, C. R., "Lightwave applications of fiber Bragg gratings," *Journal of Lightwave Technology*, Vol. 15, 1997, pp. 1391–1404.

[50] Saleh, A. A. M., et al. "Modeling of gain in erbium-doped fiber amplifiers," *IEEE Photonics Technology Letters*, 1990.

[51]   Bayart, D., B. Clesca, and L. Hamon, "Experimental investigation of the gain flatness characteristics of 1.55 μm erbium-doped fluoride fiber," *IEEE Photonics Technology Letters*, Vol. 6, 1994, pp. 613–615.

[52]   Godstein, E. L., L. Eskildsen, and V. da Silva, "Inhomogeneously broadened fiber amplifier cascades for transparent multiwavelengths lightwave networks," *Journal of Lightwave Technology*, Vol. 13, 1995, pp.782–790.

[53]   Giles, C. R., and E. Desurvire, "Modeling erbium-doped fiber amplifiers," *Journal of Lightwave Technology*, Vol. 9, 1991, pp. 271–283.

[54]   Vengsarkar, A. M., et al. "Long-period fiber-grating-based gain equalizers," *Optics Letters*, Vol. 21, 1996, pp. 336–338.

[55]   Kashyap, R., R. Wyatt, and R.J. Campbell, "Wideband gain flattened erbium fiber amplifier using a photosensitive fiber blazed grating," *Electronics Letters*, Vol. 29, 1993, pp. 154–156.

[56]   Kashyap, R., R. Wyatt, and P. F. Mckee, "Wavelength flattened saturated erbium amplifier using multiple side-tap Bragg gratings," *Electronics Letters*, Vol. 29, 1993, pp. 1025–1026.

[57]   Wysocki, P. F., et al. "Broad-band Erbium-doped fiber amplifier flattened beyond 40 nm using long-period grating filter," *IEEE Photonics Technology Letters*, Vol. 9, 1997, pp.1343–1345.

[58]   Zyskind, J. L., "Performance issues in optically amplified systems and networks," in *Proc. OFC'97*, Dallas, TX, paper TuP1, 1997.

[59]   Onaka, H. et al. "1.1 Tb/s WDM transmission over a 150 km 1.3 mm zero-dispersion single-mode fiber," *in Proc. OFC'96*, San Jose, CA Paper PD19, 1996.

[60]   Delevaque, E., et al. "Gain control in erbium-doped fiber amplifiers by lasing at 1480 nm with photoinduced Bragg gratings written on fiber ends," *Electronics Letters*, Vol. 29, 1993, pp. 1112–1114.

[61]   Massicott, J. F., et al. "1480 nm pumped erbium doped fiber amplifier with all optical automatic gain control," *Electronics Letters*, Vol. 30, 1994, pp. 962–963.

[62]   Ko Seong Yun, et al. "Gain control in erbium-doped fiber amplifiers by tuning center wavelengths of fiber Bragg grating constituting resonant cavity," *Electronics Letters*, Vol. 34, pp. 990–991, 1998.

[63]   Grubb, S. G, et al. "1.3 μm cascaded Raman amplifier in germanosilicate fibers," in *Proc. Optical Amplifiers and Their Applications*, Breckenridge, CO, Paper PD3, 1994.

[64]   Dianov, E. M., et al. "Demonstration of 1.3 μm Raman fiber amplifier gain of 25 dB at a pumping power of 300 mW," *Optic Fiber Technology*, Vol. 1, 1995, pp. 236–238.

[65]   Nielsen, T. N., et al. "8x10 Gb/s 1.3-μm unrepeatered transmission over a distance of 141 with Raman post- and pre-amplifiers," *IEEE Photonics Technology Letters*, Vol. 10, 1998, pp. 1492-1494.

[66]   Brinkmayer, E., et al. "Fiber Bragg reflector for mode selection and line-narrowing of injection laser," *Electronics letters*, Vol. 22, 1986, pp. 134–135.

[67]   Park, C. A., et al. "Single-mode behavior of multimode 1.55 μm laser with a fiber grating external cavity," *Electronics Letters*, Vol. 22, 1986, pp. 1132–1134.

[68]   Burns, D., et al. "Active modelocking of an external cavity GaInAsP laser incorporating a fiber-grating reflector," *Electronics Letters*, Vol. 24, 1988, pp. 1439-1441.

[69]   Bird, D. M., et al. "Narrow line semiconductor laser using fiber grating," *Electronics Letters*, Vol. 27, 1991, pp. 1115–1116.

[70]   Morton, P. A., et al. "Stable single mode hybrid laser with high power and narrow linewidth," *Applied Physics Letters*, Vol. 64, 1994, pp. 2634–2636.

[71]   Kashyap, R., "Wavelength uncommitted lasers," *Electronics Letters*, Vol. 30, 1994, pp. 1065-1066.

[72]   Loh, W. H., et al. "Single frequency erbium fiber external cavity semiconductor laser," *Applied Physics Letters*, Vol. 66, 1995, pp. 3422-3424.

[73]   Timofeev, F. N., et al. "Spectral characteristics of a reduced cavity single-mode semiconductor fiber grating laser for application in dense WDM systems," *Proc. 21 st European Conference Optics Commun. (ECOC'95)* Brussels, Belgium Paper Tu.P.26, 1995, pp. 477–480.

[74]   Campbell, R. J., et al. "Wavelength stable uncooled fiber grating semiconductor laser for use in an all optical WDM access network," *Electronics Letters*, Vol. 32, 1996, pp. 119–120.

[75]   Ziari, M., et al. "High speed fiber grating coupled semiconductor WDM laser," *Proc. Conf. Lasers*

[76]   Morton, P. A., et al. "Hybrid soliton pulse source with fiber external cavity and Bragg reflector," *Electronics Letters*, Vol. 28, 1992, pp. 561–562.

[77]   Giles, C. R., T. Erdogan, and V. Mizrahi, "Simultaneous wavelength stabilization of 980-nm pump lasers," *IEEE Photonics Technology Letters*, vol. 6, 1994, pp. 907–909.

[78]   Ventrudo, B. F., et al. "Wavelength and intensity stabilization of 980 nm diode lasers coupled to fiber Bragg gratings," *Electronics Letters*, Vol. 30, 1994, pp. 2147–2148.

[79]   Hargreaves, D., G. S. Lick, and B. F. Ventrudo, "High-power 980 nm pump module operating without a thermoelectric cooler," *Proc. Optc. Fiber Commun. Conf. (OFC'96)*, San Jose, CA, paper ThG3, 1996, pp. 229–230.

[80]   Morton, P. A., et al. "Mode-locked Hybrid soliton pulse source with extremely wide operating frequency range," *IEEE Photonics Technology Letters*, Vol. 5, 1993, pp. 28–31.

[81]   Tkach, R. W. and A. R. Chraplyvy, "Regimes of feedback effects in 1.5-im distributed feedback laser," *Journal of Lightwave Technology*, Vol. 4, 1986, pp. 1655–1661.

[82]   Hand, D. P. and P. St J. Russell, 7^h *international conference on integrated optics and optical fiber communication (IOOC'89)*, Kobe, 1989, Technical Digest, p. 64.

[83]   Hill, K. O. et al. "Narrow-bandwidth optical waveguide transmission filters: A new design concept and applications to optical fiber communications," *Electronics Letters*, Vol. 23, 1987, pp. 464–465.

[84]   Bilodeau, F. et al. "High-return-loss narrowband all-fiber bandpass Bragg transmission filter," *IEEE Photonics Technology Letters*, Vol. 6, 1994 , pp. 80–82.

[85]   Johnson D.C. et al. " New design concept for a narrowband wavelength-selective optical tap and combiner," *Electronics Letters*, Vol. 23, 1987, pp. 668–669.

[86]   Fielding, A. et al. "Compact all-fiber wavelength drop and insert filter," *Electronics Letters*, Vol 30, 1994, pp. 2160–2161.

[87]   Reid, D. C. et al. "Phase-shifted Moiré grating fiber resonators," *Electronics Letters*, Vol. 26, 1990, pp. 1011.

[88]   Legoubin, S., et al. "Formation of Moiré grating in core of germanosilicate fiber by transverse holographic double exposure method," *Electronics Letters*, Vol. 27, 1991, pp. 1945–1946.

[89]   Zhang, L., et al. "Wide-stopband chirped fiber moiré grating transmission filters," *Electronics Letters*, Vol. 31, 1995, pp. 477–479.

[90]   Canning, J. and M.G. Sceats, "π-phase-shifted periodic distributed structures in optical fibers by UV post processing," *Electronics Letters,* Vol. 30, 1994, pp.1344–1345.

[91]   Kashyap, R., P.F. Mckee, and D. Armes, "UV written reflection grating structures in photosensitive optical fibers using phase-shifted phase masks," *Electronics Letters*, Vol. 30, 1994, pp. 1977–1978.

[92]   Agrawal, G. P., and S. Radic, "Phase-shifted fiber Bragg grating and their applications for wavelength demultiplexing," *IEEE Photonics Technology Letters*, Vol. 6, 1994, pp. 995–997.

[93]   Farries, M. C. et al. "Very Broad reflection bandwidth (44nm) chirped fiber grating and narrow bandpass filters produced by the use of an amplitude mask," *Electronics Letters,* Vol. 30, 1994, pp. 891–892.

[94]   Mizrahi, V., et al. "Four channel fiber grating demultiplexer," *Electronics Letters*, Vol. 30, 1994, pp. 780–781.

[95]   Morey, W. W., *OFC'91*, San Diego, California PDP 20, 96, 1991.

[96]   Bilodeau, F., B.Malo, D. C. Johnson, J. Albert, and K. O. Hill, *Proc. of the European Conference on Optical Communications*, ECOC'93, p.29 ThC12.8, 1993.

[97]   Bilodeau, F., et al. "High-Return-Loss narrowband all-fiber bandpass Bragg transmission filter," *IEEE Photonics Techn. Letters*. Vol. 6, 1994, pp. 80.

[98]   Hill, K. O., et al. "Narrow-bandwidth optical waveguide transmission filters: A new design concept and applications to optical fiber communications," *Electronics Letters,*. Vol. 23, 1987, pp. 465–466.

[99]   Zhang, L., et al. "Post fabrication exposure of gap-type bandpass filters in broadly chirped fiber gratings," *Optics Letters*, Vol. 20, 1995, pp. 1927–1929.

[100] Bakhti, F. and P. Sansonetti, "Wide bandwidth, low loss and highly rejective doubly phase-shifted UV-written fiber bandpass filter," *Electronics Letters*, Vol. 32, 1996, pp. 581–582.

[101] Zengerle, R. and O. Leminger, "Phase-shifted Bragg grating filters with improved transmission

[101]   Zengerle, R. and O. Leminger, "Phase-shifted Bragg grating filters with improved transmission characteristics," *Journal of Lightwave Technology*, Vol. 13, 1995, pp. 2354–2358.

[102]   Capmany, J., R. I. Laming, and D.N. Payne, "A novel highly selective and tunable optical bandpass filter using a fiber grating and a fiber Fabry-Perot," *Microwave and Optical Technology Letters*, Vol. 7, 1994, pp. 499–501.

[103]   Town, G. E., et al. "Wide-band Fabry-Perot-like filters in optical fiber," *IEEE Photonics Technology Letters*, Vol. 7, 1995, pp. 78–80.

[104]   Othonos A., X. Lee, and R.M. Measures, "Superimposed multiple Bragg gratings", *Electronics Letters*, Vol. 30, 1994, pp. 1972–1973.

[105]   Eggleton, B. J., et al. "Long periodic superstructure Bragg gratings in optical fibers," *Electronics Letters*, Vol 30, 1994, pp. 1620–1621.

[106]   Archambault, J.-L., et al. "Grating-frustrated coupler: A novel channel-dropping filter in single mode optical fiber," *Optics Letters*, Vol. 19, 1994, pp. 180–182.

[107]   Vengsarkar, A. M., et al. "Long-period fiber grating as band-rejection filters," *Journal of Lightwave Technology*, Vol. 14, 1996, pp. 58–65.

[108]   Hill, K. O., et al. "Efficient mode conversion in telecommunication fiber using externally written gratings," *Electronics Letters,* Vol. 26, 1990, pp. 1270–1272.

[109]   Hill, K. O., et al. "Birefringent photosensitivity in monomode optical fiber: application to external writing of rocking filters," *Electronics Letters*, Vol. 27, 1991, pp. 1548–1550.

[110]   Bilodeau, F. et al. "An all-fiber dense-wavelength-division multiplexer/demultiplexer using photo-imprinted Bragg gratings," *IEEE Photonics Technology Letters*, Vol. 7, 1995, pp. 388–390.

[111]   Bilodeau, F., et al. "Low-loss highly overcoupled fused couplers: Fabrication and sensitivity to external pressure," *Journal of Lightwave Technology*, Vol. 6, 1988, pp. 1476–1482.

[112]   Bethuys, S., et al. "Optical add/drop multiplexer based on UV-written Bragg gratings in twincore fiber Mach-Zehnder interferometer," *Electronics Letters*, Vol. 34, 1998, pp. 1250–1251.

[113]   Giles, C. R. and V. Mizrahi, "Low-loss add/drop Multiplexer for WDM lightwave networks," in Tenth Int. Conf. Integrated Optics and Optical Fiber Communication, *Techn. Dig.*, Vol. 3, paper ThC2-1, 1995.

[114]   Kim Yoon Se, et al. "Highly stable optical add/drop multiplexer using polarization beam splitters and fiber Bragg grating," *IEEE Photonics Technology Letters*, Vol. 9, 1997, pp. 1119–1121.

[115]   Dong L. et al. "Novel add/drop filters for wavelength-division multiplexing optical fiber systems using a Bragg grating assisted mismatched coupler," *IEEE Photonics Technology Letters*, Vol. 8, 1996, pp. 1656–1658.

[116]   Kewitsch, A. S., et al. "All-fiber zero-insertion-loss add-drop filter for wavelength-division multiplexing," *Optics Letters*, Vol. 23, 1998, pp. 106–108.

[117]   Quetel, L., et al. "Programmable fiber grating based wavelength demultiplexer," in *Proc. OFC'96*, Paper WF6, pp. 120–121, 1996.

[118]   Mizuochi, T., K. Shimizu, and T. Ktayama, "All-fiber add/drop multiplexing of 6x10 Gbit/s using a photo-induced Bragg grating filter for WDM networks," in *Proc. OFC'96*, paper WF2, 1996.

[119]   Pan ,J. J. and Y. Shi, *Electronics Letters*, Vol. 33, 1997, pp. 1895–1896.

[120]   Winful, H. G., "Pulse compression in optical fiber filters," *Applied Physics Letters*, Vol. 46, 1985, pp. 527–529.

[121]   Ouellette, F., "All-fiber filter for efficient dispersion compensation," *Optics Letters*, Vol. 16, 1991, pp.303–305.

[122]   Williams, J. A. R., I. Bennion, K. Sugden, and N. J. Doran, "Fiber dispersion compensation using a chirped in fiber Bragg grating," *Electronics Letters*, Vol. 30, 1994, pp. 985–987.

[123]   Kashyap, R., et. al. "30ps chromatic dispersion compensation of 400fs pulses at 100Gbit/s in optical fibers using an all fiber photoinduced chirped reflection grating," *Electronics Letters*, Vol. 30, 1994, pp. 1078–1080.

[124]   Zervas, M. N., K. Ennaser, and R.I. Laming, "Design of apodised linearly-chirped fiber gratings for optical communications," in *Proc. ECOC'96*, Oslo, Norway, Paper WeP.06, 1996.

[125]   Barcelos, S., M. N. Zervas ,and R. I. Laming, "Characteristics of chirped fiber gratings for dispersion

compensation," *Optical Fiber Technology*, 1996, pp. 213–215.

[126] Atkinson, D., et al. "Numerical study of 10 cm chipped-fiber grating pairs for dispersion compensation at 10 Gb/s over 600 km of nondispersion shifted fiber," *IEEE Photonics Technology Letters*, Vol. 8, 1996, pp. 1085–1087.

[127] Krug, P. A. et al. "Dispersion compensation over 270 km at 10Gbit/s using an offset-core chirped fiber Bragg grating," *Electronics Letters*, Vol. 31, 1995, pp. 1091–1093.

[128] Loh, W.H., et al. "10 cm chirped fiber Bragg grating for dispersion compensation at 10 Gbit/s over 400 km of non-dispersion shifted fiber," *Electronics Letters*, Vol. 31, 1995, pp. 2203–2204.

[129] Loh, W. H., et al. "Dispersion compensation over distances in excess of 500 km for 10 Gbit/s systems using chirped fiber gratings," *IEEE Photonics Technology Letters*, Vol. 8, 1996, pp. 944–946.

[130] Cole, M. J., et al. "Broadband dispersion-compensating chirped fiber Bragg gratings in 10 Gbit/s NRZ 110 km non-dispersion-shifted fiber link operating at 1.55 μm," *Electronics Letters*, Vol. 33, 1997, pp. 70–71.

[131] Dong, L., et al. "40 Gbit/s 1.55 μm RZ transmission over 109 km of non-dispersion shifted fiber with long continuously chirped fiber gratings," *Electronics Letters*, Vol. 33, 1997, pp. 1563–1565.

[132] Grudinin, A. B., et al. "Straight line 10Gbit/s soliton transmission over 1000 km of standard fiber with in-line chirped fiber grating for partial dispersion compensation," *Electronics Letters*, Vol. 33, 1997, pp. 1572–573.

[133] Stephens, T., et al. "257 km transmission at 10 Gbit/s in ono-dispersion-shifted fiber using an unchirped fiber Bragg grating dispersion compensator," *Electronics Letters*, Vol. 32, 1996, pp. 1599–1601.

[134] Gnauck, A. H., et. al. "8x20 Gbit/s 315-km, 8x10 Gbit/s 480-km WDM transmission over conventional fiber using multiple broad-band fiber gratings," *IEEE Photonics Technology Letters*, Vol. 10, 1998, pp. 1495–1497.

[135] Eggleton, B. J., et al. "Dispersion compensation using a fiber gating in transmission," *Electronics Letters*, Vol. 32, 1996, pp. 1610–1611.

[136] Eggleton, B. J., et al. "Bragg grating solitons," *Physics Review Letters*, Vol. 76, 1996, pp. 1627–1630.

[137] Litchinister, N. M., B. J. Eggleton, and D. B. Patterson, "Fiber Bragg gratings for dispersion compensation in transmission: Theoretical model and design criteria for nearly ideal pulse recompression," *Journal of Lightwave Technology*, Vol. 15, 1997, pp. 1303–1313.

[138] Eggleton, B. J., et al. "Implications of fiber grating dispersion for WDM communication systems," *IEEE Photonics Technology Letters*, Vol. 9, 1997, pp. 1403-1405.

[139] Litchinister, N. M., et al. "Dispersion of cascaded fiber gratings in WDM lightwave systems," *Journal of Lightwave technology*, Vol. 16, 1998, pp. 1523–1529.

[140] Frigo, N. J., et al. "RITENet: A passive optical network architecture based on the remote interrogation of terminal equipment," *Optical Fiber Communication Conf.*, OFC'94, postdeadline Paper PD8, 1994.

[141] Giles, R. and Song Jiang, "Fiber grating sensor for wavelength tracking in single-fiber WDM access PON's," *IEEE Photonics Technology Letters*, Vol. 9, 1997, pp. 523–525.

[142] Zirngibl, M., C. H. Joyner, and L. W. Stulz, "Demonstration of a 9x200 Mbits/s wavelength division multiplexed transmitter," *Electronics Letters*, 1994 pp. 1484–1485.

[143] Yariv, A., D. Fekete, and D. M. Pepper, "Compensation for channel dispersion by nonlinear optical phase conjugation," *Optics Letters*, Vol. 4, 1979, pp. 52–54.

[144] Jopson, R. M. and R. E. Tench, "Polarization-independent phase conjugation of lightwave signals," *Electronics Letters*, Vol. 29, 1993, pp. 2216–2217.

[145] Giles, C. R., V. Mizrahi and T. Erdogan, "Polarization-independent phase conjugation in a reflective optical mixer," *IEEE Photonics Technology Letters*, Vol. 7, 1995, pp. 126–128.

[146] Yamashita, S., Sze Y. Set, and R. I. Laming, "Polarization independent all-fiber phase conjugation incorporating inline fiber DFB lasers," *IEEE Photonics Technology Letters*, Vol. 10, 1998, pp. 1407–1409.

[147] Seeds, A., "Optical technologies for phased array antennas," *IEICE Trans. Electron.*, Vol. E76-C, 1995, pp. 198–206.

[148] Frgyes, I. and A. Seeds, "Optically generated true-time delay in phased array antennas," *IEEE Transactions on microwave theory and techniques*, Vol. 43, 1995, pp. 2378–2386.

[149] Ng, W. et al. "The first demonstration of an optically steered microwave phased array antenna using true-time delay," *Journal of Lightwave Technology*, Vol. 9, 1991, pp. 1124–1131.

[150] Goutzoulis, A. P., et al. "Prototype binary fiber optic delay line," *Optical Engineering*, Vol. 28, 1989, pp. 1193–1202.

[151] Esman, R. D., et al. "Fiber-optic prism true time-delay antenna feed," *IEEE Photonics Technology Letters*, Vol. 5, 1993, pp. 1347–1349.

[152] Ball, G. A., W. H. Glenn, and W. W. Morey, "Programmable fiber optic delay line," *IEEE Photonics Technology Letters*, Vol. 6, 1994, pp. 741–743.

[153] Molony, A., C. Edge, and I. Bennion, "Fiber grating time delay element for phased array antennas," *Electronics Letters*, Vol. 31, 1995, pp. 1485–1486.

[154] Cruz, J. L., et al., "Array factor of a phased array antenna steered by a chirped fiber grating beamformer," *IEEE Photonics Technology Letters*, Vol. 10, 1998, pp. 1153–1155.

[155] Molony, A., et al. "Fiber Bragg-grating true time-delay systems: Discrete-grating array 3-b delay and chipped-grating 6-b delay lines," *IEEE transactions on microwave theory and techniques*, Vol. 45, 1997, pp. 1527–1530.

[156] Corral, J. L., et al. "Continuously variable true time-delay optical feeder for phased-array antenna employing chirped fiber gratings," *IEEE transactions on microwave theory and techniques*, Vol. 45, 1997, pp. 1531–1536.

Chapter 7

FIBER
BRAGG GRATING
SENSORS

7.1 Introduction

From the earliest stage of their development, fiber Bragg gratings have been considered excellent sensor elements, suitable for measuring static and dynamic fields, such as temperature, strain, and pressure [1]. The principal advantage is that the measurand information is wavelength-encoded (an absolute quantity), thereby making the sensor self-referencing, rendering it independent of fluctuating light levels and the system immune to source power and connector losses that plague many other types of optical fiber sensors. It follows that any system incorporating Bragg gratings as sensor elements is potentially interrupt-immune. Their very low insertion loss and narrowband wavelength reflection offer convenient serial multiplexing along a single monomode optical fiber, for which any fiber optic network can be implemented (star, series, parallel, ring) and modified over the long term, thereby increasing flexibility. There are further advantages of the Bragg grating over conventional electrical strain gauges, such as linearity in response over many orders of magnitude, from parts per billion (ppb) to several percent, and many of which that are intrinsic to the properties of optical fibers. Examples such as immunity to electromagnetic interference (EMI), light weight, flexibility, stability, high temperature tolerance, and even durability against high radiation environments (darkening of fibers), make reproducible measurements possible. The small diameter of the optical fiber also makes it compatible with applications for which small diameter probes are required, such as in the human body for temperature profiling. Moreover, Bragg gratings can easily be embedded into materials to provide local damage detection as well as internal strain field mapping with high localization, strain resolution, and measurement range. The Bragg grating is therefore an important component for the development of smart structure technology and for monitoring composite material curing and response; and, indeed, gratings have been tested with civil structures to monitor load levels, offering the promise of real-time structural measurements. Applications for fiber grating sensors are also emerging in process control and aerospace industries.

In this chapter we shall elaborate on the aforementioned advantages and describe a multitude of applications, both current and future, for Bragg grating sensors (BGSs). The key detection issue is the determination of the often small measurand-induced wavelength shifts. We shall present a number of demodulation schemes that have been studied for both dynamic and static measurand fields that may be broadly classed into passive and active (where the grating forms an intrinsic part of a laser cavity) sensor geometries and that are most often applied to point sensors. It is in applications, however, such as distributed sensing, where optical fiber sensors in general and Bragg gratings in particular have the edge over current electro-mechanical sensors. Therefore, we shall outline how BGS arrays may be formed into several architectures that may be interrogated by wavelength or time division multiplexing (WDM/TDM); techniques that have been used with "conventional" optical fiber sensors. If WDM and/or TDM are combined with spatial division multiplexing (SDM), truly large sensor arrays may be realized. That gratings can be written at well-defined wavelengths and operate over an application-specific range has made them most suitable to WDM approaches. Another important sensor category exists where the grating is used simply as a reflective marker, demodulated through conventional optical time domain reflectometry (OTDR). The realization of extremely long gratings has made possible a new BGS implementation of intragrating sensing that offers the promise of highly localized, real-time strain mapping. The general requirements for an ideal interrogation method demand high resolution with a large measurement range, from subpicometer to several picometers for most applications, offering a range to resolution ratio lying within $10^3{:}1\text{--}10^5{:}1$. The interrogating scheme should also be cost-effective and competitive against conventional optical or electrical sensors. Finally, any scheme must be compatible with multiplexing, for reducing overall costs, and taking advantage of the suitability of these sensors for quasi-distributed sensing.

7.2 Sensing External Fields

In Chapter 3 we showed how the strain response of the grating arises from the physical elongation of the sensor, leading to a fractional change in the grating pitch, with a corresponding change in the fiber index because of the photoelastic effect. The thermal response results from the inherent thermal expansion of the fiber material and the temperature dependence of the refractive index. The equations relevant to these terms and their relation to the wavelength shift may be found in Chapter 3. For the measurement of acceleration, ultrasonic waves, and force, (3.10) and (3.11) are still applicable, as these measurands are converted to strain in all practical measurement systems. We expand the discussion from Chapter 3 to include the influence of pressure and dynamic magnetic fields that have recently been measured using Bragg gratings.

7.2.1 Pressure Sensitivity

A pressure change of ΔP leads to a corresponding wavelength shift $\Delta\lambda_p$ of

$$\frac{\Delta\lambda_P}{\lambda_B} = \frac{\Delta(n\Lambda)}{n\Lambda} = \left(\frac{1}{\Lambda}\frac{\partial\Lambda}{\partial P} + \frac{1}{n}\frac{\partial n}{\partial P}\right)\Delta P \tag{7.1}$$

In the case of a single-mode fiber the fractional change in fiber diameter resulting from the applied pressure is negligible compared with the change in the physical length and refractive index for which

$$\frac{\Delta L}{L} = -\frac{(1-2\nu)P}{E} \tag{7.2a}$$

and

$$\frac{\Delta n}{n} = \frac{n^2 P}{2E}(1-2\nu)(2\rho_{12} + \rho_{11}) \tag{7.2b}$$

where E is the fiber Young's modulus. Given that $\Delta L/L = \Delta\Lambda/\Lambda$, the normalized pitch-pressure and the index-pressure coefficients are given by

$$\frac{1}{\Lambda}\frac{\partial\Lambda}{\partial P} = -\frac{(1-2\nu)}{E} \tag{7.3a}$$

and

$$\frac{1}{n}\frac{\partial n}{\partial P} = \frac{n^2}{2E}(1-2\nu)(2\rho_{12} + \rho_{11}) \tag{7.3b}$$

Therefore, the wavelength-pressure sensitivity is given by

$$\Delta\lambda_P = \lambda_B\left[-\frac{(1-2\nu)}{E} + \frac{n^2}{2E}(1-2\nu)(2\rho_{12} + \rho_{11})\right]\Delta P \tag{7.4}$$

Xu et al. [2] have measured $\Delta\lambda_P/\Delta P$ to be 3×10^{-3} nm/MPa over a pressure range of 70 MPa for a Bragg grating written at 1550 nm.

7.2.2 Dynamic Magnetic Field Sensitivity

Bragg gratings have also been used for dynamic magnetic field detection through the Faraday effect to induce a slight change in the fiber index experienced by left and right circularly polarized light in a Bragg grating. A longitudinal magnetic field applied to the grating changes the refractive index for the two circular polarizations, which results in two Bragg conditions being satisfied: $\lambda B_+ = 2n_+\Lambda$ and $\lambda B_- = 2n_-\Lambda$, where the subscripts refer to right and left circularly polarized light at the Bragg grating. The reliance on silica fiber as the host material means that this is an inherently weak effect that is proportional to the low fiber Verdet constant V of ~8×10^5 rad/Gm at 1300 nm. The index change induced by a magnetic field applied to the optical fiber is

$$n_+ - n_- = \frac{VH\lambda}{2\pi} \qquad (7.5)$$

Consequently, the resulting wavelength shift is very small, but it has been detected with an interferometric demodulation scheme by Kersey and Marrone [3]. Fields of 1 to 10^6 Gauss can be measured with high linearity, making this approach suitable for applications in nuclear magnetic resonance (NMR), plasma confinement and spectroscopy.

7.3 Wavelength Demodulation of Bragg Grating Point Sensors

For temperature and strain the key numbers show that a wavelength resolution of ~1 pm is required to resolve a temperature and strain change of ~ $0.1°C$ and 1 $\mu\varepsilon$, respectively, both at an initial Bragg wavelength of 1.3 μm. Therefore, the precision measurement of the Bragg grating wavelength shift is crucial to achieving good sensor performance. This wavelength resolution can be measured with commercially available optical spectrum analyzers (OSA) and tunable lasers, but it is a more challenging goal when using small electro-optic devices. Various attempts have spawned a multitude of grating demodulation schemes presented in the literature. We shall cover much of the key work, both in terms of

Table 7.1   Summary of BGS Interrogation Techniques

	Edge filter	Tunable filter	Interferometric	Tunable laser	CCD-spectrometer
Range to resolution	10^2–10^3	10^3–10^4	10^3–10^4	10^3–10^5	10^3–10^4
Measurement speed	high	high	high	high	high
Long-term stability	good [1]	good [1]	good [2]	good	good
WDM compatibility	low	high	high	high	high
Potential cost	low	medium	low	high	medium

(1) Requires filter stabilization.
(2) Requires a reference grating to compensate for thermal drift in the interferometer.

innovation and performance. It must be borne in mind, however, that although a wide variety of techniques have been demonstrated for monitoring the Bragg wavelength shift, only a few appear to have the potential for being realized as practical, cost-effective instrumentation systems for real-world use. We shall attempt to highlight these schemes (Table 7.1).

For high-resolution wavelength-shift detection, interferometric wavelength discrimination can be used by employing a piezoelectrically tuned, unbalanced Mach-Zehnder interferometer (MZI) that proves most suitable for demodulating high-precision single-grating sensors, although wavelength multiplexing schemes have also been demonstrated. An alternative approach uses a tunable, narrowband filter at the detector, which may be scanned through the entire wavelength range of interest using electronic control. Filters include the fiber Fabry-Perot, acousto-optic, and fiber Bragg gratings and have proven well suited to multiplexing, with arrays demonstrated that contain eight or more grating sensors. In applications where high resolution is not paramount, but for which there is a need for simple wavelength-shift detection, bulk wavelength selective filters or couplers may be used.

7.3.1 Quasi-Static Strain Monitoring

A number of schemes have been demonstrated for the recovery of the wavelength-shift information and where the majority of applications are directed to the measurement of quasi-static strain fields for smart structure applications. The most fundamental means for interrogating a Bragg grating relies on broadband illumination of the device. The bandwidth of the source covers the extent over which the grating will operate when exposed to a perturbation. The narrowband reflected light is subsequently directed to a wavelength detection system. The gratings used in sensor applications typically have bandwidths of 0.05 to 0.3 nm. A number of methods may be used to differentiate any change in the reflected wavelength, including the use of a basic spectrometer, passive or active optical filtering, and interferometric detection.

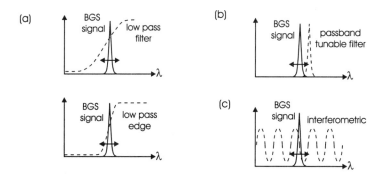

Figure 7.1 Basic optical filtering functions for processing fiber Bragg grating return signals for (a) broadband/edge filters, (b) tunable narrowband filter, and (c) unbalanced interferometers.

7.3.1.1 Passive Broadband Interrogation

Broadband or edge filters provide a wavelength-dependent loss when the cutoff is close to the signal wavelength, offering a linear relationship between the wavelength shifts and the output intensity changes of the filter, as shown in Figure 7.1(a). Comparing light transmitted through the filter with light passed along a reference path recovers the wavelength shift of the sensor (Figure 7.2(a)). This approach provides either very limited sensitivity (broadband filter) or limited range (edge filter); and because it relies on bulk optic components, alignment stability is critical, which reduces portability. Here the measurement range is inversely proportional to the detection resolution. This approach was demonstrated by Melle et al. in 1992, who removed the influences of light source fluctuations and losses in the optical fiber links via a ratiometric detection approach, with the signal intensity I_s and the reference intensity I_R related by [4]

$$\frac{I_S}{I_R} = A\left(\lambda_B - \lambda_0 + \frac{\Delta\lambda}{\sqrt{\pi}}\right)$$

(7.6)

This system offers several advantages, such as low cost and ease of use, and a low resolution of a few tens of $\mu\varepsilon$ has been demonstrated over a range of several mε. It has also been commercialized by Electrophotonics Co. of Canada.

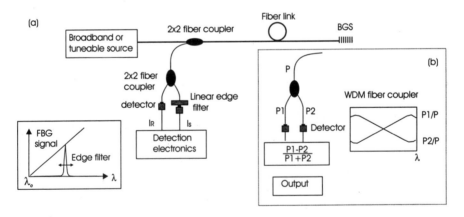

Figure 7.2   (a) Example of the bulk ratiometric scheme (*After*: [4]). (b) A variation based on a fiberized WDM coupler scheme (*After*: [5]).

7.3.1.2 Ratiometric Detection with a WDM Fiber Coupler

All fiber approaches have been demonstrated to overcome the limitations of bulk optic filters. Equivalent fiber optic devices such as fiber WDM fused-tapered couplers provide a monotonic change in the coupling ratio between the two output fiber ports. Taking the ratio of the difference and sum of the two outputs, P_1 and P_2, of the WDM coupler gives a drift compensated output for wavelength-shift detection, as shown in Figure 7.2(b). An acceptable strain and temperature resolution of $\pm$ 5 $\mu\varepsilon$ and $\pm$ 0.5°C, respectively, has been demonstrated using this approach [5]. This scheme offers an all-fiber approach that is compact, has very little power loss, is low cost, and offers a reasonably accurate measurement (~1% of full scale). The multiplexing of sensors using this scheme has yet to be addressed. A further variation of this approach uses a highly over-coupled coupler to increase the sensitivity [6].

7.3.1.3 Interrogation via Scanning Optical Filter

One of the most successful and attractive demodulation approaches is to use a scanning optical filter to track the BGS wavelength changes. Examples such as tunable Fabry-Perot (FP) [7], acousto-optic tunable filter (AOTF) [8], and Bragg grating–based filters [9] have all been demonstrated. In all cases the demodulated output is the convolution of the tunable filter spectrum with that of the grating (Figure 7.1(b)), and the output is optimized when the spectrum of the tunable filter matches that of the grating. Tracking the wavelength change associated with this maximum point leads to the wavelength shift of the BGS. The use of narrowband scanning optical filters has the drawback of sampling a narrow slice of the optical spectrum at a given time; as a result, the measured resolution is strongly dependent on the signal-to-noise ratio of the return signal and the linewidths of the tunable filter and BGS. For example, repeated scanning of a grating array at a frequency f results in an energy E_R reflected by each grating per sampling period that is equal to [1]

$$E_R = \frac{RI\Delta\lambda_g}{f} \tag{7.7}$$

where R is the grating reflectivity, I the spectral brightness of the source, and $\Delta\lambda_g$ the grating's spectral width. The scanning filter further limits the detection energy E_D [1]

$$E_D = \frac{E_R\Delta\lambda_f}{\lambda_s} \tag{7.8}$$

where $\Delta\lambda_f$ is the filter bandwidth and λ_s the width of the scanned wavelength range. Therefore, if the filter's bandwidth is ~1% of λ_s, the detectable energy E_D per scan for close to 100% reflectivity gratings is still only ~1% E_R. This demands that strong gratings and sources be used for good wavelength resolution. Nevertheless, measurements made with gratings with low reflectivities of ~2% to 4% and erbium fiber sources, with average powers of 10 mW, have demonstrated 1-$\mu\varepsilon$ resolution, which together with the large working range is sufficient for most structural monitoring applications. We now present specific examples of the above filter types.

Tunable Wavelength Fiber Fabry-Perot Filter

Tunable FP filters (FPF) have to date been employed in optical fiber communication systems to remove amplified spontaneous emission noise emanating from optical fiber amplifiers at the receiver end. Their stability and ease of use make them ideal for BGS applications. The filters are characterized by band-pass resonances of Lorentzian lineshapes and bandwidths of typically 0.3 nm, with a wide operating range of tens of nanometers, limited by the free spectral range (FSR) between resonances that depends upon the physical mirror separation. Filter finesses F of 120 are typical, $F = \text{FSR}/\Delta\lambda_f$. Filter tuning is achieved by accurately displacing the mirror separation using a piezoelectric (PZ) element and thereby changing the cavity spacing. Currently available FPF can be scanned at rates exceeding 300 Hz, and scan rates of close to 1 kHz should be possible. Figure 7.3 shows how a tunable filter is used to demodulate the wavelength shift from a single Bragg grating, in this case operating in either a tracking (closed-loop) or scanning mode. The former scheme is limited to addressing one BGS, whereas the latter is equally applicable to addressing more than one sensor. In this example, (tracking mode) light from a broadband source illuminates the system and a narrowband component is reflected from the grating and directed to the FPF through a fused-tapered coupler. The filter bandwidth is comparable to the grating bandwidth, whereas the FSR is larger than the operating range of the grating. Kersey et al. [7] have reported a resolution of ~1 pm over a working range in excess of 40 nm for a single BGS. The single sensor implementation locks the FP passband to the grating return signal, with a simple closed-loop arrangement. Dithering of the FP resonance wavelength by a fraction (~0.01 nm) of its passband (~0.3 nm) at a frequency f_m produces a modulation in the optical output of the FPF that contains components at f_m and its harmonics. When the wavelengths of the BGS return signal and FPF transmission peak are aligned, the fundamental component is nulled. The amplitude of the modulation component at f_m serves as an error signal that can be used to lock the FPF passband to the Bragg wavelength, and the control voltage is used to give a measure of the mechanical or thermal perturbation acting on the grating. Figure 7.3 (inset)

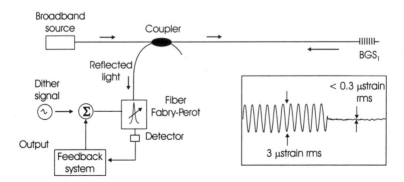

Figure 7.3   Single Bragg grating strain sensor demodulation. Inset: Output response to a 1-Hz strain perturbation of 3-με rms and system noise floor (30-Hz bandwidth) (*After:* [7,10]).

shows the output response to a low frequency 1-Hz strain perturbation of 3-με rms, indicating a strain detectability of better than 0.3-με rms. Operating the FPF in wavelength-scanning mode can be used to address several BGS elements, which are chosen such that the nominal Bragg wavelengths and operational wavelength domains do not overlap and yet fall within the spectral envelope of the source and the FSR of the FPF. For example, a maximum strain load of ± 750–1500 με at 1.3 μm requires a grating spacing of 2–4 nm, with an additional 0.5-nm guard band between gratings. With this constraint a maximum of 12–20 gratings can be written to lie within the bandwidth of a typical broadband source of 30–40 nm. As the filter is tuned, the passband scans over the return signals from the gratings, resulting in the resolution of peaks associated with each grating, as one would observe using a conventional spectrometer. The wavelengths can be determined and recorded from the voltage applied to the filter as the return signals are detected. This approach gives a low-resolution output as the observed peaks are line-broadened by the convolution of the FPF passband and the grating signals. If, however, the dither signal is maintained and the detector output is passed to an electrical mixer and low-pass filter arrangement that detects the output component signal at the dither frequency, a derivative response to the spectral components is obtained. This produces a zero crossing at each of the grating center wavelengths and offers improved measurement resolution [7, 10]. The generation of the scanning voltage for the FPF via a 16-bit digital to analog converter can produce a minimum resolvable wavelength shift of ~0.8 pm, or equivalent strain of 0.8 με. Figure 7.4 compares the strain monitored with a scanning filter demodulated grating and resistive strain gauge (RSG) when both are subject to a strain level of ~2000 με, with resolutions of ~ ± 1 με being possible for a bandwidth from DC to ~360 Hz. The use of optical switches allows for the addressing of several serial arrays; a system for tracking 60 sensors has recently been demonstrated [11], as shown in Figure 7.5. The separate fibers are illuminated using a single 1.3-μm edge-emitting light-emitting-diode (ELED) source of power ~150 μW. The sensors in the arrays are addressed with 50 averages/sensor within a 2.5-second interval. Increasing the number of switches or having a multi-input/output port FPF can substantially increase the number of sensors that are multiplexed.

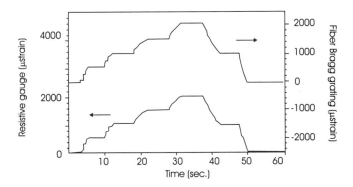

Figure 7.4 Comparison between a resistive strain and a Bragg grating strain sensor (*After:* [1]).

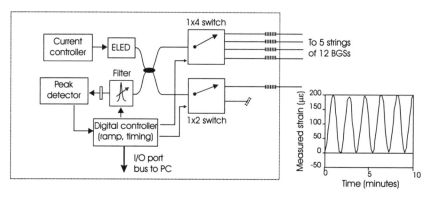

Figure 7.5   Schematic diagram of a 60 channel Bragg grating sensor electro-optic system. Inset: Single channel output to sine-wave of 2-minute periodicity (*After*: [11]).

Figure 7.5 (inset) shows the output of one sensor subject to sine-wave modulation of periodicity 2 minutes, displaying a short-term system resolution of $\pm$ 1 $\mu\varepsilon$, with a temperature-limited drift over 30 minute of $< \pm$ 3 $\mu\varepsilon$.

Tunable Fiber Bragg Grating Filters

The wavelength shift from a BGS element can be tracked with a tunable Bragg grating filter (TBGF) at the receiver end. The receiver-end gratings may be connected to interrogate sensor signals in parallel [9] (in which case each sensor requires a separate detector), or in series (here only a single detector is required) [12, 13]. The receiver-end and sensor gratings are wavelength matched in an unstrained state and have similar, preferably identical, bandwidths. Connecting a receiver-end grating onto a PZ element or another form of fiber stretcher leads to the TBGF. When a voltage is applied across the PZ element, the receiver-end grating is stretched, changing its mean reflecting wavelength. When this is tuned to the sensor grating (using closed-loop signal processing), a voltage output is obtained that is proportional to the wavelength change of the receiver-end grating that is itself equal to the wavelength shift of the sensing grating. The measurement range and resolution are determined by the PZ element used. Jackson et al. [9] first demonstrated this approach in a reflectometric geometry, as depicted in Figure 7.6, with a network of two sensor-receiver-grating pairs. In the reflectometric configuration the receiver grating acts as a notch filter and strongly rejects the light reflected from the sensor element for which it behaves as a matched filter. The matched receiver grating is either "locked" to or scanned across the sensor-grating spectrum and a maximum in the signal strength is detected. Light from a broadband source, in this case an ELED, with a bandwidth of 70 nm, provided 10 μW of launched power into the network described in [9]. The minimum detectable strain under closed-loop signal processing for quasi-static signals was 4.12 $\mu\varepsilon$, which corresponded to the minimum induced strain necessary to destroy the matched condition and is dictated by the linewidth of the grating. Gratings with a linewidth of 0.05 nm can potentially improve the minimum strain resolution to ~1 $\mu\epsilon$. However, as the linewidths of the gratings are narrowed, the level of signal power returned decreases, increasing light loss and limiting resolution. It is relatively straightforward to extend this topology for multiplexing, by increasing the number of sensors wavelength-matched to a fixed number of receiver gratings by using additional sensing fibers with time division multiplexing.

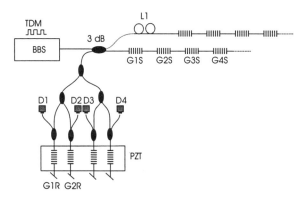

Figure 7.6   Schematic diagram of a reflectometric fiber Bragg grating–based tunable filter scheme (*After*: [9]).

Brady et al. [12] have modified the parallel topology by using instead a serial array of receiving gratings, again in a reflectometric approach. This has the advantage of using the power reflected from the sensing gratings in an efficient manner, while also reducing the number of system components required. The measured strain and temperature resolutions were 2.8 $\mu\epsilon$ and 0.2°C, respectively. Davis et al. [13] have further modified the above approach by using the receiving gratings in an efficient transmissive mode, thereby minimizing the effects of light loss and improving sensor sensitivity (Figure 7.7). The power loss is reduced by a factor of four and, because a minimum is detected at the detector output, this approach is inherently more sensitive. In the demonstration described in [13] light from a superfluorescent fiber source (SFFS) was incident on an array of six gratings operating in the wavelength range 1530–1557 nm. To track the sensor signal changes, the receiving gratings are mounted in PZ actuators and a small dither signal is applied to the actuators. The light transmitted through the filter-grating array is detected by a single output photodetector, whose signal is fed to lock-in amplifiers referenced to the dither signals applied to the filter gratings. The lock-in outputs are summed with the dither signals and are fed back to the filter gratings, thereby closing the loop. This provides an output voltage directly proportional to the applied strains. Clearly, the dither frequency must be applied to each receiving grating, permitting simultaneous measurement from all strained sensor gratings. No static strain resolution was specified but it can be assumed to be

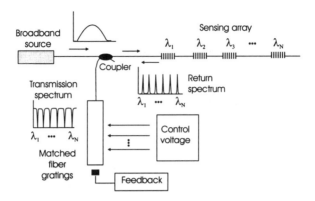

Figure 7.7   Experimental arrangement for transmissive matched-filter interrogation of Bragg grating sensors (*After*: [13]).

comparable or better than the results quoted in [9] and [12]; however, a dynamic strain resolution of ~10 nε/Hz$^{1/2}$ rms was recorded for low-frequency vibrations. The linearity, resolution, and range of the PZ tuning device determine the measurement accuracies for all of these topologies, calling for the use of expensive precision elements if consistent results are to be obtained.

Acousto-Optic Tunable Filters

The acousto-optic tunable filter (AOTF) is a solid-state optical filter in which the wavelength of the diffracted light is selected by applying a corresponding RF drive frequency. The AOTF has a very large wavelength tuning range that can extend to several micrometers, fast access times exceeding 5 kHz, and narrow spectral bandwidth. The device can be operated in a variety of modes, such as a spectrometer, dithered filter, and tracking filter. Therefore, the AOTF appears particularly attractive for wavelength multiplexing very large Bragg grating arrays, with the proviso that a suitable broadband source, or array of sources, is available to cover the same working range. The device uses few components and like the FPF it is suitable for connecting to optical fibers. An electronic feedback loop provides lock-in operation and it may be controlled by a personal computer (PC). Small changes in the AOTF wavelength are a linear function of the change in the applied RF frequency, a convenient parameter to measure allowing for accurate measurements and good long-term stability. Unlike the FPF, the AOTF can be driven at multiple wavelengths simultaneously by applying multiple RF signals of different frequencies, allowing in principle parallel interrogation of many gratings, in addition to TDM offered by using a FP. There are two principal modes of operation: scanning mode and lock-in mode [14, 15]. In scanning mode the PC tunes the AOTF via a voltage-controlled oscillator (VCO) over the wavelength range of interest, with the power reflected from the grating recorded. In lock-in mode the system tracks the instantaneous wavelength

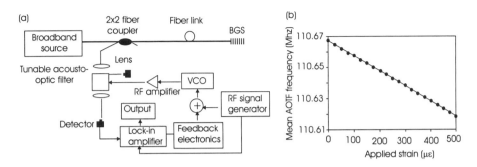

Figure 7.8 (a) Schematic of AOTF Bragg grating demodulation scheme (*After*: [16]). (b) AOTF system response to applied strain (1-second measurement period) (*After*: [15]).

of a particular grating using the feedback loop (Figure 7.8(a)). For a given grating and AOTF bandwidth there is an ideal frequency deviation for maximizing the tracking error signal given by

$$\Delta\lambda_{opt} = \sqrt{\frac{\Delta\lambda^2_{AOTF} + \Delta\lambda^2_g}{8\ln 2}} \qquad (7.9)$$

where $\Delta\lambda_{AOTF}$ is the AOTF bandwidth. Note that (7.9) is independent of the filter transmission, grating reflectivity, and intensity noise. This technique may track the wavelength of multiple gratings with the system initially scanning the wavelength range of interest to identify suitable gratings, in either transmissive or reflective configurations, followed by wavelength tracking a selected grating. The switching time between gratings is ~50 ms.

In a practical demonstration by Geiger et al. [15], the mean AOTF frequency due to applied strain was calibrated against a resistive strain gauge and Bragg grating in close contact. Figure 7.8(b) shows the change in the mean AOTF frequency with strain, corresponding to a scale factor of −96.7 Hz/με. The measurement resolution improves when the counter gating time is increased, thereby measuring the AOTF mean frequency over a longer period, but this is at the expense of the system response time. Therefore, this measurement period determines the system response and is software controlled to allow both fast and slow measurements from the same system. It is found that transmissive configurations deliver the best resolution for high reflectivity gratings of similar bandwidth to the AOTF. For low reflectivity gratings and when the grating bandwidth is much smaller that that of the AOTF, the reflective configuration shows the better noise performance [16]. For example, a transmissive configuration gives a theoretical resolution of ~3.5 nε/Hz$^{1/2}$, compared with ~5.5 nε/Hz$^{1/2}$ for a reflective configuration. Xu et al. [16] measured a standard deviation of 0.4 με for a 5-Hz measurement bandwidth (100-ms measurement period) for a reflective configuration where a fiber-pigtailed ELED operating at 1300 nm, producing ~6-μW output power over a 56-nm bandwidth, illuminated a 90% reflectivity grating having a 0.4-nm bandwidth. This compares favorably with a predicted noise-limited strain resolution of 0.26 µε. The AOTF had a wavelength tuning range of 1.2–1.4 µm and a 3.3-nm bandwidth.

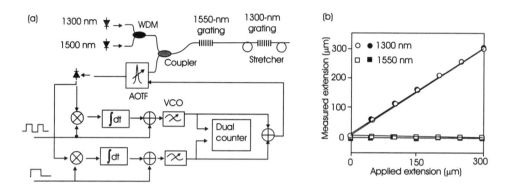

Figure 7.9   (a) Schematic diagram of AOTF electronics demultiplexing scheme. (b) Results of optical and electronic demultiplexing techniques (open) optical, (filled) electronic, squares: 1550 nm, circles: 1300 nm (*After*: [17]).

One of the most notable advantages of the AOTF over all other filters is the possibility of driving the device at multiple RF signals to allow for true parallel processing of multiple wavelength signals using a single filter and detector. To that end Volanthen et al. [17] have demonstrated the simultaneous monitoring of two fiber Bragg gratings written at 1300 and 1550 nm using a single AOTF (Figure 7.9(a)). Two demodulating schemes were presented. The first, termed "optical demultiplexing technique", separates the return signals from each grating using optical filtering directly after the AOTF. In this case a separate optical receiver provides an input to a separate lock-in loop and detects every wavelength band. The second scheme is termed "electronic demultiplexing" and uses a single receiver that provides the same amplitude information but different dithering frequencies to both lock-in loops that lock the mean interrogation wavelength to that of their chosen grating. The accuracy of the optical demultiplexing method was measured by monitoring the standard deviation of the output frequency of each loop with a constant strain applied to both gratings. At 1300 nm the rms error was 3 µε, whereas at 1550 nm the error was 6 µε, both for a 100-ms measurement period. Similar experiments were performed using the electronic demultiplexing approach and produced rms errors of 2 µε at both wavelengths. Figure 7.9(b) compares the results obtained via both demultiplexing approaches by straining the 1300-nm grating and recording the outputs corresponding to both wavelengths. Very low cross-talk levels of −52 and −56 dB between different wavelengths were observed for the electronic and optical demultiplexing techniques, respectively. The electronic technique is limited by the available dithering frequencies that maintain low cross-talk within the receiver bandwidth, whereas the optical technique is limited by the availability of optical filters. When combined together, however, these approaches can offer demultiplexing to a very large number of Bragg gratings.

We close by noting that the AOTF performance can be dramatically improved by using a far smaller bandwidth. For example, Dunphy et al. [18] have presented an AOTF with a bandwidth of 0.2 nm that offers <1-pm resolution, over a smaller operating range of ~120 nm. The AOTF reported in [16] (bandwidth 3.3 nm) had a temperature dependence of 2.68 kHz/°C or 0.03 nm/°C; therefore, a temperature change of 1°C produces an equivalent strain change of ~28 με. However, the temperature stability of very high-resolution devices has yet to be investigated.

Other Tunable Filter Types

Recently, Coroy and Measures [19] have reported a wavelength measurement system based on the use of a semiconductor quantum well electroabsorption filtering detector (QWEFD). The absorption edge of the QWEFD is tuned, in a manner similar to that reported for AOTFs, using the quantum confined Stark effect as a wavelength tunable filter [20, 21]. A strain resolution of ± 8 με, or ± 9.7 pm, was measured at 1550 nm. The QWEFD has a power-dependent response and the system in [19] was able to measure the Bragg grating wavelength independent of the input power levels for signals as weak as 3 nW. This device offers the potential of small, robust, and low-cost wavelength-shift demodulation for BGSs.

7.3.1.4 Bragg Grating Interrogation Using Wavelength Tunable Source

A calibrated, narrow-linewidth, single-frequency, continuously wavelength-tunable erbium fiber laser has been used to interrogate an array of three Bragg grating temperature sensors at 1550 nm [22]. The laser cavity was formed between two 98% Bragg gratings and was pumped to produce 100 μW of output power in a 20-kHz linewidth (Chapter 6). A PZ transducer was used to alter the cavity length and therefore the lasing wavelength. An immediate advantage of this approach is the improved signal-to-noise ratio as the measurement determines a maximum in grating reflected power and it dispenses with the need for optical filtering. The system operating range, however, is limited by the laser wavelength-tuning range of ~2.3 nm, corresponding to a maximum temperature excursion of 180°C, and the frequency accuracy is limited by the precision of the PZ transducer, in this case to 2.3 pm or ~0.18°C. As with tunable filter demodulation the PZ transducer properties limit system bandwidth. A commercially available gain-coupled distributed feedback tunable laser (Nortel) has been recently used to provide continuous tuning from 1536.5 to 1544.5 nm. Strain measurements over 2000 με gave a resolution of ± 0.76 με in a 10-ms sampling period (± 0.076 με/Hz$^{1/2}$) [23]. As with the AOTF there is a trade-off between measurement range and bandwidth for a fixed acquisition rate, in this case an improved resolution to ± 0.024 με/Hz$^{1/2}$ over a range limited to 200 με.

7.3.1.5 Recovery of Bragg Grating Wavelength Shift Using CCD Spectrometer

Thus far we have presented schemes based on various forms of scanning optical filters, for which the optical throughput power penalty is the major demodulation drawback. Laser-based interrogation can circumvent this but is limited by tuning range and system bandwidth. These issues are overcome through the use of parallel detection of the entire recovered-light spectrum using a charge-coupled device (CCD) (Figure 7.10(a)). A dispersive optical element, most often a finely ruled bulk, diffraction grating directs the incident light across a linear array of detector pixels. This approach collects all the light returned by each fiber Bragg grating over the entire scan period, hence $E_D = E_R$, and a 1% grating provides as much signal through parallel detection as does a 95% grating via scanned detection. The spectrometer returns the wavelength with a resolution equal to the product of the grating's linear dispersion at the detector plane with the pixel width, defined as nanometers per pixel (nm/pixel), with the received wavelength converted into position information along a line imaged onto the array of detector elements. For example [1], by dispersing a 24-nm bandwidth over a 256-pixel CCD, as many as 22 fiber gratings spaced by 1 nm can be resolved, with > 0.4-nm overlap-free range (corresponding to ~600-$\mu\varepsilon$ range). The center-to-center pixel spacing often corresponds to ~0.1 nm, whereas a strain resolution of 1 $\mu\varepsilon$ requires a wavelength resolution of better than 7×10^{-4} nm or less than 1/100 of a pixel. As the image from each Bragg grating is spread across many adjacent pixels, a weighted average of the center wavelengths of the illuminated pixels scaled according to the incident signal level gives a computed center of the reflected grating wavelength, the centroid detection algorithm [24]. Using this approach, strain sensitivities of better than 1 $\mu\varepsilon$, with signal averaging, at repetition rates exceeding 3.5 kHz have been demonstrated with 20 1% to 3% grating reflectors, illuminated by several hundred microwatts (μW) of light from a broadband source. CCDs are Si-detector-based, developed to work below 1 μm, therefore, their use requires the coupling of optical fibers

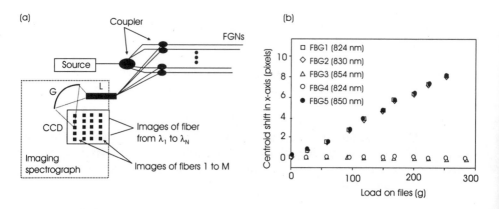

Figure 7.10 (a) Schematic of fiber Bragg grating demodulation employing CCD spectrometer-based system. (b) Shift of centroid position as a function of grating loading (*After*: [26]).

with bulk optic spectrometers. Fortunately, spectrometer technology has rapidly decreased in size to the point that PC-based spectrometer cards are available. Although these do not apparently offer the wavelength resolution for the most demanding grating applications, software post-processing algorithms can dramatically improve performance, making this form of demodulation highly attractive for many applications. For example, in aerospace applications mechanically scanned instrumentation or passive optical filters that are highly susceptible to environmental fluctuations are prohibited. The use of CCD spectrometer is a form of SDM/WDM.

Conventional spatial multiplexing splits sensors into many fiber channels and employs separate electronic signal processing units for each channel. This is not strictly multiplexing and there are inevitable penalties associated with the use of multiple single sensors in terms of cost, durability, and weight. Davis et al. [25] have demonstrated an optical fiber switch that circumvents this problem by allowing for separate sensors on multiple fiber channels to share the same processing electronics. The use of a mechanical switch, however, severely limits the speed of the system and its applicability; for example, the reported system could address a large array of 60 sensors but at a sample rate of 0.4 Hz. This sampling rate is too low for sensor systems monitoring, say, aerospace structures. Chen and co-workers [26, 27] have reported combined digital spatial domain multiplexing (DSDM) with traditional WDM to interrogate BGSs, allowing for high-resolution measurements with a good sample rate for each sensor. Figure 7.10(a) shows the system, with light from a broadband source incident upon a number of BGS, through a star coupler. BGSs on the same fiber have different predetermined wavelengths, whereas BGSs on different fibers can have the same wavelengths. Light reflected from the BGSs is sent down different down-lead fibers to the processing unit, where a bulk grating (fabricated on the surface of a mirror lens) produces a real image of the fiber array on a two-dimensional CCD pixel array. The output of the CCD is digitized by a frame grabber and processed by a computer. The light reflected at different wavelengths is diffracted into different directions onto the pixels of the CCD array. Therefore, if the system has M fiber channels and N BGSs of different wavelengths along each fiber, one produces an MxN array, with spatial positions of the fiber channels encoded, for example, along the y-axis of the CCD, while wavelengths are encoded along the x-axis.

An experimental system employed two fiber channels of five gratings in total, a CCD camera of 512x512 pixels, and a frame rate of 30 frames per second, the output of which is shown in Figure 7.10(b). To obtain subpixel resolution, the centroid of each peak is calculated, with the system resolution being substantially improved from 125 $\mu\varepsilon$ for a one-pixel shift to ~1.5 $\mu\varepsilon$, at 850 nm. This system is potentially capable of accommodating 25 channels (the size of each light spot is < 20x20 pixels), with seven sensors in each channel for a strain sensor range of $\pm$ 3500 $\mu\varepsilon$, leading to a potential multiplexing capacity of 175 BGSs, with each being sampled at 30 Hz. This is superior to that achieved with any other form of multiplexing technique. Ezbiri et al. [28] have also presented a readout system capable of repeatable, high-resolution measurements using a PC-controlled spectrometer. For the low-cost spectrometer of resolution 0.1 nm, initial results indicate that a resolution of ~1 pm can easily be achieved with post-processing, translating into a sensing resolution of ~1 $\mu\varepsilon$ for gratings at 820 nm.

Peak-Wavelength-Detection Algorithms

A number of algorithms can be used to locate the peak of the grating-reflected wavelength when the grating image is spread over several adjacent pixels. These include the two-pixel-ratiometric technique, whereby the grating spectrum is divided into two areas along a vertical line; the ratio of the two areas is then used to track the evolution of the peak wavelength. This technique is optimized when the grating's bandwidth covers a maximum of two pixel's spectral windows and the wavelength shifts to be monitored are smaller than the pixel spectral window.

The centroid detection algorithm (CDA) uses the following algorithm to determine the Bragg wavelength [24]:

$$\lambda_B = \frac{\sum_j \lambda_j I_j}{\sum_j I_j} \tag{7.10}$$

where I_j and λ_j represent the intensity and center wavelength of the jth CCD pixel, respectively. This expression represents a moving average that points to the Bragg grating maximum to within a fraction of the pixel spectral window. The number of pixels chosen to perform the processing should both offer stable wavelength resolutions inside the pixels as well as prevent discontinuities at the pixel boundaries. The improvement to pixel resolution is limited by the optical power reflected from the grating; therefore, gratings having different reflectivities exhibit different accuracies [24]. Alternatively, a least squares method (LSM) may be implemented, fitting a quadratic polynomial to sequential pixel outputs. In this case a polynomial of second order $\sum c_k \lambda_j^k$ approximates the peak region of the Bragg grating reflection such that the error ε_j in approximating the jth point satisfies [28]

$$\frac{\partial \left(\sum_j (\varepsilon_j)^2 \right)}{\partial c_k} = 0 \quad k = 0, 1, 2... \tag{7.11}$$

where ε_j is given by $\varepsilon_j = I_j - \sum c_k \lambda_j^k$ for which I_j is the measured intensity. Figure 7.11(a) shows the ability of each technique to return the exact Bragg wavelength as a function of the number of pixels used in the computation. The LSM maintains a quasi-flat response as the pixel number varies, whereas the CDA proves to be more error prone, leading to a large deviation from the initial wavelength. Ezbiri et al. [28] have tested the suitability of both algorithms to locating the peak Bragg wavelength. An ELED source (20-nm bandwidth, 820-nm center wavelength) illuminated several Bragg gratings with 0.6-nm bandwidths corresponding to ~9 CCD pixels. The light returned from the gratings was imaged onto a linear 1024-element CCD array, following diffraction from a 1800 groove/mm bulk grating (70-nm operating range, centered at 835 nm). Cooling cycle data for the LSM and

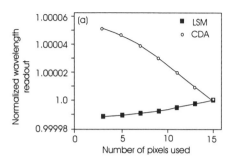 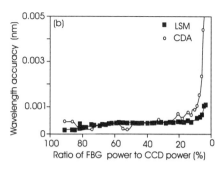

Figure 7.11 (a) Normalized Bragg wavelength readout versus number of pixels used (upper trace: CDA; lower trace: LSM). (b) Relation between power and optical wavelength resolution of the CDA and LSM methods (*After*: [28]).

CDA approaches returned good comparative results with a deviation from the decay curve in both cases of <2 pm. The best performance was achieved using 21 pixels for the CDA and 9 pixels for the LSM techniques. Figure 7.11(b) shows the relation between the ratio of grating reflected power with CCD saturation power and optical wavelength resolution, displaying that both approaches offer subpicometer resolution down to 50% of saturation power. After that point the CDA technique begins to degrade exponentially, whereas the LSM technique gives a flat response down to ~10% of saturation power. The resolution readout instrument based on the LSM method can therefore be maintained for a much wider range of available power, providing the most stable response of the two approaches. Consequently, the requirement for powerful sources and tailored gratings is minimized. These data indicate that high-resolution wavelength readout can be performed with a low-cost spectrometer by deconvolving the Bragg grating spectrum and the pixel resolution. For example, with a 0.1-nm resolution, shifts of ~1 pm can be measured with power levels of ~1 nW per grating at 25 Hz.

7.3.1.6 Analysis of Bragg Grating Wavelength Shift Using Fourier Transform Spectroscopy

Another form of direct spectroscopic analysis of the Bragg grating wavelength is through Fourier transform spectroscopy (FTS), as reported by Davis and Kersey [29, 30] (Figure 7.12). Light from an array of gratings illuminates a Michelson interferometer in which the length of one arm is linearly scanned to change the relative optical path lengths. When the optical path difference is zero, a beat signal derived from the optical components is generated at the detector. In this way each grating, reflecting at a discrete wavelength, produces a distinct signal at audio frequencies. This frequency component is then modulated by the application of an external perturbative field. The Michelson optical path difference (OPD) is scanned through the full coherence length of the grating's reflection spectrum to yield the interferogram. Physical path scans of 10 cm were induced,

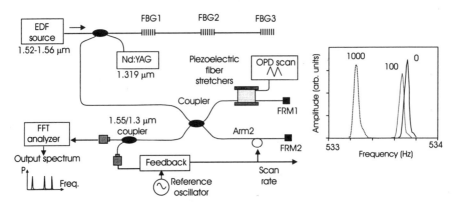

Figure 7.12   Schematic of the fiber Bragg grating sensor system demodulated via fiber Fourier transform spectroscopy. Inset: spectrum recovered for 0, 100, and 1000 με (*After:* [29, 30]).

corresponding to an OPD of ~30 cm (i.e., $2n\Delta L$). When the interferometer OPD is linearly scanned over a range $\pm D$, an interferogram is generated at the output symmetrically around zero OPD. The width of the interferogram depends on the coherence length l_c of the incident light, and the Fourier transform of the interferogram reproduces the optical spectrum of the light. The effective coherence length of a typical Bragg grating at 1550 nm is ~0.5 to 1 cm (for bandwidths of 0.2 to 0.1 nm, respectively). Therefore, the interferometer OPD must be scanned over a minimum of 2 cm to observe the interferogram. So as not to be limited by the bandwidth of the grating, even longer scans are necessary, with $D \gg l_c$. To maintain the linearity of the scan a narrow-line single-frequency Nd:YAG laser, operating at 1319 nm, is used to monitor the rate at which fringes are produced, correcting the scan using a feedback loop when necessary. Further complication results from birefringence-induced polarization fading that randomly apodizes the signal, necessitating the use of Faraday rotator mirrors. Figure 7.12 (inset) shows the Fourier transform of the interferogram and the shift in the component for strain levels of 100 με and 1000 με applied to a grating element with a nominal Bragg wavelength of 1548.2 nm. In [30] the resolution of the FFT analyzer was ~6 mHz, or a fractional change in frequency of ~$1:10^5$ that represents an equivalent wavelength shift resolution of 0.015 nm, that translates to a strain resolution of ~12 με at 1550 nm. This scheme overcomes the 2π measurement range limitation of the interferometric wavelength-shift detection (Section 7.3.2.1) and offers reasonable wavelength resolution. It is also suitable for multiplexing, with each wavelength corresponding to a different frequency component; however, it is complicated and expensive compared with the schemes discussed thus far. Of course, the FFT can be implemented using a computer. Finally, this technique offers sufficient resolution that it could potentially be used to monitor the distortion of the actual grating profile under loading, so-called intragrating sensing.

Flavin et al. [31] have reported an alternative short-OPD scan, covering only the central part of the interferogram. The grating center wavelength is determined by measuring the periodicity of the interferogram with respect to the group-delay scan rate in the interferometer, obtained from a high coherence interferogram derived from a He-Ne laser. The analytic signal of both sampled interferograms is calculated with a phase value assigned to every sampled point. The periodicities of these interferograms can be compared to a fraction of a period over the entire scan, without the need for a uniform OPD scan. Consider a scanning interferometer illuminated by light reflected from a grating of center wavelength λ_1. When the effective coherence length of the reflected light exceeds the OPD of the scanning interferometer, an interferogram is generated of the form [32]

$$I(\tau) = I_0 + I_1(\tau)\cos\varphi_1(\tau) \tag{7.12}$$

where $I_1(\tau)$ is a slowly varying envelope function, τ is the difference in group delay between interferometer arms, and the phase $\varphi_1(\tau) = \omega_1\tau$, where $\omega_1 = 2\pi c/\lambda_1$. The grating center wavelength can be found from the periodicity of this interferogram, which can be measured by finding an analytic signal $A_1(\tau) = I_1(\tau)\exp(\varphi_1(\tau))$ of the interferogram, extracting and unwrapping the phase $\varphi_1(\tau)$ to remove the 2π discontinuities and finding the best fit gradient of $\varphi_1(\tau)$ versus τ [31]. The analytic signal is the Hilbert transform of the AC part of the interferogram

$$A_1(\tau) = H[I(\tau) - I_0] \tag{7.13}$$

This gives an extremely high-resolution measurement of the Bragg grating center wavelength for a short-OPD scan, without the need for a precise, constant velocity OPD scan. This approach, however, cannot be applied directly to a multiplexed system of gratings. In that case $A_1(\tau)$ is the superposition of signals from the individual gratings

$$A(\tau) = \sum_i A_i(\tau) = \sum_i I_i(\tau)\exp(\varphi_i(\tau)) \tag{7.14}$$

The individual phase functions $\varphi_i(\tau)$ cannot be easily extracted form the composite signal; however, separating the signals at an intermediate stage allows for the accurate measurement of the center wavelengths. The numerical computation of the Hilbert transform is achieved by taking the FT of a real signal, zeroing the signal at zero frequency and all negative frequencies, and then taking the inverse FT, giving a complex signal whose real part is the original real valued signal. Individual gratings are represented as sharp lines in this spectrum and so can be selected by windowing in the appropriate region. Applying an inverse FT of the windowed signal leads to a complex signal that is the analytic signal of the component of the original interferogram due to a single grating. Each of the analytic signals can thus be derived independently, with each grating wavelength found by unwrapping the phase and finding the gradient $\omega_i = d\varphi_i/d\tau$ by linear regression, as per Figure 7.13(a). The measurements of λ_i are obtained from $d\varphi_i/d\tau$, and one requires the high-resolution measurement of the group delay τ. This is achieved by using a stable laser source that produces a reference interferogram in the Michelson during the OPD scan, and the phase φ_R is extracted from the analytic signal. If λ_R is stable and known to high

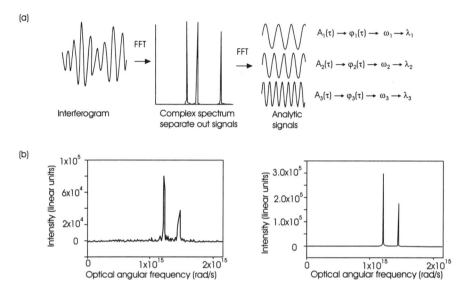

Figure 7.13   (a) Outline of short scan Fourier transform technique. (b) Recovered FT spectra without and with group delay calibration (*After*: [32]).

precision, an accurate delay can be attributed to each sampled value of the interferogram by using $\tau = \varphi_R \lambda_R / 2\pi c$. These delay values also apply to the sample values of the multiple grating interferogram, as both signals are sampled simultaneously. A nonuniform OPD scan does not affect the case of a single grating; however, in the case of multiplexed arrays, the nonuniformity will cause a broadening of the spectral components recovered after the first FT, limiting the ability to separate the individual grating components. This can be overcome by resampling the calibrated interferograms using a cubic spline interpolation.

Measurements have been made on gratings at 1556 and 1301 nm [32], illuminated by ELEDS, while a He-Ne laser provides a reference interferogram and the interferograms are scanned through a total OPD of 1.2 mm, a delay of 4 ps. The group delay of the sampled interferograms must be calibrated and resampled in equal delay increments. Figure 7.13(b) shows examples before and after calibration, with scan rate variations initially distorting the peaks. Individual peaks are then FT and a least squares fit of the analytic signal phase φ_i against delay τ yields ω_i and hence λ_i. The frequency accuracy of this technique depends ultimately on the stability of the He-Ne interferogram that gives an effective wavelength error of 0.02 pm, which is well below the measurement resolution and is therefore negligible. The center-grating wavelength is derived from the phase of the analytic signal of the interferogram, obtained by Fourier analysis. Signals from different gratings are separated out in the frequency domain. For an OPD scan of 1.2 mm, measurements of the grating center wavelength are independent to within 0.007 nm, which is similar to the 0.005-nm single grating resolution, which compares well to the FTS technique, providing an improved resolution for a shorter OPD scan. When applied to the measurement of absolute strain, the wavelength resolution corresponds to $\pm$ 3.5 $\mu\varepsilon$. Scanning about a point of non-zero path imbalance may allow for higher wavelength resolution or shorter OPD scans, while retaining the advantages of high accuracy and software-based grating demultiplexing.

7.3.1.7 Mode-Locked Fiber Laser

Grating sensor demodulation has also been implemented with broadband, ultrashort pulses generated with a passively mode-locked fiber laser (Figure 7.14(a)) [33, 34]. Nonlinearly chirped pulses of tens of picosecond duration and corresponding to an 85-nm optical bandwidth are obtained through self-phase modulation. Average output powers of ~60 mW are available from 1500 to 1650 nm and the repetition rate of 6.48 MHz implies average pulse energies of close to 10 nJ. The pulsed nature of the source is not relevant to its operation, with the grating sensor system responding much as it would were it to be illuminated with a broadband, continuous source (Figure 7.15(a)). A demodulation scheme, however, has also been developed to specifically take advantage of the short pulse output (Figure 7.14(b)). The pulses are injected into a 3.25-km length of highly dispersive fiber (dispersion of 83 ps/nm km at 1550 nm) before being incident on the sensor array. The reflected light traverses back through the highly dispersive fiber and is sampled with a high-speed detector and sampling oscilloscope. The highly dispersive fiber serves as a wavelength-dependent optical delay line, with strain- and temperature-induced wavelength shifts being converted into a shift in the arrival time of the pulse at the detector (Figure 7.15(b)). A 3.67-nm shift in the peak wavelength produces a 1.96-ns advance in the pulse arrival time. Therefore, in the time domain an array of Bragg gratings would generate a sequence of pulses separated by the time of flight between the gratings, with an additional wavelength-dependent delay resulting from the double-pass through the highly dispersive fiber. If the physical spacing between the gratings is constant, only changes in the Bragg wavelength will shift the relative time of the reflected pulses. Sensitivities of $\pm$ 20 $\mu\varepsilon$ have been demonstrated over a strain range exceeding 3500 $\mu\varepsilon$ (7-nm wavelength range), primarily limited by the total fiber dispersion of only -540ps/nm and the timing jitter of the scope. Higher dispersion fiber or a 30-cm chirped Bragg grating could lead to 1-$\mu\varepsilon$

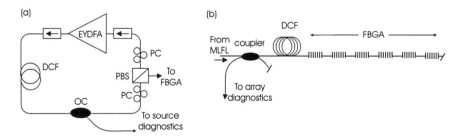

Figure 7.14   (a) Passively mode-locked fiber laser source, using an Er:Yb co-doped fiber amplifier and dispersion compensating fiber (D = -95 ps/nm km). (b) Fiber Bragg grating array interrogation scheme (*After*: [34]).

sensitivity. The time-domain approach also removes the limitation of wavelength-domain demodulation that requires that the gratings never overlap in wavelength. The time-domain demodulation measures only the induced delay; therefore, wavelengths of different gratings may cross over or overlap in wavelength. Finally, one should be able to derive both the local strain at the gratings and the integrated strain between the gratings by comparing standard OTDR with the dispersive OTDR. The mode-locked fiber laser source offers useful advantages over superfluorescent fiber sources, superluminescent diodes, and tunable narrow-line lasers, with possible bandwidths of 200 nm that could cover the complete low-loss communications window. The source is bright (>0.5 mW/nm), and the large pulse energy and high repetition rate permit real-time damage detection, such as explosions and mechanical failure of structural elements, when time-domain demodulation is used. Broadband square-pulse operation of a passively mode-locked fiber laser for Bragg grating interrogation has also been demonstrated for bandwidths greater than 60 nm [35].

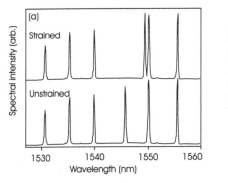 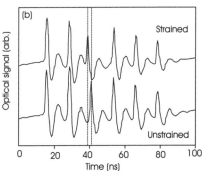

Figure 7.15 Response to 1850-με applied strain with (a) 3.67-nm spectral shift of strained grating. (b) Corresponding 1.96-ns temporal shift resulting from propagation through a dispersive delay (*After*: [34]).

7.3.1.8 Other Lasers

Yun and co-workers [36] have demonstrated a 0.1-nm linewidth, wavelength-swept erbium-doped fiber laser scanned over a 28-nm range used to demodulate a BGS array with a strain resolution of 0.47 με rms at 250 Hz (42 nε/Hz$^{1/2}$). Finally, there is also the potential of using super-continuum sources, for which 140-nm bandwidths have been demonstrated using a mode-locked semiconductor laser source [37].

7.3.2 Dynamic Strain Sensing

One can of course use fiber interferometers, such as the Michelson and Mach-Zehnder, that convert the wavelength shift from the grating into a phase change at the interferometer output. This too is a form of filtering, in this case with a transfer function of the form 1 + cos(φ) (Figure 7.1(c)). This approach is open to many signal-processing schemes that offer very high phase resolution measurements. However, the detection of quasi-static (absolute) strains is precluded by temperature-induced phase drift in the interferometer itself and by the limited unambiguous range, corresponding to a phase change of 2π radians in the interferometer, with the consequent loss of the absolute wavelength measurement. Although schemes have been presented that circumvent these issues with varying degrees of success (for example, Section 7.3.2.2), interferometric wavelength-shift detection remains most suited to recovering dynamic phase signals, with a dynamic strain range limited to 100 dB.

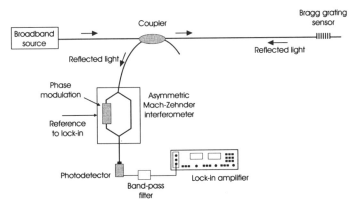

Figure 7.16   Grating sensor system using fiber interferometer wavelength discriminator for dynamic strain measurements (*After*: [38]).

7.3.2.1 Interferometric Wavelength-Shift Detection

In this approach, light from a broadband source is coupled to the BGS and reflected off to an unbalanced optical fiber Mach-Zehnder interferometer (MZI). Figure 7.16 shows the principle of this technique. The reflected light in effect becomes a tunable light source into the interferometer, with wavelength shifts induced by perturbation of the Bragg grating sensor resembling a frequency- (wavelength-) modulated source. The unbalanced interferometer serves to filter the light with a raised cosine function; the wavelength dependence on the interferometer output can be expressed as [10, 38]

$$I(\lambda_B) = A[1 + k\cos(\psi(\lambda_B) + \varphi)] \tag{7.15}$$

where $\psi(\lambda_B) = 2\pi nd/\lambda_B$ and A is proportional to the input intensity and system losses, d is the physical length imbalance between the fiber arms, n is the effective index of the core, λ_B is the wavelength of light reflected from the BGS (sensor signal), k is the interference fringe visibility, and φ is a bias phase offset of the MZI resulting from slowly varying and random environmental perturbations acting on the product nd. This concept has been extensively used for interferometric fiber sensors to introduce a phase carrier signal into an unbalanced interferometer using direct laser emission modulation for the purpose of phase demodulation [39]. In this case the converse is done and the interferometer is used as the discriminator, detecting the wavelength shifts of the effective source formed by the strained grating element.

For a dynamic strain-induced modulation in the reflected wavelength $\Delta\lambda_B \sin\omega t$ from the BGS, the change in the phase shift $\Delta\psi(t)$ is [10, 38]

$$\Delta\psi(t) = -\frac{2\pi\Delta L_{MZI}}{\lambda_B^2}\Delta\lambda_B \sin\omega t = -\frac{2\pi\Delta L_{MZI}}{\lambda_B}\xi_g\Delta\varepsilon \sin\omega t \tag{7.16}$$

with $\Delta L_{MZI} = nd$, the optical path difference (OPD) of the Mach-Zehnder scanning interferometer and $\Delta\varepsilon$ is the dynamic strain subjected to the grating. ξ_g is the normalized strain-to-wavelength-shift responsivity of the Bragg grating, with

$$\xi_g = \frac{1}{\lambda_B}\frac{\delta\lambda_B}{\delta\varepsilon} \tag{7.17}$$

Of course, (7.16) and (7.17) are equally applicable to the recovery of temperature-induced wavelength shifts, with $\Delta\varepsilon$ replaced by ΔT. By measuring $\Delta\psi(t)$ with pseudo-heterodyne processing, the strain or temperature change can be determined. It is clear that the phase sensitivity in response to strain or temperature is directly proportional to the OPD in the scanning interferometer. It is also apparent that the appropriate choice of interferometer OPD can lead to an extremely high sensitivity to dynamic strain-induced Bragg wavelength shifts. As an example, consider an interferometer having an OPD of 1 cm (~6.7-mm fiber length, $n = 1.5$), the output wavelength-to-phase conversion factor is ~26 rad/nm at 1.55 μm (using (7.16)). The grating strain response is ~1.2 nm/mε at 1.55 μm, therefore giving a strain-to-phase system response of ~0.031 rad/με. Interferometric systems can routinely make phase shift measurements down to μrad/Hz$^{1/2}$, for which a strain resolution of ~0.020 nε/Hz$^{1/2}$ (20 pε/Hz$^{1/2}$) is possible. The strain-induced phase shift may be determined by holding the interferometer in quadrature using a low gain feedback loop. Dynamic strain-induced phase shifts are then detected at the output of a difference amplifier to provide balanced detection of the interferometer output. Alternatively, pseudo-heterodyne signal processing may be implemented by ramping the phase in one arm of the MZI over 2π radians (serrodyne modulation), thereby generating an electrical carrier at the receiver. With the phase deviation set to 2π radians (depending on the fly-back time) the output signal of the difference amplifier after band-pass filtering at the fundamental ramp modulation frequency is of the form

$$S(\lambda) = A\cos(\omega_0 t + \psi(\lambda)) \tag{7.18}$$

Strain variations of the grating lead to phase modulation of the carrier $S(\lambda)$ that can be detected using a lock-in amplifier. In the case of harmonic variations the carrier modulation leads to sidebands in the frequency domain, the amplitudes of which are given by Bessel functions of the first kind. This approach is particularly suited for tracking large phase shifts resulting from high quasi-static strain levels.

One caveat of interferometric wavelength-shift detection is that the interferometer path difference must be less than the effective coherence length of the light reflected from the grating if temporal coherence is to be maintained between the interfering beams traversing the two arms of the MZI. Typically, an effective grating coherence length of 1 cm results for a 0.3-nm grating operating at 1.55 μm. The operational system range is inversely proportional to the OPD of the MZI and is set by the FSR of the scanning interferometer

$$FSR_{MZI} = \frac{\lambda_B^2}{\Delta L_{MZI}} \qquad (7.19)$$

Hence, there is a trade-off between sensitivity and operational range. In this method of wavelength signal recovery the unambiguous measurement range is limited to a 2π change in the scanning interferometer. Figure 7.17 shows the strain to phase shift conversion responsivity for different values of the OPD between the interferometer arms. To enable absolute, unambiguous sensing with this approach, the OPD would have to be set such that at the maximum anticipated strain level applied to the grating sensor the induced phase shift is less than $\pm \pi$ radians. For example, with a 300-μm OPD the interferometer wavelength to phase shift responsivity is 0.785 rad/nm, resulting in a strain to phase shift responsivity of 9×10^{-4} rad/με for a 1550-nm grating. For a limit of $\pm \pi$ rad, this would lead to a maximum strain range of ± 3500 με.

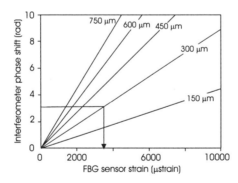

Figure 7.17 Strain to phase shift conversion for an integrated optic Mach-Zehnder detection scheme for various OPDs (*After:* [10]).

Kersey et al. [38] demonstrated interferometric wavelength-shift detection for dynamic strain measurements. An Er-doped superfluorescent fiber source producing ~300 μW of output power within a 35-nm bandwidth centered at 1548 nm illuminated the optical network. A BGS at nominal wavelength of 1545 nm was contacted to a PZ transducer for the application of known dynamic strains signals. The returned light was coupled into a

MZI of 10-mm fiber path imbalance that was held in quadrature by feedback applied to a PZ transducer in one arm of the Mach-Zehnder to compensate for low frequency drifts. When a 2π-phase shift was applied to one arm of the MZI using a PZ cylinder at 500 Hz (in place of the feedback signal), the output visibility was ~0.3, consistent with the interferometer path imbalance and effective coherence length of the 0.2-nm bandwidth of the grating. A signal-to-noise ratio of 46 dB in a 1-Hz bandwidth was recorded for a dynamic strain signal of 0.12 $\mu\epsilon$ rms, applied to the fiber grating at 500 Hz. The minimum detectable strain was ~0.6 $n\epsilon/Hz^{1/2}$ for frequencies above 100 Hz to >100 kHz. This corresponds to an average wavelength shift of the grating of 1.5×10^{-4} nm rms, or ~20 MHz rms, and the noise floor corresponds to a shift of ~7.5×10^{-7} $nm/Hz^{1/2}$ rms, or ~100 $kHz/Hz^{1/2}$ rms in optical frequency, or better than 1 part in 10^9. The actual phase noise floor was ~20 $\mu rad/Hz^{1/2}$ at frequencies above 100 Hz, limited by detector noise. The increase in noise below 100 Hz is accounted for by the MZI sensitivity to environmental perturbations. Pseudo-heterodyne signal processing has also been investigated for this system [40]. This approach is clearly sensitive to dynamic strain changes; however, it is less suited to quasi-static or static strain measurements due to drifts in the interferometer bias phase $\varphi(t)$. This can be compensated for by incorporating a local reference grating to synthesize a reference wavelength that directly reflects information regarding the instability of the interferometer. The phase difference between the sensing grating and reference grating can be used to compensate for thermal drift. This approach has also allowed for accurate quasi-static strain and differential temperature measurements [41, 42]. An alternative solution is to use an unbalanced integrated optic interferometer; this would provide a stable bias and negate the need for a reference grating [10]. An OPD of several hundred microns can be realized using a planar integrated optic Mach-Zehnder.

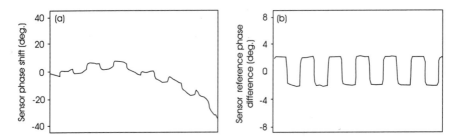

Figure 7.18   Output response to a strain perturbation of 3-$\mu\epsilon$ pk-to-pk in (a) uncompensated (direct carrier) and (b) drift compensated (sensor-reference phase difference) modes (*After*: [41]).

Kersey and co-workers [41] have also presented a fiber Bragg grating strain sensor system that uses interferometric determination of the strain-induced wavelength shifts and that incorporates a reference channel to compensate for random thermally induced drift in the interferometric discriminator. This system was capable of resolving quasi-static strain changes at 1 Hz with a resolution of ~ 6$n\epsilon/Hz^{1/2}$. Figure 7.18 shows the effectiveness of the drift compensation in the output response under 3-$\mu\epsilon$ strain steps: (a) gives the uncompensated single-channel carrier phase shift, and (b) the compensated sensor-reference phase difference. The fivefold increase in sensitivity and the stable baseline are noteworthy. The long-term stability in phase difference output of the lock-in amplifier of $\pm\ 1°$ corresponds to a rms random drift of ~0.5 µε/h for this system. Kersey and Berkoff [42] have implemented the same setup to realize a differential temperature sensor having a temperature resolution of <0.05°C or 0.5% of the grating bandwidth (<6x10^{-4} nm) over a 60°C range. In the case of temperature-induced shifts in the Bragg wavelength, (7.16) and (7.17) become $\Delta\psi = (-2\pi\Delta L_{MZI}/\lambda_B)\chi_g\ \Delta T$ and $\chi_g = [1/\lambda_B]\cdot[\partial\lambda_B/\partial T]$, respectively.

7.3.2.2 Enhanced Range Interferometric Wavelength-Shift Detection

We have discussed how the operational range of the Bragg grating is limited by the FSR of the receiving interferometer that is itself inversely proportional to the OPD; therefore, it is apparent that by varying the OPD one can control the range and resolution of the sensor. In the examples given a Mach-Zehnder interferometer was used, but the arguments are equally applicable to a Michelson interferometer. The change in OPD is limited for a fiber interferometer by the breaking strain of the glass; however, a bulk receiving interferometer can potentially offer large, controlled changes in the path imbalance. Recently, a novel method has been demonstrated that allows for two sets of interferometric fringes to be obtained by stepping a bulk Michelson receiving interferometer from a long to a short cavity, thereby enhancing the unambiguous measurement range [43]. The optical phase output from the cavity with the larger OPD (i.e., range 1) gives the high resolution measurement, while the shorter OPD (i.e., range 2) determines the number of fringes obtained with the long cavity, as in Figure 7.19. Therefore, the total absolute value of the phase change is given by $(\Delta\psi + 2\pi N)$, where N is the fringe number. If the interferometer is limited to one range and the phase excursion exceeds the FSR of the interferometer, N becomes ambiguous. Reducing the OPD and hence the measurement resolution increases the FSR allowing for N to be determined. We may define an enhancement factor M in the unambiguous range that is simply the ratio of the cavity lengths in the receiving interferometer

$$M = \frac{FSR_{short}}{FSR_{long}} = \frac{\Delta L}{\Delta L'} \qquad (7.20)$$

In principle M can be very large since a bulk cavity can be increased in size from a few hundred microns to tens of centimeters, although there are stability issues that need to be considered. In practice it is unlikely that M will exceed 100, nevertheless, this still represents a substantial improvement in range. Recently, this approach has been extended [44] by using two cascaded interferometers with outputs differentiated using frequency division multiplexing, with each interferometer modulated and filtered at discrete carrier frequencies. The system was demonstrated for a maximum enhancement factor of 40. An unambiguous measurement range for the first scanning interferometer (optical path imbalance ~0.78 mm, corresponding to a FSR of ~0.98 nm) was 0.0375 mε. The path imbalance of the second interferometer was varied from 5 to 30 mm, corresponding to a

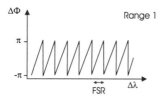 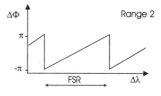

Figure 7.19 Principle of dual cavity, dual range interferometric scanning scheme.

FSR of 0.017 to 0.14 nm (i.e., an enhancement factor of 7 to 57). For $M = 40$, the overall unambiguous range was increased to 1.5 mε while maintaining a resolution of ~37 nε/Hz$^{1/2}$, corresponding to a range to resolution of ~ 4x10^4:1 and making BGSs competitive with conventional fiber optic interferometric sensors and strain gauges. If a bulk cavity can be maintained in alignment and isolated from thermal and low frequency vibrations, with a suitable reference channel, this approach appears quite feasible for the realization of large range, quasi-static measurements.

7.3.3 Simultaneous Interrogation of Bragg Gratings and Interferometric Sensors

There is considerable interest in the concomitant use of different forms of optical fiber sensors, particularly Bragg gratings and interferometric sensors, and there is a need for common demodulation instrumentation. The first dual, high-resolution measurements were made by Brady and co-workers [45] using a combination of interferometric wavelength-shift detection (Section 7.3.2.1) for the BGS and path-matching (coherence tuned) interferometry for a Fizeau cavity sensor [46], an approach that is suitable for multiplexing. Satisfactory performance figures indicated that the combined demodulation scheme did not affect the performance of the individual sensors. Typical resolutions of ~50 nε/Hz$^{1/2}$ and 1 mrad/Hz$^{1/2}$ at 20 Hz, were recorded for the BGS and Fizeau cavity, respectively. These figures compare well with previously reported data for single-sensor systems [10, 46]. A number of common demodulation schemes have been demonstrated [47–51].

7.4 Simultaneous Measurement of Temperature and Strain

Fiber Bragg gratings, in common with many types of fiber optic sensors, are subject to both strain and temperature fields simultaneously. Measurement of the perturbation-induced wavelength-shift from a single grating does not facilitate the discrimination of the sensor's response to these two variables. This inability to discriminate between temperature and strain is possibly the most significant limitation of Bragg gratings as sensors and has serious implications for strain sensors designed to measure quasi-static signals, as any temperature variation along the fiber will be indistinguishable from strain. For dynamic

Table 7.2   Comparison of Temperature and Strain Separation Measurement Schemes

Compensating method	Relative error		Extrinsic vs. Intrinsic	Reference
	Strain	Temperature		
Dual overlaid Bragg gratings at different center wavelengths	17με/pm	1.7K/pm	Intrinsic	[62]
Bragg grating in highly birefringent fiber	20με	2°C	Intrinsic	[65]
Two Bragg gratings of different cladding diameters	17με/pm	1K/pm	Intrinsic	[66]
Two Bragg grating diffraction orders	17με/pm	1.7K/pm	Intrinsic	[64]
Double core, dual wavelength Bragg gratings	Unknown [1]	Unknown [1]	Intrinsic	[73]
One rocking filter and one Bragg grating	40με/0.1nm	0.25K/0.1nm	Intrinsic	[69]
Two Bragg gratings and a long period grating	9με	1.5°C	Intrinsic	[68]
Fabry-Perot interferometer and Bragg grating	1.25με/pm	0.35K/pm	Extrinsic	[164]
Fiber Bragg grating rosette	3με/pm + 2.5με	0.14K/pm	Intrinsic	[72]
Two Bragg gratings mounted on opposite surfaces of a cantilever	1με/pm	Not measured	Extrinsic	[59]
Passive temperature compensating package	70με error on 120°C range	Not measured	Extrinsic	[57]

(1) Similar performance to dual overlaid/dual diffraction order gratings expected.

strain measurements this is not an issue, since the thermal fluctuations occur at low frequencies that tend not to coincide with the resonance frequencies of interest. We shall discuss schemes that separate strain- and temperature-induced wavelength shifts (ε, T) in the sections that follow.

Briefly, the elimination of cross-sensitivity may be achieved by measurements at two different wavelengths or two different optical modes, for which the strain and temperature responsivity is different. Temperature-compensating methods may be classified as intrinsic (relying on the fiber properties) or extrinsic (combining the grating with an external material of suitable properties and dimensions). The simplest approach is to use two sensors with one isolated from either of the unwanted perturbations; however, in applications where sensors must be embedded with minimal intrusion, a second sensor may not be practicable. Regardless of the exact details, any scheme should ideally be compatible with a large number of gratings to facilitate multiplexing. Table 7.2 summarizes these results.

7.4.1 Principle of Operation

For an ideal sensor recovering two observables, φ_1 and φ_2, that represent the changes induced by (ε,T) for two system eigenmodes (in this case the two Bragg grating wavelengths λ_1 and λ_2), each observable corresponds to either ε or T alone [52]

$$\begin{pmatrix} \varphi_1 \\ \varphi_2 \end{pmatrix} = \begin{pmatrix} K_{1T} & 0 \\ 0 & K_{2\varepsilon} \end{pmatrix} \begin{pmatrix} T \\ \varepsilon \end{pmatrix} \qquad (7.21)$$

assuming that the strain- and thermally-induced perturbations are linear. This is plotted in the (ε,T) plane in Figure 7.20(a) for a specific φ_1 and φ_2, with (ε,T) corresponding to the intersection of φ_1 and φ_2 lines, with an angle of 90 degrees at the intersection indicating the independence of the observables. In practice, both observables each show some sensitivity to ε and T as a result of the fiber material dependence on temperature and strain; that is,

$$\begin{pmatrix} \varphi_1 \\ \varphi_2 \end{pmatrix} = \begin{pmatrix} K_{1T} & K_{1\varepsilon} \\ K_{2T} & K_{2\varepsilon} \end{pmatrix} \begin{pmatrix} T \\ \varepsilon \end{pmatrix} \qquad (7.22)$$

The measurement of φ_1 and φ_2 and inversion of the matrix determine ε and T. Equation 7.22 assumes that the strain and temperature are essentially independent (i.e., the related strain-temperature cross-term is negligible); this applies well for small perturbations [53]. For this approach to be successful one must accurately know K_ε and K_T, the strain and temperature coefficients, respectively. Assuming this is the case, the limitation becomes the conditioning of the matrix, demanding that the determinant is non-zero

$$\begin{pmatrix} T \\ \varepsilon \end{pmatrix} = \frac{1}{(K_{1T}K_{2\varepsilon} - K_{2T}K_{1\varepsilon})} \begin{pmatrix} K_{2\varepsilon} & -K_{1\varepsilon} \\ -K_{2T} & K_{1T} \end{pmatrix} \begin{pmatrix} \varphi_1 \\ \varphi_2 \end{pmatrix} \qquad (7.23)$$

Figure 7.20(b) shows the loci corresponding to φ_1 and φ_2 in the (ε,T) planes, again the intersection yields the derived (ε,T). Clearly, when the angle between the lines at the intersection point becomes zero (i.e., the lines are parallel), the following condition exists: $K_{1T}/K_{2T} = K_{1\varepsilon}/K_{2\varepsilon}$ and (7.23) tends to infinity, invalidating the measurement. Therefore, a fundamental tenet is that the ratio of the strain responses of two gratings be different from the ratio of their temperature responses. An equivalent description has been provided in [54], with the errors in φ_1 and φ_2 translated to an error-ellipse in the (ε,T) plane. One can compare the errors in the derived measurands with those in the ideal case. Ideally

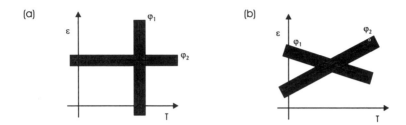

Figure 7.20 (a) Loci of φ_1 and φ_2 in the (ε, T)-plane for when $\varphi_1 = \varphi_1(T)$ only and $\varphi_2 = \varphi_2(\varepsilon)$ only, with shaded areas corresponding to measurement errors $\delta\varphi_1$ and $\delta\varphi_2$. (b) Nonideal intersection case with cross-coupled strain and temperature information and larger measurement error (*After*: [52]).

δT and $\delta\varepsilon$, the errors in temperature and strain are given by [46]

$$\begin{pmatrix} \delta T \\ \delta\varepsilon \end{pmatrix} = \begin{pmatrix} \delta\varphi_1 / K_{1T} \\ \delta\varphi_2 / K_{2\varepsilon} \end{pmatrix} \tag{7.24}$$

where $\delta\varphi_i$ are errors in φ_1 and φ_2. The effect of the phase errors is indicated in Figure 7.20(a), where the shaded region around the intersection indicates the uncertainty in (ε, T). The errors are increased in the non-ideal case where the (φ_1, φ_2) responses are not orthogonal [55]

$$|\delta\varepsilon| = \frac{|K_{2T}\|\delta\varphi_1| + |K_{1T}\|\delta\varphi_2|}{|K_{1T}K_{2\varepsilon} - K_{2T}K_{1\varepsilon}|} \tag{7.25}$$

where the domain of uncertainty is illustrated in Figure 7.20(b) (with a similar equation for T). Any uncertainties in K correspond to variations in the gradients of the φ_i-lines [56].

7.4.2 Extrinsic Temperature Compensation

At least three extrinsic methods have been proposed and demonstrated. The first utilizes a passive temperature-compensating package to nullify the temperature to wavelength coefficient [57]. In this case the grating is mounted under tension in a package comprised of two materials with different thermal expansion coefficients; as the temperature rises, strain on the grating is progressively released. Packaging the Bragg grating in a liquid crystalline polymer tube has resulted in an improvement in temperature stability by a factor of ten [58]. In another scheme, a cantilever is used with two Bragg gratings mounted on opposite surfaces (top and bottom), with one grating stretched while the other is compressed. The difference in the Bragg grating wavelengths is temperature independent because both Bragg gratings have the same temperature sensitivity [59]. The work described in [57–59] strictly involves mechanical compensation and does not provide a separate temperature measurement.

7.4.3 Intrinsic Temperature Compensation

All the methods below can provide both temperature and strain information provided that the two linear equations of (7.23), both functions of temperature and strain, can be solved.

7.4.3.1 Reference Grating

The relative contributions for temperature and strain may be ascertained if a reference exists for the temperature information. The most obvious approach is to use a separate grating as a temperature sensor, being in thermal contact with local sensor environment but shielded from strain changes (e.g., using Teflon sleeving). The compensation results from subtracting the wavelength shift induced by the temperature excursion in the reference grating from the total wavelength shift recorded by the strain sensor [41, 60].

7.4.3.2 Tapered Grating

Xu et al. [61] have shown that a chirped grating in a tapered fiber can be temperature independent; when heated the grating spacing changes uniformly, modulating the center wavelength. The taper profile is designed such that the grating is linearly chirped when tension is applied, creating a strain gradient along the fiber. Measuring the increase in the return signal, as the effective spectral bandwidth increases with strain, rather than the Bragg wavelength change, gives an intensity reflected signal that is independent of temperature. Having a tapered fiber section, however, weakens the structure. Furthermore, because the information is now converted to an intensity measurement, one needs to consider the effect of system losses producing measurement errors. An intensity-measurement strain resolution of 4 $\mu\varepsilon$ over a range of 4000 $\mu\varepsilon$ was recorded. As for extrinsic methods reported in Section 7.4.3, this approach does not provide a separate temperature measurement. This is not suitable for multiplexing.

7.4.3.3 Dual-Wavelength Superimposed Gratings

Xu et al. [62] have measured the responses of two overlapping gratings written at 848 and 1298 nm and have found that their responses are 6.5% higher for strain and 9.8% lower for temperature at ~1300 nm compared with ~850 nm. The two wavelengths derive a temperature and strain response analogous to (7.23). The technique exploits the differential dispersion of the strain and temperature coefficients with wavelength. The matrix conditioning results in a predicted strain and temperature error of $\pm$ 10 $\mu\varepsilon$ and $\pm$ 5°C, respectively. Udd et al. [63] proposed a similar scheme by writing two overlaid gratings at different wavelengths in birefringent optical fiber; because of the fiber birefringence four gratings are actually written. Temperature measurements result through solving four equations having four unknowns. Dual-wavelength techniques require two broadband sources to address each grating and suitable wavelength filtering at the system output.

7.4.3.4 Multiple Bragg Grating Orders

In practical gratings the sinusoidal axial variation of the refractive index is not perfect; therefore the gratings also reflect at harmonics of the fundamental wavelength. The strength of the higher order reflections is determined by the magnitude of their respective Fourier coefficients that describe the index perturbation. For conventionally written gratings, the second-order signal is very weak at 0.1% reflectivity, compared with 50% reflectivity at the primary wavelength, and the second harmonic wavelength does not occur at exactly half the Bragg wavelength due to material and waveguide dispersion. Kalli and co-workers [54, 64] have demonstrated the viability of this approach, offering similar strain and temperature resolutions to the dual-wavelength superimposed grating method, in this case $\pm$ 17 $\mu\varepsilon$/pm and $\pm$ 1.7 °C/pm. However, both this and the dual-wavelength superimposed grating approach will be difficult to implement with more than one sensor because of the incompatibility of the two operating wavelengths and the need for expensive WDM components.

7.4.3.5 Simultaneous Measurement of Temperature and Strain Using PANDA Fiber Grating

Polarization-maintaining fiber has different propagation constants for the slow and fast axes; therefore, a fiber Bragg grating written in such a fiber has two distinct Bragg wavelengths corresponding to the fast and slow axes. Of course, each Bragg wavelength has a different dependence on temperature and strain, in accordance with (7.22). Sudo et al. [65] have written a grating in PANDA fiber with corresponding Bragg wavelengths of 1535.32 and 1535.78 nm for the slow and fast axes, respectively. Wavelength to temperature responsivities of 9.5 and 10.1 pm/°C and strain sensitivities of 1.342 and 1.334 pm/$\mu\varepsilon$ were measured, with cross-sensitivity errors between temperature and strain of $\pm$ 2°C and $\pm$ 20 $\mu\varepsilon$.

7.4.3.6 Gratings in Dissimilar Diameter Fibers

A novel composite sensor has been formed from two gratings of similar center wavelength, photowritten into fiber having two different cladding diameters (80 and 120 μm) that are spliced together [66]. When subjected to strain, the gratings experience different individual strains (greater in the fiber with the thinner cladding), producing different changes in the Bragg center wavelength. The temperature sensitivity, however, is similar for the two gratings. Thus, the relative wavelengths provide a measurement of the strain, while the absolute wavelength shifts include information on both temperature and strain. Maximum errors of $\pm$ 17 $\mu\varepsilon$ and $\pm$ 1°C were measured over ranges of 2500 $\mu\varepsilon$ and 120°C, respectively. An interferometric scheme has been developed to directly measure the wavelength difference between the gratings, thereby leading to ε [67].

7.4.3.7 Hybrid Bragg Grating/Long Period Grating

The long period grating also has rather different temperature and strain response coefficients compared with the Bragg grating. This is because the Bragg grating wavelength is linearly proportional to the grating period multiplied by the effective refractive index of the core, whereas the long period grating wavelength is proportional to the grating period multiplied by the difference in index of refraction between the core and the cladding. The respective strain and temperature responses depend on the changes in these terms. Therefore, simultaneous strain and temperature measurement can be achieved, as has recently been verified by Patrick et al. [68]. For the long period gratings the wavelength shifts for strain and temperature were 0.5 pm/$\mu\varepsilon$ and 60 pm/°C, compared with 1 pm/$\mu\varepsilon$ and 9 pm/°C, giving a well-conditioned transformation and resulting in strain and temperature errors of $\pm$ 9 $\mu\varepsilon$ and $\pm$ 1.5°C, respectively. This is a particularly attractive approach as it retains the use of components that are designed to operate at a single wavelength, thereby avoiding costly WDM devices, as would be required in the scheme presented in [62] and [64]. There are, however, several potential problems. For example, the bandwidth of the long period grating is relatively large, limiting the accuracy and the number of sensors that could be used in a WDM system. The physical length of the grating at a few centimeters far exceeds that of a Bragg grating, posing problems if the physical length exceeds the separation between two Bragg grating sensors and for where very high spatial resolution is required. In smart structure applications, the gratings may experience a significant nonuniform strain field along a length of a few centimeters. Finally, the long period grating is highly bend-sensitive, and separating the relative wavelength contributions from bend and longitudinal strain would be problematic.

7.4.3.8 Superimposed Gratings and Polarization-Rocking Filters

A rocking filter, where the periodicity of the refractive index perturbation coincides with the polarization beat length in the birefringent fiber in which it is written, can be combined with a Bragg grating to realize a compensating scheme. Again, the devices operate in the same wavelength band and exploit the very different wavelength sensitivities for the two grating types. The temperature sensitivity of the wavelength shift in the rocking filter is about 100 times greater than for the Bragg grating. For example, Kannellopoulos et al. [69] have demonstrated the use of a fiber Bragg gratings and a long period rocking filter in the 800-nm band with errors of $\pm$ 165 $\mu\varepsilon$ and $\pm$ 1.5°C. For practical reasons there are advantages to avoiding polarization-based techniques and special birefringent fibers. First, the physical length of the rocking filter is large compared with the grating, limiting the application range. Second, the bandwidth of the rocking filter is far larger than that of the grating, restricting the measurement accuracy and the number of sensors that can be multiplexed using WDM for quasi-distributed measurement. Third, the use of polarization-maintaining fiber and polarization-related components increases the cost and difficulty of multiplexing, particularly for a large number of sensors.

7.4.3.9 Bragg Gratings and Brillouin Scattering

Davis and Kersey have demonstrated that a combined interrogation system based on Bragg gratings and a distributed Brillouin network, operating at different wavelengths, can decouple temperature and strain [70]. This leads to a modified matrix form

$$\begin{pmatrix} T \\ \varepsilon \end{pmatrix} = \begin{pmatrix} K_{B,T} & K_{B,\varepsilon} \\ K_{g,T} & K_{g,\varepsilon} \end{pmatrix}^{-1} \begin{pmatrix} \Delta v_B \\ \Delta \lambda_g \end{pmatrix} \tag{7.26}$$

where $K_{B,T}$ and $K_{B,\varepsilon}$ are the responsivities of the Brillouin frequency shift Δv_B to temperature and strain, respectively. The Brillouin frequency shifts near 1300 nm are 14 times more sensitive to temperature than the Bragg grating element at 1550 nm. Strain sensitivity is also higher by a factor of 5.6, and it is this increase in relative sensitivity due to temperature against that of strain that permits the decoupling. Strain and temperature resolutions of 22 $\mu\varepsilon$ and 1.9°C were obtained from this approach. This is suitable for use with a large number of gratings.

7.4.3.10 Combined Grating and In-Line/Extrinsic Fiber Etalon Sensor Methods

An in-line fiber etalon sensor formed by splicing a short segment of silica hollow-core fiber between two sections of a single-mode fiber has very low temperature sensitivity and high mechanical strength, making it suitable for use in smart structure applications [71]. A serial combination of grating and in-line etalon has enabled simultaneous strain and temperature measurements, resulting from their different strain and temperature coefficients [47]. As the return signals from the cascaded grating and etalon sensors mix together in the spectral domain, it is impossible to separate them completely using only WDM, resulting in relatively high cross-talk. Recently, Liu et al. [48] have demonstrated a combined sensor configuration consisting of a single-mode lead-in/lead-out optical fiber and multimode reflector, incorporating a Bragg grating in the single-mode fiber section. This design essentially consists of two sensors in series, a Bragg grating in a strain-free environment and an air cavity extrinsic Fabry-Perot strain sensor. When illuminated with a broadband source, a narrow band of the optical power is reflected from the grating with transmitted light illuminating the etalon sensor, whose cavity modulates the reflection spectrum. The reflected light from both sensors is directed to a CCD spectrometer for measurement. Temperature changes cause shifts in the peak Bragg grating wavelength, while exhibiting negligible response to applied strain (less than 13 pm for a strain range of 1200 με). Applied strain, however, causes a change in the periodicity of the interference fringes. Strain and temperature resolutions of 32 με (mean strain of 605 με) and 2.4°C (mean temperature change of 34.5°C) were measured.

7.4.3.11   Bragg Grating Strain Rosettes

Bragg grating rosettes made of two or three noncollinear strain gauges mounted on a common substrate at 45 or 60 degrees to form a rectangular or delta rosette, respectively, have been shown to provide temperature and strain discrimination [72]. They are used extensively in stress analysis to measure the two principal strains and stresses and the orientation of the principal axis, whenever it is not known a priori. An innovative way has also been described for use as a uniaxial strain gauge rigorously independent of temperature effects as well as the orientation of the sensor onto the structure. The uniaxial strain and temperature are accurately measured to 3 με/pm and 0.14°C/pm.

7.4.3.12   Dual-Core Fiber Bragg Gratings

The Boeing Company has developed and patented a double-core Bragg grating fiber optic sensor transmitting different wavelengths of light. Bragg gratings at these two different wavelengths are written at the same location onto the optical fiber, and the response from these Bragg gratings can be used to simultaneously determine both the strain and temperature at the sensing location [73].

7.5   Polarization Stability of Interrogation Schemes

An issue that remains to be addressed for all the demodulation systems is Bragg grating polarization-dependent behavior. This can result in systematic errors when taken in combination with any optical component having polarization-dependent emission (ELEDs), transmission (fiber couplers or acousto-optic tunable filters), or detection responsivity (grating spectrometer). The polarization-dependent behavior occurs either intrinsically (due to intrinsic fiber birefringence, or for non-cylindrically-symmetric grating inscription) or is externally induced during use as a sensor (e.g., in extreme environments or through lateral mechanical pressure applied to a carbon fiber matrix during the production of laminate sheets).

Ecke et al. [74] have found that laterally compressing or bending a fiber lead produces dynamic changes due to polarization-mode conversion, resulting in severe noise and systematic errors of 30 με or 4°C. Scrambling the state of polarization (SOP) in either the wavelength or time domain will average the Bragg wavelength shift, regardless of polarization-mode conversion in the leads. A Lyot filter [75] consisting of two cascaded birefringent elements with a relative polarization axis rotation of 45 degrees can be used to

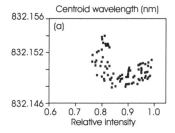 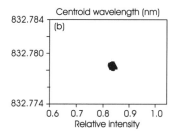

Figure 7.21 Variation in peak CCD intensity (a) without (40% degree of polarization) and (b) with polarization scrambling (1.7% residual degree of polarization) (*After*: [74]).

polarization scramble in the wavelength domain. Under broadband light illumination the net effect is to give an output SOP that varies cyclically with optical wavelength, effectively averaging all possible SOPs over the finite bandwidth of the system. The AOTF is particularly polarization sensitive, while a mixed SOP input to a FPF will broaden the measured response, generating large errors, unless the device has sufficient resolution to resolve the individual peaks. Time domain scrambling [76] is achieved by modulating the length of hi-bi fiber (two 4-m lengths spliced at 45 degrees [74]) using PZ transducers, such that all SOPs are swept through rapidly and effectively averaged over time. Figure 7.21 shows that for a simple broadband source and spectrometer-CCD arrangement, the passive Lyot depolarizer gives a dramatic improvement to the system stability, reducing wavelength errors to less than 1 pm. Since this device operates by averaging in the wavelength domain, it is suitable for use in systems having fast dynamic responses, but are not suitable for use with narrowband (e.g., laser) interrogation systems. For time domain scrambling the system can still be used with narrowband and laser interrogation schemes, providing a wavelength stability error of <1.5 pm p-p (x5 improvement).

7.6 Multiplexing Techniques

Multiplexing is a very important issue for optical fiber sensors, as it is in the domain of distributed sensing where their greatest impact is anticipated. Therefore, by sharing the source and processing electronics, the cost per sensor is reduced, improving the competitiveness of optical-based sensors against conventional electro-mechanical sensors. Minimizing the number of electronic components substantially reduces the overall system weight while enhancing durability. The following multiplexing techniques have been reported and applied to optical fiber sensors: time [77], spatial [78], wavelength and frequency [13], and coherence domain multiplexing [79]. The multiplexing capacities of any of these techniques, however, are limited to approximately 10 sensors due to various factors, including speed, cross-talk, signal-to-noise ratio, and wavelength bandwidth. The most popular formats for increasing sensor numbers combine spatial domain multiplexing with other techniques, as these combinations do not generally degrade system performance.

7.6.1 Wavelength Division Multiplexing (WDM)

WDM encodes each BGS with a unique spectral slice of the available source spectrum, which defines the sensor's operating range and is also associated with a specific spatial location along the optical fiber. An immediate advantage of wavelength discrimination between sensors is that the physical spacing between individual gratings may be as short as desired and the need for high-speed electrical signal processing, as is often required in TDM, is removed. Several approaches have been reported, as we shall outline below.

7.6.1.1 Parallel and Serial WDM Topologies

As demonstrated by Jackson et al. [9], this is a simple, parallel method that permits the simultaneous interrogation of all sensors. The outputs, however, are split into many fiber channels, employing separate electronic signal processing units for each channel, which increases potential cost and weight. Both Brady et al. [12] and Davis and Kersey [13] have presented similar serially based multiplexing schemes. The former scheme suffers a power penalty because of its reflectometric approach. Further experimental details may be found in Section 7.3.1.3.

7.6.1.2 WDM Schemes with Tunable Filters

It was shown in Section 7.3.1.3 that tunable filters (FPF, AOTF, or FBGF) could be used to realize compact and simple multiplexing schemes, with typical resolutions of a few picometers over ranges of tens of nanometers [11, 14, 17]. Systems based on AOTF demodulation have no moving parts and are therefore particularly suitable to rapid scanning of sensor arrays. The level of sensor cross-talk is limited by the extinction ratio of the filter and is expected to vary, increasing as operating wavelengths of perturbed gratings tend to be in close proximity. Nevertheless, a principal advantage of these WDM-based schemes is the minimal impact that cross-talk has on practical sensor demonstrations.

7.6.1.3 Combined WDM and Interferometric Detection

It has been shown that wavelength-shift interferometric detection offers extremely high strain resolution, and, in principle, single sensor performance can be realized for individual elements in large Bragg grating sensor arrays. Sensor elements are written at different nominal wavelengths, and an additional wavelength-selective element is incorporated into the system for selective sensor demodulation. Figure 7.22 shows two possible approaches [78, 80]. The tunable FPF spectrally slices the output of a broadband source for wavelength addressing of individual sensors. An unbalanced MZI serves to further spectrally filter the light with a sinusoidal transfer function, and an electrical carrier is generated via the detector through serrodyne modulation of one of the interferometer arms. The carrier

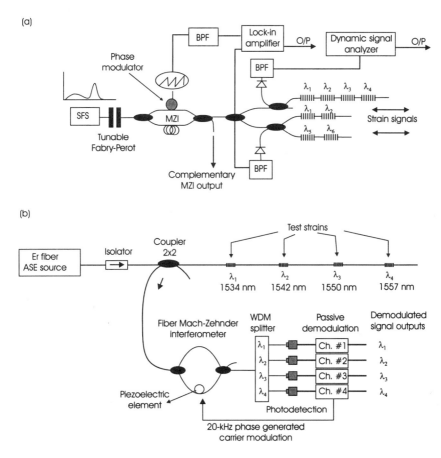

Figure 7.22 Two schemes for multiplexing Bragg sensors using band-pass elements for WDM and interferometric detection: (a) active band-pass, and (b) passive band-pass filter (*After*: [78] and [80]).

signal is phase modulated by any perturbation acting on the chosen grating that is addressed sequentially, producing the high-resolution wavelength-shift detection. Kalli and co-workers [81] have demonstrated an eight-sensor system for dynamic and static phase measurements (Figure 7.22(a)) that was extended through spatial multiplexing. A high-resolution, low-frequency dynamic strain resolution of ~90 nε/Hz$^{1/2}$ at 7 Hz was achieved in a system operating at 1550 nm. An identical system operating at 800 nm, incorporating a reference grating, recovered static phase measurements via the detection of the differential phase shift between grating pairs and produced a noise-limited static strain resolution of 1.8 με. Berkoff and Kersey have used a similar approach (Figure 7.22(b)), however, the wavelength filtering is achieved passively, before the detector, using a band-pass wavelength division demultiplexer (BWDM). In this case all sensor signals are summed and filtered through the MZI simultaneously, producing a composite signal that is demultiplexed with the BWDM. A four-sensor system was demonstrated with a minimum strain resolution of 1.5 nε/Hz$^{1/2}$ at ~600 Hz. A drawback of these systems is the cyclic phase error that occurs because the serrodyne modulation is not exactly 2π radians for all the different Bragg wavelengths. If a pure frequency shift is generated using an acousto-optic modulator, this issue is mitigated [51].

7.6.2 Time Division Multiplexing (TDM)

For TDM the return pulses between adjacent BGSs are recovered with the two pulses separated by a distance equal to the round-trip delay time between gratings. The temporally separated output pulses that are then fed into a compensating interferometer of OPD equivalent to the round-trip time delay between the reflective markers. In this way portions of light in adjacent pulses interfere to produce a high-resolution interferometric output.

7.6.2.1 Combined TDM and Interferometric Detection

In the case of TDM, Weis et al. [77] have demonstrated a four-element sensor array demodulated using a single, unbalanced Mach-Zehnder interferometer. A strain resolution of ~2 nε/Hz$^{1/2}$ was demonstrated for frequencies exceeding 10 Hz, with the cross-talk between channels less than 30 dB for a system at 1300 nm illuminated by an ELED source of power ~250 μW. The main shortcoming of TDM is that as the number of sensors increases the signal-to-noise ratio deteriorates as the pulse duty cycle increases. Berkoff and Kersey [82] have achieved a static strain resolution of 2 με (random drift 3.5 με/hr without using a reference grating) for an eight-sensor TDM system with the MZI replaced by an integrated optic wavelength discriminator, which provided wide range and good stability for quasi-static measurements.

7.6.2.2 TDM and WDM

It is apparent that WDM allows for the interrogation of several sensors per fiber; however, because the reflectivity of the Bragg gratings is rarely more than a few percent, a large part of the source energy remains unused. When combining WDM with TDM, reusing the source spectrum can scale the number of sensors. Figure 7.23 shows three principal configurations: serial, parallel, and branching. A serial configuration using several concatenated wavelength-stepped arrays is shown in Figure 7.23(a), with each at a greater distance along the fiber. A pulsed light source, whose spectral bandwidth covers the sensors' operating range, is reflected off consecutive gratings at successively more distant positions along the fiber and these pulses are consecutively returned to the detector. The detection instrumentation responds to the reflected signals only during a selected time window after the pulse is launched, so that a single WDM set of sensors is selected for detection. This approach has the potential of addressing very large arrays (e.g., a 100-

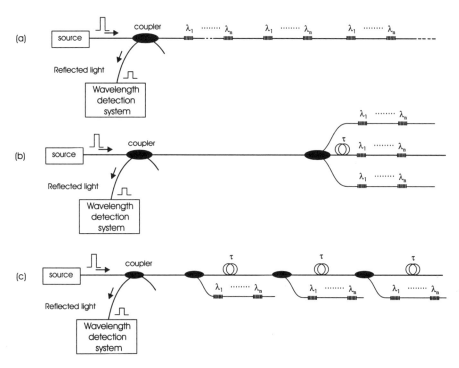

Figure 7.23 Combined WDM/TDM multiplexing topologies for Bragg grating sensor arrays: (a) serial system with low reflectivity gratings, (b) parallel network, and (c) branching network (*After*: [1]).

element system is quite feasible and that may also be combined with CCD detection). There are drawbacks, however, to using such concatenated arrays since gratings whose reflected light signals are separated in time but which overlap in wavelength can experience cross-talk, termed "multiple-reflection" and "spectral-shadowing cross-talk" [1]. Multiple-reflection cross-talk results if light reflected from a grating arrives in the detection time window allotted to another downstream grating, because of delays associated with multiple reflection paths. Low reflectivity gratings can minimize this cross-talk. Spectral-shadowing cross-talk is a distortion of a downstream grating spectrum that results from light having to pass twice through an upstream grating. If the two gratings' center wavelengths are slightly offset, it appears as though the downstream grating is shifted further in the direction of the actual offset. Large errors exceeding 1 με are expected for a pair of interfering Bragg gratings with reflectivities of greater than 5%. The parallel and branching networks in Figure 7.23 (b) and (c) eliminate these deleterious effects at the expense of optical efficiency and the need for additional couplers and stronger reflecting gratings. Davis et al. [83] have addressed a nine-sensor Bragg grating array for mapping the strain levels in a plate subject to deflection forces, using an approach based on Figure 7.23(a).

7.6.3 Spatial Division Multiplexing (SDM)

In applications where point measurements are required, such as in the aerospace industry and security applications, it is often necessary for sensors to be operated independently, interchangeable, and replaceable without substantial recalibration. Sensors with identical characteristics are quite feasible with BGSs. However, whereas the serial WDM and TDM schemes prove unsuitable for the interchangeability of sensors, SDM based on a parallel sensor topology are ideal. Recently, a SDM approach has been demonstrated for demodulating a large number of Bragg sensors (potentially 32) with a strain resolution of ~0.36 $\mu\varepsilon/Hz^{1/2}$ [84]. The implementation of SDM typically results in negligible cross-talk between sensors.

7.6.3.1 SDM and WDM

Conventional spatial multiplexing splits the sensor outputs into separate detection channels, the combination of SDM and WDM given by the topologies in Section 7.5.1.3 are examples of this [78, 80, 81]. The increased number of detection units is a drawback, increasing system weight and cost. Davis et al. [25] used a multichannel optical fiber switch, allowing for separate sensors on multiple fiber channels to share the same processing electronics, while using a single FPF to achieve both the WDM and wavelength-shift detection. A large array of 60 sensors was addressed for a low sampling rate of 2.2 seconds, which may be suitable for applications where speed is unnecessary, such as for civil structures. Chen et al. [26, 27] have reported a SDM/WDM system based on using a two-dimensional CCD array to generate an MxN array, with spatial positions of the fiber channels encoded along one axis of the CCD, while wavelengths are encoded along the orthogonal axis.

7.6.3.2 SDM and TDM

A combined SDM and TDM topology with drift compensation based on wavelength-shift detection has been reported in [85], for the demodulation of eight sensors and a quasi-static strain resolution of ~ 0.22 $\mu\varepsilon/Hz^{1/2}$ for a range to resolution of 1250:1.

7.6.4 Combined SDM/WDM/TDM

The serial multiplexing schemes of WDM and TDM, and combinations thereof, make efficient use of the source power. Parallel topologies such as SDM allow for the independent operation and interchangeability of sensors. Combining both serial and parallel multiplexing types can offer extremely large and efficient two-dimensional quasi-distributed sensing networks, as reported by Rao and co-workers [86]. This is essentially an extension of the topology in [78, 81, 87]. A reported static strain resolution of ~1 $\mu\varepsilon/Hz^{1/2}$

Table 7.3 Summary of BGS Multiplexing Techniques

	TDM	WDM	SDM	SDM/WDM
Multiplexing capacity	medium	good [1]	good	very good
Spatial resolution	low	high	high	high
Usage of optical power	good [2]	good	medium [3]	good [3]
Interchange-ability	low	low	high	medium
Potential cost	low	medium	medium	medium

(1) In combination with TDM.
(2) Poor signal-to-noise ratio may occur for large numbers of sensors.
(3) Optical power usage is improved by optical switching between channels, but at the expense of sampling rate.
After: [88].

was reported for this system. We conclude this section by referring to Table 7.3, which provides a summary of Bragg grating sensor-multiplexing techniques [88].

7.7 Sensors Based on Chirped Bragg Gratings

In Chapters 3 and 4 we discussed details regarding chirped grating properties and manufacture. Their intended application is for dispersion compensation in high-bit-rate transmission systems (Chapter 6); however, they have recently received attention as sensing elements. Unlike conventional Bragg gratings, novel sensing devices can be constructed, as we shall review in the following sections.

7.7.1 Broadband Chirped Grating Sensor

We have thus far demonstrated a number of schemes for demodulating conventional, narrowband BGSs that rely on some form of optical filtering. Chirped Bragg gratings possessing a broadband, step-function chirp profile may also be used as sensors and can offer several advantages, particularly as an alternative to matched-grating demodulation. Fallon et al. [89] have presented a demodulation relying on the mismatching of two identical, broadband chirped gratings, called identical chirped grating interrogation (ICGI). This scheme dispenses with the need for high-quality piezoelectric actuators or

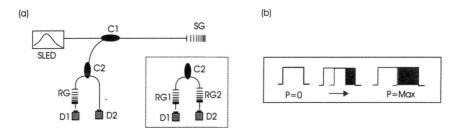

Figure 7.24 (a) Schematic diagram of ICGI interrogation scheme with source fluctuation compensation (inset), and (b) principle of the ICGI technique (*After:* [90]).

tunable filters and is particularly attractive for aerospace applications because of the passive demodulation with no moving parts. Figure 7.24 illustrates the basic approach with two initially matched filters. In this state light from a broadband source, incident on the sensor grating, is reflected to the reference grating, which acts as a rejection filter, minimally transmitting light to the detector. When a perturbation acts on the sensing grating, its spectral profile is linearly shifted, resulting in a fraction of the light reflected from the sensing grating falling outside the rejection band of the rejection filter and being transmitted to the detector. It is the quasi-square reflection profiles that permit a linear relationship between the change in strain or temperature encoded in the Bragg wavelength and the intensity of the light transmitted by the reference grating. The grating bandwidth clearly determines the system range. For example, a 10-nm bandwidth grating gives a sensing range of 10 mε (~1pm/με). Such broadband chirped gratings with near-square-like reflection profiles are achievable in practice using the dissimilar wavefront, two-beam holographic or chirped phase-mask method (Chapter 4). Figure 7.24 also shows how the output of the system may be linearized to account for the nonuniform intensity output from most broadband sources by measuring its power from the second arm of the output coupler and taking the ratio

$$\text{Output} = \frac{P_1}{P_2} = \frac{\int I(\lambda)T(\lambda)R(\lambda,\varepsilon)d\lambda}{\int I(\lambda)R(\lambda,\varepsilon)d\lambda} \tag{7.27}$$

where $I(\lambda)$ is the spectral intensity of the light source, $T(\lambda)$ is the transmission coefficient of the receiving grating, and $R(\lambda,\varepsilon)$ is the reflectivity of the sensing grating as a function of wavelength and strain. Furthermore, the sensitivity of the system can be increased by inserting another identical grating into the second arm of the output coupler and by measuring the ratio between the difference and the sum of the two outputs [90]. The ICGI can be extended to a multiplexing scheme where multiple sensors are arranged in serial, parallel, or a combination of both. The sensors are simultaneously interrogated, and therefore interrogation speed can be very high, unlike tunable filter or switching approaches. The disadvantage here is that multiple broadband sources are required to increase total bandwidth and light intensity. Zhang et al. [90] have demonstrated a spatial and wavelength division multiplexing architecture for multiplexing four parallel sensors. In this system four identical 30-nm broadband chirped grating sensors (centered at 1285 nm) are used to facilitate a sensing range of $\pm$ 15000 µε. Similarly, four identical 30-nm broadband chirped reference gratings are used (centered at 1270 nm). There is a 15-nm difference between the central wavelengths of the sensor and reference gratings so that zero strain corresponds to the mid-range of the system output signal. The superluminescent light emitting diode (SLED) source (output power 0.15 mW) exhibited a $\pm$ 10% intensity modulation that did not prove problematic because of the use of a reference, which permitted the use of low-quality and low-cost sources. This approach is particularly suitable for monitoring extreme strain and offers a resolution of ~5 µε. Combining such a system with the sensing scheme recently reported by Davis et al. [11], interrogating 60 BGSs, would form a combination architecture offering very high resolution and range.

7.7.2 Tapered Chirped Grating Sensor

The tapered grating sensor utilizes the strain gradient realized when a tapered fiber is under tension, with only the outer diameter of the fiber tapered in the region of the grating [91]. This removes effects associated with changing the effective index, while allowing control over the spectral shape and location of the chirp. The fiber is first tapered and then the grating is written as the fiber is held under a preset tensile load. The unstrained fiber now contains a chirped grating, and as strain is applied the reflectance peak narrows and shifts until the tension at which the grating was written is applied, in this case 1750 µε. The grating width and location are now the same as for an untapered fiber exposed under zero strain. Under additional strain the grating shifts further and broadens (Figure 7.25(a)). An alternative approach results by exposing an untapered fiber and then etching it to produce a linear taper. This now results in a narrow grating in unstrained fiber that broadens under loading, along with a shift in the spectrum (Figure 7.25(b)). The more pronounced chirp per unit applied strain is a consequence of a more pronounced taper. Such devices are

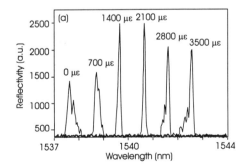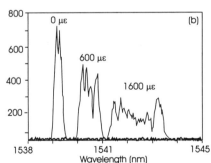

Figure 7.25   (a) Reflectance spectra of chirped tapered grating (from 115 to 105 µm). Grating written while fiber subjected to 1750 µε. (b) Reflectance spectra of chirped grating tapered from 115 to 85 µm. Grating written in unstrained fiber (*After*: [91]).

interesting as they have unique responses to temperature and strain fields. Strain results in a broadening and shifting of the Bragg condition, whereas temperature affects only the location of the centroid through the temperature-dependent refractive index. Through careful calibration and by measurement of the spectral shift and broadening, one can use such devices to simultaneously measure temperature and strain [91].

7.7.3 Asymmetrically Chirped Grating Sensor

An alternative strain sensor concept makes use of a tailored Bragg grating element in which the grating reflectivity is monitored as a function of strain for a given probing wavelength. In this instance a chirped grating having an asymmetric broadband spectral response is utilized as a strain-sensitive reflective filter [92]. A large number of such weakly reflecting gratings could be addressed by OTDR via this technique. A particularly suitable shape for this type of application is a strongly asymmetric ramp profile that produces a gradual change in reflectivity on one side of the grating response. If the probing wavelength is located in the middle of this ramp, one can determine whether compressive or tensile strain is applied to the sensor through an increase or decrease in reflectivity. Figure 7.26(a) shows such a spectrum for a chirped and strongly apodized grating, along with a shift resulting from an applied strain of 2000 µε. Figure 7.26(b) shows the change in reflectivity with applied strain to the grating for a probe wavelength of 1534 nm. This technique, however, is sensitive to system losses between the sensor and detection system. In principle one could overcome this through OTDR by using the intrinsic Rayleigh scattering of the fiber as a reference, thereby reducing the effects of optical losses. Kersey has proposed that this type of sensor could be used as a pressure-sensing probe, with pressure-altered strain on the grating yielding reflectivity as a function of pressure.

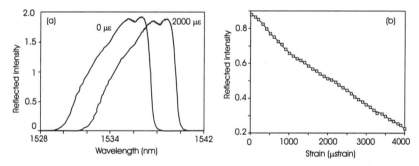

Figure 7.26   (a) Spectrum of asymmetric profile Bragg grating at 0 and 2000 µε. (b) Change in reflectivity of asymmetric Bragg grating for a fixed probing wavelength of 1534 nm (*After*: [92]).

7.7.4 Intragrating Sensing

The measurement of the peak Bragg wavelength provides information only on the average value of the measurand over the length of the grating. In many sensing applications, however, (e.g., crack detection) one may require multiple, adjacent, high-resolution measurements. A fully distributed profile of the measurand along a grating may be derived from the Bragg wavelength as a function of distance, this is known as intragrating sensing. The profiles of gratings having lengths from 5 mm to 10 cm have been reported [93–96] with a spatial resolution as low as 0.8 mm [93]. The technique uses the fact that different grating sections contribute to the reflectivity at different wavelengths when the grating modulation periodicity is nonuniform along its length. Each grating section contributes at the "local" Bragg wavelength given by $\lambda_B(z) = 2n(z)\Lambda(z)$. Hence, a uniform grating subjected to a nonuniform strain or temperature distribution will have different parts of the grating contributing to different wavelengths according to the local condition of the measurand. This broadens the reflected spectrum, while reducing the peak reflectivity. This nonuniformity affects both the intensity and phase response of the grating; therefore, one may analyze either one or both components to determine the Bragg wavelength with position. By subtracting the modified Bragg wavelength from the original Bragg wavelength, both as a function of position along the grating length, one finds a difference that is dependent on either the nonuniform temperature or strain. The grating can either be uniform or chirped to start with; the latter has obvious advantages.

7.7.4.1 Intensity Reflection Spectrum Analysis

This is the simplest approach. It requires knowledge of the intensity reflection spectrum $R(\lambda)$ measured using an optical filter. The distribution in $\lambda_B(z)$ is found from the following integral equation [93]:

$$\int_{\lambda(z=-(L/2))}^{\lambda(z)} -\ln[1 - R(\lambda')]\, d\lambda' = \pm\frac{\pi^2}{2} \int_{-(L/2)}^{L/2} \frac{\Delta n^2(z')}{n(z')}\, dz' \tag{7.28}$$

where $\Delta n(z)$ is the modulated refractive index that must be accurately known. A further limitation is that only profiles that result in monotonic $\lambda_B(z)$ can be interpreted with this technique. For an initially uniform grating, only continuously increasing or decreasing strain or temperature profiles along the grating can be measured and the technique does not give the sign of the gradient. If, however, one uses an initially chirped grating, measurand fields that are nonmonotonic can be measured, as long as the total effect of the measurand gradient is limited so as not to cancel the pre-chirp bias of the grating. Chirped gratings also increase the spatial resolution of the measurement. The position dependant resolution is given by [93]

$$\delta z(z) = \frac{1}{2} \sqrt{\frac{2n_0\Lambda_0^2}{\left|\dfrac{d\lambda_B(z)}{dz}\right|}} \tag{7.29}$$

Figures 7.27 (a) and (b) illustrate the use of this approach using a chirped grating 19 mm long and with a 33-nm bandwidth. A 70°C hot air jet results in a nonuniform temperature profile along the grating. The final result shows that the changes in $\lambda_B(z)$ are small compared to the initial chirp, giving a calculation of $\Delta\lambda(z)$ and $T(z)$ with reasonable accuracy [1].

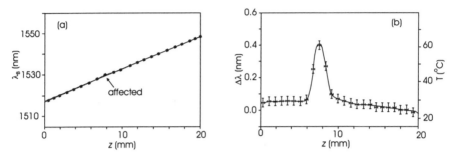

Figure 7.27   (a) Calculated $\lambda_B(z)$ and assumed $\lambda_B^0(z)$ with (b) evaluated $\Delta\lambda(z)$ and $T(z)$ using an intensity-based intragrating approach (*After*: [1]).

7.7.4.2 Intragrating Sensing Through Phase Measurement

The phase response of the grating can be obtained by incorporating the grating as a broadband reflector in a Michelson interferometer configuration. The group delays of the light propagating in the two interferometer arms differ by [97]

$$\Delta\tau = \frac{\partial\phi}{\partial\omega} = \frac{2nz_p}{c} \tag{7.30}$$

where φ is the measured phase difference, $\omega = 2\pi c/\lambda$ is the angular frequency of the light, and z_p is a penetration depth that defines along the sensing fiber where the optical path imbalance with the reference arm is zero. Ignoring the optical fiber material dispersion, the effective penetration depth with wavelength is given by

$$z_p(\lambda) = -\left(\frac{\lambda^2}{4\pi n}\right)\frac{\partial\phi}{\partial\lambda} \tag{7.31}$$

When the grating is monotonic and has a large enough chirp, the center of the reflection of light at a given wavelength is the point where the local Bragg condition is satisfied; therefore, $z_p(\lambda)$ is just the inverse of $\lambda_B(z)$. The advantage of this approach compared with the intensity approach is that the sign of the gradient can be unambiguously determined. Additionally, one does not need to accurately know the refractive index profile, as long as it is sufficiently high and varies relatively smoothly so that its variation does not affect the phase of the reflected light. Figure 7.28(a) shows the change in phase with wavelength derived from the raw interferometric response of a chirped grating Michelson interferometer [98]. The grating has a 33-nm bandwidth centered at 1535 nm and a physical length of 19 mm; the interferometric balance point occurs ~80% along the length of the chirped grating. Figure 7.28(b) shows the derived Bragg wavelength shift both with and without the grating subject to a thermal gradient over a ~2-mm section.

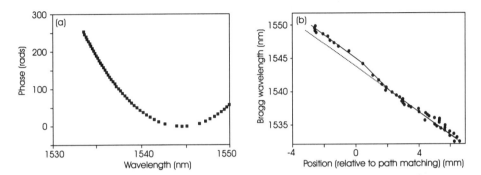

Figure 7.28   (a) Interferometer phase response versus wavelength. (b) Bragg wavelength shift for the grating subject to a thermal gradient over a 2-mm fiber section near the center of the fiber grating (*After*: [98]).

7.7.4.3 Combined Reflection Spectrum and Phase Measurement

Huang et al. [97] have exploited the fact that for low grating reflectivity, intensity and phase responses are essentially equivalent to the Fourier transform of the grating's structure and this can lead to the determination of an arbitrary measurand profile based on the combined measurement of phase and intensity. This approach removes the monotonicity of the fields and gives a spatial resolution that is not limited by the effective length of the BGS. This

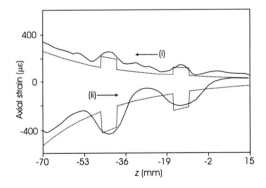

Figure 7.29   Comparison of measured and expected strain profile along a fiber Bragg grating in tension (−300 με) and compression (600 με) (*After*: [99]).

does require the simultaneous measurement of intensity and phase, making this approach complex. A practical demonstration of this approach is presented in [99] using an ~86-mm long grating sensor for the measurement of a nonmonotonic strain variation along a nonuniform cross-section cantilever beam. A theoretical strain accuracy of ~25 με was anticipated with a measured spatial accuracy of 2.58 mm. Figure 7.29 compares the measured and expected strain profile along the Bragg grating under tension and compression.

7.7.4.4 Low Coherence Reflectivity Measurement

Volanthen et al. [100] have demonstrated the first real-time measurements of strain fields along a fiber grating, allowing for the measurement of arbitrary strain fields, to an accuracy of 12 με. The experimental setup is similar to the phase measurement and the setup described in [101]. By using low coherence interferometry, a small section of the grating is selected and the Bragg wavelength of the selection is measured using a tunable filter. The sequential interrogation of many small sections gives a fully distributed wavelength profile. A single-mode Michelson interferometer incorporates the grating-under-test in one arm and a narrowband grating in the reference arm to provide a wavelength-tunable reflection. A stretcher and PZT are used to stretch and modulate the OPD of the reference arm. This generates a rapidly varying optical signal at the detector that has peak amplitude when the OPD of the interferometer is optimized for the wavelength selected by the reference grating. In this way, a plot of optimal wavelength versus OPD is obtained with the wavelength change being proportional to the measurand strain or temperature, $\lambda_B(z)$ [102]. The interrogated grating ($L = 3$ cm, bandwidth = 3 nm, $\lambda = 1531$ nm, $R = 40\%$) was characterized by the reference grating (bandwidth = 0.18 nm, $\lambda = 1529.5$ nm, $R = 92\%$). Figure 7.30 shows the extension applied to the reference grating as a function of the position of the fiber stretcher used to vary the delay, for the interrogated grating when unstrained (solid line) and strained (dashed line) to 1500 με in the center section of the

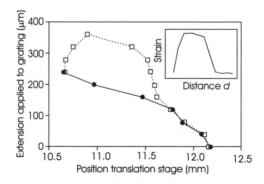

Figure 7.30 Characterization of interrogated grating using low coherence reflectometry when strained and unstrained. Inset: measured strain against distance for interrogated grating (*After*: [100]).

grating. The plot of the strained grating shows the underlying chirp of the grating with an additional strain superimposed over a certain region. The measured strain was calculated by subtracting the extensions for the unstrained plot from those for the strained plot and is shown in the inset, for a measurement accuracy of ~50 μɛ.

An AOTF (bandwidth = 3 nm) has also been used for the receiver to perform the wavelength selection with a broadband mirror replacing the reflective grating [100]. The location of the interrogated region in the grating is selected by balancing its path length with that of the mirror in the reference arm. The AOTF is then tuned to find the Bragg wavelength at the chosen point. Measurement of the original grating profile gave a noise level of 18 pm (~12 μɛ) for a 100-ms measurement time. High spatial resolution requires a wide filter bandwidth to give a small coherence length, at the expense of spectral resolution; nevertheless, the spectral accuracy can be improved using a lock-in system. LeBlanc and Kersey [103] have presented a similar approach to Volanthen [100]; however, a FPF provides wavelength tuning, offering the possibility of improving the spatial resolution by changing the FPF bandwidth. An Erbium-doped fiber source is filtered through a FPF that sets the effective coherence length, the chirped grating sensor is interrogated by scanning the OPD of the reference arm of the Michelson interferometer, and the FPF determines the wavelength coinciding with maximum fringe visibility. Measurement of the original grating profile gives a rms deviation of 0.015 nm for a 0.24-nm FPF bandwidth. The system spatial resolution is limited by the effective coherence length, in this case limited by the FPF bandwidth to 4.3 mm ($L_{coh} = 2\lambda^2/(\pi n_{eff}\Delta\lambda^{FP})$) for better than 12-μɛ strain resolution. The spatial resolution can be improved by increasing the FPF bandwidth. The ultimate limit, however, will be the effective interaction length of the light reflected from the grating, at 1.5 mm, corresponding to a 0.69-nm FPF bandwidth. Therefore, increasing the bandwidth of the FPF beyond this effective interaction length does not improve the spatial resolution; it is now determined by the grating and not the FPF.

7.8 Distinguishing Bragg Grating Strain Effects

In the previous section we presented several methods for determining the nonmonotonic strain profiles acting along the grating sensor length, broadening the reflection spectrum, and developing multiple reflection peaks, eventually leading to the breakdown of the Bragg condition. In this section we make the distinction between axial and transverse strain loads and their effect on the recovered wavelength spectrum. The action of unequal diametric strain can also produce the aforementioned spectral response [104]. It is important to be able to distinguish between these two strain effects.

The reflection distribution from a grating experiencing a general strain state is given by [105]

$$I(\lambda) = \frac{I_0\sqrt{\alpha/\pi}}{L}\left[P_x \int_{\lambda_x}\int_0^L S(\lambda)\, e^{-\alpha(\lambda-\lambda_{Bx}(z))^2}d\lambda dz + P_y \int_{\lambda_y}\int_0^L S(\lambda)\, e^{-\alpha(\lambda-\lambda_{By}(z))^2}d\lambda dz \right] \quad (7.32)$$

with P_x and P_y given by

$$P_x = \frac{\left\langle E_{0x}^2 \right\rangle}{\left\langle E_{0x}^2 \right\rangle + \left\langle E_{0y}^2 \right\rangle} \qquad\qquad P_y = \frac{\left\langle E_{0y}^2 \right\rangle}{\left\langle E_{0x}^2 \right\rangle + \left\langle E_{0y}^2 \right\rangle} \qquad (7.33)$$

where $\left\langle E_{0x}^2 \right\rangle$ and $\left\langle E_{0y}^2 \right\rangle$ are the time-averaged scalar magnitudes of the square of the electric fields in the $\hat{x}$ and $\hat{y}$ directions respectively. The grating length is L, reflected intensity I_0, α is related to the bandwidth of the grating, and $S(\lambda)$ is the source envelope spectrum. Equation 7.33 takes into account the variation of the reflection spectrum along the length of the grating from strain gradients and for the two polarization directions under unequal diametric strain. For a grating experiencing the most general thermo-mechanical strain conditions, the change in the reflection wavelength is given by [106]

$$\Delta\lambda_{Bx}(z) = \lambda_0 \varepsilon_z(z) - \frac{\lambda_0}{2} n_0^2 (P_{11}\varepsilon_x(z) + P_{12}[\varepsilon_z(z) + \varepsilon_y(z)]) + \lambda_0 \xi \Delta T \qquad (7.34a)$$

and

$$\Delta\lambda_{By}(z) = \lambda_0 \varepsilon_z(z) - \frac{\lambda_0}{2} n_0^2 (P_{11}\varepsilon_y(z) + P_{12}[\varepsilon_z(z) + \varepsilon_x(z)]) + \lambda_0 \xi \Delta T \qquad (7.34b)$$

where the component of the light with its polarization vector in the $\hat{x}$ and $\hat{y}$ direction, respectively. Here, n_0 is the mean refractive core index, P_{11} and P_{12} are the strain-optic coefficients of the fiber, ΔT is the local temperature change, ξ is the corresponding fiber thermal phase coefficient, and $\varepsilon_i(z)$ are the three orthogonal normal thermo-mechanical strain components (sum of mechanical and thermal strains) experienced by the grating ($i = x,y,z$).

To examine the strain effects, a grating is illuminated by a broadband source, with the grating reflection spectrum passing through a polarizer before being measured with an OSA. Reflection spectra very similar in form are measured under unequal transverse strains and gradient axial strains due to debonding, even though the origin of the applied strain is very different in each case. The polarizing element shows that the output spectrum of the grating, when subjected to unequal transverse strains varies as the polarizer is rotated, whereas the gradient axial strain that results from debonding does not. These results are shown in Figure 7.31 (a) and (b). To simulate a real-world situation, a Bragg grating embedded in kevlar epoxy was also studied. Force was applied to induce both a transverse and a strain gradient using a combination of bending and contact stresses. The insertion of the polarizing element allows for the origin of the strain inducing the reflection peaks to be distinguished (Figure 7.31(c)). The transmitted intensity of the two shortest wavelength peaks varies as the polarization axis is rotated, while the remaining three reflection peaks maintain the same transmitted intensity level. Therefore, the two short wavelength peaks result from unequal transverse strain, while the remaining peaks are due to the gradient axial strain, removing the measurement ambiguity.

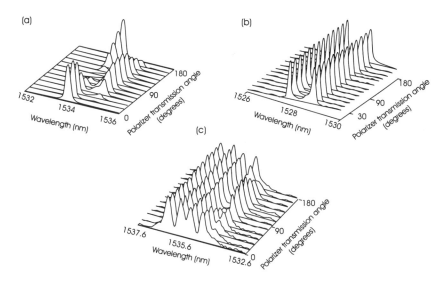

Figure 7.31 Effect of (a) unequal transverse strain distribution, (b) gradient strains resulting from debonding, and (c) combined strain state for wavelength versus polarizer angle (*After:* [105]).

Jin et al. [107] have reported the combined use of an in-line fiber etalon (ILFE) and Bragg grating for the simultaneous measurement of axial and transverse strain, resulting from the different sensitivities of both sensors. ILFE sensors are sensitive to axial but insensitive to transverse strain, whereas BGS are intrinsically sensitive to both axial and transverse strain. The precision of the ILFE and BGS measurements is ~1% and 5% for the axial and transverse strains, respectively. Udd et al. [108] have demonstrated the overlaying of dual-fiber Bragg gratings into polarization-preserving fibers for the measurement of the two axes of transverse strain by measuring the peak to peak spectral separation between fiber gratings along each of the birefringent axes. When the grating is subject to transverse strain, the closely spaced grating peaks associated with the different polarization axes separate, as the relative birefringence of the two axes is affected. Under longitudinal strain both peaks corresponding to the two birefringent axes will be affected in a similar manner. It is found that the transverse strain sensitivity is approximately a factor of 6 less than the longitudinal strain sensitivity for a commercially available fiber. Bjerkan et al. [109] have investigated the effects of transverse compressive loading of Bragg gratings by investigating the induced birefringence for gratings written in fibers with different buffer coatings, acrylate, and polyimide. For high loads (>0.5 N/mm) the birefringence behavior is nonlinear and coating dependent.

7.9 Bragg Grating Fiber Laser Sensors

Fiber Bragg gratings may also be used as narrowband reflectors for forming in-fiber laser cavities. The basic Bragg grating laser sensor employs two Bragg gratings of matched

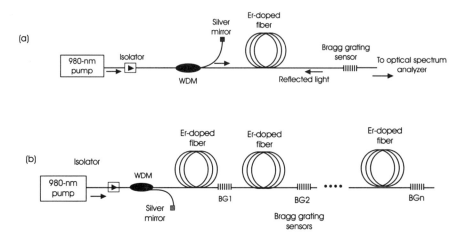

Figure 7.32 Two basic fiber Bragg grating laser sensor systems: (a) short cavity fiber Bragg grating-pair lasers, and (b) extended cavity fiber Bragg grating lasers.

wavelength to create an in-fiber cavity, or one grating combined with a broadband reflector, with Er-doped fiber as the usual gain medium (Figure 7.32(a)). Changes in the lasing wavelength occur in response to external perturbations acting on the Bragg reflector, located at a distal sensing point. Recovery of the laser wavelength using filtering techniques, as per the demodulation schemes for passive gratings, can then be used to determine the induced strain/temperature shifts with any temperature and strain perturbation acting on the grating alone producing the same responsivity as measured for passively monitored devices. With grating laser sensor configurations, however, one may also detect effects such as the beating between different longitudinal cavity modes or polarization modes in the system [110]. The device can be implemented in a variety of ways and may be operated in either a single or multimode fashion [111–115]. Similar to basic Bragg grating sensing, the inherent wavelength division addressing capabilities of gratings also allows distributed laser sensors to be implemented (Figure 7.32(b)) [114, 116]. The main advantage to laser-based sensors is an increased signal-to-noise ratio over the passive fiber grating sensors described in Section 7.3, while also providing a line-narrowed signal compared with light reflected from a passive grating element, therefore offering greater measurement resolution. Of course, a calibrated, tunable fiber laser may also serve as a source and demodulator to a passive grating array. The most immediate disadvantage is an incompatibility in extending to a multielement, multipoint sensing array, as one cannot simply concatenate additional fiber gratings to share the same length of active fiber, although we shall present methods that overcome this issue. The many facets of tunable single-frequency devices incorporating Bragg grating elements are dealt with in Chapter 6.

7.9.1 Single and Multipoint Bragg Grating Laser Sensors

The basic Bragg grating laser sensor is suitable for the inclusion of only a single grating; addressing more elements requires a suitable tuning element within the cavity that sequentially tunes to each output coupler, maximizing the cavity gain at the newly selected wavelength. The insertion of a tunable, narrowband, intracavity FPF has been shown by Kersey and Morey [116] to permit the operation of a four-sensor array in a linear cavity incorporating four series-connected grating reflectors (Figure 7.33). Selective laser oscillation occurs at one of the Bragg wavelengths when the transmission passband of the filter coincides with the wavelength of one Bragg grating element. The section containing the Er-doped fiber gain section is a loop reflector that operates in a unidirectional manner through the inclusion of an isolator. The other cavity reflection point is defined by one of a series of Bragg grating elements at different wavelengths connected to the cavity via a conventional fiber link of arbitrary length. Selective lasing allows for each of the grating sensor elements to be interrogated by assigning each to a particular operational wavelength domain, none of which are overlapping and all of which fall into the gain bandwidth. In this way tuning the intraloop transmission wavelengths over the bandwidth of the gain allows for sequential addressing of each Bragg grating with the ring laser forced to lase only at wavelengths determined by the Bragg grating center wavelengths. The system strain resolution was limited by the wavelength discrimination of the OSA to $\pm$ 25 $\mu\epsilon$.

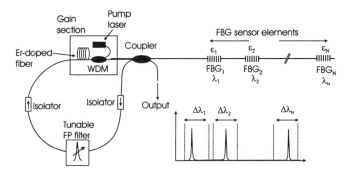

Figure 7.33   Fiber laser sensor configuration with multiple fiber Bragg grating sensor elements (*After*: [116]).

The linear resonator with a series of grating reflectors has also been constructed with an intracavity, acousto-optic modulator serving as a modelocker (Figure 7.34) [113]. Here each grating forms a cavity of a different length measured from the common broadband reflector; therefore, each cavity is modelocked to a different modulator drive frequency, permitting selective interrogation of the individual sensors, if (with reference to Figure 7.34)

$$f_1 = \frac{c}{2nL_0} \qquad \text{or} \qquad f_2 = \frac{c}{2n(L_0 + \Delta L)} \qquad (7.35)$$

However, it is important that the gain-grating reflectivity product at the two Bragg wavelengths should be approximately equal to avoid competition between the modes of the different cavities.

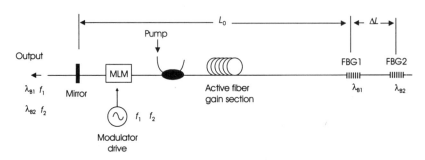

Figure 7.34   Mode-locked fiber Bragg grating/fiber laser cavity with collinear cavities (*After*: [113]).

Alavie et al. [114] have overcome the single-sensor limitation with up to three grating elements by inserting an additional length of active fiber between successive gratings (Figure 7.32(b)). The Bragg wavelengths must be sufficiently separated, and a pump source with sufficient power is required to induce simultaneous lasing at the three wavelengths defined by these gratings. The advantages here are the inherently low cross-talk and that the outputs can be analyzed simultaneously. In a single-sensor version of this setup, a strain resolution of ~5 με was measured with a 13-kHz frequency response [112].

7.9.2 Ultra-High Resolution Bragg Grating Laser Sensor Demodulation

Another laser sensor approach utilizes a linear-cavity fiber laser formed by a pair of matched in-fiber gratings as the sensor element with several sensors connected in series in a single length of Er-doped fiber for multisensor operation [117]. The grating laser may be as short as ~3 cm [118] and behaves as a sensor with a gauge length equal to the length of the cavity, offering the same spectral response as a normal grating element, assuming that the entire cavity is exposed to uniform strain. One advantage of the laser sensor system is that the bandwidth of the cavity sets the wavelength resolution of the system, which can be much narrower than the passive grating bandwidth. The drawback to this system is that each grating element of the sensor pair must be subject to the same perturbation. The final point is important, as nonuniform strain would break down the lasing condition. One of the most important attributes of this sensor type is the extremely narrow linewidth, from 10–50 kHz, which permits the detection of very weak dynamic strain signals, particularly if coupled with phase-sensitive demodulation schemes. Figure 7.35(a) shows the basic form of a fiber-grating laser with light directed to the unbalanced Mach-Zehnder interferometer. As with the passive demodulation scheme of Section 7.3.2.1, the strain essentially frequency modulates the laser output and the resulting small wavelength changes are demodulated using an unbalanced Mach-Zehnder interferometer as output phase shifts.

For a combined broadband source/passive grating arrangement, the bandwidth of the reflected signal is typically greater than 0.1 nm, restricting the OPD of the receiving interferometer to short lengths that lie within the effective coherence length of the reflected light. Conversely, the grating laser can have a coherence length of several kilometers, allowing for a very large OPD in the receiving interferometer that enhances system sensitivity to [117]

$$\frac{\Delta\Phi_{MZ}}{\Delta\Phi_{FL}} = \frac{\Delta L_{MZ}}{2L_{FL}} = G \qquad (7.36)$$

which compares the phase shift that would occur in a resonant linear cavity of length L_{FL} at a fixed wavelength, for a particular strain level, to the MZI output phase shift, for a physical path imbalance of ΔL_{MZ}, with the phase at the output of the MZI being G times higher than that produced in a passive cavity. The enhancement G applies equally to single-mode lasers (with the proviso that G can be any value to a limit determined by the coherence length of the laser) and multimode lasers (here G is restricted to integer values within the coherence limit of individual modes). This ensures the MZI is biased to provide coherent addition of the phase response from each laser mode [117]. For example, for an imbalance of 100m, the optical frequency to phase response is ~3000 rad/GHz. If we assume a not unreasonable 1-μrad/Hz$^{1/2}$ phase detection sensitivity, using an interferometer, this leads to a strain resolution of ~2.5×10^{-15}/Hz$^{1/2}$, potentially exceeding that obtained by using direct, interferometric sensing. Figure 7.35(b) shows results from a strain resolution measurement made with a single-mode 1554-nm Bragg grating laser sensor (~2.5 cm Er/Yb cavity) with a 96-m path imbalance in the readout interferometer, corresponding to a phase amplification factor $G = 1920$. The data indicate a phase noise level of ~26 μrad/Hz$^{1/2}$ at 7 kHz, corresponding to a dynamic strain resolution of ~5.6×10^{-14}/Hz$^{1/2}$, which is equivalent to a wavelength jitter of 8×10^{-11} nm/Hz$^{1/2}$.

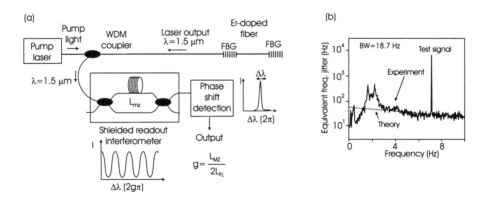

Figure 7.35 (a) Schematic of a fiber Bragg grating laser sensor interrogated by a Mach-Zehnder interferometer that converts the laser emission wavelength changes to interferometric phase changes. (b) Sensitivity of Bragg grating laser sensor system (*After*: [117]).

A multilongitudinal-mode laser sensor (longitudinal-mode linewidth <10 kHz), having a 3-m cavity length (L_{FL}), with a receiving interferometer OPD matched to an integer multiple of $2L_{FL}$, for a $G = 16$ ($\Delta L_{MZ} = 32L_{FL}$) has also been examined. Figure 7.36 shows that the interferometer noise floor is 13 dB better than that obtained with a single frequency laser (at 7 kHz), translating to a dynamic strain resolution of $7\times10^{-15}/\text{Hz}^{1/2}$, or equivalent wavelength jitter of 8.5×10^{-12} nm/Hz$^{1/2}$. The strain resolutions are limited by thermal cavity fluctuations in the fiber laser cavity. The Bragg grating lasers may also be interrogated by using their inherent WDM capabilities, for which four, single-frequency fiber laser sensors have been interrogated with a single readout interferometer [119]. A strain resolution of $7\times10^{-14}/\text{Hz}^{1/2}$ was measured that is comparable to that for a single fiber laser sensor, with a cross-talk level between channels of < -60 dB. The extremely high sensitivity, comparable to that obtained with direct interferometric sensing, could be used to realize a strain, acoustic pressure, or magnetic field sensors.

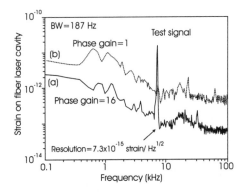

Figure 7.36   Spectral response of the Mach-Zehnder interferometer to the fiber Bragg grating multimode laser output with a cavity modulation signal. Trace (a) $L_{FL} = 3$ cm and MZI OPD = 96m. Trace (b) $L_{FL} = 3$ m and MZI OPD = 6m (*After*: [117]).

7.10   Bragg Gratings as Interferometric Sensors and Reflective Markers

Fiber Bragg gratings have also been used to form interferometric sensor elements, where the gratings serve as in-fiber reflectors, defining the interferometer dimensions and hence the properties of the in-fiber Fabry-Perot or Michelson. Alternatively, they can be used purely as reflective markers, with their position monitored through OTDR techniques.

7.10.1   Reflectometric Sensing Arrays Using Bragg Reflectors

Among the first demonstrations of optical fiber multiplexing topologies was the use of reflective markers (in this case mechanical splices) to define sensing segments along a continuous length of optical fiber, forming an array of Fizeau cavities, interrogated using TDM [120, 121]. In this system the magnitude of the cross-talk is proportional to the

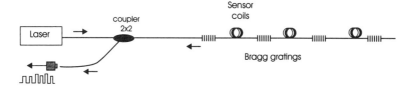

Figure 7.37 Serially multiplexed interferometric sensors based on in-fiber Bragg gratings as reflectors.

marker reflectivity, and therefore, low and consistent reflectivities are required. Using mechanical splices, however, severely limits the successful application of this technique. If, on the other hand, Bragg gratings are used, a reproducible, low insertion loss, in-fiber partial reflector, may be realized. Bragg gratings are particularly promising as their wavelength selective nature offers the possibility of implementing WDM/TDM arrays with low sensor cross-talk [122]. Figure 7.37 illustrates the basic configuration for serially multiplexing Bragg grating reflectors. The input light is square-wave modulated to produce a double-input pulse, with the two pulses separated by a distance equal to the round-trip delay time between the partial reflectors. The output pulses reflected from consecutive partial reflectors can now overlap, allowing for interferometric mixing. One may produce beat frequencies from the overlapping pulses at the output by setting the optical frequencies of the two input pulses to be different. Alternatively, one may input a single pulse, producing a train of temporally separated output pulses that are then fed into a compensating interferometer of OPD equivalent to the round-trip time delay between the reflective markers. In this way portions of light in adjacent pulses interfere, producing the desired interferometric output. An inherent feature of reflectometric and tapped serial array topologies is that multiple reflection paths give rise to cross-talk through multiple pulse interactions. By capitalizing on the wavelength dependence Vohra et al. [122] have presented an approach to circumvent this issue through a dual wavelength, four sensor TDM array (i.e., combined WDM/TDM). The array, operating on a single fiber, is composed of four high-sensitivity Fabry-Perot interferometric sensors, 40 m in length, each formed by a pair of in-fiber Bragg gratings acting as reflection elements. Wavelength filters at the output reduce the cross-talk level between the different wavelength channels, while radio frequency (RF) modulation of the laser sources is implemented to heterodyne any excess phase-induced intensity noise out of the baseband, spreading the excess noise power over a larger bandwidth at the detectors. Cross-talk between the sensors operating at dissimilar wavelengths is measured to be below −60 dB (sensor noise limited) when wavelength filtering is used and ~35 dB without wavelength filtering. The cross-talk between sensors operating at the same wavelength is measured to be <−30 dB. The RF modulation of the laser frequency was shown to improve the array noise by >15 dB.

Geiger and co-workers [123] have presented a novel OTDR approach for addressing the optical path length to multiple reflective markers, based on simple optics and sophisticated signal processing. The OTDR interrogates the distance between the gratings, and any perturbation acting on the grating changes the time delay between the reflected signals. Using a modified electrically coherent receiver (correlator) to detect the reflections from the fiber has been found to enhance the range resolution over conventional OTDR. In conventional OTDR a delay is used that can only be discretely stepped, continuously sweeping the delay leads to the range resolution improvement, as the resulting triangular autocorrelation function of the pulse gives a precision measurement of the optical time delay. A pseudo-random bit sequence is used to illuminate the system, improving the duty cycle and hence the signal-to-noise ratio. The technique relies solely on delay information and as such is insensitive to amplitude and polarization changes.

7.10.2 Nested Fiber Interferometers Using Bragg Reflectors

An approach has also been demonstrated for configuring multiple nested interferometric sensor elements based on Fabry-Perot and Michelson interferometers with common fiber paths. This allows for the possibility of forming adaptive sensor arrays, or to implement specialized sensor configurations such as differential and vector based sensors; Figure 7.38 illustrates this idea [124, 125]. First, we have multiple coexisting interferometers within a single Michelson configuration (Figure 7.38(a)). The multiple interferometers are formed using a series of Bragg gratings between sensor and respective reference coils. The wavelength selectivity of the gratings produces an interferometer signal, the phase response of which is determined by the input source wavelength. Interrogation of the system at one of the wavelengths corresponding to one of the grating pairs in the system produces an interferometer output, the phase of which can be described by [124, 125]

$$\varphi_{\lambda j} = \sum_{i=1}^{j} \varphi_i \qquad (7.37)$$

where φ_i is the phase induced in the ith sensor coil ($i = 1$ to N). The phase at the jth wavelength is the sum of the phase induced at each sensor coil up to the last addressed element. Hence, the phase of any sensor coil, or combination of coils, can be assessed via

$$\varphi_j = \varphi_{\lambda j} - \varphi_{\lambda (j-1)} \qquad (7.38)$$

As shown, the reference coils can be included to maintain the optical path imbalance of each interferometer section constant. Second, a Michelson may be implemented to form a $\pm$ magnitude/gradient sensor formed by two sensing coils and a single reference length (Figure 7.38(b)). In this case, interrogation of the system at λ_1 provides a measure of the phase of the sensor coil S_1, λ_2 provides a measure of the phase of sensor coil S_2, and λ_3 provides a differential, or gradient, measurement. Finally, we have a serial array of sensor coils that can be read independently, in groups, or as an effective single sensor element by choice of the interrogation wavelength (Figure 7.38(c)). For an interrogation wavelength λ_1 the gratings at λ_1 define a series of sensors comprising single sensing coils that can be interrogated using a pulse source [120]. At other interrogation wavelengths, the sensors comprise groups of sensing coils, up to λ_4 in this case, which sees only one sensor comprising all the sensing coils. This is a basic form of adaptive sensor array. It can operate

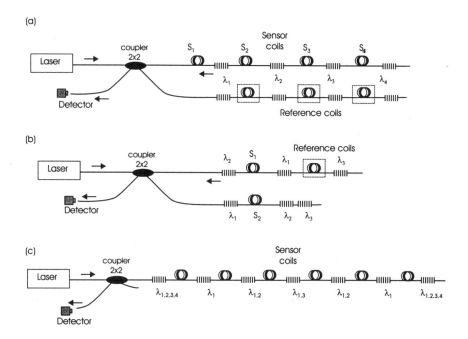

Figure 7.38 Nested interferometer configurations: (a) serial Michelson, (b) implementation for providing magnitude and gradient, and (c) nested Fabry-Perot array.

with different spatial resolution depending on the interrogation wavelength. This is a very basic geometrically nested array. The nesting of multiple interferometers using common fiber paths provides certain flexibility in designing interferometric sensors, especially for differential, vector, and spatially varying measurands. Nested interferometric arrays based on this concept are possible and may be useful for forming adaptive arrays where the array spatial properties are controlled via the interrogation wavelength.

A three-element interferometer (Figure 7.38(b)) has been constructed using gratings at 1535, 1540, and 1545 nm to demonstrate the potential utility of this arrangement [124, 125]. Signals are applied to S_1 and S_2 of 6π and 4π, respectively, resulting in a sensor output of three fringes at λ_1 and two full fringes at λ_2, whereas λ_3 results in a differential output of one full fringe. To further demonstrate the differential operation, the sensor coil S_1 is driven by a strong noise signal and a weak tonal signal at 700 Hz, whereas coil S_2 was driven by noise alone. The recovered spectra for wavelengths at λ_1 and λ_2 are predominantly noise, whereas the output observed for λ_3 provides cancellation of the noise, and yields the weak tonal signal.

7.10.3 Bragg Grating-Based Fabry-Perot Sensors

To provide high sensitivity to temperature and strain effects and to avoid inter-grating cross-talk, BGSs are made with very narrow bandwidths. This is at the expense of the total reflected light level, and in the extreme case of infinitely narrow bandwidth (and in principle maximized wavelength-shift resolution) the returned light level tends to zero. One may follow another approach and use two narrowband Bragg gratings as in-fiber reflectors to form a Fabry-Perot (FP) cavity. The reflectivity and bandwidth of the gratings determine the exact FP properties. For example, a pair of identical, high-reflectivity (95.5%) in-line Bragg gratings with a 10-cm center-to-center spacing have been shown to produce an etalon with characteristic FP resonance peaks when illuminated with a suitable laser source. A finesse of 67 was recorded for a 15-MHz band-pass [126]. The bandwidth of the Bragg gratings is ~53GHz and the FP FSR ~1 GHz; therefore, more than 53 fringes occur within the grating band-pass. The linewidth of the etalon is more than 3500 times smaller than that of the grating, which gives a corresponding improvement to measurand sensitivity. One may therefore anticipate that monitoring the shift in the FP band-pass, or the phase difference between light signals from the two grating reflectors, can result in a very high-resolution sensor. Alternatively, the mirror reflectivities may be kept low and not necessarily the same, producing a low-finesse FP etalon. In this case the transmission spectrum is cosinusoidal in nature. For practical sensor demonstrations, low coherence interrogation of low finesse FPs is highly attractive. This entails having a path-matching condition necessary to observe fringes from a composite sensor network and requires two unbalanced interferometers, one receiver and one sensor (FP), having approximately the same OPD. Illumination with a source whose coherence length is far shorter than the path imbalance in either individual interferometer does not result in fringes, but if the interferometer path imbalances are matched to within the coherence length of the source composite cosinusoidal fringes are observed with reduced visibility (30% maximum). The fringe packet is modulated by a Gaussian envelope of width inversely proportional to the spectral width of the light reflected from the FP mirrors (Bragg gratings). The advantage here is for absolute, interrupt-immune measurements. Narrowband Bragg grating–based FPs are unsuitable for use as low coherence sensors because even when illuminated with broadband light the reflected signal is highly coherent; chirped gratings have a broader bandwidth that overcomes these limitations. Otherwise, long gauge length sensors must be used with $OPD_{FP} >> l_{geff}$. A multiplexing scheme based on this principle has been presented in [127]. Two Bragg gratings are used as the mirrors of each FP cavity with the same mean reflecting wavelengths but different reflectivities (38% and 100%) in order to maximize the contrast of the signal back-reflected from the FP sensor. Additionally, the mean reflecting wavelength of each FP sensor is different, allowing for WDM interrogation of each sensor. A fiber Mach-Zehnder interferometer with a 2-m path imbalance is used as the path-matched filter. The round-trip optical path length within each FP is also 2m. Although the superfluorescent fiber source has a low coherence length of ~60 μm, the path-matching condition is actually determined by the effective coherence length of the light reflected by the FP at ~1.2 cm. As an example of the system performance, the sensors demonstrated very high respective strain resolutions of 0.2 and 0.09 nε/Hz$^{1/2}$, in response to

12- and 25-kHz strain signals, for a 100-kHz carrier and a 1-kHz measurement bandwidth. Cross-talk is buried in the system noise. Kaddu et al. [128] have demonstrated an improved low coherence FP sensor using three in-line Bragg gratings (a triple FP sensor) for improved identification of the center of the recovered fringe pattern.

7.10.4 Collocated Fabry-Perot Cavity with Wavelength Addressable Cavity Lengths

Single FP cavities have a fixed FSR, and for a given finesse, a predetermined wavelength resolution (bandwidth dependent). Therefore, a large FSR can only be obtained at the expense of resolution. In sensor applications a high resolution translates to high sensitivity, whereas a large FSR translates to a large dynamic range; these are both desirable quantities in practice. With this in mind, Koo et al. [129] have proposed and demonstrated a wavelength-encoded FP sensor with a range of addressable free spectral ranges allowing multiple dynamic ranges and resolutions to be realized in a single sensor head. The FP sensor is made of two collocated chirped gratings arranged such that their chirped direction is opposite to each other, thereby forming multiple wavelength-dependent FP cavities with different cavity lengths (Figure 7.39(a)). If the chirp directions are the same, then each multiple grating will have equal FSRs; this is simply the case of a chirped grating used as a broadband mirror [130]. For the collocated chirped gratings the largest FSR corresponds to the shortest cavity length L_{min} associated with an interrogation wavelength at the center of the passband of the grating. Conversely, the smallest FSR or largest cavity length L_{max} is associated with interrogation wavelengths near the edges of the grating passband. The

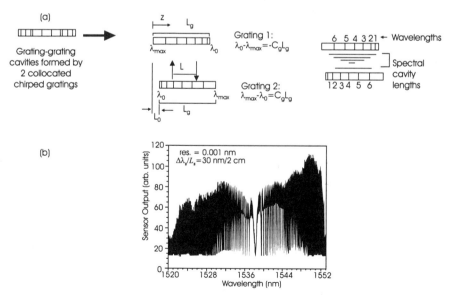

Figure 7.39 (a) Illustration of a chirped fiber grating Fabry-Perot sensor with wavelength-dependent cavity lengths; (b) spectral response of the sensor measured with a tunable laser (*After:* [129]).

opposite wavelength chirp orientation produces reflection symmetry in the device spectral response, resulting in two interrogation wavelengths corresponding to each cavity larger than L_{min}. Figure 7.39(b) shows the spectral response from 1520 to 1552 nm. Responses at 1551 nm have smaller FSR and higher resolution, whereas 1539 nm has a larger FSR and lower resolution. These FP cavities provide effective FSRs from 0.04 to 1.3 nm with a 26-dB dynamic range for static strain, from 3 to 1300 µε. The strain resolution is limited by the wavelength resolution, here derived from a tunable laser source. It is expected that interferometric techniques could result in far higher resolutions. Therefore, by probing the chirped grating sensor at both 1539 and 1551 nm, both large dynamic range and high resolution have been achieved.

7.11 Other Bragg Sensor Types

There are also new developments utilizing specially modified or tailored gratings. These include π-phase shifted devices, for which a strain sensitivity of 0.5 nε/Hz$^{1/2}$ has been measured [131], multimode gratings [132], and superstructure gratings [133].

7.12 Applications of Bragg Grating Sensors

In this section we provide examples of applications for Bragg grating sensors and attempt to present the myriad applications that have most recently been discussed in the open literature.

7.12.1 Introduction to Aerospace Applications

The aerospace industry is a potentially important user of optical fibers, particularly as data links, which are considered a mature technology following a decade of intense development. Suppliers of military aircraft, such as Westland Helicopters, already use such components in their Lynx helicopters as part of replaceable elements in their communication links, with projects such as the Boeing 777 and Eurofighter 2000 being potential candidates. Although research projects have shown that optical fiber sensors can operate within prescribed limits for use in aircraft, they are still considered an immature technology. To date efforts are directed towards the sensor development for harsh environments unsuitable for conventional electro-mechanical sensors, taking advantage of radiation resistance and EMI immunity. The aerospace industry is very conservative and new sensors must undergo long testing periods before major changes in instrumentation systems can be accepted in both civil and military aircraft. Given the economy of scales, however, optical sensors can only be successful if they "buy" their way onto the aircraft. Practically this means a sensor system weight or cost reduction compared with electrical sensors. Increases in sensor reliability and ease of installation and maintenance with little training and without special handling are demanded, ideally leading to the so-called "fit and forget" systems. Finally, standardizing sensors and instrumentation reduces part numbers and overall costs.

Sensing strategies for aerospace applications (airplanes, helicopters, missiles, and space vehicles) broadly follow the same conditions. The most important requirements are to have passive, low weight, and ideally common sensors that may be multiplexed over optical links. By carefully defining sensor requirements, it may be possible to specify a range of optical sensors, satisfying the majority of avionics applications that are either interchangeable or at least use common interrogation instrumentation (Section 7.3.3). Currently, many sensor types perform similar functions but are not interchangeable. Any new optical fiber sensors cannot exceed the size and weight of conventional electro-mechanical sensors; therefore, a sensor intrinsic to the core of standard fiber optic cable, for our purposes a Bragg grating, is highly attractive. The Bragg grating sensor solves one of the major drawbacks to optical fiber sensors: the lack of a standard demodulation approach, while maintaining a completely passive network. The largest class of sensors measures the position of flight control elements such as landing gear status, flap and rudder position, and so forth. When taking into account high levels of system redundancy (triple level for key equipment functions), well in excess of 100 sensors are employed. As a result, size and weight savings become important. Flight control sensors monitor fuel levels, vibration, and so forth. The sensor types used on the landing gear system of an Airbus Industries A340-type aircraft may be found in [134]. This is an area of high sensor density, in a critical application area where sensors are systematically exposed to harsh environments. There are a number of sensor categories, such as temperature, pressure, speed and proximity sensors, fluid level, torque and debris monitoring sensors.

7.12.1.1 Bragg Gratings for Use in Aircraft

Bragg grating sensors have potential aerospace applications as temperature, pressure, and strain sensors, and they are likely to find success as embedded sensors that monitor the performance and fabrication of reinforced carbon fiber composites and advanced testing of gas turbine engines, for which there is no acceptable competing technology.

Monitoring of Reinforced Carbon Fiber Composites–Curing and Impact Detection

A formidable task for any sensor is monitoring the fatigue life and failure of new aerospace materials, particularly reinforced carbon fiber composites (RCFCs). These materials are finding increasing use in airframes because of their extremely good mechanical properties. The composites and optical fibers share a similar material structure, and this complementary nature has led to their combination, forming a sensing structure called the "smart skin." When the sensor signal is used to drive an actuator modifying the structure for active damping and control, the system is known as a "smart structure." Advanced RCFCs are finding use on civil and military aircraft, and more recently, on space vehicles. These materials offer a strength-to-weight ratio greater than steel and yet can be molded into complex shapes at reasonable cost (e.g., through resin transfer molding). They are also radar transparent, giving aircraft very low radar cross sections. The materials consist of layers of plastic-impregnated fibers lying in parallel. Successive layers are stacked together with parallel, crossed, or 45-degree orientations between the plies, forming a sheet whose strength characteristics may be tailored to suit the specific application. The inherent anisotropy of these materials means their physical properties are strongly dependent on their orientation; therefore, it is of paramount importance that the final composite material undergoes complete curing. The fully cured composite is extremely flexible and can undergo very high dynamic strain levels of 1% over many cycles, without an apparent loss of strength. A major concern is the difficulty in detecting internal damage as surface-mounted strain gauges cannot readily be applied to this problem. Delamination can occur within the structure without any exterior evidence to the fact.

Moreover, impact damage is usually only visible underneath the point of impact, which is almost always enclosed within the airframe structure. Furthermore, the composite is susceptible to laser-induced damage through optical absorption of light and its subsequent conversion to heat. There are also compatibility issues between the RCFC and optical fibers; for example, fiber diameters are ~10 μm and 100 μm (thin clad fibers), respectively, leading to a significant mechanical disturbance to the lay of the carbon fibers. This disturbance does lead to a weakening of the local structure, but experimental data suggest that this is marginally less than that of the surrounding matrix. The manufacture of the RCFC consists of pressing the oriented fiber layers together with a bonding agent and heating from anywhere between 180°C and 400°C and with pressures up to 100 psi. For applications where the composite panel is thick (e.g., the roots of wing sections), the thermal treatment must be completed throughout the entire structure. An embedded sensor system could provide information in real time, enabling intelligent manufacturing process monitoring, such as in-process temperature profiling and strain release monitoring of the composite directly in the autoclave, thereby improving real-time manufacturing and product quality. Moreover, the sensor can improve the design of smart structures and skins for structural monitoring for distributed strain and temperature measurements. This applies to wings, fuselage, etc. for health monitoring aging aircraft and modern composite structures as well as for the prediction of residual lifetime of repaired systems, thereby

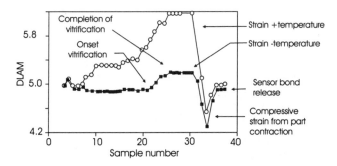

Figure 7.40   Cure monitoring of two multiplexed Bragg grating sensors at 824 and 829 nm located within 6-layer graphite/epoxy prepreg (*After*: [135]).

enabling true control of the structural integrity and maintenance. Furthermore, the detection of impacts and cracks and active structural control to provide information to actuators, such as vibration damping of flexible structures, shape control of structures under thermal or mechanical loads. These applications are equally important to aeronautics and naval applications.

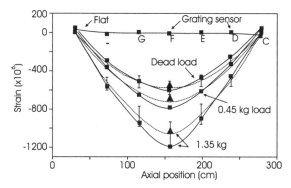

Figure 7.41 Strain measured in CRTM™ beam with Bragg grating sensors (■) and compared to finite element analysis (dashed line) and theoretical model (▲) (*After*: [136]).

Dunphy et al. [135] identified a number of benefits in combining optical fibers and composites. Gratings manufactured at ~825 nm were embedded in the plies of 6-layer graphite/epoxy prepreg material in a hot press for curing. One sensor was located within the uncured composite, with the other used as a reference, and the gratings' response were monitored during the 175°C/3500lb traditional autoclave curing program. Figure 7.40 confirms the onset of vitrification along with the build-up of residual strain during cooling. This was the first indication that curing program optimization and parts testing could successfully result from the use of optical fiber sensors. Friebele et al. [136] have embedded Bragg gratings fabricated during fiber drawing in composite panels made by the alternative continuous resin transfer molding process. The performance under axial strains is shown in Figure 7.41 along with finite element and theoretical analysis, indicating reasonable agreement. The measurement scheme assumes uniaxial stress along the grating, which is unlikely for this type of structure (Section 7.8). Tatam et al. have studied real-time internal strain development during composite curing and impact testing of RCFCs [137, 138]. Measurements made with embedded dielectric microsensors and BGSs show that it is possible to monitor strain levels resulting from the onset of liquification, gelation, and vitrification within the surrounding resin matrix. Figure 7.42 (a) and (b) show results for two sequential curing cycles. The strain information from the gratings demonstrates the difference in curing regimes with the rapid ramp down in temperature in Figure 7.42(a), resulting in a significantly higher strain than the final strain in Figure 7.42(b). The final strain recorded 2 weeks after curing induced a 0.16-nm wavelength shift, implying a residual strain of ~330 με (for a response of 0.48 pm/με at 836 nm), which is consistent with the results of Friebele [136]. There is also the possibility of compressive

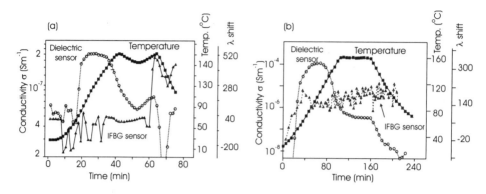

Figure 7.42 (a) First cure cycle for 934 carbon fiber composite, for a 3°C/min ramp to 180°C. (b) Second curing cycle for a 1°C/min ramp to 177°C (*After*: [137]).

strains contributing to the final output from accumulative resin shrinkage in directions not reinforced with carbon fiber [109]. The vulnerability of embedded optical fibers to impact damage indicates that the fiber location and orientation influence the acceptable impact level. Coated fibers embedded near the top surface survive 35-J impacts, unlike stripped fibers that fail regardless of sensor orientation. Coated fibers embedded at the lower surface fail and survive when oriented at 90 and 0 degrees to the plies, for the same impact level, respectively. Stripped fibers survive at 0-degree orientation to 15-J impacts. Table 7.4 summarizes the results and suggests locating the BGS on the upper surface. Figure 7.43 shows the result from 1-J impact measurements. The grating response closely matches that of the impact tower transducer, displaying the same high-frequency irregularities that are missing form the strain gauge data. The demodulation bandwidth exceeds 25 kHz compared to 2 kHz for the strain gauge. Finally, there is clear evidence that the distance of the fiber from the impact site is more important than the impact energy.

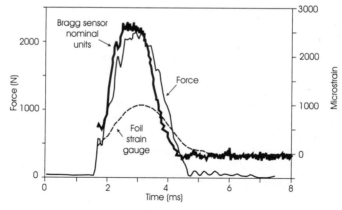

Figure 7.43 Bragg grating response to a 1-J impact with time and compared to strain gauge and force meter (*After*: [138]).

Table 7.4   Vulnerability of Stripped/Coated Optical Fibers to Impact Energies 2 mm from Embedding Point

Fiber orientation/ type	Surface location	Integration	Failure energy for coated	Failure energy for stripped
0° coat and strip	upper	embedded	no failure > 35 J	between 25 and 35 J
0° coat and strip	lower	embedded	no failure > 35 J	between 10 and 15 J
90° coat and strip	upper	embedded	no failure > 35 J	between 25 and 35 J
90° coat and strip	lower	embedded	between 25 and 35 J	not available
0° coat and strip	upper	surface mounted	no failure > 35 J	between 25 and 35 J
0° coat and strip	lower	surface mounted	no failure > 35 J	no failure > 35 J
90° coat and strip	lower	surface mounted	no failure > 35 J	between 5 and 10 J

After: [138].

Large-scale projects are underway and we give some interesting examples. In France the DRET/AIA-Cuers-Pierrefeu/ONERA-L3C and CEA/LETI consortium are assessing the integrity of a fighter aircraft radome through the detection of invisible impacts within the composite structure, such as delamination and cracks. Aerospatiale are largely involved in the design and/or manufacturing of the advanced composite structures for new engines and high-pressure tanks and have recently started development with CEA-LETI in order to equip a high-pressure tank with embedded Bragg grating strain gauges [139]. The Photonics Group at the NASA Ames Research Center is developing a nondestructive, sensitive pressure sensor for helicopter airfoils using Bragg grating sensors embedded in a specially designed glove, no greater than 1/16" thick, providing two-dimensional, real-time pressure readings in flight or wind tunnel tests [140]. Currently, helicopter wing or rotor pressure sensing uses specially drilled airfoils with individual pressure sensors located within the holes. This destructive approach requires the design and production of special purpose and expensive wings and rotors, costing in excess of $1 million for a single wing or rotor. Other applications include measurements of tail rotor loads, rotor/fuselage interaction, inboard rotor aerodynamics, rotor/rotor interactions for tilt-rotors, hub aerodynamics, and wake/tail interactions. In Sweden the Institute of Optical Research is involved in a national project named SMART together with the FFA to develop time multiplexed strain and temperature measurement systems for composite monitoring in military aircraft [139]. There is also interest in developing on-line, health and usage monitoring systems based on advance load monitoring and damage detection technologies. British Aerospace heads an 11 partner cooperation whose aim is to reach a reduction of 20% in inspection, within 5 years of the project completion date (1999) (i.e., ~$2 million per aircraft if related to a 20-year life). To achieve these goals, several sensing technologies are under investigation, one of which is BGSs [139].

Adaptive Structures

Active damping of a composite structure embedded with grating sensors was first reported by Dunphy and co-workers in 1990 by using a feedback loop-controlled PZ actuator pair to counteract the displacement of the structure under loading through monitoring the Bragg reflecting wavelength [135]. Recent adaptive structure approaches are more ambitious. For example, since 1996 in Germany, the Daimler-Benz Research Center, together with Daimler-Benz Aerospace Airbus and the DLR (Institute for Aerospace Studies), is investigating the development of an adaptive wing loaded with Bragg gratings. The technical aim of this 7-year project is to find a structural dynamic solution to optimize the aerodynamic performance of aircraft, while the economic objective is to incur lower fuel consumption, higher payload, and lower operating costs. These goals are to be realized by local contour modifications at the wing surface and with variable trailing edges. The structural changes are to be monitored by quasi-distributed embedded strain and temperature BGSs scanned by a tunable laser in the vicinity of 1550 nm [139].

Application to Gas Turbine Engines

The gas turbine engine environment is characterized by very high temperatures and pressures compounded by rapid gas flow and high-velocity, high-intensity acoustic waves. The thermal environment varies from extreme cold at the intake (-50°C) to emissions from the jet-pipe, at temperatures exceeding 1500°C. The various compression chambers require pressure measurements from 15000 kPa. We are unaware of the use of BGSs in this field but the potential certainly exists, with measurements largely centered upon high pressure and temperature, demodulated with schemes having no moving parts [90].

7.12.1.2 Space Vehicular Utilization

There is also interest in using BGSs in space-bound craft, where all structures are subject to high acoustic and vibration loads to simulate launch conditions. The Photonics Group at the NASA Ames Research Center supports the Reusable Launch Vehicle (RLV) Technology Program, designing specialized fiber Bragg sensors that can monitor the condition of the cryogenic tanks aboard an RLV [141]. This work is part of the X-33 technology program with Lockheed Martin, in partnership with NASA, the US Air Force, and industry directed at world leadership in low-cost space transportation. A critical aspect for RLVs is a health management system, one aspect of which is the maintenance of the cryogenic tanks that hold the liquid hydrogen and oxygen fuel. The current goal is a 6-hour recertification time for the tanks between each flight. Fiber Bragg gratings have been identified by Lockheed Martin as ideal candidate embedded sensors for measuring strain loads on the cryogenic tanks (range 2500 $\mu\varepsilon$, resolution 10 $\mu\varepsilon$). Fiber sensors are ideal since they provide accurate data with a small lightweight sensor that can be embedded into the composite walls and they present no spark hazard. The sensors are expected to operate in a wide temperature range from -250°C to 380°C.

Friebele et al. [142] have used BGSs for measuring the dynamic loads on a lightweight antenna reflector made of graphite-epoxy composite and having metal-epoxy fittings. The reflector surface is a thin composite membrane attached to supporting rings and struts. The Naval Research Laboratory attached Bragg gratings at the four corners of two struts in five different locations around the inner ring to measure longitudinal strain, bending, and torsion. The small dimensions of struts make it impossible to use conventional resistive strain gauges. The demodulation of 40 grating sensors has been achieved; an equivalent number of resistive strain gauges would result in a severe structural weight penalty. Comparisons between the electrical and optical sensors indicate excellent agreement, but the optical sensors display none of the electrical noise recorded by resistive strain gauges. In another application the dynamic strains of the advanced tether experiment (ATEX) composite support deck were measured with embedded gratings. Two four-element sensor arrays were embedded in one graphite-epoxy composite face sheet of an aluminum-core honeycomb support deck for the tether canister, with a 16-element array embedded in the other. The embedded grating sensors agreed with surface-mounted RSGs during vibration testing, demonstrating a system noise floor of <20 $\mu\varepsilon/\text{Hz}^{1/2}$ at 100 Hz, thereby easily detecting a 0.2-$\mu\varepsilon$ rms strain associated with a 92-Hz resonant mode of the supporting deck.

7.12.2 Bragg Grating Sensors in Marine Applications

To date the use optical fibers for marine applications has essentially been for military use, such as subsea acoustic measurements using fiber optic hydrophones. Applications to naval vessels are perhaps the only area where sensor instrumentation is commensurate to that used for aircraft, although the sensor demands are considerably less. Naval vessels are becoming increasingly computerized, utilizing complex data handling and transmission for which the light weight, space, and low cross-talk advantages offered by optical fibers for both communications and sensing are appealing. Most of the sensor types are generic and find applicability to other areas, such as in civil engineering. Moreover, there is great potential for down-hole instrumentation using fiber optics for exploration of natural oil resources, where the extremely harsh environments cannot be tolerated by conventional electronic sensors.

7.12.2.1 Transient Hull-Loading Effects and Composite Panel Slamming

Recently, trials have been conducted using Bragg gratings for the characterization of scaled marine vehicle models [143]. Measurements made on a scaled catamaran model recorded slamming forces between the wet deck (between the sea and two hulls) and sea waves, dynamic loads, and bending moments. The system was designed to measure slamming forces by locking a laser diode to the peak wavelength of a BGS mounted in a watertight test device (slam patch) made of a hollow aluminum box. This device is mounted in the wet deck of the catamaran model with the grating axis facing the sea loads. The sensor system had a dynamic range of 27 dB for a 1-kHz frequency range. A 16-channel system has been developed for composite panel slamming tests [144] to simulate transient loading forces based on the system reported in [80]. In this case the BGSs are bonded to the wet and dry side of the composite panel, mounted on a mechanical rig loaded with weights exceeding 100 kg, and dropped from a range of heights. This type of testing can also provide information regarding the survival characteristics of the grating sensors. The system displayed sub-10 $n\varepsilon/Hz^{1/2}$ resolution and bandwidth exceeding 5 kHz for $\pm$ 1000 $\mu\varepsilon$ maximum strain levels. The composite panel and sensors both survived testing, where an impact speed of 2 m/s and 20-cm drop height produced a 67-Hz primary frequency response. Increasing the height ninefold and the impact velocity to 6 m/s leads to higher order modes up to 1 kHz.

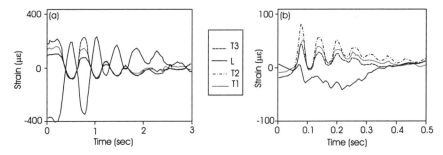

Figure 7.44 (a) Longitudinal and transverse strain induced by global oscillation of vessel hull. (b) Longitudinal and transverse strain induced by local vibrations (*After*: [145]).

A joint Naval Research Laboratory–Norwegian Navy program, Composite Hull Embedded Sensor System (CHESS), has the objective of instrumenting an in-service fiberglass mine countermeasure vessel with over 100 sensors by exploiting the static and dynamic strain measuring capabilities of Bragg gratings [145]. In high seas it is important to monitor dynamic strain loading of critical hull locations, and active warning should be given under extreme loading conditions by surface-mounting an array of Bragg grating sensors to the lower and upper hulls and the wet deck of the vessel. The sensor system is also based on [80]. Figure 7.44(a) shows transients induced by global oscillation of the hull at 2 Hz, from which we observe that the longitudinal strain and the strain in the transverse directions are 180 degrees out of phase. Transient local vibrations at ~20 Hz in the bottom panel are shown in Figure 7.44(b), with transverse strain being dominant. Semi-periodic impacts at ~0.2 Hz result from the hull contacting the surface waves. Signal processing is achieved using wavelet transforms that prove well suited for the detection and characterization of transients (Figure 7.45).

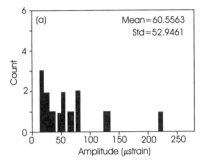 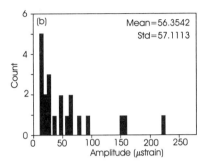

Figure 7.45 (a) Amplitude distribution of 2-Hz transient during 330-second time interval; (b) matched filter approximation (*After*: [145]).

7.12.2.2 Application to Lock Gates

A glass reinforced plastic (GRP) lock gate has been developed by the Direction des Constructions Navales (DCN-Lorinet-France) for use on French waterways to replace conventional steel gates that require heavy maintenance schedules [139]. Each set consists of two "mitre-type leaves" rotating around a vertical axis, with the leaves made from thick-skin GRP panels, strengthened by horizontal torsion beams. A five Bragg grating-element sensor system (one for temperature compensation and four for strain measurement) has been embedded by CEA-LETI into a lock gate. Real-time strain measurements indicate results in good agreement with changing water levels and with expected strain at specific sensor locations. The data can be used by designers to scale finite element models and for system optimization.

7.12.2.3 Deep Bore-Hole Applications

There is also some potential in the area of down-hole instrumentation for using fiber optics. Well-monitoring systems are designed for permanent down-hole installation for measuring temperature and pressure under extreme conditions, in excess of 150°C and 1000 bar, respectively. Optical fibers are attractive as they enable the elimination of all down-hole electronics and drift and faults associated with them. Typical sensor specifications are shown in Table 7.5 [146].

Bragg gratings are potentially useful for this application, operating well in this high-temperature environment, over the long working distances and lifetimes required; however, they currently lag behind conventional fiber systems in terms of pressure

Table 7.5 Sensor Specification for Deep Bore-Hole Applications

Pressure range	1-1000 bar
Temperature range	20-200°C
Accuracy	< 0.4 bar (0.2-bar target)
Resolution	< 0.1 bar (0.02-bar target)
Working distance	7 km (20-km target)
Lifetime	> 5 years (10-year target)

After: [146].

resolution and require ultra-high wavelength resolution to be successful sensor replacements. There has been recent interest in "measurement-while-drilling" (MWD) systems that measure drilling and formation parameters (e.g., weight and torque on the drill bit) without interrupting the drilling operation [147]. A BGS has been incorporated between a circular bimorph and fixed support, forming a down-hole Bragg grating modulator that tracks changes in the amplitude of the drumhead-like action of the bimorph at its natural resonant frequency, during drilling. Furthermore, recent geophysical research includes measurement of seismic events in very deep bore-holes, and vibration measurements to 300°C using a fiber-coupled Michelson interferometer for seismic sensing have been demonstrated by Ecke et al. [148]. At temperatures above 250°C, interferometric detection and fiber optic signal transmission appears the only viable approach for performing long-term measurement of seismic waves. The application of a three-dimensional seismic network requires frequency operating ranges of 0.05 to 30 Hz and amplitudes of 0.1 nm to 100 μm, which is comparable to surface-based seismometers.

7.12.3 Applications to Civil Engineering Structural Monitoring

There is growing concern over the state of civil infrastructure in both the United States and Europe. It is essential that mechanical loading be measured for maintaining bridges, dams, tunnels, buildings, and sport stadiums. By measuring the distributed strain in buildings, one can predict the nature and grade of local loads, for example, after an earthquake. The mechanical health of bridges, however, is increasingly under scrutiny, as old structures are often excessively loaded leading to a real possibility of increased structural failure rates. In fact, a 1996 US Department of Transportation survey estimates that 40% of all bridges in the United States are seriously deteriorated. A recent structural failure example relates to the Forth Road Bridge in Scotland (built in 1964) when in December 1997 one of the steel bolts securing the suspension cables to the bridge deck snapped. The snapped bolt, 2 ft long with a diameter of 2.5 inches, was one of 768 that helped secure the ends of suspension ropes to the bridge deck. Wires were also found to have broken in steel hanger ropes the previous year. Four bolts, in pairs of two, anchor each of the 192 hanger assemblies on the bridge. They were due to be replaced as part of a £9 million refit expected to last 3 years. There is concern with 50-year old railroad bridges in the United States as regulatory limits on railcar loads are relaxed.

The current inspection routine depends on periodic visual inspection. The use of modern optical-based sensors can lead to real-time measurements, that monitor the formation and growth of defects. Additionally, optical fibers sensors allow for data to be transmitted over long distances to a central monitoring location. The advantage of optical fibers is that they may either be attached to an existing structure or embedded into concrete decks and supports prior to pouring, thereby monitoring the curing cycle and the condition of the structure during its serviceable lifetime. One of the most important applications of Bragg gratings as sensors is for smart structures, where the grating is embedded directly into the structure to monitor its strain distribution; however, for error-free, quasi-static strain measurement, temperature compensation of thermal fluctuations is required. This could lead to structures that are self-monitoring or even self-scheduling of their maintenance and repair through the union of optical fiber sensors and artificial intelligence with material science and structural engineering. Several types of fiber optic sensor are capable of sensing structural strain, such as, the intrinsic and extrinsic fiber Fabry-Perot sensor. For example, Lee et al. [149] have used a multiplexed array of 16 fiber Fabry-Perot sensors to monitor strain on the Union Pacific Bridge that crosses the Brazos River at Waco, Texas. The fiber sensors are located at fatigue-critical points for measuring dynamic loads induced by trains crossing the bridge, with the recorded data correlating well with those recovered by resistive strain gauges. Nevertheless, the general consensus is that fiber Bragg gratings are presently the most promising and widely used candidates for smart structures. The instrumentation for multiplexing large grating sensor arrays can be the same, offering a potentially low-cost solution for monitoring structural strain. Fiber sensors are robust with a tensile strength exceeding that of steel; therefore, with suitable protection of the fiber surface, the mechanical integrity of the fiber can be maintained. Under the correct conditions catastrophic failure of the grating will only follow the failure of the structure as a whole. As the wavelength shift with strain is linear and with zero offset, long-term measurements are possible, and because the measurement can be interrupt-immune one can avoid perpetual monitoring of a structure, instead performing periodic measurements when necessary.

7.12.3.1 Applications to Bridges

There are numerous applications of Bragg gratings to structural monitoring. One of the first uses (1993) of fiber Bragg gratings was for stress measurement in the Beddington Trail Bridge near Calgary, Canada, where 16 Bragg grating sensors were attached to steel reinforcing bars and carbon-fiber composite tendons used in the prestressed concrete girders that support the bridge. The project demonstrated the feasibility for long-term monitoring of bridge structures in a manner previously considered impractical [142]. The I-10 bridge project in Las Cruces, New Mexico, uses an array of 67 fiber optic sensors attached to the steel I-beams under the bridge and measures the deformation-induced strain resulting from vehicular movement over the bridge. The sensors determine the number of heavy or overweight trucks crossing the bridge and also the structural resonances caused by dynamic loading. Projects such as this may also be used to identify damage or deterioration by placing sensors in critical, high-stress locations and by interpreting long-term changes in the response of the bridge to traffic. A current and ongoing research program, led by Fuhr and Huston of the University of Vermont, has established the widespread use of optical fiber-based sensors, including BGSs, for monitoring the Waterbury Vermont Steel Truss Bridge. Bragg strain sensors are cast into plastic and placed within the rebar matrix such that they are directly encased in concrete, whereas other BGSs are epoxied directly onto the bridge rebar and covered in a protective coating. Measurements from the fiber sensors are made remotely and then transferred to a central computer that analyzes the information and displays it on a World Wide Web site [150].

Meissner et al. [151] have embedded gratings into concrete prisms and loaded to 150 kN, measuring an essentially linear response under loading. The sensors have been tested at a prestressed 72-m long concrete bridge spanning the A4 motorway near Dresden, Germany. A sensor carrier protects the gratings from damage during the construction of the bridge (Figure 7.46(a)) and also prevents distortion of the measurements that may result from microcracks. The fiber is embedded in a groove of a loop of reinforcement steel (6 mm diameter), and two cross-ribs guarantee good contact between the sensor carrier and concrete. Two sensor carriers with two fiber Bragg gratings attached to each carrier were installed in the middle of the bridge under construction. The grating sensors were placed near the tension ducts because the bending stress shows maximal values at that point. Before the sensors were embedded, the strain was set to zero and the strain was monitored during construction loading of the bridge (Figure 7.46(b)). After prestressing, negative strains (compressive strains) were measured. As the weather cooled the bridge,

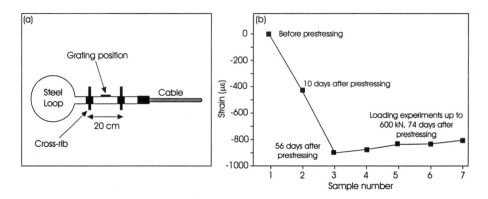

Figure 7.46 (a) Bragg grating sensor carrier. (b) Histogram of one Bragg grating sensor (*After*: [151]).

creeping and shrinkage of the concrete resulted in a higher strain 56 days after prestressing of the bridge than directly after prestressing. Loading experiments up to 600 kN were made 74 days after prestressing of the bridge. The wavelength measurements were made using a long period grating as a broadband edge filter for the back-reflected light from the sensing grating. Future applications include traffic control and axle load control.

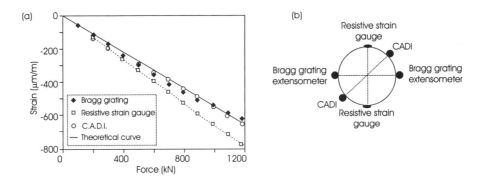

Figure 7.47 (a) Measured strain for the three sensors types, and (b) their locations (*After*: [152]).

Finally, Dewynter-Marty et al. [152] have made strain and crack detection measurements with surface-mounted and embedded Bragg grating "extensometers," which are dedicated components having a central metallic transducing rod anchored to the building, onto which is bonded a BGS. The device has a base length of 0.1–0.2m. It is likely that in many practical applications the BGS will experience strain transduced through a sensor carrier rather than by direct anchoring to the civil structure surface, unless direct embedding is utilized. We shall present a series of results for strain and crack propagation demodulated using portable instrumentation based on an FPF and 40-nm bandwidth superluminescent diode operating at 1.3 μm. Results are taken for cylindrical concrete samples and rectangular flooring as may be used on any generic civil structure. Figure 7.47(a) and (b) show the strain measurement on a concrete cylindrical surface, comparing BGS, resistive strain gauge, and inductive sensors located at opposite points on the perimeter of the sample that was subjected to crescent stress in compression up to 1200 kN in 100-kN steps. A theoretical curve is calculated from Hooke's law and indicates that, of the three sensors, the BGS and resistive strain gauge behave well from 0–700 με with ± 2.8% and ± 1.3 % full-scale resolution, respectively. In a second experiment 10 repetitive stress cycles up to 1000 kN with a fast return to zero were used to quantify the zero point fluctuation due to potential shifting of the anchoring point. The optical sensors demonstrated a return to zero of ± 3 με, good reproducibility of ± 3 με, and a ± 4-με repeatability (30 averages) in the Bragg sensor measurement. Inductive and Bragg sensors were also located on a concrete floor (1.5m long, 1m wide, and 0.26m high) and metallic plates were used to create areas of weakness. The extensometer under loading has a strain threshold of 2 με compared with 30 με for the inductive sensor. Figure 7.48 shows the

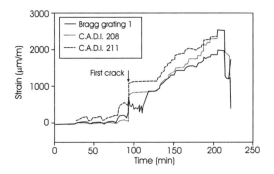

Figure 7.48   The three sensors' responses to a concrete floor fracture (*After*: [152]).

results for an applied force of 1000 kN and indicates that the inductive sensor located at the periphery measures the largest strain, which is also where the crack initiates. Bragg grating 1 indicates a 500-μm crack. Measurements were also made for an embedded Bragg grating extensometer, commercial vibrating wire extensometer, and externally mounted inductive sensors, with sensors embedded above or parallel to one another. Strain cycling the BGS shows good linearity to $<\pm$ 2 % full scale, whereas all sensor types display a residual offset, between + 2 and + 50 με for the Bragg grating extensometer, with a reproducibility of 7 με and repeatability of less than 5 με (30 averages). For embedded measurements in the concrete floor, loading is increased from 0 to 600 kN in 50-kN steps. At a load of 350 kN two cracks appear. The Bragg grating measures a strain of 228 με for a 45.6-μm crack and the applied force is slowly reduced to 0 kN, with the grating in the cracked area measuring a resultant positive strain of 90 με for an 18-μm wide crack (Figure 7.49). These experiments clearly show that Bragg gratings are able to detect and monitor early structural failures for a measurement range of 2500 με and a strain threshold of 2 με.

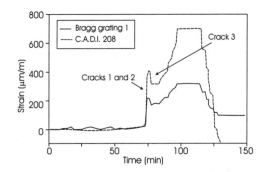

Figure 7.49   The three sensors' responses to a concrete floor fracture (*After*: [152]).

7.12.3.2   Structural Maintenance and Applications to Tunnels and Underground Civil Works

There is a growing use of novel materials such as fiber-reinforced polymers (FRP) in the construction industry. These materials are corrosion resistant and have a greater serviceable lifetime than their steel counterparts. There is, however, little practical experience with FRPs and long-term monitoring is necessary. A recent demonstration of 65 BGSs is on-going at the Taylor bridge near Winnipeg, Canada for monitoring FRP reinforcement and prestressing bars in the concrete girders and FRP material in the bridge deck and barrier wall. Carbon fiber cables have also been used to substitute steel supporting cables in the Storck's cable-stayed bridge in Winterthur, Switzerland. BGSs were either inserted in the periphery of two FRP cables or bonded on a 2-m long steel rod for embedding in concrete. The BGS-based system was demodulated with a 0.5-m CCD-equipped grating spectrometer to give a 1-$\mu\varepsilon$ strain resolution at 10 Hz. Furthermore, fiber sensors can also be used to monitor strain during construction, as witnessed by the use of an array of 32 fiber Bragg gratings at the Vaux Viaduct Bridge near Lausanne, Switzerland. In this case the strain during the movement of large steel box beams pushed and rested on piers was measured [142].

Another application within the European Union STABILOS project concerns monitoring the Mont-Terri tunnel in Switzerland, conducted by CSEM. The sensing network is composed of bare BGSs strained between two anchoring points and connected in series, demodulated using a rotating interference filer [139]. Finally, a European Union–funded Brite project COSMUS was launched in December 1996 for 3 years, to improve safety in civil works. The aim is to assess and control subterranean movements to within 1 mm during the development of underground transport systems. Sensors currently used in civil engineering cannot achieve this goal and hence high accuracy instrumentation and control systems are currently under development in order to control the grouting plant every 30 seconds. The BGSs are used as temperature-compensated, quasi-distributed strain sensors with a 10-$\mu\varepsilon$ accuracy, along with high-resolution tilt meters (10 arc. sec).

7.12.4   Bragg Gratings for Medical Applications

One of the first applications of optical fibers was in producing endoscopes, which act purely as light guides to previously inaccessible areas of the human body. With the advent of optical fiber-based sensors there has been interest in their employment for physiological parameter measurement, of which extrinsic fiber sensors are most commonly used (e.g., the Luxtron Corporation and extrinsic Fabry-Perot sensors [153]). The intrinsic electrical isolation and immunity to electromagnetic interference of optical fibers is particularly useful in avoiding the risk of electrical shock when high voltages are applied to a patient, such as for resuscitation through defibrillation after sudden cardiac arrest. Furthermore, the small size of optical catheter probes enables the in vivo measurement of organ functions through minimally invasive procedures, providing accurate local information on temperature, pressure, and acoustic fields. There is often minimal tissue trauma as a result of the dielectric properties of the optical fiber and the small sensor size, which is small enough so as to not interfere with routine medical procedures such as hemodialysis. Sensor materials must also be inert to prevent an undesirable immunity response from the human defense system leading to the aggregation of plaque and coagulation. Therefore, it is considered advantageous if the sensor can be intrinsic to the fiber host.

There is increasing interest in thermal therapies and methods in medicine as minimally invasive alternatives to surgery, using RF currents, microwaves and lasers to induce temperatures several tens of degrees Celsius above normal body temperature for periods of 20–60 minutes. There is also a trend towards increasing acoustic outputs from diagnostic ultrasonic systems and a widening in the therapeutic applications of high-intensity ultrasound. Some examples include the treatment of benign prostatic diseases, ablation therapy of cardiac arrhythmia, microwave induced hyperthermia for radiotherapy in the treatment of recurrent breast cancer, and the removal of some inoperable tumors. In the local environment, high electric or ultrasonic pressure fields can result in temperature gradients of tens of degrees Celsius per centimeter that need to be accurately mapped. BGSs are amenable to the measurement of ultrasonic, temperature and pressure fields for in vivo and in situ study of the acoustic and thermal properties of diseased tissue [154]. As minimally invasive sensors they are attractive and can be inserted into catheters and hypodermic needles. The only real issue is a question of accuracy and stability. Bragg gratings also solve the problem of reproducibility in production, which is perhaps the biggest failing of mass-produced, miniaturized, and low-cost disposable sensors.

7.12.4.1 Temperature Sensors

Optical fiber thermometers are particularly suitable for competing with current thermocouple and thermistor devices that are electrically active and therefore inappropriate for a number of medical applications. Conventional electrical temperature sensors can perturb any incident field and can lead to localized points of heating, where high magnetic or electromagnetic fields are present. For medical applications a resolution of $0.1°C$ over a range from $\sim35–50°C$ is required, with a definite need for sensor multiplexing when temperature profiling measurements are made. These requirements make Bragg gratings appear particularly useful for this purpose.

Feasibility of In Vivo Bragg Grating Sensor Measurements in NMR Machines

A scheme presented by Rao and co-workers [155], based on that reported in [7], uses Bragg gratings for remote temperature measurements in an NMR machine. A Bragg grating drift-compensated FPF demodulated the BGS calibrated against a conventional thermocouple and limited by a 16-bit digital-to-analog converter to 1.2-pm resolution, equivalent to $0.2°C$ at 800 nm. The processing measurement time of 0.4-second limits the system response, but is within the 1-second requirement for medical applications. The probe with four sensing gratings was placed inside a NMR machine with high magnetic field of ~4.7

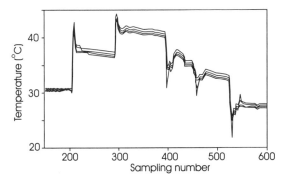

Figure 7.50 Experimental results of temperature measurement using Bragg gratings operated in an NMR machine (*After*: [155]).

Tesla and linked to the instrumentation over a 25-m distance. The actual measured resolution was $\pm$ 0.2°C with an accuracy of $\pm$ 0.8°C, limited by the quality of the filter PZ transducer element. Averaging for 1.6 second improved the resolution to $\pm$ 0.1°C. The data in Figure 7.50 arise by adding warm water followed by ice into a container holding the BGSs and shows the suitability of the gratings for this application.

Bragg Grating Flow-Directed Thermodilution Catheter for Cardiac Monitoring

Measurement of the heart's efficiency plays a key role for cardiac monitoring, and physicians currently inject patients with a cold solution to measure their heart's output, using a flow-directed thermodilution catheter inserted directly into the right atrium. The catheter also enables blood temperature measurement in the pulmonary artery. Catheters with conventional thermistor and thermocouple devices have been commercially available for many years and are still widely used. A typical flow-directed thermodilution catheter is used to measure the amount of blood pumped by the heart. For a known heart rate the time delay t_1 in the temperature profile can be used to determine the amount of blood pumped, and a thermistor is used to obtain the temperature profile. Any blockage of blood vessels increases the time delay to t_2, where $t_2 > t_1$ (i.e., the amount of blood pumped by the heart in a given time interval is reduced). The amplitude of the temperature curve also decreases because the cold solution moves slowly and warms gradually. The temperature and volume of the solution, the speed of injection, and the catheter size affect the temperature distribution measured by the thermistor; therefore, careful calibration is required, with a typical reproducibility of 5%. Rao et al. [156] have used a test rig to simulate blood flow using a peristaltic pump (from 60 to 120 r/min), utilizing an unbalanced, bulk Michelson scanning interferometer to track changes in the grating response with a resolution of 0.2°C over a 10-Hz bandwidth. A thermocouple compared the optical and electrical measurements, both of which displayed a linear relationship between the pump rate and time delay. The change of blood flow for a constant pump rate and different sizes

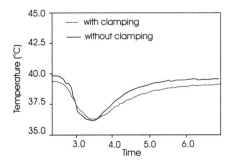 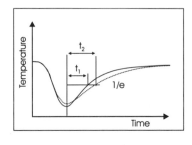

Figure 7.51 Experimental results for blood-vessel blocking simulation measured with a Bragg grating sensor (*After:* [156]).

of vessels is shown in Figure 7.51. There is good agreement between the grating and thermocouple sensors, indicating a relative change in time delays in both cases of 10% from the clamped to the unclamped state.

7.12.4.2 Bragg Grating Response to Focused Ultrasonic Fields

The safety of ultrasound for medical applications requires assessment because of a growing trend toward increasing output powers from diagnostic ultrasound equipment and the widening use of high-intensity ultrasonic fields in a range of therapeutic applications. Currently, fields are assessed using complex theoretical modeling or direct detection with piezoelectric devises that are susceptible to EMI and signal distortion, along with reduced sensitivity due to the electrical loading effects of the transducer leads. Fisher et al. [157] have demonstrated that Bragg gratings can in principle be used to detect high frequency (MHz) ultrasonic fields. There is, however, a complex interaction of acoustic coupling from the ultrasonic field to the grating leading to standing wave formation in the fiber, obscuring the system response and reducing the grating effectiveness. Matters are improved by desensitizing the fiber. Pressure fields in the region of the grating of 10–20 atms were produced at 1.9 MHz and a noise-limited pressure resolution of ~4.5×10^{-3} atm/Hz$^{1/2}$ was deduced. Scanning the acoustic focal spot longitudinally along the grating results in multiple peaks and troughs in the system response with displacement observed over a distance greater than the physical grating length and results from the compressional standing waves in the fiber. Jacketing the fiber in PVC sleeving with a 1-mm aperture significantly attenuates the fiber acoustic modes, allowing for controlled grating exposure to the acoustic fields (Figure 7.52). Alternatively, the grating length can be shortened to less than the half the acoustic wavelength; therefore, for most medical applications frequencies of 500 kHz to 4 MHz are used, implying that grating lengths should be less than 0.5 mm at the highest frequency. There is also evidence that a *single* grating exposed to the acoustic field can produce a linear response and a noise-limited pressure resolution of 2×10^{-3} atm/Hz$^{1/2}$, without the need for interferometric setups.

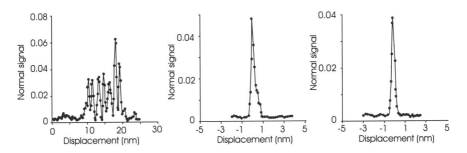

Figure 7.52   Longitudinal scan for 5-mm long grating and for shielded 1-mm grating (*After*: [157]).

7.12.5 Bragg Sensors Within the Nuclear Power Industry

In countries where a large part of the energy production is based on the nuclear power industry (USA, UK, Japan, Germany and France), there is a paramount need to improve safety in the operation and monitoring of equipment. Recent studies have shown that optical fiber sensors can potentially improve the safety of strategic equipment, such as transformers, generators and structures, such as containment shells, cooling towers and steam pipes. There are two main systems, the nuclear steam supply system includes the plant's reactor, related steam-generating equipment and pumps and pipes to transfer water and steam. The turbo-generator system is driven by steam from the nuclear steam supply system to produce electricity.

7.12.5.1 Nuclear Shield Monitoring

Nuclear power plants are very large structures covering 200–300 acres. There are separate buildings for the reactor and the plant's turbines and electricity generators. The reactor building or containment structure has a thick concrete floor and walls of steel or concrete-lined steel, designed to prevent accidental leakage from the reactor. It is the ultimate protective barrier. It is designed to withstand a potential H_2 risk in the case of a nuclear accident. The maximum pressure level is ~12 bar for a 900-MW nuclear power plant with a single shield and 9 bar for a 1300-MW double-shielded plant. Structural integrity tests are performed periodically; however, the deployment of a quasi-distributed sensing system as offered by optical fiber-based remote sensors would significantly enhance reliability, safety, and reduce maintenance costs. In 1995 CEA-LETI, EDF, and Framatome in France launched a joint project for the development of Bragg grating-based extensometric instrumentation for concrete measurements. The extensometers are surface-mounted onto or embedded into large walls of the nuclear shield made of high-performance prestrained concrete. The Bragg grating extensometer operates under elongation and compression over a range of ~2500 µε with a 2-µε detection limit [139].

7.12.5.2 Steam Pipe Monitoring

The reactor vessel is a tank-like structure (~700 metric tons) with 15-cm thick steel walls that holds other reactor parts and is installed at the base of the reactor building, with steel pipes carrying water and steam in and out of the vessel. In a pressurized-water reactor nuclear power plant, water is used to absorb heat but does not boil because it is kept under high pressure. Therefore, there is the potential for breaks to occur between elements of a complex circuitry (vessel, pump, pressurizer, steam generator) with a large number of solder joints aging under thermo-mechanical cycling. Current safety rules do take into account complete pipe breakage; nevertheless, early leakage detection is an important safety issue. A leak or break in a reactor water pipe can have serious consequences if it is accompanied by a loss of coolant, in an extreme case this could result in the reactor melting down and the widespread release of radiation. Current approaches that detect such leaks do so globally (water balance, dose level), but local detection is highly desirable for redundancy. The placement of BGSs at critical points along pipes such as soldering joints and elbows is an attractive and cost-effective solution. Recognizing this, a consortium of nine partners led by BICC Cables Ltd. (UK) launched in early 1996 a 3-year Brite project called "Fiber optic strain monitoring at elevated temperature" (FOSMET, BE 1432). The primary objective is to develop a quasi-distributed monitoring system with fully integrated temperature compensation able to multiplex several strained BGSs and provide real-time (1-second scanning per sensor) lifetime prediction for high temperature components (~550°C). The demonstration will take place in an operating fossil fuel plant at the end of 1999 [139]. ERA Technology Ltd. and 10 partners launched a similar project in 1995, called "Distributed temperature monitoring on high temperature pipework" (FORMS, BRE 20965).

7.12.5.3 Waste Conditioning and Disposal Monitoring

Highly radioactive waste must be kept in underground storage for long time periods. The French National Radioactive Waste Management Agency (ANDRA) is currently studying underground repository concepts and monitoring equipment with the CEA for a network of disposal interconnected galleries leading to a decommissionable site. Another example is a concept initiated by SCK-CEN Mol Research Center in Belgium, with research undertaken at the underground laboratories of HADES. In Germany future work will investigate salt mines, where Bragg gratings could ensure strain and temperature measurements in underground repositories [139].

7.12.6 Applications to Power Transmission Lines

The electrical power industry demands highly reliable on-line equipment with low maintenance schedules. Often equipment is located in difficult-to-access points, such as in gas or oil tanks, between the coil windings of power transformers, or even in vacuum. In most cases sensors must have small dimensions and be passive, operating on low power consumption. The extremely high voltages and currents demand high electrical insulation, and there is also a need for immunity to EMI that results from the noisy electromagnetic environment at a power plant. To date a number of optical fiber-based devices have been tested for power system applications, examples of which are current and voltage sensors aimed at replacing current and voltage transformers on high voltage lines, or to perform direct current and high frequency transient measurements. Other applications are magnetic and electric field sensors for measuring stray flux in electrical machinery and insulators; temperature sensors to probe internal overheating of transformers; and generators and wind speed and strain sensors for monitoring long-haul conductor lines. Measurement of single and 3-phase ac and dc currents are of interest, with a performance for fixed installations equivalent to conventional, power management current transformers (i.e., a measurement range of 1–10000 A with 1% accuracy and 0.1-second response time). For fault and transient diagnostics a larger 100-kHz bandwidth is required (mechanical switching devices operate quickly in order to protect the grid from large current spikes) with reduced 5% accuracy.

Optical current sensors have been extensively researched and developed for use in the power distribution industry for more than a decade. Most designs are based on devices that behave magneto-optically, with the enclosed current determined from the Faraday effect through measurement of plane of polarization rotation of linearly polarized light, as the magnetic field induces material birefringence. Any stress-induced birefringence resulting from temperature inhomogeneities and vibrations may influence the SOP of the light, quenching the Faraday effect. The temperature and wavelength sensitivity of the Verdet constant and the vulnerability of the scale factor to vibration continue to limit the application of this technique. Therefore combining the benefits offered by optical fiber leads with more conventional electrical sensors, the so-called hybrid approach, appears very attractive. For example, the current transformer (CT) (based on Ampere's law) is independent of both the position of the current within a closed path and the presence of magnetic fields associated with current external to the closed path. The enclosed current induces a magnetic flux change in the CT's core that generates an alternating current in a solenoid (the secondary) that is wrapped around the core. This alternating current may be detected as a voltage signal by including a resistor in series with the solenoid. Ning et al. [158] have demonstrated the measurement of large currents at high voltages by using a combination of conventional CT and PZ element, transducing the voltage developed across the transformer secondary into a phase modulation in a fiber interferometer. In this approach external vibrations acting on the fiber leads corrupt the signals used to determine the currents.

7.12.6.1 Combined Bragg Grating/Current Transformer Current Metering Sensor

Henderson et al. [159] have presented an alternative Bragg grating/CT approach with the BGS mounted on a PZ element, combining the CT's high resolution with the electrical insulation offered by the fiber optic system. At low signals (<50V for ac signals) the variation in the diameter of the PZ cylinder is proportional to the applied voltage, over more than 5 orders of magnitude, for frequencies less than the first mechanical resonance. Hence, the variation in strain applied to the grating is linearly related to the current induced by the secondary of the CT. A maximum effective current of 700A was applied. Interferometric wavelength-shift detection was used to demodulate the Bragg shift from which a current resolution of $0.7 A/Hz^{1/2}$ was deduced. The frequency dependence of the optical system, for a constant amplitude current, showed good fidelity to 10 kHz. By providing wavelength-encoded signals, Bragg gratings are ideal candidates for incorporation into WDM schemes. In the power distribution industry it is often necessary to measure three phase currents simultaneously. With the hybrid approach birefringence in the fiber leads, leading to polarization noise corrupting the encoded Faraday signal, is not an issue. The influence of temperature is also not such a problem as this effect occupies a lower frequency regime than the signal of interest.

7.12.6.2 Multiplexed Load Monitoring System for Power Transmission Lines Using Fiber Bragg Grating

Ogawa et al. [160] have developed a multiplexed Bragg grating load monitoring system for power transmission lines in heavy snows, based on the performance of 10 sensing points over 30 km. The BGSs are bonded to a metal plate that joins a power transmission cable to a pylon. The strain is directed to the grating through the metal plate, which therefore determines the sensing characteristics, in this case a dual resolution sensing mode. The resolution below 3000 kgf is ~50 kgf with a reduction above this point, as strain is redistributed to outer edges of the plate. (Note: 25 kgf ~ 2000 $\mu\varepsilon$ / 10 $\mu\varepsilon$.)

7.12.6.3 Magnetic Field Sensors

Bragg gratings coated with a suitable magnetostrictive or resistive coating could be used for magnetic and current sensing; however, the metallic coating modifies the magnetic field and is electrically conducting, essentially negating the benefits of optical fibers.

7.12.7 Other Applications

In addition to the above uses, Bragg grating sensors have been proposed for a diverse number of applications, such as in intruder detection schemes [161], as accelerometers [162], and strain sensors measuring transient loading of gun barrels [163]. Clearly these sensors demonstrate excellent versatility.

References

[1]   Kersey, A. D., et al. "Fiber grating sensors," *IEEE Journal of Lightwave Technology*, Vol. 15, 1997, pp. 1442–1463.

[2]   Xu, M. G., et al. "Optical in-fibre grating high pressure sensor," *Electronics Letters*, Vol. 29, 1993, pp. 389–399.

[3]   Kersey, A. D., and M. J. Marrone, "Fiber Bragg grating high-magnetic-field probe," 10th Optical Fibre Sensors Conference, Glasgow, Scotland, Oct. 11–13, 1994, pp. 53–56.

[4]   Melle, S. M., K. Liu, and R. M. Measures, "A passive wavelength demodulation system for guided-wave Bragg grating sensors," *IEEE Photonics Technology Letters*, Vol. 4, 1992, pp. 516–518.

[5]   Davis, M. A., and A. D. Kersey, "All-fibre Bragg grating strain-sensor demodulation technique using a wavelength division coupler," *Electronics Letters*, Vol. 30, 1994, pp. 75–77.

[6]   Zhang, Q., et al. "Use of highly overcoupled couplers to detect shifts in Bragg wavelength," *Electronics Letters*, Vol. 31, 1995, pp. 480–481.

[7]   Kersey, A. D., T. A. Berkoff, and W. W. Morey, "Multiplexed fiber Bragg grating strain-sensor system with a fiber Fabry-Perot wavelength filter," *Optics Letters*, Vol. 18, 1993, pp. 1370–1372.

[8]   Xu, M. G., et al. "Novel interrogation system for fibre Bragg grating sensors using an acousto-optic tunable filter," *Electronics Letters*, Vol. 29, 1993, pp. 1510–1511.

[9]   Jackson, D. A., et al. "Simple multiplexing scheme for fiber-optic grating sensor network," *Optics Letters*, Vol. 18, 1993, pp. 1192–1194.

[10] Kersey, A. D., "Interrogation and multiplexing techniques for fiber Bragg grating strain-sensors," Society of Photo-Optical Instrumentation Engineers, Distributed and Multiplexed Fiber Optic Sensors III, Boston, Sept. 1993, Vol. 2071, pp. 30–48.

[11] Davis, M. A., et al. "Interrogation of 60 fibre Bragg grating sensors with microstrain resolution capability," *Electronics Letters*, Vol. 32, 1996, pp. 1393–1394.

[12] Brady, G. P., et al. "Demultiplexing of fibre Bragg grating temperature and strain sensors," *Optics Communications*, Vol. 111, 1994, pp. 51–54.

[13] Davis, M. A., and A. D. Kersey, "Matched-filter interrogation technique for fibre Bragg grating arrays," *Electronics Letters*, Vol. 31, 1995, pp. 822–823.

[14] Geiger, H., et al. "Progress on grating interrogation schemes using a tunable filter," *Proceedings of the Optical Fiber Sensors Conference* (OFS-11), Sapporo, Japan, 1996, pp. 376–379.

[15] Geiger, H., et al. "Electronic tracking system for multiplexed fibre grating sensors," *Electronics Letters*, Vol. 31, 1995, pp. 1006–1007.

[16] Xu, M. G., H. Geiger, and J. P. Dakin, "Modeling and performance analysis of a fiber Bragg grating interrogation system using an acousto-optic tunable filter," *IEEE Journal of Lightwave Technology*, Vol. 14, 1996, pp. 391–396.

[17] Volanthen, M., et al. "Simultaneous monitoring of multiple fibre gratings with a single acousto-optic tunable filter," *Electronics Letters*, Vol. 32, 1996, pp. 1228–1229.

[18] Dunphy, J. R., et al. "Instrumentation development in support of fiber grating sensor array," Society of Photo-Optical Instrumentation Engineers, Distributed and Multiplexed Fiber Optic Sensors III, Boston, Sept. 1993, Vol. 2071, pp. 2–11.

[19] Coroy, T., and R. M. Measures, "Active wavelength demodulation of a Bragg grating fibre optic strain sensor using a quantum well electroabsorption filtering detector," *Electronics Letters*, Vol. 32, 1996, pp. 1811–1812.

[20] Miller, D. A. B., et al. "Electric field dependence of optical absorption near the band gap of quantum well structures," *Physical Review B*, Vol. 32, 1985, pp. 1043–1060.

[21] Coroy, T., et al. "Active wavelength demodulation of Bragg fibre-optic strain sensor using acousto-optic tunable filter," *Electronics Letters*, Vol. 31, 1995, pp. 1602–1603.

[22] Ball, G. A., W. W. Morey, and P. K. Cheo, "Fiber laser source/analyzer for Bragg grating sensor array interrogation," *IEEE Journal of Lightwave Technology*, Vol. 12, 1994, pp. 700–703.

[23] Coroy, T., et al. "Peak detection demodulation of a Bragg fiber optic sensor using a gain-coupled distributed feedback tunable laser," *Proceedings of the Optical Fiber Sensors Conference* (OFS-12),

Williamsburg, VA, USA, 1997, pp. 210–212.

[24]   Atkins, C. G., M. A. Putnam, and E. J. Friebele, "Instrumentation for interrogating many-element fiber Bragg grating arrays embedded in fiber/resin composites," Society of Photo-Optical Instrumentation Engineers, Smart Sensing Processing and Instrumentation, 1995, Vol. 2444, pp. 257–266.

[25]   Davis, M. A., et al. "A 60 element fiber Bragg grating sensor system," *Proceedings of the Optical Fiber Sensors Conference* (OFS-11), Sapporo, Japan, 1996, pp. 100–103.

[26]   Chen, S., et al. "Digital spatial and wavelength domain multiplexing of fiber Bragg grating based sensors," *Proceedings of the Optical Fiber Sensors Conference* (OFS-12), Williamsburg, VA, USA, 1997, pp. 448–451.

[27]   Hu, Y., et al. "Multiplexing Bragg gratings using combined wavelength and spatial division techniques with digital resolution enhancement," *Electronics Letters*, Vol. 33, 1997, pp. 1973–1975.

[28]   Ezbiri, A., S. E. Kanellopoulos, and V. A. Handerek, "High resolution instrumentation system for demodulating of Bragg grating aerospace sensors," *Proceedings of the Optical Fiber Sensors Conference* (OFS-12), Williamsburg, VA, USA, 1997, pp. 456–459.

[29]   Davis, M. A., and A. D. Kersey, "Fiber Fourier transform spectrometer for decoding Bragg grating sensors," *Proceedings of the Optical Fiber Sensors Conference* (OFS-10), Glasgow, Scotland, 1994, pp. 167–170.

[30]   Davis, M. A., and A. D. Kersey, "Application of a fiber Fourier transform spectrometer to the detection of wavelength encoded signals from Bragg grating sensors," *IEEE Journal of Lightwave Technology*, Vol. 13, 1995, pp. 1289–1295.

[31]   Flavin, D. A., R. McBride, and J. D. C. Jones, "Short optical path scan interferometric interrogation of a fibre Bragg grating embedded in a composite," *Electronics Letters*, Vol. 33, 1997, pp. 319–321.

[32]   Flavin, D. A., R. McBride, and J. D. C. Jones, "Absolute measurements of wavelengths from a multiplexed in-fibre Bragg grating array by short-scan interferometry," *Proceedings of the Optical Fiber Sensors Conference* (OFS-12), Williamsburg, VA, USA, 1997, pp. 24–27.

[33]   Putnam, M. A., et al. "Sensor grating demodulation using a passively mode locked fiber laser," Technical Digest Optical Fiber Communications Conference, Optical Society of America, Vol. 6, 1997, p. 156.

[34]   Dennis, M. L., et al. "Grating sensor array demodulation by use of a passively mode-locked fiber laser," *Optics Letters*, Vol. 22, 1997, pp. 1362–1364.

[35]   Putnam, M. A., et al. "Broadband square-pulse operation of a passively mode-locked fiber laser for fiber Bragg grating interrogation," *Optics Letters*, Vol. 23, 1998, pp. 138–140.

[36]   Yun, S. H., D. J. Richardson, and B. Y. Kim, "Interrogation of fiber grating sensor arrays with a wavelength-swept fiber laser," *Optics Letters*, Vol. 23, 1998, pp. 843–845.

[37]   Takushima, Y., F. Futami, and K. Kikuchi, "Generation of over 140-nm wide super-continuum from a normal dispersion fiber by using a mode-locked semiconductor laser source," *IEEE Photonics Technology Letters*, Vol. 10, 1998, pp. 1560–1562.

[38]   Kersey, A. D., T. A. Berkoff, and W. W. Morey, "High-resolution fibre-grating based strain sensor with interferometric wavelength-shift detection," *Electronics Letters*, Vol. 28, 1992, pp. 236–238.

[39]   Jackson, D. A., A. D. Kersey, and M. Corke, "Pseudo-heterodyne detection scheme for optical interferometers," *Electronics Letters*, Vol. 18, 1982, pp. 1081–1083.

[40]   Kersey, A. D., T. A. Berkoff, and W. W. Morey, "Fiber-grating based strain sensor with phase sensitive detection," *Proceedings 1st European Conference on Smart Structures and Materials*, Glasgow, Scotland, 1992, Session (2), pp. 61–67.

[41]   Kersey, A. D., T. A. Berkoff, and W. W. Morey, "Fiber-optic Bragg grating strain sensor with drift-compensated high-resolution interferometric wavelength-shift detection," *Optics Letters*, Vol. 18, 1993, pp. 72–74.

[42]   Kersey, A. D., and T. A. Berkoff, "Fiber-optic Bragg-grating differential-temperature sensor," *IEEE Photonics Technology Letters*, Vol. 4, 1992, pp. 1183–1185.

[43]   Rao, Y. J., et al. "Dual-cavity interferometric wavelength-shift detection for in-fiber Bragg grating sensors," *Optics Letters*, Vol. 21, 1996, pp. 1556–1558.

[44]   Rao, Y. J., et al. "Dynamic range enhancement of in-fibre Bragg grating sensors with two cascaded scanning interferometers," *Proceedings of the Optical Fiber Sensors Conference* (OFS-12), Williamsburg, VA, USA, 1997, pp. 512–515.

[45]   Brady, G. P., et al. "Simultaneous interrogation of interferometric and Bragg grating sensors," *Optics Letters*, Vol. 20, 1995, pp. 1340–1342.

[46]   Jackson, D. A., and J. D. C. Jones, "Interferometers" in *Optical Fiber Sensors: Systems and Applications, Vol. 2*, B. Culshaw and J. Dakin (eds.) Norwood, MA, Artech House, 1989: Ch. 10, pp. 239–280.

[47]   Singh, S., and J. S. Sirkis, "Simultaneous measurement of strain and temperature using optical fiber sensors: two novel configurations," *Proceedings of the Optical Fiber Sensors Conference* (OFS-11), Sapporo, Japan, 1996, pp. 108–111.

[48]   Liu, T., et al. "Simultaneous strain and temperature measurement using a combined fibre Bragg grating/extrinsic Fabry-Perot sensor," *Proceedings of the Optical Fiber Sensors Conference* (OFS-12), Williamsburg, VA, USA, 1997, pp. 40–43.

[49]   Rao, Y. J., et al. "Spatially-multiplexed fibre-optic interferometric and grating sensor system for quasi-static absolute measurements," *Proceedings of the Optical Fiber Sensors Conference* (OFS-11), Sapporo, Japan, 1996, pp. 666–669.

[50]   McGarrity, C., and D. A. Jackson, "A multi-purpose, self-calibrating network for large numbers of interfereomtric and grating sensors," *Proceedings of the Optical Fiber Sensors Conference* (OFS-12), Williamsburg, VA, USA, 1997, pp. 460–463.

[51]   Webb, D. J., et al. "Signal recovery technique for in-fibre Bragg grating and interferometric sensors using a Mach-Zehnder interferometer incorporating a fibre-pigtailed acousto-optic modulator," *Proceedings of the Optical Fiber Sensors Conference* (OFS-12), Williamsburg, VA, USA, 1997, pp. 508–511.

[52]   Jones, J. D. C., "Review of fibre sensor techniques for temperature-strain discrimination," *Proceedings of the Optical Fiber Sensors Conference* (OFS-12), Williamsburg, VA, USA, 1997, pp. 36–39.

[53]   Morey, W. W., G. Meltz, and W. H. Glenn, "Fiber optic Bragg grating sensors," Society of Photo-Optical Instrumentation Engineers, Fiber Optic and Laser Sensors VII, 1989, Vol. 1169, pp. 98–107.

[54]   Brady, G. P., et al. "Simultaneous measurement of strain and temperature using the first- and second-order diffraction wavelengths of Bragg gratings," *IEE Proceedings in Optoelectronics*, Vol. 144, 1997, pp.156–161.

[55]   Jin, W., et al. "Simultaneous measurement of strain and temperature: Error analysis," *Optical Engineering*, Vol. 36, 1997, pp. 598–609.

[56]   Jin, W., et al. "Geometric representation of errors in measurements of strain and temperature," *Optical Engineering*, Vol. 36, 1997, pp. 2272–2278.

[57]   Yoffe, G. W., et al. "Passive temperature-compensating package for optical fiber gratings," *Applied Optics*, Vol. 34, 1995, pp. 6859–6861.

[58]   Iwashima, T., et al. "Temperature compensation technique for fibre Bragg gratings using liquid crystalline polymer tubes," *Electronics Letters*, Vol. 33, 1997, pp. 417–419.

[59]   Xu, M. G., et al. "Thermally-compensated bending gauge using surface mounted fibre gratings," *International Journal of Optoelectronics*, Vol. 9, 1994, pp. 281–283.

[60]   Morey, W. W., G. Meltz, and J. M. Weiss, "Evaluation of a fibre Bragg grating hydrostatic pressure sensor," *Proceedings of the Optical Fiber Sensors Conference* (OFS-8), Monterey, CA, USA, 1992, Postdeadline paper PD-4.4.

[61]   Xu, M. G., et al. "Temperature-independent strain sensor using a chirped Bragg grating in a tapered optical fibre," *Electronics Letters*, Vol. 31, 1995, pp. 823–825.

[62]   Xu, M. G., et al. "Discrimination between strain and temperature effects using dual-wavelength fibre grating sensors," *Electronics Letters*, Vol. 30, 1994, pp. 1085–1087.

[63]   Udd, E., et al. "Three axis strain and temperature sensor," *Proceedings of the Optical Fiber Sensors Conference* (OFS-11), Sapporo, Japan, 1996, pp. 224–247.

[64]   Kalli, K., et al. "Possible approach for the simultaneous measurement of temperature and strain via

first and second order diffraction from Bragg grating sensors," *Proceedings of the Optical Fiber Sensors Conference* (OFS-10), Glasgow, Scotland, 1994, Postdeadline paper.

[65]   Sudo, M., et al. "Simultaneous measurement of temperature and strain using PANDA fiber grating," *Proceedings of the Optical Fiber Sensors Conference* (OFS-12), Williamsburg, VA, USA, 1997, pp. 170–173.

[66]   James, S. W., M. L. Dockney, and R. P. Tatam, "Simultaneous independent temperature and strain measurement using in-fibre Bragg grating sensors," *Electronics Letters*, Vol. 32, 1996, pp. 1133–1134.

[67]   Song, M., et al. "Interferometric temperature-insensitive strain measurement with different-diameter fiber Brag gratings," *Optics Letters*, Vol. 22, 1997, pp. 790–792.

[68]   Patrick, H. J., et al. "Hybrid fiber Bragg grating/long period fiber grating sensor for strain/temperature discrimination," *IEEE Photonics Technology Letters*, Vol. 8, 1996, pp. 1223–1225.

[69]   Kannellopoulos, S. E., V. A. Handerek, and A. J. Rogers, "Simultaneous strain and temperature sensing with photogenerated in-fiber gratings," *Optics Letters*, Vol. 20, 1995, pp. 333–335.

[70]   Davis, M. A, and A. D. Kersey, "Simultaneous measurement of temperature and strain using fibre Bragg gratings and Brillouin scattering," *IEE Proceedings in Optoelectronics*, Vol. 144, 1997, pp. 151–155.

[71]   Sirkis, J. S., et al. "In-line fiber etalon for strain measurement," *Optics Letters*, Vol. 18, 1993, pp. 1973–1975.

[72]   Magne, S., et al. "State-of-strain evaluation with fiber Bragg gratings rosettes: Application to discrimination between strain and temperature effects in fiber sensors," *Applied Optics*, Vol. 36, 1997, pp. 9437–9447.

[73]   The Boeing Company, "Fiber with multiple overlapping gratings," Patent number 5,627,927. Issued 1997.

[74]   Ecke, W., et al. "Improvement of the stability of fiber grating interrogation systems using active and passive polarization scrambling devices," *Proceedings of the Optical Fiber Sensors Conference* (OFS-12), Williamsburg, VA, USA, 1997, pp. 484–487.

[75]   Burns, W. K., et al. "Degree of polarization in the Lyot depolarizer," *IEEE Journal of Lightwave Technology*, Vol, LT-1, 1983, pp. 475–479.

[76]   Kersey, A. D., A. Dandridge, and M. J. Marrone, "Single-mode fiber pseudo-depolarizer," Society of Photo-Optical Instrumentation Engineers, Fiber Optic and Laser Sensors V, 1987, Vol. 838, pp. 360–364.

[77]   Weis, R. S., A. D. Kersey, and T. A. Berkoff, "A four-element fiber grating sensor array with phase-sensitive detection," *IEEE Photonics Technology Letters*, Vol. 6, 1994, pp. 1469–1472.

[78]   Kalli, K., et al. "Wavelength division and spatial multiplexing using tandem interferometers for Bragg grating sensor networks," *Optics Letters*, Vol. 20, 1995, pp. 2544–2546.

[79]   Dakin, J. P., et al. "New multiplexing scheme for monitoring fiber optic Bragg grating sensors in the coherence domain," *Proceedings of the Optical Fiber Sensors Conference* (OFS-12), Williamsburg, VA, USA, 1997, pp. 31–34.

[80]   Berkoff, T. A., and A. D. Kersey, "Fiber Bragg grating array sensor system using a bandpass wavelength division multiplexer and interferometric detection," *IEEE Photonics Technology Letters*, Vol. 8, 1996, pp. 1522–1524.

[81]   Kalli, K., et al. "Wavelength division and spatial multiplexing of Bragg grating sensor networks using concatenated interferometers," *Proceedings of the Optical Fiber Sensors Conference* (OFS-11), Sapporo, Japan, 1996, pp. 522–525.

[82]   Berkoff, T. A., and A. D. Kersey, "Eight element time-division multiplexed fiber grating sensor array with integrated-optic wavelength discriminator," Second European Conference on Smart Structures and Materials, Glasgow, Scotland, 1994, Session 10, pp. 350–353.

[83]   Davis, M. A., D. G. Bellemore, and A. D. Kersey, "Structural strain mapping using a wavelength/time division addressed fiber Bragg grating array," Second European Conference on Smart Structures and Materials, Glasgow, Scotland, 1994, Session 10, pp. 342–345.

[84]   Rao, Y. J., et al. "Spatially-multiplexed fibre-optic Bragg grating strain and temperature sensor

based on interferometric wavelength-shift detection," *Electronics Letters*, Vol. 31, 1995, pp. 1009–1010.

[85] Rao, Y. J. et al. "Combined spatial- and time-division-multiplexing scheme for fibre grating sensors with drift compensated phase-sensitive detection," *Optics Letters*, Vol. 20, 1995, pp. 2149–2151.

[86] Roa. Y. J., et al. "Simultaneous spatial, time and wavelength division multiplexed in-fibre grating sensor network," *Optics Communications*, Vol. 125, 1996, pp. 53–58.

[87] Lobo Ribeiro, A. B., et al. "Time-and-spatial-multiplexing tree topology for fiber-optic Bragg-grating sensors with interferometric wavelength-shift detection," *Applied Optics*, Vol. 35, 1996, pp. 2267–2273.

[88] Rao, Y. J., "In-fibre Bragg grating sensors," *Measurement Science and Technology*, Vol. 8, 1997, pp. 355–375.

[89] Fallon, R. W., et al. "Identical broadband chirped grating interrogation technique for temperature and strain sensing," *Electronics Letters*, Vol. 33, 1997, pp. 705–706.

[90] Zhang, L., et al. "Spatial and wavelength multiplexing architectures for extreme strain monitoring system using identical-chirped-grating-interrogation technique," *Proceedings of the Optical Fiber Sensors Conference* (OFS-12), Williamsburg, VA, USA, 1997, pp. 452–455.

[91] Putnam, M. A., G. M. Williams, and E. J. Friebele, "Fabrication of tapered, strain-gradient chirped fiber Bragg gratings," *Electronics Letters*, Vol. 31, 1995, pp. 309–311.

[92] Kersey, A. D., M. A. Davis, and T. Tsai, "Fiber optic Bragg grating strain sensor with direct reflectometric interrogation," *Proceedings of the Optical Fiber Sensors Conference* (OFS-11), Sapporo, Japan, 1996, pp. 634–637.

[93] LeBlanc, M., et al. "Distributed strain measurement based on a fiber Bragg grating and its reflection spectrum analysis," *Optics Letters*, Vol. 21, 1996, pp. 1405–1407.

[94] Huang, S., et al. "A novel Bragg grating distributed-strain sensor based on phase measurements," Society of Photo-Optical Instrumentation Engineers, Smart Sensing, Processing and Instrumentation, 1995, Vol. 2444, pp. 158–169.

[95] Huang, S., et al. "Fiber optic intra-grating distributed strain sensor," Society of Photo-Optical Instrumentation Engineers, Distributed and Multiplexed Fiber Optic Sensors IV, 1994, Vol. 2294, pp. 81–92.

[96] Huang, S., et al. "Bragg intragrating structural sensing," *Applied Optics*, Vol. 34, 1995, pp. 5003–5009.

[97] Huang, S., M. M. Ohn, and R. M. Measures, "Phase-based Bragg intragrating distributed strain sensor," *Applied Optics*, Vol. 35, 1996, pp. 1135–1142.

[98] Marrone, M. J., A. D. Kersey, and M. A. Davis, "Fiber sensors based on chirped Bragg gratings," *Proceedings of the Optical Society of America Annual Meeting*, Rochester, NY, October 1996, paper WGG5.

[99] Ohn, M. M., et al. "Arbitrary strain profile measurement within fibre gratings using interferometric Fourier transform technique," *Electronics Letters*, Vol. 33, 1997, pp. 1242–1243.

[100] Volanthen, M., et al. "Distributed grating sensor for real-time monitoring of arbitrary strain fields," *Proceedings of the Optical Fiber Sensors Conference* (OFS-11), Sapporo, Japan, 1996, pp. 6–9.

[101] Volanthen, M., et al. "Low coherence technique to characterise reflectivity and time delay as a function of wavelength within a long fibre grating," *Electronics Letters*, Vol. 32, 1996, pp. 757–758.

[102] Volanthen, M., et al. "Measurement of arbitrary strain profiles within fibre gratings," *Electronics Letters*, Vol. 32, 1996, pp. 1028–1029.

[103] LeBlanc, M., and A. D. Kersey, "Distributed, intra grating sensing by Fabry-Perot wavelength tuned low-coherence interferometry," *Proceedings of the Optical Fiber Sensors Conference* (OFS-12), Williamsburg, VA, USA, 1997, pp. 52–55.

[104] Wagreich, R. B., et al. "Effects of diametric load on fibre Bragg gratings fabricated in low birefringent fibre," *Electronics Letters*, Vol. 32, 1996, pp. 1223–1224.

[105] Wagreich, R. B., and J. S.Sirkis, "Distinguishing fiber Bragg grating strain effects," *Proceedings of the Optical Fiber Sensors Conference* (OFS-12), Williamsburg,VA, USA, 1997, pp. 20–23.

[106] Sirkis, J. S., "Unified approach to phase-strain-temperature models for smart structure

interferometric optical fiber sensors: Part 1, development," *Optical Engineering*, Vol. 32, 1993, pp. 752–761.

[107] Jin, X. D., J. S. Sirkis, and V. S. Venkateswaran, "Simultaneous measurement of two strain components in composite structures using embedded fiber sensors," *Proceedings of the Optical Fiber Sensors Conference* (OFS-12), Williamsburg, VA, USA, 1997, pp. 44–47.

[108] Udd, E., D. Nelson, and C. Lawrence, "Multiple axis strain sensing using fiber gratings written onto birefringent single mode optical fiber," *Proceedings of the Optical Fiber Sensors Conference* (OFS-12), Williamsburg, VA, USA, 1997, pp. 48–51.

[109] Bjerkan, L., K. Johannessen, and X. Guo, "Measurement of Bragg grating birefringence due to transverse compressive forces," *Proceedings of the Optical Fiber Sensors Conference* (OFS-12), Williamsburg, VA, USA, 1997, pp. 60–63.

[110] Ball, G. A., G. Meltz, and W. W. Morey, "Polarimetric heterodyning Bragg-grating fiber-laser sensor," *Optics Letters*, Vol. 18, 1993, pp.1976–1978.

[111] Ball, G. A., W. W. Morey, and P. K. Cheo, "Single- and multi-point fiber-laser sensors," *IEEE Photonics Technology Letters*, Vol. 5, 1993, pp. 267–270.

[112] Melle, S. M., et al. "A Bragg grating-tuned fiber laser strain sensor system," *IEEE Photonics Technology Letters*, Vol. 5, 1993, pp. 263–266.

[113] Kersey, A. D., and W. W. Morey, "Multiplexed Bragg grating fibre-laser strain-sensor system with mode-locked interrogation," *Electronics Letters*, Vol. 29, 1993, pp. 112–114.

[114] Alavie, A. T., et al. "A multiplexed Bragg grating fiber laser sensor system," *IEEE Photonics Technology Letters*, Vol. 5, 1993, pp. 1112–1114.

[115] Othonos, A., et al. "Fiber Bragg grating laser sensor," *Optical Engineering*, Vol. 32, 1993, pp. 2841–2846.

[116] Kersey, A. D., and W. W. Morey, "Multi-element Bragg-grating based fibre-laser strain sensor," *Electronics Letters*, Vol. 29, 1993, pp. 964–966.

[117] Koo, K. P., and A. D. Kersey, "Bragg grating-based laser sensors systems with interferometric interrogation and wavelength division multiplexing," *IEEE Journal of Lightwave Technology*, Vol. 13, 1995, pp. 1243–1249.

[118] Ball, G. A., "60mW 1.5 μm single-frequency low-noise fiber laser MOPA," *IEEE Photonics Technology Letters*, Vol. 6, 1994, pp. 192–194.

[119] Koo, K. P., and A. D. Kersey, "Noise and crosstalk of a 4-element serial fiber laser sensor array," *Technical Digest of the Conference on Optical Fiber Communications* (OFC'96), San Jose, CA, USA, Vol. 2, 1996, pp. 266–267.

[120] Dakin, J. P., et al. "Novel optical fiber hydrophone array using a single laser source and detector," *Electronics Letters*, Vol. 20, 1984, pp. 14–15.

[121] Kersey, A. D. et al. "Analysis of intrinsic crosstalk in tapped serial and Fabry-Perot interferometric fiber sensor arrays," Society of Photo-Optical Instrumentation Engineers, Fiber Optic Laser Sensors VI, 1988, Vol. 985, pp. 113–116.

[122] Vohra, S., et al. "An hybrid WDM/TDM reflectometric array," *Proceedings of the Optical Fiber Sensors Conference* (OFS-11), Sapporo, Japan, 1996, pp. 534–537.

[123] Geiger, H., et al. "Multiplexed fibre-optic system for both local and spatially-averaged strain monitoring," Second European Conference on Smart Structures and Materials, Glasgow, Scotland 1994, Session 10, pp. 366–370.

[124] Kersey, A. D., and M. J. Marrone, "Nested interferometric sensors utilizing fiber Bragg grating reflectors," *Proceedings of the Optical Fiber Sensors Conference* (OFS-11), Sapporo, Japan, 1996, pp. 618–621.

[125] Kersey, A. D., and M. J. Marrone, "Bragg grating based nested fibre interferometers," *Electronics Letters*, Vol. 32, 1996, pp. 1221–1223.

[126] Morey. W. W., J. R. Dunphy, and G. Meltz, "Multiplexing fiber Bragg grating sensors," Society of Photo-Optical Instrumentation Engineers, Distributed and Multiplexed Fiber Optic Sensors, 1991, Vol. 1586, pp. 216–224.

[127] Henderson, P. J., et al. "Simultaneous dynamic-strain and temperature monitoring using multiplexed

fibre-Fabry-Perot array with low-coherence interrogation," *Proceedings of the Optical Fiber Sensors Conference* (OFS-12), Williamsburg, VA, USA, 1997, pp. 56–59.

[128] Kaddu, S. C., et al. "Intrinsic fibre Fabry-Perot sensors based on co-located Bragg gratings," *Optics Communications*, Vol. 142, 1997, pp. 189–192.

[129] Koo, K. P., et al. "Fiber-chirped grating Fabry-Perot sensor with multiple-wavelength-addressable free-spectral ranges," *IEEE Photonics Technology Letters*, Vol. 10, 1998, pp. 1006–1008.

[130] Town, G. E., et al. "Wide-band Fabry-Perot-like filters in optical fiber," *IEEE Photonics Technology Letters*, Vol. 7, 1995, pp. 78–80.

[131] LeBlanc, M., A. D. Kersey, and T. E. Tsai, "Sub-nanostrain strain measurements using a pi-phase shifted grating," *Proceedings of the Optical Fiber Sensors Conference* (OFS-12), Williamsburg, VA, USA, 1997, pp. 28–30.

[132] Wanser, K. H., K. F. Voss, and A. D. Kersey, "Novel fiber dives and sensors based on multimode fiber Bragg gratings," 10th Optical Fibre Sensors Conference, Glasgow, Scotland, 1994, pp. 265–268.

[133] Eggleton, B. J., et al. "Long superstructure Bragg gratings in optical fibers," *Electronics Letters*, Vol. 30, 1994, pp. 1620–1621.

[134] Huntley, A. R., "Optical sensor networks," Avionics Conference and Exhibition"Systems Integration: Is The Sky The Limit?" November 1994.

[135] Dunphy, J. R., et al. "Multi-function, distributed optical fiber sensor for composite cure and response monitoring," Society of Photo-Optical Instrumentation Engineers, Fiber Optic Smart Structures and Skins III, 1990, Vol. 1370, pp. 116–118.

[136] Friebele, E. J., et al. "Distributed strain sensing with fibre Bragg grating arrays embedded in CRTM™ composites," *Electronics Letters*, Vol. 30, 1994, pp. 1783–1784.

[137] O'Dwyer, M. J., et al. "Relating the state of cure to the real-time internal strain development in a curing composite using in-fibre Bragg gratings and dielectric sensors," *Measurement of Science and Technology*, Vol. 9, 1998, pp. 1153–1158.

[138] Dykes, N. D., et al. "Mechanical and sensing performance of embedded in-fibre Bragg grating devices during impact testing of carbon fibre reinforced polymer composite," Workshop on Smart Systems Demonstrators: Concepts and Applications, Harrogate, UK, July 1998, pp. 168–175.

[139] Ferdinand, P., et al. "Applications of Bragg grating sensors in Europe," *Proceedings of the Optical Fiber Sensors Conference* (OFS-12), Williamsburg, VA, USA, 1997, pp. 14–19.

[140] Internet reference: ic.arc.nasa.gov/ic/projects/photonics/OS/Proposed/Bragg/bragg.html

[141] Internet reference: ic.arc.nasa.gov/ic/projects/photonics/OS/HealthSensors/health.html

[142] Friebele, E. J., "Fiber Bragg grating strain sensors: present and future applications in smart structures," *Optics and Photonics News*, Vol. 9, 1998, pp. 33–37.

[143] Hjelme, D. R., et al. "Application of Bragg grating sensors in the characterization of scaled marine vehicle models," *Applied Optics*, Vol. 36, 1997, pp. 328–336.

[144] Vohra, S. T., et al. "Sixteen channel WDM fiber Bragg grating dynamic strain sensing system for composite panel slamming tests," *Proceedings of the Optical Fiber Sensors Conference* (OFS-12), Williamsburg, VA, USA, 1997, pp. 662–665.

[145] Wang, G., et al. "Digital demodulation and signal processing applied to fiber Bragg grating strain sensor arrays in monitoring transient loading effects on ship hulls," *Proceedings of the Optical Fiber Sensors Conference* (OFS-12), Williamsburg, VA, USA, 1997, pp. 612–615.

[146] Lequime, M., "Fiber sensors for industrial applications," *Proceedings of the Optical Fiber Sensors Conference* (OFS-12), Williamsburg, USA, 1997, pp. 66–71.

[147] Weis, R. S., and B. D. Beadle, "MWD telemetry system for coiled-tubing drilling using optical fiber grating modulators downhole," *Proceedings of the Optical Fiber Sensors Conference* (OFS-12), Williamsburg, VA, USA, 1997, pp. 416–419.

[148] Ecke, W., et al. "First field tests with the fiber optic high-temperature borehole seismometer," *Proceedings of the Optical Fiber Sensors Conference* (OFS-11), Sapporo, Japan, 1996, pp. 630–633.

[149] Lee, W., et al. "Railroad bridge instrumentation with fiber optic sensors," *Proceedings of the Optical Fiber Sensors Conference* (OFS-12), Williamsburg, VA, USA, 1997, pp. 412–415.

[150] Internet reference: bob.emba.uvm.edu/Waterbury.html

[151]   Meissner, J., et al. "Strain monitoring at a prestressed concrete bridge," *Proceedings of the Optical Fiber Sensors Conference* (OFS-12), Williamsburg, VA, USA, 1997, pp. 408–411.

[152]   Dewynter-Marty, V., et al. "Concrete strain measurements and crack detection with surface-mounted and embedded Bragg grating extensometer," *Proceedings of the Optical Fiber Sensors Conference* (OFS-12), Williamsburg, VA, USA, 1997, pp. 600–603.

[153]   Harmer, A., and A. Scheggi, "Interferometers" in *Chemical, Biochemical and Medical Sensors, Vol. 2*, B. Culshaw and J. Dakin (eds.), Norwood, MA: Artech House, 1989, Ch. 16, pp. 599–651.

[154]   Rao, Y. J., et al. "In-fiber Bragg-grating temperature sensor system for medical applications," *IEEE Journal of Lightwave Technology*, Vol. 15, 1997, pp. 779–785.

[155]   Rao, Y. J., et al. "In-situ temperature monitoring in NMR machines with a prototype in-fibre Bragg grating sensors system," *Proceedings of the Optical Fiber Sensors Conference* (OFS-12), Williamsburg, VA, USA, 1997, pp. 646–649.

[156]   Rao, Y. J., et al. "In-fibre Bragg grating flow-directed thermodilution catheter for cardiac monitoring," *Proceedings of the Optical Fiber Sensors Conference* (OFS-12), Williamsburg, VA, USA, 1997, pp. 354–357.

[157]   Fisher, N. E., et al. "Ultrasonic field and temperature sensor based on short in-fibre Bragg gratings," *Electronics Letters*, Vol. 34, 1998, pp. 1139–1140.

[158]   Ning, Y. N., et al. "Interrogation of a conventional current transformer by a fiber-optic interferometer," *Optics Letters*, Vol. 16, 1991, pp. 1448–1450.

[159]   Henderson, P. J., N.E. Fisher, and D. A. Jackson, "Current metering using fibre-grating based interrogation of a conventional current transformer," *Proceedings of the Optical Fiber Sensors Conference* (OFS-12), Williamsburg, VA, USA, 1997, pp. 186–189.

[160]   Ogawa, Y., et al. "A multiplexed load monitoring system of power transmission lines using fiber Bragg grating," *Proceedings of the Optical Fiber Sensors Conference* (OFS-12), Williamsburg, VA, USA, 1997, pp. 468–471.

[161]   Bryson, C., "Interferometric sensor system for security applications," 10th Optical Fibre Sensors Conference, Glasgow, Scotland, 1994, pp. 485–488.

[162]   Todd, M. D., et al. "Flexural beam-based fiber Bragg grating accelerometers," *IEEE Photonics Technology Letters*, Vol. 10, 1998, pp. 1605–1607.

[163]   James, S. W., et al. "Transient strain monitoring on a gun barrel using optical fibre Bragg grating sensors," Society of Photo-Optical Instrumentation Engineers, Laser Interferometry IX: Applications, San Diego, CA, USA, July 1998, Vol. 3479, pp. 216–221.

Selected Bibliography

Liu, T., et al. "Simultaneous strain and temperature measurements in composites using a multiplexed fibre Bragg grating sensor and an extrinsic Fabry-Perot sensor," Society of Photo-Optical Instrumentation Engineers, Smart Sensing, Processing and Instrumentation, 1997, Vol. 3042, pp. 203–212.

Chapter 8

IMPACT OF
FIBER
BRAGG GRATINGS

8.1 Introduction

In this chapter we close the book with a general discussion of the impact of fiber Bragg gratings in both the arenas of optical fiber communications and sensing. This is not an easy task, as the emergence of a new and rapidly growing technology is notoriously difficult to track. Nevertheless, we aim to provide growth indicators to look for in both the aforementioned areas. It is particularly noteworthy that no other single device has had a significant impact on both sensing and optical communications applications and markets simultaneously.

8.2 Importance of Fiber Bragg Gratings to Global Communications

The communications industry has developed rapidly over the past two decades from an orderly and steadily growing enterprise to a chaotic marketplace with continually altered regulations, complex business relationships, and explosive optical technology leading to dynamic growth. The increasing installation of optical fibers, which is directly linked to telecommunications, has grown consistently this decade. Market estimates project that this steady growth is set to continue from a global level of 44 million fiber-km in 1998, with an annual growth rate of 21%, approaching 100 million fiber-km by 2002. The demands for global broadband communications are driving fiber optical–based technologies towards wavelength division multiplexing (WDM). In particular WDM applications are aimed at relieving congestion on previously installed fiber cables in long distance and trunk backbones. Existing long distance routes in service from the 1980's need overhauling in response to demands for increased capacity. However, it is particularly attractive for companies such as AT&T, MCI, and Sprint to increase capacity without overlaying new fiber cable on older routes, where unused fibers are unavailable. Based on this, network distance operators have opted for the deployment of WDM systems, and this necessarily involves the use of thousands of optical fiber laser sources and amplifiers. The spectral characteristics of these components, each operating over very distinct wavelength bands, can easily be tailored using fiber Bragg gratings. Furthermore, these long haul applications require a suitable means of compensating for fiber dispersion. This issue may similarly be overcome using specialized Bragg gratings. For example, in 1998 Nortel manufactured probably the world's longest dispersion compensating grating for fiber optic networks at 2.4m in length, for use in dense WDM (DWDM). In addition, the increased design flexibility of chirped gratings allows for virtually any chirp characteristic to be implemented. These grating types may offer network designers alternatives to the use of tens of kilometers of dispersion compensating fiber, provided that technical problems associated with delay-ripple can be overcome.

Optical access networks (OANs) are a rapidly emerging market where fiber Bragg grating technology will have a great impact. OANs are of particular importance in the United States and show significant promise in western Europe and other nations within the Asia-Pacific region. This emergence lies principally where telecommunications competition is prevalent, demanding that new interactive services be provided to system customers. OANs extend beyond the traditional star distribution topology, which is more or less fiber optic, by adding novel all-optical components and subsystems. For example, optical add drop multiplexers, optical cross-connects for wavelength management and restoration, and optical amplifiers, in addition to other devices which increase wavelength capacity expansion, create an optical stratum within the access network of the DWDM systems. This allows for routing, switching, and cross-connect functions to be completely implemented in the optical domain, improving the quality and reliability of services while maintaining signal formats, for compatibility reasons. Optical add drop multiplexers having already been deployed and optical cross-connects are expected by 1999 or 2000. According to Pioneer Consulting (Cambridge, MA), an increase from $76 million in 1998 to $1 billion by 2003 will occur for the deployment of WDM systems into access networks in the United States alone. The forecast includes end-to-end WDM systems for enterprise; metropolitan area, local exchange carrier, and local access networks.

As seen in Chapter 6, the development of fiber Bragg gratings has opened the way to all-fiber devices in communications, facilitating the basic functions of very high quality, selective filtering and reflecting components. The associated properties directly influence systems such as WDM and have led to developments in add drop filters, gain flattened fiber amplifiers, and high-quality fiber lasers. This has facilitated another means for the important transition to rugged, well-defined, and high-performance in-fiber, in-line components. Fiber Bragg gratings have revolutionized the means by which light is processed within the fiber. It is clear that the ability to tailor fiber Bragg gratings to any desired spectral characteristics, along with their ease of fabrication, will make them a fundamental building block of any fiber optic based communication system. Fiber Bragg gratings will continue to play a crucial role in the future of fiber optic communications.

Currently, the most promising uses of fiber Bragg gratings in telecommunications lie with fiber-based devices in DWDM, which is considered to be the communication format of the future. DWDM lightwave transmission systems can provide fiber networks with a very high capacity range (40 to 1000 Gbps). Initially, wavelength division multiplexing meant increasing capacity of any single-fiber line from one to two wavelengths, without making major technological changes. But now, just a few years after the introduction of DWDM, the demand for higher channel densities and data rates is exerting tremendous pressure on underlying components such as fiber amplifiers and sources. A critical factor in providing DWDM systems is delivering the necessary optical components in large volume and to the desired specifications. One year ago, transmitting 20 Gbps worth of data from point A to point B with two 10-Gbit signals running at 10 Gbps would have required a regenerator every 80 to 100 km. Today, sending at 2.5 Gbps with eight OC-48 signals, the same speed and clarity are accomplished without any regeneration over more than 600 km. Therefore, the components determine the DWDM capabilities. The demands on components such as fiber amplifiers have pushed them beyond their normal design capabilities, for example their limited operating wavelength range from 1530 to 1560 nm. Gain flattening and equalization using gratings has extended their useable wavelength range. Clearly, current amplifier designs do not take advantage of the low loss window of the fiber, which is about ten times wider. Therefore, any means of extending the usable optical amplification is highly attractive.

The DWDM market is emerging as one of the fastest growing markets in the telecommunications industry. The phenomenal growth of the Internet and other applications such as e-mail, large fiber transfer, and video-conferencing, as well as emerging technologies in the local loop such as digital subscriber lines and integrated services digital network are consuming massive bandwidth in the fiber backbone. As an example of the great performance gains, in 1995 the total capacity of a single-mode fiber was 2.5 Gbps, whereas in 1998, this has scaled to 96 times 2.5 Gbps. The DWDM network has developed to the point that systems supporting 16 to 40 channels are common place with add/drop capabilities that are based on the combined use of fiber Bragg gratings with dielectric-coated band pass filters. This proves to be a cost-effective and performance-leading solution. The most recent multiple-channel systems demand very narrow channel spacing, towards which CIENA Corporation has developed a unique optical architecture utilizing 50-GHz channel spacing (MultiWave 4000 DWDM system). The key to this success has been the use of fiber Bragg gratings to enable precise channel spacing at half the width of current commercial systems. The excellent cross-talk performance and high wavelength margins exhibited by the Bragg gratings and their compatibility with existing erbium-doped fiber amplifiers allow for 40 channels. The 50-GHz grid is a superset of the 100-GHz grid currently being recommended by the International Telecommunications Union (ITU). It is anticipated that other vendors will join in the process of developing a 50-GHz standard. Under the currently defined 100-GHz grid, a maximum of 41 channels are supported. The 50-GHz spacing is already in the process of being approved by Bellcore for systems with 64 and higher count wavelengths. The advantage in systems costs can be dramatic as a 16-channel system uses half the optical amplifiers of two 8-channel systems, and a 96-channel system uses one-twelfth the number of amplifiers of parallel 8-channel systems. CIENA has received a contract from Sprint to supply the new 40-channel MultiWave 4000 DWDM system, based on an in-fiber Bragg grating design. The new equipment will increase network capacity by 250% immediately, with a total capacity increase of 600% being possible as additional channels come on line. In this way scalable foundations for future telecommunications are being laid today. The continued standardization of products should allow DWDM devices to be easily deployed in all Synchronous Optical Network/Synchronous Digital Hierarchy networks, or as stand-alone equipment. This will include test-and-measurement standards. Another trend is that companies are migrating toward providing total product lines rather than one product for systems vendors. Cases in point are the CIENA Corporation and Lucent Technologies, which represent the two primary competing WDM market specialists that have literally cornered the DWDM markets to date with very different technologies.

All-optical network technology is forging and driving the component market upwards. The market for DWDM systems is expected to grow at a breakneck pace of 65% annually until 2000, then taper off to about 33%. This market will grow to $4.3 billion by the year 2000, according to the latest market research reports by Information Gatekeepers Inc. (IGI). The cost of components is expected to drop from approximately 10% to 50% annually, depending on demand, quality, and competition. The use of DWDM in telecommunications networks is challenging component manufacturers to design a variety of new devices that can be integrated into these systems. The most promising advances for this are optical amplifiers, add/drop multiplexers, and optical cross-connects. A key element to this growth is the fiber Bragg grating, a device that is easy to manufacture and cost-effective, which can be further used to fashion special components and hybrid devices performing unique functions. For example, Uniphase Corporation has invested in Bragg grating technology through the company Indx, a developer of fiber optic reflection filters for WDM applications. This move is based on the synergy between the technologies of Indx and Uniphase's modulator, amplifier, and transmitter products. Optical networking is a reality and an all-optical network is a reasonable goal. Today DWDM provides an instantaneous burst of capacity without the expense of laying new fiber, and the number of channels will certainly not stop here. Advances in optical network technology will dictate how the industry will grow, and we can predict with some certainty that the fiber Bragg grating will be at the technological forefront.

8.3 Commercial Prospects of Fiber Bragg Grating Sensors

The field of optical fiber sensing is highly diverse, and this diversity is perceived as the great advantage over conventional electromechanical sensors in the ability to tailor an optical sensor to measure any one of a myriad of physical parameters to high resolution. This in turn serves many different application areas, and therefore, the potential for a rapid growth in the optical fiber sensors market is well founded. Most sensor types, however, do not share a common instrument base, limiting sensor interchangeability and increasing system costs, as specialized components are used to meet specific sensor characteristics. A situation where new instrumentation costs significantly exceed those of the sensor does not offer any economic benefits to the adoption of new sensor types. When this is coupled with issues such as technical risk, measurand cross sensitivity, specialized operator handling, and noninterrupt-immune instrumentation, one may appreciate the limited adoption of fiber sensors and their highly fragmented niche-driven marketplace. This slow advancement into a new market is by no means unusual and constitutes the traditional pattern for the adoption of new technologies in the measurement industries. To date a large proportion of optical sensor development has been directed by universities and small business ventures. This has resulted in the majority of sensors being low-volume, high manufacturing cost devices. Therefore, the adoption of a new sensor type must offer an improved price/performance ratio over existing technologies, excelling as accurate but expensive, or competitive, meaning not as accurate but economical. In the latter case secondary performance criteria become important, such as resistance to extreme temperatures, radiation environments, or shock, or improved ruggedness and simplicity in manufacture. In the case of a low price/performance ratio, inroads into new application areas are a must if a foothold is to be accomplished and maintained. If there is no direct competition then it is easy to play on the component strengths, winning in a new market sector by default. Perhaps the best example to date is represented by the distributed optical fiber sensor (based on Brillouin scattering), for which the most advanced systems have demonstrated 1 m spatial resolution over 50-plus-km distances. No other sensor type can compete.

The market trend for the adoption of optical fiber sensors has been for large companies to sit in the wings waiting for the emergence of new technologies compatible with a large-scale application, such as the special needs of aircraft sensor systems. Moreover, compatibility with high-volume manufacturing has always been demanded from the final sensor prototype before resources are committed. Therefore, all the benefits envisioned by workers in this field, and there are many, have not yet resulted in low-cost, high-volume manufacturing. The emergence of a single technology platform and ideally a single sensor type, produced from a relatively small number of component building blocks, is needed. This concept is not new and it is widely held that the components of optical fiber communications can (and indeed do) provide the necessary building blocks; nevertheless, to date this emergence has not happened. Research trends indicate that the most promising sensors are interferometric, fiber Bragg grating, and evanescent wave sensors. In the last decade, one sensor type above all others has generated a great deal of excitement, not seen since the fiber optic gyroscope in the 1980s, the fiber Bragg grating sensor.

In the United States, Europe, and Asia there is a fast growing interest in optical sensing, with a predicted growth potential reaching approximately $5 billion by the first decade of the next century in the United States alone. Currently, the largest activities in the sensor market are chemical gas sensing, medical pressure sensing, automotive rotation and direction sensing, and temperature sensing. The main growth areas are expected to be environmental chemical sensing and biomedical sensing. However, strain sensing associated with smart structures is also expected to be an area of considerable growth in both the United States and Europe. It is interesting to note that the United States has commercial leadership in strain, chemical, biomedical, and pressure sensing while maintaining a larger R&D effort in all sensing areas, while Japanese companies lead in commercial rotation and electric and magnetic sensing. There is not this level of specificity in Europe, and this emerging market area has demonstrated interest in all of the above sensor types. The United States is populated with many small companies heading state-of-the-art optics research, whereas in Europe and Japan, the optical sensor effort comes mostly from universities, divisions of large companies such as British Aerospace (UK), government centers such as CEA-LETI (France), Hitachi Cable, and Sumitomo Electric (Japan). Japanese companies have focused on potentially high-volume applications with lower performance requirements; a notable example is Hitachi's fiber optic gyroscope, designed for high-volume automotive applications.

This brief summary of research directions in the most significant of markets gives our first clear indicator of where fiber Bragg grating market potential exists. It appears that the greatest inroads will be made by the entrepreneurial business base in the USA and a multitude of large-scale collaborations in Europe between industry and universities, funded by the European Union. In what follows we will draw on the specific areas where we believe Bragg grating sensors will have the greatest impact; however, we shall not list explicit projects as those that are current and pending have been adequately dealt with in Chapter 7.

Perhaps the following properties, more than any others, display why the Bragg grating element can excel as a sensor. (1) It displays versatility as a point or quasi-distributed sensor element that encodes its information as an absolute quantity and can be demodulated via a myriad of schemes, using active or completely passive approaches. The former are well suited to structural monitoring, whereas the latter are particularly relevant to aerospace sensing. (2) The compatibility of Bragg gratings with "conventional" interferometric optical fiber point sensors, using a common instrument base, has an importance that should not be underestimated. (3) The relative simplicity, low cost, and consistency in grating manufacture, directly into the fiber core is advantageous. Therefore, the inherent properties of optical fiber as a dielectric material host are exploited, offering a small overall sensor size and excellent long term sensor stability. Considering each point in turn, we may conclude as follows.

1. In civil infrastructure sensing there is no universal failure criteria for concrete and composite materials. Therefore, thousands of constantly aging systems of bridges, highways, buildings, railroads, dams, pipelines, etc., that are failing due to material fatigue must be monitored in real time, particularly as scientists do not fully understand the nature and control of material fatigue. Furthermore, this is warranted by the increasing use of new, advanced materials that are promising great strength and longevity in a light structure. There are numerous examples of structural failures. For example, the United States Army Corps of Engineers spends in excess of $20 million to meet its responsibility for 580 United States dams; during the first half of the 1980s, 36 dam failures in the United States cost seven lives and $48 million in damages. Increasingly, embedded optical fiber sensors hold promise for routine monitoring of aging effects and offer the possibility of taking remedial action prior to catastrophic failure. On-line concrete and composite curing monitoring and long term maintenance scheduling run parallel to this. Issues associated with the technology for this market mostly have to do with packaging and attachment of the sensors. The other issue is cost. Currently, optical sensors are as much as 100 times more expensive than corresponding mechanical or electronic sensors; however, the increased performance of Bragg grating sensors and low manufacturing costs, coupled with the multiplexing of large sensor arrays tilts the cost/performance ratio towards these sensors. In the other predominant field making extensive use of composite materials in aerospace applications,

light, sensitive, interrupt-immune sensors with good redundancy are essential. The easy maintenance and long-term stability of sensors are principal requirements, along with no moving parts. More than any other optical fiber sensor, the Bragg grating fulfills these criteria, particularly in the guise of the chirped grating sensor.

As part of a distributed sensor array, the Bragg grating arguably performs better than any interferometric or electromechanical sensors, but how do the real-time sensing capabilities of Bragg grating quasi-distributed arrays compare with the established and truly distributed sensing topologies based on Brillouin, Raman, and Rayleigh backscatter? Actually, quite well with regard to spatial resolution, but the distributed sensor range cannot be matched. A spatial resolution of 1 meter can easily be met and surpassed during grating inscription; however, one cannot reasonably expect to multiplex more than 1000 sensors, using the most advanced multiplexing techniques to date. Therefore, applications such as intruder detection schemes or distributed temperature sensors for monitoring fire hazards remain the realm of the traditional distributed optical sensors, unless a limited range is acceptable, in which case the lower system cost of the Bragg grating sensor system is unbeatable.

2. The multitude of ways available to sensing has meant that no single sensing technique has emerged to become the large-volume leader. Some techniques seem to be more prominent than others for sensing a given measurand; for example, the extrinsic Fabry-Perot interferometer is well suited for use in hostile, high pressure and temperature environments as found in bore-hole applications. A case in point is a Fabry-Perot sensor available from FPPI measuring dynamic pressure to 2500 psi and 10 kHz. However, more often than not sensitivity to unwanted measurands is the limiting factor for successful sensor performance. A network able to incorporate both interferometric point and Bragg grating sensors can prove useful in situations where temperature and strain need to be measured separately or for which one measurand needs to be compensated. For example, the Bragg grating can provide complementary temperature information to the Fabry-Perot pressure sensor, with both sensors capable of surviving temperatures of several hundred degrees, this performance is well beyond the capabilities of conventional electrical sensors. The potential for the oil exploration market is very large with tens of thousands of bore-hole locations worldwide demanding constant monitoring.

3. Many market surveys for optical sensors have been performed over several years in the United States, Japan, and Europe. While the estimates vary considerably, owing to a variety of assumptions, they nevertheless agree in predicting a large and growing optical fiber sensor market over the next two decades, with anticipated markets exceeding $1.2 billion by the year 2000, followed by rapid growth into the next century. A very important parameter is the unit price of sensors, which is currently about $300 and is expected to decline to about $150 by the year 2000 for a typical optical fiber sensor. The cost of Bragg gratings is potentially far lower. When the lower costs are coupled with the ability to exactly reproduce the grating characteristics through the use of phase masks, applications utilizing disposable sensors become commercially feasible. The most obvious of which lies in medical temperature probes where sensors are generally disposable. In this way instrumentation unit costs are amortized over many sensor units, improving the long-term commercial viability. In fact, temperature, pressure, and chemical sensors probably

constitute the biggest segment of the sensor market. Of these three, chemical sensing, mostly associated with medical and biomedical applications, is showing the fastest growth.

To summarize, the future of the Bragg grating sensor is looking very promising and certainly has the potential to finally push optical fiber sensors to the forefront, resolutely challenging conventional forms of electromechanical sensors. We have projected that areas of distributed strain and point temperature sensing are where the strongest challenge can be advanced. These encompass aerospace, civil infrastructure, medical, and oil exploration sensing fields. Strain sensing is clearly dominated by the United States and is growing rapidly, causing great interest in Europe, as a result of smart structure efforts. At present, activities in point sensing of temperature also appear to be stronger in the United States and Europe, while distributed sensing of temperature is stronger in Japan, using conventional optically based systems. That trend is expected to continue. The one significant market area of considerable interest that has yet to be penetrated, even at an R&D level, is automotive sensing. Potentially, the automotive sensor market could grow to several billion dollars alone over the next few decades if all the parameters to be sensed could be done cost-effectively in an optical format. This market, however, is extremely price conscious, and it is not clear that optical techniques, even those as promising as Bragg gratings, will be suitable to penetrate this market.

ABOUT THE AUTHORS

Andreas Othonos, received his B.Sc. in theoretical physics (1984) and M.Sc. in high energy physics (1986) from the University of Toronto. Between 1986-1990 he conducted research in the area of ultrafast laser semiconductor interactions at the the National Research Council of Canada (NRC) and the University of Toronto where he was awarded a Ph.D. degree. From 1990 to 1996 he was with the Ontario Laser and Lightwave Research Center at the University of Toronto working on ultrafast dynamics in semiconductors and fiber photosensitivity. Since 1996 he has been an assistant professor in the Department of Natural Sciences (Physics) at the University of Cyprus and a principle investigator of the Photonics Research Group. His current research interests involve ultrafast laser induced dynamics, laser-matter interactions, nonlinear optics, laser physics, fiber optic photosensitivity and telecommunications. He has published more than 50 journal papers along with several invited reviews and chapters in books.

Kyriacos Kalli, C.Phys., M.Inst.P., received the B.Sc. (Hons) in Theoretical Physics (1988) and Ph.D. in Physics (1992) from the University of Kent at Canterbury. His Ph.D. thesis investigated linear and non-linear optical phenomena using high finesse ring resonators. In 1993 he joined the Fiber and Electro-Optics Research Center, Virginia Tech as a visiting research scholar, where he investigated optical fiber-based intruder detection schemes and temperature sensors for on-board-ship fire monitoring. From 1994–1996 he undertook research at the Applied Optics Group, University of Kent into the use of fiber Bragg gratings in multiplexed sensor arrays and Raman spectroscopy for pollution monitoring of ground water. He joined the Photonics Research Group, University of Cyprus, in 1996, engaged in research into integrated gas flow and gas sensors based on porous silicon micromachining, fluorescence spectroscopy for environmental pollution studies, and non-destructive evaluation of semi-conductors using photoreflectance and photothermal measurements. His research interests are in Bragg grating and optical fiber sensors, laser material interactions, and photosensitivity and environmental pollution monitoring. He has 50 journal and conference publications. Dr. Kalli is a member of the Optical Society of America.

Email: othonos@ucy.ac.cy
Email: kkalli@ucy.ac.cy
Internet site http://photonics.ucy.ac.cy

Index